For Product Safety Concerns and Information please contact our
EU representative GPSR@taylorandfrancis.com Taylor & Francis
Verlag GmbH, Kaufingerstraße 24, 80331 München, Germany

# Advances in Probiotic Technology

# Advances in Probiotic Technology

*Editors*
**Petra Foerst**
Lehrstuhl für Verfahrenstechnik disperser Systeme (VDS)
Maximus-von-Imhof-Forum 2
Freising · Germany

**Chalat Santivarangkna**
Institute of Nutrition, Mahidol University
Nakhon Pathom
Thailand

CRC Press
Taylor & Francis Group
Boca Raton London New York

CRC Press is an imprint of the
Taylor & Francis Group, an **informa** business
A SCIENCE PUBLISHERS BOOK

CRC Press
Taylor & Francis Group
6000 Broken Sound Parkway NW, Suite 300
Boca Raton, FL 33487-2742



Printed on acid-free paper
Version Date: 20150706

International Standard Book Number-13: 978-1-4987-3453-0 (Hardback)



**Visit the Taylor & Francis Web site at**
**http://www.taylorandfrancis.com**

**and the CRC Press Web site at**
**http://www.crcpress.com**

# Preface

The development and application of probiotics in the past was oriented toward pharmaceuticals or limited to traditional probiotic foods such as yoghurt. However, these days application of probiotics has been extended to novel functional foods that are targeted at the general population. We both have been working at the Technische Universität München together for a decade in the field of fermentation and drying technology of probiotics and are convinced that these fields of research are very vital and will increase in the future along with this trend. Novel probiotic foods require, for example, novel probiotic strains, appropriate cultivation process, and efficient preservation methods that fit to the new food matrices or food production processes. Despite the availability of huge amount of scientific literature related to health benefits, little effort has been made to put together insight related to these technological and biotechnical aspects of probiotics. This is why the book "Advances in Probiotic Technology" was initiated by us.

The aim of this book is to attract and engage key experts in the field of probiotic technology to assemble the latest advances and trends in technology and biotechnology of probiotics into a book. The focus of this book is to present the state-of-the art technology for isolation and characterisation of probiotics as well as the production process of probiotic cell concentrate including fermentation, freezing and drying, storage and microencapsulation. A part of the book is dedicated to the use of probiotics in different food products. We are convinced that the probiotic technology is still in its infancy and will become more important in future when the role of microorganisms in the gut is better understood and specific probiotic effects of individual strains are elucidated.

The book comprises 18 chapters, that are divided in four parts. The chapters are arranged in such a way that it resembles the probiotic production process, i.e., starting from upstream to downstream processing, and ending with today's application of probiotics in food matrices.

Part I deals with screening and isolation of probiotics, novel probiotic strains, and cultivation and fermentation processes. Part II deals with molecular techniques used for the identification and enumeration of probiotics, analysis of functionalities of probiotics and of gut microbiota, and genetic engineering of probiotics. Part III deals with drying and storage of probiotics, as well as control of the processes. Next to the state-of-the art freeze drying process for drying probiotics, alternative, more economical processes and industrial aspects of probiotic production are addressed. Part IV will deals with application of probiotics in dairy and non-dairyproducts as well as on microencapsulation technology.

Each chapter is written by leading experts and will provide not only up-to-date developments and information but also viewpoints on the topics.

We would like to express our gratitude to all authors who contributed to this book and hope that the book will prove to be essential to researchers, technologists and industry professionals in the field of probiotics.

**Petra Foerst**
**Chalat Santivarangkna**

# Contents

## Part III: Preservation of Probiotics



## Part IV: Applications of Probiotics

# PART I

# Probiotic Cells and The Cultivation

CHAPTER 1

# Isolation, Identification and Characterisation of Potential New Probiotics

*Akihito Endo*[1,*] and *Miguel Gueimonde*[2]

## Introduction

Probiotics are defined by FAO/WHO (2002) as live microorganisms, which when administered in adequate amounts confer a health benefit on the host. A number of bacterial strains (and sometimes yeast strains) have been characterised, proposed and applied in food matrices as human probiotics. These probiotics have been shown to be beneficial for treatment and prevention of allergies, irritable bowel syndrome, inflammatory bowel disease, viral diarrhoea and constipation, immune modulation and promotion of oral health by intervention studies (reviewed by Collado et al. 2009). The majority of probiotics in use are classified in the genera *Bifidobacterium* and *Lactobacillus* (Ouwehand et al. 2002a). *Lactobacillus* spp. are members of the lactic acid bacteria (LAB) group, and other members of the group, e.g., *Leuconostoc, Pediococcus, Enterococcus*, *Lactococcus* and *Oenococcus*, are also assessed on the basis of their probiotic properties mainly by *in vitro* studies. *Bifidobacterium* spp. share some similar characteristics to LAB but differentiate from LAB on their phylogenetic positions and high GC content in their genomes. These organisms usually originate in human faeces, breast-milk, and fermented and non-fermented foods. However, these origins usually have complex microbiota, and sometimes target organisms are not

[1] Department of Food and Cosmetic Science, Department of Bioindustry, Tokyo University of Agriculture, Hokkaido, Japan.
E-mail: a3endou@bioindustry.nodai.ac.jp
[2] Department of Microbiology and Biochemistry of Dairy Products, Instituto de Productos Lácteos de Asturias (IPLA-CSIC), Villaviciosa, Spain.
E-mail: mgueimonde@ipla.csic.es
* Corresponding author

a major part of the microbiota. In such cases, enrichment culturing techniques that enhance only a certain group of bacteria is needed for specific isolation. Following the isolation, the isolates are identified by phenotypic and molecular approaches and characterised by their beneficial and risk factors. This chapter discusses the isolation, identification at species level and characterisation of potential probiotics.

## 1. Isolation of Bifidobacteria and LAB

### *1.1 Isolation of Bifidobacteria from Human Faeces and Breast Milk*

Human fecal and breast milk samples are rich and well-used sources for isolation of bifidobacteria as probiotic candidates. *Bifidobacterium adolescentis, B. bifidum, B. breve, B. catenulatum, B. longum* subsp. *longum, B. longum* subsp. *infantis* and *B. peudocatenulatum* are generally regarded as human bifidobacteria (Ventura et al. 2004). Other bifidobacteria, e.g., *Bifidobacterium angulatum* and *B. dentium*, can sometimes be seen in human faeces however, they are not commonly regarded as probiotic candidates. To isolate bifidobacteria from the faecal and breast milk samples, the samples are usually brought to the laboratory at 4°C to stop bacterial growth and to keep the viability of bacteria. Anaerobic transport media are sometimes used to keep viability of anaerobic bacteria including bifidobacteria (Hartemink and Rombouts 1999, Makino et al. 2011). The transported samples are serially diluted to make adequate dilution with buffers, and pre-reduced buffers or broths are usually used for the dilution. Pre-reduced buffers which are frequently used are listed in Table 1. Non-pre-reduced buffers, e.g., saline and PBS are also used sometimes (Ferraris et al. 2010, Makino et al. 2011). The serially diluted samples are spread on a *Bifidobacterium*-selective or a

**Table 1.** List and composition of pre-reduced dilution buffers.

| Buffers | Base | Composition or additives | References |
|---|---|---|---|
| Anaerobic dilution buffer | | 0.02925% $K_2HPO_4$, 0.017625% $KH_2PO_4$, 0.04425% $(NH_4)_2SO_4$, 0.0045% CaCl2, 0.00825% $MgSO_4$, 0.0001% resazurin, 0.05% L-cysteine, 0.05% L-ascorbic acid, 0.4% $Na_2CO_3$ and 0.05% agar at pH 7.4–7.6 | Mitsuoka et al. (1965) Kataoka et al. (2007) |
| 0.25 X Ringer solution supplemented with cysteine | 0.25 X Ringer solution | 0.05% L-cysteine | Nebra and Blanch (1999) Mullié et al. (2006) |
| Wilkins–Chalgren broth | Wilkins-Chalgren broth | | Thitaram et al. (2005) Rada et al. (2006) |
| Saline supplemented with cysteine | Saline | 0.2% L-cysteine | Alp and Aslim (2010) |
| Reduced physiological salt solution | | 0.1% peptone, 0.05% L-cysteine and 0.8% NaCl | Hartemink and Rombouts (1999) |
| Reduced medium | BHI broth | 0.5% glucose, 0.5% yeast extract, 0.25% L-cysteine, 0.001% vitaminK1 and 0.002% hemin | Solís et al. (2010) |

non-selective medium and are usually cultured at 37°C for 2 to 4 d under anaerobic conditions, using gas generating systems or anaerobic workstations. de Man, Rogosa and Sharpe (MRS) agar supplemented with L-cysteine (ranging from 0.05 to 0.5%, w/v) is a well-used non-selective medium for isolation of bifidobacteria from the samples, and LAB are also target organisms when this medium is used. Selective media for the specific growth of *Bifidobacterium* spp. have been developed by researchers. BS agar, Beerens agar, improved Petuely's medium and BFM might be originally used as the selective media (Mitsuoka et al. 1965, Tanaka and Mutai 1980, Benno et al. 1989, Beerens 1990, Nebra and Blanch 1999, Hartermink and Rombouts 1999). These media contain one or a mixture of selective agents, including propionic acid, neomycin, lithium chrolide, paromomycin and nalidixic acid. Selective media supplemented with mupirocin seem to be more commonly in use during the last decade. These are BSM (*Bifidobacterium*-selective media), TOS-AM, WCBM and MTPY agar (Simpson et al. 2004a, Thitaram et al. 2005, Rada et al. 2006, Ferraris et al. 2010). Mupirocin, originally characterised as pseudonomic acid A (Fuller et al. 1971), is an antibiotic produced by *Pseudomonas fluorescens* and blocks bacterial protein synthesis by the inhibition of bacterial isoleucyl-tRNA synthetase (Hughes and Mellows 1978). Mupirocin inhibits the growth of *Lactobacillus* spp., *Lactococcus lactis* and *Bacillus* spp. but not of *Bifidobacterium* spp. (Rada 1997, Simpson et al. 2004a). Growth inhibition of *Lactobacillus* spp. is important as they are usually found in the same habitat as *Bifidobacterium* spp. including human faeces and breast milk. *Pediococcus* spp. and *Propionibacterium freudenreichii* produce colonies on MRS agar in the presence of 50 mg/l Mupirocin, but their colonies are very small (<0.5 mm) and thus can be distinguished from colonies of *Bifidobacterium* spp. (Simpson et al. 2004a). Supplements of glacial acetic acid might be further helpful (Thitaram et al. 2005, Rada et al. 2006). Major *Bifidobacterium*-selective media are summarized in Table 2. Several articles have compared and discussed levels of selectivity of the selective media (Hartemink and Rombouts 1999, Apajalahti et al. 2003, Mikkelsen et al. 2003, Mullié et al. 2006, Ferrais et al. 2010). Kawasaki et al. (2007) studied the growth of several *Bifidobacterium* spp. under different atmospheric conditions and reported that *Bifidobacterium* spp. requires $CO_2$ for colony development. Colonies on the selective agar are transferred into non-selective culture broth, e.g., MRS broth supplemented with 0.05% L-cysteine, and incubated at 37°C for 24–48 hr. Impact of $O_2$ on the growth in liquid broth varies between *Bifidobacterium* spp. (Kawasaki et al. 2006). The cultures are then used for further research or are stocked in broth supplemented with glycerol at –80°C. The described method can also be applicable to isolate bifidobacteria from animal faeces.

### *1.2 Isolation of LAB from Human Faeces and Breast Milk*

Isolation of LAB from human samples is similar to the procedure for bifidobacteria but might require less attention to produce anaerobic conditions. Samples are usually transported to laboratory at 4°C, and serially diluted with non-reduced buffers, e.g., saline or PBS, plated on non-selective or selective mediums and incubated at 37°C for 2–4 d under anaerobic conditions. MRS agar supplemented with 0.05% L-cysteine is usually used as a non-selective medium, and Rogosa agar, Rogosa SL agar, LBS agar, modified LBS agar and LAMVAB agar are major agar media for the selective isolation of Lactobacillus spp. from human samples (Table 3). These selective media generally have low pH around 5 to 5.5 for growth inhibition of clostridia, enterobacteria and bifidobacteria, however, the low pH might inhibit the growth of gastrointestinal acid-sensitive LAB *Leuconostoc* and *Weissella* spp. although they are also major LAB in infant faeces and in breast milk (Dal Bello et al. 2003, Cabrera-Rubio et al. 2012). Culturing at 37°C is suitable for most gastrointestinal LAB but might have a negative impact on certain mesophilic species (Dal Bello et al. 2003). Vancomycin used in

**Table 2.** List of *Bifidobacterium*-selective media developed for fecal microbiota.

| Media | Base | Additives and selective agents | Reference |
|---|---|---|---|
| BS agar | BL | 5% BS solution composed of 30% sodium propionate, 0.1% paromomycin, 0.4% neomycin and 6% Lithium chloride | Mitsuoka et al. 1965 |
| Beerens agar | Columbia agar | 0.5% propionic acid | Beerens 1990 |
| BFM | | 0.2% meat extract, 0.7% yeast extract, 0.2% starch, 0.05% L-cysteine, 0.5% sodium chloride, 0.5% peptone, 0.2% tryptone, 0.5% lactulose, 0.0001% riboflavin, 0.0001% thiamine, 0.0016% methylene blue, 0.2% lithium chloride and 0.5% propionic acid. | Nebra and Blanch 1999 |
| MTPY agar | TPY | 0.1% Tween 80, 0.1% glacial acetic acid, 0.05% L-cysteine and 0.005% mupirocin | Rada and Petr 2000 |
| BSM | MRS | 0.05% cysteine and 0.005% mupirocin | Simpson et al. 2004a |
| TOS-AM | TOS | 1% glacial acetic acid and 0.005% mupirocin | Thitaram et al. 2005 |
| WCBM | Wilkinse-Chalgren agar | 1% glucose, 0.5% agar, 0.5% Tween 80, 0.05% L-cysteine and 0.005% mupirocin | Ferrais et al. 2010 |

**Table 3.** List of *Lactobacillus*-selective media developed for fecal microbiota.

| | Base | Additives and selective agents | Reference |
|---|---|---|---|
| LAMVAB agar | MRS | 0.05% L-cysteine, 0.005% bromocresol green, 0.002% vancomycin and 2% agar (pH 5.0) | Hartemink et al. 1997 |
| modified LBS agar | LBS | 0.8% Lab-lemco powder, 1.5% sodium acetate and 0.37% acetic acid | Mitsuoka 1980 |

Commercially available media are not included in this table

LAMVAB agar is effective for selective isolation of *Lactobacillus* spp. (Hartemink et al. 1997) but might have a risk of loosening typical lactobacilli in human faeces. Jackson et al. (2002) compared the reliability of Rogosa and LAMVAB agars to isolate lactobacilli from human faeces and suggested that LAMVAB agar did not support the growth of *L. acidophilus* and *L. gasseri,* whereas bifidobacteria were unexpectedly isolated on Rogosa agar. *L. acidophilus* and *L. delbrueckii* subsp. *bulgaricus* have been shown to be susceptible to low concentrations of vancomycin (Pacher and Kneifel 1996). Colonies on the selective agar are transferred into a non-selective culture broth and incubated at 37°C for 24–48 hr. The cultures are used for further research or are stocked in the broth supplemented with glycerol at –80°C. The described method can be also applicable to isolate bifidobacteria from animal faeces.

### 1.3 Isolation of Bifidobacteria from Animal Milk and Fermented Milk

Bifidobacteria found in raw animal milk and fermented animal milk might be possible sources of probiotics. *Bifidobacterium animalis* subsp. *lactis, B. mongoliense* and *B. psychraerophilum* have been found in fresh or fermented milks (Meile et al. 1997, Watanabe et al. 2009, Hsieh et al. 2012). *Bifidobacterium crudilactis* has also been isolated from fresh milk and cheese (Delcenserie et al. 2007). Isolation of bifidobacteria from such food samples is generally the same as that from human samples however, bifidobacteria are sometimes isolated from the milk samples with LAB by standard techniques for the isolation of LAB, e.g., MRS agar incubated at 30°C (Watanabe et al. 2009, Hsieh et al. 2012). Such species are usually oxygen tolerant (Simpson et al. 2004b). Bifidobacteria found in animal milks are sometimes regarded as contaminates from other sources (Ventura et al. 2004, Delcenserie et al. 2011), as some of them are usually inhabitants of gastrointestinal tracts of animals (Scardovi and Trovatelli 1974, Simpson et al. 2004b).

### 1.4 Isolation of LAB from Fermented and Non-fermented Foods

A number of fermented foods have been studied to look for new probiotics candidates; these include fermented milk, cheese, fermented vegetables, fermented meat, fermented fish and fermented cereals (reviewed by Anukam and Reid 2009, Rivera-Espinoza and Gallardo-Navarro 2010). In such fermented foods, LAB are usually the predominant bacteria, and thus strict selective media are not needed for isolation. Commercially available MRS medium, M17 medium and Rogosa medium are usually used for isolation of LAB from the fermented foods, however, dependent on the food product, slight modification might be needed. For instance, lactose should be used to isolate dairy LAB as a carbon source and maltose is a suitable carbon source for certain sourdough LAB. Cycloheximide and sodium azide are sometimes useful to inhibit the growth of concomitant fungi and aerobes (Endo and Okada 2005). Vera et al. (2009) compared the impacts of culture media for isolation of LAB from sourdough using 11 elective/selective media and found that each medium has different merits, e.g., Maltose MRS and MRS media recovered the highest bacterial counts and MRS5 agar isolated

**Table 4.** Primers used for identification of lactobacilli and bifidobacteria.

| Name | Sequence (5'-3') | Target gene | Target organisms | Reference |
|---|---|---|---|---|
| BI8F | GGGTTYCGATTCTGG CTCAGGATG | 16S rDNA | eubacteria including bifidobacteria | Miyake et al. 1998 |
| 15R | AAGGAGGTGATCCAR CCGCA | 16S rDNA | eubacteria including bifidobacteria | Giovannoni 1991 |
| 8F (27F) | AGAGTTTGATCMTGG CTCAG | 16S rDNA | eubacteria excluding certain bifidobacteria | Giovannoni 1991 |
| DnaJ1-uni | GAGAAGTTCAAGGAC ATCTC | *dnaJ1* | bifidobacteria | Ventura et al. 2006 |
| DnaJ1-rev | GCTTGCCCTTGCCGG | | | |
| ClpC-uni | GAGTACCGCAAGTAC ATCGAG | *clpC* | bifidobacteria | Ventura et al. 2006 |
| ClpC-rev | CATCCTCATCGTCGAA CAGGAAC | | | |
| H60F | GGNGAYGGNACNAC NACNGCNACNGT | *hsp60* | bifidobacteria | Jian et al. 2001 |
| H60R | TCNCCRAANCCNGGN GCYTTNACNGC | | | |
| recEXT-F | GGCTATGAAACAAAT TGAAAAACAATWYG GNAARGG | *recA* | lactobacilli and some other LAB | Dellaglio et al. 2005 |
| recEXT1-R | TGTTTAAACGGTGGA GCAACTTTRTTYTTNA C | | | |
| pheS-21-F | CAYCCNGCHCGYGAY ATGC | *pheS* | lactobacilli and some other LAB | Naser et al. 2005 |
| pheS-22-R | CCWARVCCRAARGCA AARCC | | | |
| rpoA-21-F* | ATGATYGARTTTGAA AAACC | *rpoA* | lactobacilli and some other LAB | Naser et al. 2005 |
| rpoA-22-R* | ACYTTVATCATNTCW GVYTC | | | Naser et al. 2007 |
| rpoA-23-R* | ACHGTRTTRATDCCD GCRCG | | | Naser et al. 2005 |

*If an amplicon is not obtained with the primer combination of rpoA-21-F/rpoA-23-R, reverse prime can be replaced with rpoA-22-R (Naser et al. 2007).

a specific population found only in this medium. Enrichment culturing is sometimes helpful for an effective isolation of a wide variety or a minor population of LAB in fermented food (Endo and Okada 2005, Lundström and Björkroth 2007). Moreover, this technique might be useful to isolate LAB with specific characteristics. For instance, an enrichment broth set at a low pH might be suitable for selective isolation of acid-tolerant probiotic candidates, and an enrichment broth containing bile salt might be suitable for bile-tolerant strains. Carbohydrates used for culturing have a big impact on the diversity of enrichment isolation of LAB (Endo et al. 2011a).

Fresh vegetables and fruits are also possible sources of probiotic LAB (Vitali et al. 2012a). LAB are usually a minority in the microbiota of fresh plant materials (Buckenhüskes 1997, Di Cango et al. 2008), and thus enrichment culturing in the presence of selective agents is useful. Cycloheximide and sodium azide are very helpful for selective culturing of LAB in the fresh plants (Endo et al. 2009). MRS medium is usually used for this isolation. Mediums containing fructose as a carbon source enhance the growth of a special group of LAB called fructophilic LAB (Endo et al. 2009 and 2011b). They are found in several plant materials and fermented foods, and moreover, several uncommon characteristics in the LAB group have been reported for fructophilic LAB (Endo and Okada 2008, Endo et al. 2009). Such characteristics include tolerance to oxygen and high osmotic pressure, suggesting their application potential.

### *1.5 Other Probiotic Candidates*

Microorganisms other than bifidobacteria and LAB are also studied and used as potential probiotics. They are yeasts, propionibacteria, *Bacillus* spp., *Escherichia coli*, etc. Several yeasts, e.g., *Saccharomyces cerevisia*, *S. boulardii* and *Kluyveromyces* spp., originating from fermented food or animal faeces have been characterised as probiotics (Kumura et al. 2004, Rajkowska and Kunicka-Styckzyńska 2010). To obtain yeast strains with a good tolerance to the human digestive tract conditions, Pennacchia et al. (2008) pre-cultured food samples (fermented meat, wine and cheese) in series of simulated gastric juices, simulated intestinal fluid in duodenum and simulated upper intestinal fluid and plated on Malt extract agar. Several isolates identified as *S. cerevisiae* and *Candida* spp. showed good survivability under the simulated human gastrointestinal tract (Pennacchia et al. 2008). *Propionibacterium freudenreichii* is the mainly used propionibacteria in probiotics but others are also sometimes used (Darilmaz and Beyatli 2012). Probiotic propionibacteria are mostly found in dairy products and are usually used in combination with other probiotic lactobacilli and bifidobacteria (Kekkonen et al. 2011). Metabolite of *P. freudenreichii* has a bifidogenic effect but suppresses the growth of *Helicobacter pyroli* (Nagata et al. 2010). *Bacillus* spp. including *Bacillus subtilis* and *B. cereus*, found from several sources, have also been characterised as probiotics. Ahire et al. (2011) isolated siderphoregenic *Bacillus* spp. from human faeces, and Sánchez et al. (2009) characterised adhesion ability of several *B. cereus* probiotics to model human intestinal surfaces. Apart from human probiotics, an interesting study isolated fish probiotics with the ability to degrade quorum sensing molecules produced by fish pathogens. *Bacillus* spp. which possesses quorum sensing molecules degrading activity was isolated from the fish gut using enrichment culturing techniques with the quorum sensing molecules as the sole carbon and nitrogen source (Chu et al. 2011). Several bacilli have been isolated as candidates of animal probiotics (Barbosa et al. 2005, Hong et al. 2005, Guo et al. 2006, Capkin and Altinok 2009, Klose et al. 2010). Reissbrodt et al. (2009) found *Escherichia coli* isolate inhibiting the growth and toxin production of Shiga-toxin producing *Escherichia coli* (STEC).

Various microorganisms are introduced here as probiotic candidates. However, some of the organisms are not food-grade microorganisms but are rather reported to be associated with human infectious diseases. In fact commercially available *B. cereus* probiotic strains have haemolysis activity and produce enterotoxin (Duc et al. 2004). Therefore, whether the microorganisms are safe or not should be carefully assessed.

## 2. Identification of Probiotic Candidates

Appropriate strain identification constitutes the starting point for the assessment of microbial properties. This information not only confirms the identity of the strain, but it is also needed for proper labelling. In addition, a proper identification may also be very helpful to predict the properties of the microorganism on the basis of the previous knowledge about other members of the same microbial group. Moreover, accurate and reliable identification of probiotic strains is also necessary to evaluate both the documented health benefits and the safety of probiotic products. However, a number of studies have reported that the identity of the probiotics included in some products does not correspond to the information stated on the product label (Gueimonde et al. 2004, Huys et al. 2006).

Identification of probiotic candidates is usually carried out by phylogenetic analysis based on 16S rRNA gene. Consensus primers for amplification of 16S rRNA gene sometimes do not amplify the genes of bifidobacteria, and thus alternative primers are usually used (Miyake et al. 1998, Watanabe et al. 2009). Certain species are known to share high sequence similarities based on 16S rRNA genes, resulting in difficulties of identification. In such cases, housekeeping genes or 16S–23S rRNA genes intergenic spacer region is often used for identification and classification of bifidobacteria and LAB, as these have more discriminatory power for identification of bifidobacteria and LAB. For bifidobacteria, *hsp60, dnaJ1* and *clpC* are usually used (Jian et al. 2001, Ventura et al. 2006, Watanabe et al. 2009, Endo et al. 2012, Junick and Blaut 2012), and *recA, pheS* and *rpoA* are used for LAB (Torriani et al. 2001, Dellaglio et al. 2005, Naser et al. 2007). Phenotypic methods, such as sugar fermentation profiles, still provide relevant physiological data, but given the low reliability observed for these methods, it has been indicated that they should not be used as stand-alone methods for identification of probiotics (Vankerckhoven et al. 2008a). Rather this technique is now used for characterisation of the isolates. A FAO-WHO expert group recommends that phenotypic tests should be done first, followed by genetic identification using methods such as DNA-DNA hybridization, 16S rRNA gene sequence analysis or other well established techniques (FAO/WHO 2002). Establishing the identity of a microorganism is thus the first crucial step for any further studies.

Probiotic effects have been found to be strain specific and, therefore, it is also necessary to identify the probiotic microorganisms at strain level. According to the FAO-WHO expert group strain typing has to be performed with a reproducible genetic method. DNA macro restriction followed by PFGE is considered the gold standard for strain identification and it has been extensively used for differentiating probiotic strains, although other molecular methods are also available to this end.

In conclusion, a proper species identification and characterisation at strain level (genetic typing), by using internationally accepted molecular methods constitutes the first tests that should be conducted for any new probiotic bacteria. In addition, the strain should be properly named according to the International Code of Nomenclature and it would be recommendable to deposit it in an internationally recognized culture collection (FAO/WHO 2002). In the immediate future, the increasing availability of genome sequences may allow genome-wide and/or multilocus phylogenetic analysis. Although complete genome assembling is still a time consuming task, the number of finished bacterial genomes, especially that of draft genomes is increasing rapidly (Chain et al. 2009).

Several probiotic strain genomes have been sequenced and in some cases the sequences have been deposited in databases. In the future, the deposit of genomes from commercial probiotic strains, in public or restricted access databases, would overcome all the current limitations regarding identification at species and strain level. The genome sequence constitutes the best possible genetic fingerprint of a given strain.

## 3. *In vitro* Selection and Characterisation of Potential Probiotic Strains

Several health benefits have been attributed to specific probiotic bacteria and recommendations for probiotic use have recently been issued (Floch et al. 2011). However, the lack of a homogenous regulation has led to the coexistence in the market of well-studied strains and products together with non-tested ones. In Europe the recent enforcement of a strict *Regulation on Nutrition and Health Claims Made on Foods* (Regulation (EC) nº1924/2006), which, for approval of any health claim requires a rigorous scientific demonstration, has resulted in that, so far, with the sole exception of yogurt improving the symptoms of lactose intolerance, no health claim on probiotics has been approved. This underlines the need for probiotic products with an increased functionality in order to reach the high degree of evidence required to substantiate any future health claim applications. It is thus, critical to identify the strain/s with the highest potential for a specific application. To this regard, different *in vitro* tests can be performed to screen strains, for selecting the ones best suited for further assessment, before beginning expensive clinical trials. It is also important to underline that in addition to the specific strain properties, some population groups may require specific and nutritionally adequate products. Therefore, it would be advisable to consider the specific needs of these population groups, together with appropriate target-specific probiotic strains, for the development of new and improved functional probiotic products (Arboleya et al. 2012a).

The starting point for the development of such new probiotic products with increased functionality should be a careful strain characterisation. This would allow identifying and selecting the probiotic strains with appropriate properties for the desired application. To this end, the first step must always be a proper identification and typing of the strains being used. Then, the use of *in vitro* tests to assess the potential functionality of the strains may be extremely helpful for selecting the most appropriate one/s.

## 3.1 In vitro Safety Assessment

*In vitro* safety assessments should always precede the inclusion of a microorganism in the food chain. For the most widely used probiotics, LAB and *Bifidobacterium* strains, the current evidence indicate that they are safe to use in the general population. The long history of safe consumption, together with the epidemiological data, supports their safety. In Europe these commonly used probiotic species have been included in the Qualified Presumption of Safety (QPS) list by the European Food Safety Authority (EFSA) and in the USA they have received the Generally Recognised As Safe (GRAS) status. However, for other microorganisms used as probiotics the situation may be more complicated, even when they have been used for some time. Moreover, our current knowledge is limited to currently used probiotics whilst new species and strains are constantly being sought for novel probiotic products and applications. For these new and future probiotics the safety assessment may be challenging, and the specific risks related to each strain should be carefully identified. There are many *in vitro* tests that can be used for evaluating the safety of probiotics (for extensive review see Lahtinen et al. 2009). It is important, however, to point out that most of our knowledge on microbial safety comes from the study of pathogens and these may not be appropriate for probiotic strains such as lactobacilli and bifidobacteria, which are normal members of the human healthy intestinal microbiota (Borriello et al. 2003). Perhaps the most typical example of this is the ability to adhere to the human intestinal mucosa; this is considered a virulence factor for pathogens but a desirable trait for probiotic strains. Thus, although the safety assessment of probiotics may result challenging there are some properties that are frequently evaluated. These include:

### *3.1.1 Presence of Antibiotic Resistance Determinants*

During the last years the focus on the safety-related issues in the study of the antibiotic resistance (AR) determinants of probiotic strains has been receiving greater interest. This is likely due to the rise of the number of pathogens showing multi-resistances to different anti-microbials and the risk of AR genes transfer. Microorganisms intended for use as probiotics should always be screened for the presence of transferable AR genes in order to avoid their transfer to other intestinal microbes. The presence of AR genes in the probiotic genomic content is not by itself a safety concern, as long as the genes are not transferable. However, it is not always possible to demonstrate the absence of risk of transfer and, therefore, the study on whether or not probiotics can act as reservoirs for AR genes is a subject of active investigation. To this regard, the EFSA considers that the nature of any antibiotic resistance determinant present in a candidate microorganism for QPS status evaluation needs to be determined (EFSA 2012). The risk of transfer is dependent on the genetic basis of the resistance; whether it is intrinsic, acquired by chromosomal mutation(s), or acquired by horizontal gene transfer. The first two cases are considered to pose a low risk of transfer, but horizontally acquired AR genes, particularly those carried within mobile genetic elements, are likely to be transmitted between different microbes constituting an important risk

(Lahtinen et al. 2009). Thus, the ability of a probiotic strain to act as a reservoir of AR genes and the risk of them being transferred to pathogenic bacteria in the food and gut environment should always be considered.

### *3.1.2 Virulence Genes and Toxic Metabolite Production*

The potential presence of virulence genes constitutes another important point on the safety assessment of probiotics. To date, virulence factors similar to those associated with pathogenic microorganisms have not been identified for lactobacilli or bifidobacteria (Ouwehand et al. 2004, Vesterlund et al. 2007, Vankerckhoven et al. 2008a). However, for other microorganisms used as probiotics, such as enterococci or *Bacillus*, the situation is different. In general, *Enterococcus faecium* strains found in foods and used as probiotics are free from virulence determinants (Vankerckhoven et al. 2008b), while strains of *E. faecalis* typically possess multiple determinants (Eaton et al. 2001). Certain strains of *Bacillus* are also marketed as probiotics and it has been demonstrated that some of them are able to produce enterotoxins (Duc et al. 2004).

The presence of potentially harmful metabolic activities in probiotic strains has also been frequently tested. Several methods are available to study the physiological and metabolic characteristics of probiotic strains. Carbohydrate fermentation and enzymatic activity profiles have been widely used (Bianchi-Salvadori 1996, Gueimonde et al. 2004). These physiological characteristics may also be of interest for the functionality assessment of the strains. In this regard, it is important to select the specific substrates or enzymatic activities relevant to the expected functional effects of the strain. Specific tests such as the ability to hydrolyse bile salts (Lim et al. 2004), to produce biogenic amines (Muñoz-Quezada et al. 2013) or to produce antimicrobial substances (Toure et al. 2003), among others, have been frequently used.

### *3.1.3 Resistance of the Probiotic Strains to Host Defence Mechanisms*

Different *in vitro* tests can be used to determine the resistance of the strain to host defences. The ability to aggregate human platelets is considered a pathogenic trait among pathogens and it has been used for the safety assessment of probiotics (Zhou et al. 2005). However, many lactobacilli present this property, independent of their ability to cause bacteremia (Kirjavainen et al. 1999) suggesting that platelet aggregation is not a good marker for the ability of the strains to cause bacteremia. Haemolysis is another virulence factor among pathogenic microorganisms and, therefore, assessment of haemolytic activity has also been used in the *in vitro* evaluation of probiotic safety. However, this property does not seem to be present among lactobacilli and bifidobacteria (Ouwehand et al. 2004, Vesterlund et al. 2007). The test of resistance to intracellular killing by macrophages in cell culture has also been used to compare probiotic strains with clinical strains of the same species (Asahara et al. 2003). Interestingly, probiotics were found to be less resistant, although differences exist among strains. Similar findings were also reported about the ability of probiotic and clinical strains to induce respiratory burst in peripheral blood mononuclear cells and the resistance to human serum (Ouwehand et al. 2004, Vesterlund et al. 2007).

In conclusion, several different *in vitro* approaches have been used in the safety assessment of probiotics. These studies have not identified virulence determinants for the most commonly used probiotic species, suggesting that such determinants do not exist. Nowadays, *in vitro* tests assessing resistance to antibiotics and the presence of mobile antibiotic resistance determinants are commonly carried out. Probiotic intervention studies have been conducted in compromised populations, such as pre-term newborns (Luoto et al. 2010, Rougé et al. 2009) or HIV-infected patients (Wolf et al. 1998) without observing adverse effects. These together with the widespread and long-term consumption of certain probiotic products clearly demonstrate the safety of probiotic administration in general population. However, care should be taken when aiming at high risk populations, as underlined by a study carried out in severely ill patients with acute pancreatitis in which adverse events were observed in the probiotic group (Besselink et al. 2008).

## *3.2 In vitro Selection of Potentially Probiotic Strains*

The beneficial health effects of a probiotic strain cannot be assumed for other strains, even if they are from the same species, making it necessary to study several different potentially probiotic strains. This underlines the importance of *in vitro* screening for identifying and selecting the strains showing a higher potential for the further and expensive preclinical and clinical studies. Different tests can be used to this end, being highly dependent on the purpose of application of the probiotic product. These include, among others, tolerance to the gastrointestinal tract (GIT) conditions, adhesion to human mucosa, inhibition of pathogens, interaction with the gut epithelia and ability of modulating the gut microbiota or immune modulation capability.

### *3.2.1 Tolerance to GIT Conditions*

An effective probiotic should reside at the target site sufficiently long and at sufficiently high concentration to elicit the probiotic effects. Therefore, in order to survive, probiotic strains have to overcome different GIT conditions, such as pH in the stomach and bile and digestive enzymes in the intestine as well as to reach their site of action. It is known that different probiotics differ in their levels of tolerance to GIT conditions and that these gastrointestinal conditions may also modify the properties of probiotics. For instance, the adhesion properties of both *Lactobacillus* and *Bifidobacterium* have been found to be modified in the presence of bile (Ouwehand et al. 2001, Gueimonde et al. 2005). This has also been confirmed at gene expression level (Ramiah et al. 2007).

According to the FAO/WHO (2002) expert group, the survival to the GIT challenges and the ability to colonize the colon are some of the *in vitro* criteria for the selection of probiotics for food applications. Different *in vitro* models simulating the GIT transit are available to this end. These include the use of both simulated gastrointestinal juices and human origin fluids. Several publications have reported the survival of different probiotic strains under simulated conditions of the GIT using synthetic gastrointestinal juices (Masco et al. 2007, Arboleya et al. 2011, Muñoz-Quezada et al. 2013). Juices obtained from human volunteers have also been used to this end (del Piano et al. 2008), as well as simulations of the GIT transit with sequential use of human isolated gastric and duodenal juices (de los Reyes-Gavilan et al. 2011). Nevertheless, the lack of standard procedures for evaluation of tolerance to gastrointestinal conditions makes comparison among strains and studies difficult.

### *3.2.2 Adhesion to the Human GIT Mucosa*

Adhesion to the intestinal mucosa has often been considered a selection criterion for probiotic strains. It may increase the retention time of the probiotic in the gut and would bring bacteria and epithelial cells in close contact. Although still a controversial issue, some studies appear to indicate a relationship between *in vitro* adhesion and colonization ability (Castagliuolo et al. 2005) and modulation of the immune system (Schiffrin et al. 1997), and these have been related to the beneficial effects of probiotics (Castagliuolo et al. 2005). Moreover, adhesion at the target sites would result in increased levels of the probiotic at the place of action (Ouwehand et al. 2002b).

Several different methods and models have been developed for the assessment of the adhesion ability of probiotic strains (Vesterlund et al. 2005). The most frequently used ones include adhesion to intestinal mucus and adhesion to human epithelial cells in culture, wherein the most often used cell lines are Caco-2 and HT-29. The effects of gastrointestinal conditions (pH, bile and digestive enzymes) and the effects of acid and bile resistance acquisition on the adhesion of probiotic bacteria have also been documented (Gueimonde et al. 2005, Collado et al. 2006), as well as the effect of strains combination (Collado et al. 2007). However, attention should be paid to the fact that adhesion may be different depending on the target population (He et al. 2001, Arboleya et al. 2011). Thus, assessing the adhesion of the probiotic candidate strains using intestinal mucus or tissue obtained from the specific target population, for which the probiotic product is being developed, would be advisable. As an example, in a recent study Arboleya and coworkers (2011) reported important differences between two adhesion models (HT-29 cells in culture vs. infant mucus). In fact, the strain showing the highest adhesion to HT-29 cells was found to be the less adhesive to infant mucus, and differences in adhesion were also observed for some strains depending on the age of the infant at the time of the intestinal mucus collection (Arboleya et al. 2011). Moreover, adhesion may also vary in different parts of the GIT or depending on the health status of the mucosa (Ouwehand et al. 2003) and, therefore, the specific site of action could also be considered for strain selection. In this regard the use of intestinal tissue samples, allowing assessment of the adhesion to different sections along the intestine or to diseased mucosa sites, offers great possibilities for selecting site-specific probiotics (Ouwehand et al. 2002b, Ouwehand et al. 2003).

### *3.2.3 Competitive Exclusion of Pathogens*

Another one of the main selection criteria for probiotics has been the ability to competitively exclude pathogens. Probiotics may exclude pathogens by competing for nutrients or producing antimicrobial substances and by reducing the adhesion of pathogens to the gut mucosa through competition for binding sites in the GIT surface.

Different *in vitro* tests have been used to evaluate the ability of probiotic strains to inhibit pathogens, among them the use of co-culture in liquid media, agar diffusion tests, and inhibition of adhesion to both intestinal epithelial cells and human mucus have been extensively used (Collado et al. 2005, Gueimonde et al. 2007, Arboleya et al. 2011, Bermudez-Brito et al. 2013, Muñoz-Quezada et al. 2013). Different studies have found that the inhibition of the adhesion of pathogens is a rather specific process, depending on both the probiotic strain and the pathogen used (Collado et al. 2005, Gueimonde et al. 2007), thus underlining the importance of strain selection. Moreover, increases in the adhesion of pathogens to human mucus in the presence of probiotics have also been reported (Collado et al. 2005, Gueimonde et al. 2006).

### *3.2.4 Potential to Modulate Microbiota*

The intestinal microbiota has an enormous diversity and complexity, making it a very challenging subject to study. This complexity makes the *in vitro* determination of the ability of a probiotic strain to modulate the gut microbiota a difficult task. However, given the important role of the gut microbiota modulation as a mediator of probiotic function, it is important to develop a test aiming at *in vitro* assessment of this microbiota modulation ability, as this would allow screening of different strains and selecting those with the desired microbiota modulation abilities. Therefore, *in vitro* models based on the culture of faecal slurries have been used to this end (Rahman et al. 2012, Vitali et al. 2012b, Arboleya et al. 2013). In such models specific faecal samples, from certain populations or disease groups, can be used allowing for the screening and selection of probiotic strains on the basis of the purposed use. This approach has been used to identify potential probiotic strains to be used in pre-term babies. To this end, first the researchers identified the differences between the microbiota establishment processes in pre-term and full-term babies (Arboleya et al. 2012b) and then, by using an *in vitro* faecal slurry model with faeces of pre-term infants, assessed the ability of different strains to promote changes in the gut microbiota, reverting those microbiota aberrations (Arboleya et al. 2013). Most often, due to its simplicity, batch culture models are used (Vitali et al. 2012b, Arboleya et al. 2013) but dynamic gut models have also been used for determining the probiotic effects on the gut microbiota (Alander et al. 1999, Rahman et al. 2012).

### *3.2.5 Interaction with the GIT Mucosa*

In the GIT probiotics interact with intestinal epithelial cells. Although limited by the mucus layer present along the gut mucosa, this interplay may affect the local environment through modulation of gene expression by the mucosa. *In vitro* co-culture of intestinal cells with probiotics provides an easy model for studying such interactions. However, the extent to which commensal bacteria regulates the expression of immune molecules by epithelial cells *in vivo* is still poorly understood.

Although the *in vitro* studies on transcriptional responses of human epithelial cells to probiotic bacteria, especially for bifidobacteria, are scarce and lack *in vivo* confirmation (Mack et al. 2003, Riedel et al. 2006, Schlee et al. 2008, Shima et al. 2008, Paolillo et al. 2009), some studies observed interesting effects. Using co-culture models it has been reported that certain probiotics may affect the local immune environment by inducing the production of chemotactic chemokines and receptors by intestinal epithelial cells, which would attract specific immune cell subsets to the gut mucosa (Lopez et al. 2012), and by inducing the expression of antimicrobial peptides and other immune molecules (Haller et al. 2000, Schlee et al. 2008). Moreover, an increased expression of CCL20 (macrophage inflammatory protein-3α, MIP-3α) by epithelial cells in *in vitro* culture has been observed after exposure to some probiotic strains (Kingma et al. 2011, Lopez et al. 2012). During normal development and immune homeostasis, CCL20 selectively attracts CCR6-expressing lymphocytes and DCs (Baba et al. 1997, Greaves et al. 1997) contributing to the recruitment of these cells to the epithelial mucosa and to the organization of lymphoid tissue (Iwasaki and Kelsall 2000). Interestingly, these models of probiotic-epithelial cells co-culture can also be combined with immune cells to give more complex, and likely realistic models. For instance, supernatants of epithelial cells exposed to probiotics have also been used in *in vitro* experiments evaluating their ability to induce the polarization of naïve lymphocytes (Lopez et al. 2012).

In addition to the beneficial activities on the immune system, *in vitro* models of interaction between probiotics and epithelial cells have also been used to assess the potential effect of the strains in the physical resistance of the mucosa. To this regard different probiotic strains have been reported to enhance the integrity of the epithelial cells monolayer, thus contributing to strengthen the gut barrier (Jensen et al. 2012, Lopez et al. 2012).

### *3.2.6 Immune Modulation Potential*

The immune modulation ability of the strains also constitutes an important selection criterion. Elucidation of the mechanisms by which intestinal microorganisms, including potential probiotics, modulate the immune system may facilitate the development of probiotic products tailored for their immunoregulatory properties. The most frequently used *in vitro* models have been based on the determination of the production of cytokines by immune cells, most often PBMCs, after co-culture with probiotics. The observed cytokine profiles were found to be strain dependent (He et al. 2002, Christensen et al. 2002, Lopez et al. 2010). However, this model may not provide the best representation of the *in vivo* situation at the intestine.

At the gut mucosal level, the innate immune response provides the first line of defence against pathogenic microorganisms. The immune cells localized in the GALT, such as dendritic cells (DCs) and T lymphocytes, constitute the first contact point between gut commensals, or orally ingested probiotics, and the immune system. In this context, DCs are considered the link between innate and adaptive immunity. These antigen presenting cells send dendrites through the tight junctions to directly sample bacteria from the intestinal lumen (Rescigno et al. 2001). This contact with antigens will induce the maturation of DCs and their cytokine production (Joffre et al. 2009), thus acquiring the ability to induce naive T cell proliferation and polarization towards effector or regulatory cells (Zhu and Paul 2008). Several studies have reported that bacteria administered orally have the potential to modulate and regulate the immune response, through their effects on intestinal mucosa DCs. By using *in vitro* co-culture models different probiotic strains have been found to induce different maturation and cytokine production patterns on DCs, generating even opposite T-cell responses

(Christensen et al. 2002, Fujiwara et al. 2004, Hart et al. 2004, Mohamadzadeh et al. 2005, Niers et al. 2007, Baba et al. 2008, Latvala et al. 2008, Lopez et al. 2010). DCs exposed to probiotic bacteria may also acquire tolerogenic properties and then T cell activation could be directed towards the generation of regulatory T cells (Smits et al. 2005, Lopez et al. 2011). Furthermore, *in vitro* probiotics-exposed DCs can be used for stimulation of isolated naïve autologous lymphocytes to determine the effect of probiotics in a more realistic setting. Recent studies using such models suggest the ability of certain probiotic bacteria to induce regulatory T cells (Treg) from naïve precursors (de Roock et al. 2010, Lopez et al. 2011).

In conclusion, several models have been used for *in vitro* assessment of probiotic properties. However, to date no clear conclusion can be drawn on the reliability of such models and the extrapolation of the results to the *in vivo* situation. Thus, at the moment these models are used to screen for the most promising strains, with the hope that such *in vitro* screening processes will increase the possibilities for high functionality, when the chosen strains are included and further characterized for their health promoting properties in human intervention studies. Moreover, they provide valuable mechanistic information that allows both, understanding probiotic action and improving their own *in vitro* models. In the future, once the mechanisms of action of probiotics are fully unravelled and understood, it will be possible to develop *in vitro* models, or select the best suited ones, with a high predictive value in terms of *in vivo* behaviour and effects. Meanwhile, the only possible demonstration of the probiotic strain effect relies on the human studies.

## 4. Acknowledgement

The authors thank Professor Seppo Salminen for his valuable comments on the chapter.

## References

Ahire, J.J., K.P. Patil, B.L. Chaudhari and S.B. Chincholkar. 2011. *Bacillus* spp. of human origin: a potential siderophoregenic probiotic bacteria. Appl. Biochem. Biotechnol. 164: 386–400.

Alander, M., I. De Smet, L. Nollet, W. Verstraete, A. von Wright and T. Mattila-Sandholm. 1999. The effect of probiotic strains on the microbiota of the Simulator of the Human Intestinal Microbial Ecosystem (SHIME). Int. J. Food Microbiol. 46: 71–79.

Alp, G. and B. Aslim. 2010. Relationship between the resistance to bile salts and low pH with exopolysaccharide (EPS) production of *Bifidobacterium* spp. isolated from infants feces and breast milk. Anaerobe 16: 101–105.

Anukam, K.C. and G. Reid. 2009. African traditional fermented foods and probiotics. J. Med. Food 12: 1177–1184.

Apajalahti, J.H., A. Kettunen, P.H. Nurminen, H. Jatila and W.E. Holben. 2003. Selective plating underestimates abundance and shows differential recovery of bifidobacterial species from human feces. Appl. Environ. Microbiol. 69: 5731–5735.

Arboleya, S., P. Ruas-Madiedo, A. Margolles, G. Solís, S. Salminen, C.G. de Los Reyes-Gavilán and M. Gueimonde. 2011. Characterization and *in vitro* properties of potentially probiotic *Bifidobacterium* strains isolated from breast-milk. Int. J. Food Microbiol. 149: 28–36.

Arboleya, S., S. González, N. Salazar, P. Ruas-Madiedo, C.G. de los Reyes-Gavilán and M. Gueimonde. 2012a. Development of probiotic products for nutritional requirements of specific human populations. Eng. Life. Sci. 12: 368–376.

Arboleya, S., A. Binetti, N. Salazar, N. Fernández, G. Solís, A.M. Hernández-Barranco, A. Margolles, C.G. de los Reyes-Gavilán and M. Gueimonde. 2012b. Establishment and development of intestinal microbiota in preterm neonates. FEMS Microbiol. Ecol. 79: 763–772.

Arboleya, S., N. Salazar, G. Solís, N. Fernández, A.M. Hernández-Barranco, I. Cuesta, M. Gueimonde and C.G. de Los Reyes-Gavilán. 2013. Assessment of intestinal microbiota modulation ability of *Bifidobacterium* strains in *in vitro* fecal batch cultures from preterm neonates. Anaerobe 19: 9–16.

Asahara, T., M. Takahashi, K. Nomoto, H. Takayama, M. Onoue, M. Morotomi, R. Tanaka, T. Yokokura and N. Yamashita. 2003. Assessment of safety of *Lactobacillus* strains based on resistance to host innate defense mechanisms. Clin. Diagn. Lab. Immunol. 10: 169–173.

Baba, M., T. Imai, M. Nishimura, M. Kakizaki, S. Takagi, K. Hieshima, H. Nomiyama and O. Yoshie. 1997. Identification of CCR6, the specific receptor for a novel lymphocyte-directed CC chemokine LARC. J. Biol. Chem. 272: 14893–14898.

Baba, N., S. Samson, R. Bourdet-Sicard, M. Rubio and M. Sarfati. 2008. Commensal bacteria trigger a full dendritic cell maturation program that promotes the expansion of non-Tr1 suppressor T cells. J. Leukoc. Biol. 84: 468–476.

Barbosa, T.M., C.R. Serra, R.M. La Ragione, M.J. Woodward and A.O. Henriques. 2005. Henriques. Screening for bacillus isolates in the broiler gastrointestinal tract. Appl. Environ. Microbiol. 71: 968–978.

Beerens, H. 1990. An elective and selective medium for *Bifidobacterium* spp. Lett. Appl. Microbiol. 11: 155–157.

Benno, Y., K. Endo, T. Mizutani, Y. Namba, T. Komori and T. Mitsuoka. 1989. Comparison of fecal microflora of elderly persons in rural and urban areas of Japan. Appl. Environ. Microbiol. 55: 1100–1105.

Bermudez-Brito, M., J. Plaza-Díaz, L. Fontana, S. Muñoz-Quezada and A. Gil. 2013. *In vitro* cell and tissue models for studying host-microbe interactions: a review. Br. J. Nutr. 109(Suppl. 2): S27–34.

Besselink, M.G., H.C. van Santvoort, E. Buskens, M.A. Boermeester, H. van Goor, H.M. Timmerman, V.B. Nieuwenhuijs, T.L. Bollen, B. van Ramshorst, B.J. Witteman, C. Rosman, R.J. Ploeg, M.A. Brink, A.F. Schaapherder, C.H. Dejong, P.J. Wahab, C.J. van Laarhoven, E. van der Harst, C.H. van Eijck, M.A. Cuesta, L.M. Akkermans and H.G. Gooszen; Dutch Acute Pancreatitis Study Group. 2008. Probiotic prophylaxis in predicted severe acute pancreatitis: a randomised, double-blind, placebo-controlled trial. Lancet 371: 651–659.

Bianchi Salvadori, B. 1996. Selection of probiotic lactic acid bacteria (LAB) on the basis of some physiological activities. IDF Nutrition Newsletter 5: 19–23.

Borriello, S.P., W.P. Hammes, W. Holzapfel, P. Marteau, J. Schrezenmeir, M. Vaara and V. Valtonen. 2003. Safety of probiotics that contain lactobacilli or bifidobacteria. Clin. Infect. Dis. 36: 775–780.

Buckenhüskes, H.J. 1997. Fermented vegetables. pp. 595–609. *In*: P.D. Doyle, L.R. Beuchat and T.J. Montville (eds.). Food Microbiology: Fundamentals and Frontiers, 2nd ed. ASM Press, Washington DC, USA.

Cabrera-Rubio, R., M.C. Collado, K. Laitinen, S. Salminen, E. Isolauri and A. Mira. 2012. The human milk microbiome changes over lactation and is shaped by maternal weight and mode of delivery. Am. J. Clin. Nutr. 96: 544–551.

Capkin, E. and I. Altinok. 2009. Effects of dietary probiotic supplementations on prevention/treatment of yersiniosis disease. J. Appl. Microbiol. 106: 1147–1153.

Castagliuolo, I., F. Galeazzi, S. Ferrari, M. Elli, P. Brun, A. Cavaggioni, D. Tormen, G.C. Sturniolo, L. Morelli and G. Palu. 2005. Beneficial effect of auto-aggregating *Lactobacillus crispatus* on experimentally induced colitis in mice. FEMS Immunol. Med. Microbiol. 43: 197–204.

Chain, P.S.G., D.V. Grafham, R.S. Fullonn, M.G. FitzGerald, J. Hoestler, D. Muzny, J. Ali, B. Birren, D.C. Bruce, C. Buhay, J.R. Cole, Y. Ding, S. Dugan, D. Field, G.M. Garryty et al. 2009. Genome projects standards in a new era of sequencing. Science 326: 236–237.

Christensen, H.R., H. Frøkiaer and J.J. Pestka. 2002. Lactobacilli differentially modulate expresion of cytokines and maduration surface markers in murine dentritic cells. J. Immunol. 168: 171–178.

Chu, W., F. Lu, W. Zhu and C. Kang. 2011. Isolation and characterization of new potential probiotic bacteria based on quorum-sensing system. J. Appl. Microbiol. 110: 202–208.

Collado, M.C., M. Hernández and Y. Sanz. 2005. Production of bacteriocin-like inhibitory compounds by human fecal *Bifidobacterium* strains. J. Food Prot. 68: 1034–1040.

Collado, M.C., M. Gueimonde, Y. Sanz and S. Salminen. 2006. Adhesion properties and competitive pathogen exclusion ability of bifidobacteria with acquired acid resistance. J. Food Prot. 69: 1675–1679.

Collado, M.C., J. Meriluoto and S. Salminen. 2007. Development of new probiotics by strain combinations: is it possible to improve the adhesion to intestinal mucus? J. Dairy Sci. 90: 2710–2716.
Collado, M.C., E. Isolauri, S. Salminen and Y. Sanz. 2009. The impact of probiotic on gut health. Curr. Drug Metab. 10: 68–78.
Dal Bello, F., J. Walter, W.P. Hammes and C. Hertel. 2003. Increased complexity of the species composition of lactic acid bacteria in human feces revealed by alternative incubation condition. Microb. Ecol. 45: 455–463.
Darilmaz, D.O. and Y. Beyatli. 2012. Acid-bile, antibiotic resistance and inhibitory properties of propionibacteria isolated from Turkish traditional home-made cheeses. Anaerobe 18: 122–127.
Delcenserie, V., F. Gavini, H. Beerens, O. Tresse, C. Franssen and G. Daube. 2007. Description of a new species, *Bifidobacterium crudilactis* sp. nov., isolated from raw milk and raw milk cheeses. Syst. Appl. Microbiol. 30: 381–389.
Delcenserie, V., F. Gavini, B. China and G. Daube. 2011. Bifidobacterium pseudolongum are efficient indicators of animal fecal contamination in raw milk cheese industry. BMC Microbiol. 11: 178.
Dellaglio, F., G.E. Felis, A. Castioni, S. Torriani and J.E. Germond. 2005. *Lactobacillus delbrueckii* subsp. *indicus* subsp. nov., isolated from Indian dairy products. Int. J. Syst. Evol. Microbiol. 55: 401–404.
de los Reyes-Gavilán, C.G., A. Suárez, M. Fernández-García, A. Margolles, M. Gueimonde and P. Ruas-Madiedo. 2011. Adhesion of bile-adapted *Bifidobacterium* strains to the HT29-MTX cell line is modified after sequential gastrointestinal challenge simulated *in vitro* using human gastric and duodenal juices. Res. Microbiol. 162: 514–519.
de Roock, S., M. van Elk, M.E.A. van Dijk, H.M. Timmerman, G.T. Rijkers, B.J. Prakken, M.O. Hoekstra and I.M. de Kleer. 2010. Lactic acid bacteria differ in their ability to induce functional regulatory T cells in humans. Clin. Exp. Allergy 40: 103–110.
del Piano, M., P. Strozzi, M. Barba, S. Allesina, F. Deidda, P. Lorenzini, L. Morelli, S. Carmagnola, M. Pagliarulo, M. Balzarini, M. Ballarè, M. Orsello, F. Montino, M. Sartori, E. Garello and L. Capurso. 2008. *In Vitro* sensitivity of probiotics to human pancreatic juice. J. Clin. Gastroenterol. 42: S170–S173.
Di Cagno, R., R.F. Surico, S. Siragusa, M. De Angelis, A. Paradiso, F. Minervini, L. De Gara and M. Gobbetti. 2008. Selection and use of autochthonous mixed starter for lactic acid fermentation of carrots, French beans or marrows. Int. J. Food Microbiol. 127: 220–228.
Duc, le.H., H.A. Hong, T.M. Barbosa, A.O. Henriques and S.M. Cutting. 2004. Characterization of Bacillus probiotics available for human use. Appl. Environ. Microbiol. 70: 2161–2171.
Eaton, T.J. and M.J. Gasson. 2001. Molecular screening of Enterococcus virulence determinants and potential for genetic exchange between food and medical isolates. Appl. Environ. Microbiol. 67: 1628–1635.
EFSA. 2012. European Food Safety Authority Panel on Biological Hazards. Scientific Opinion on the maintenance of the list of QPS biological agents intentionally added to food and feed (2012 update). EFSA J. 10: 3020.
Endo, A. and S. Okada. 2005. Monitoring the lactic acid bacterial diversity during *shochu* fermentation by PCR-denaturing gradient gel electrophoresis. J. Biosci. Bioeng. 99: 216–221.
Endo, A. and S. Okada. 2008. Reclassification of the genus Leuconostoc, and proposals of Fructobacillus fructosus gen. nov., comb. nov., Fructobacillus durionis comb. nov., Fructobacillus ficulneus comb. nov. and Fructobacillus pseudoficulneus comb. nov. Int. J. Syst. Evol. Microbiol. 58: 2195–2205.
Endo, A., Y. Futagawa-Endo and L.M.T. Dicks. 2009. Isolation and characterization of fructophilic lactic acid bacteria from fructose-rich niches. Syst. Appl. Microbiol. 32: 593–600.
Endo, A., Y. Futagawa-Endo and L.M.T. Dicks. 2011a. Influence of carbohydrates on the isolation of lactic acid bacteria. J. Appl. Microbiol. 110: 1085–1092.
Endo, A., T. Irisawa, Y. Futagawa-Endo, K. Sonomoto, K. Itoh, K. Takano, S. Okada and L.M.T. Dicks. 2011b. *Fructobacillus tropaeoli* sp. nov., a novel fructophilic lactic acid bacterium isolated from a flower. Int. J. Syst. Evol. Microbiol. 61: 898–902.
Endo, A., Y. Futagawa-Endo, P. Schumann, R. Pukall and L.M.T. Dicks. 2012. *Bifidobacterium reuteri* sp. nov., *Bifidobacterium callitrichos* sp. nov., *Bifidobacterium saguini* sp. nov., *Bifidobacterium stellenboschense* sp. nov. and *Bifidobacterium biavatii* sp. nov. isolated from faeces of common marmoset (*Callithrix jacchus*) and red-handed tamarin (*Saguinus midas*). System. Appl. Microbiol. 35: 92–97.
FAO/WHO. 2002. Guidelines for the Evaluation of Probiotics in Food. Working Group Report 2002. Rome and Geneva: Food and Agricultural Organization of the United Nations/World Health Organization.
Ferraris, L., J. Aires, A.J. Waligora-Dupriet and M.J. Butel. 2010. New selective medium for selection of bifidobacteria from human feces. Anaerobe 16: 469–471.

Floch, M.H., W.A. Walker, K. Madsen, M.E. Sanders, G.-T. Macfarlane, H.J. Flint, L.A. Dieleman, Y. Ringel, S. Guandalini, C.P. Kelly and L.J. Brandt. 2011. Recommendations for probiotic use-2011 update. J. Clin. Gastroenterol. 45(Suppl.): S168–171.

Fujiwara, D., S. Inoue, H. Wakabayashi and T. Fujii. 2004. The anti-allergic effects of lactic acid bacteria are strain dependent and mediated by effects on both Th1/Th2 cytokine expression and balance. Int. Arch. Allergy Immunol. 135: 205–215.

Fuller, A.T., G. Mellows, M. Woolford, G.T. Banks, K.D. Barrow and E.B. Chain. 1971. Pseudomonic acid: an antibiotic produced by *Pseudomonas fluorescens*. Nature 234: 416–417.

Giovannoni, S. 1991. The polymerase chain reaction. pp. 175–201. *In*: E. Stackebrandt and M. Goodfellow (eds.). Nucleic acid Techniques in Bacterial systematics. John Wiley & Sons, New York, USA.

Greaves, D.R., W. Wang, D.J. Dairaghi, M.C. Dieu, B. Saint-Vis, K. Franz-Bacon, D. Rossi, C. Caux, T. McClanahan, S. Gordon, A. Zlotnik and T.J. Schall. 1997. CCR6, a CC chemokine receptor that interacts with macrophage inflammatory protein 3alpha and is highly expressed in human dendritic cells. J. Exp. Med. 186: 837–844.

Gueimonde, M., A. Margolles, C.G. de los Reyes-Gavilan and S. Salminen. 2007. Competitive exclusion of enteropathogens from human intestinal mucus by *Bifidobacterium* strains with acquired resistance to bile- A preliminary study. Int. J. Food Microbiol. 113: 228–232.

Gueimonde, M., L. Noriega, A. Margolles, C.G. de los Reyes-Gavilan and S. Salminen. 2005. Ability of *Bifidobacterium* strains with acquired resistance to bile to adhere to human intestinal mucus. Int. J. Food Microbiol. 101: 341–346.

Gueimonde, M., S. Delgado, B. Mayo, P. Ruas-Madiedo, A. Margolles and C.-G. los Reyes-Gavilan. 2004. Viability and diversity of probiotic *Lactobacillus* and *Bifidobacterium* populations included in commercial fermented milks. Food. Res. Int. 37: 839–850.

Gueimonde, M., L. Jalonen, F. He, M. Hiramatsu and S. Salminen. 2006. Adhesion and competitive inhibition and displacement of human enteropathogens by selected lactobacilli. Food. Res. Int. 39: 467–471.

Guo, X., D. Li, W. Lu, X. Piao and X. Chen. 2006. Screening of *Bacillus* strains as potential probiotics and subsequent confirmation of the *in vivo* effectiveness of *Bacillus subtilis* MA139 in pigs. Antonie Van Leeuwenhoek 90: 139–146.

Haller, D., S. Blum, C. Bode, W.P. Hammes and E.J. Schiffrin. 2000. Activation of human peripheral blood mononuclear cells by nonpathogenic bacteria *in vitro*: evidence of NK cells as primary targets. Infect. Immun. 68: 752–759.

Hart, A.L., K. Lammers, P. Brigidi, B. Vitali, F. Rizzello, P. Gionchetti, M. Campieri, M.A. Kamm, S.C. Knight and A.J. Stagg. 2004. Modulation of human dendritic cell phenotype and function by probiotic bacteria. Gut 53: 1602–1609.

Hartemink, R. and F.M. Rombouts. 1999. Comparison of media for the detection of bifidobacteria, lactobacilli and total anaerobes from faecal samples. J. Microbiol. Methods 36: 81–192.

Hartemink, R., V.R. Domenech and F.M. Rombouts. 1997. LAMVAB—A new selective medium for the isolation of lactobacilli from faeces. J. Microbiol. Methods 29: 77–84.

He, F., A.C. Ouwehand, E. Isolauri, M. Hosoda, Y. Benno and S. Salminen. 2001. Differences in composition and mucosal adhesion of bifidobacteria isolated from healthy adults and healthy seniors. Curr. Microbiol. 43: 351–354.

He, F., H. Morita, H. Hashimoto, M. Hosoda, J. Kurisaki, A.C. Ouwehand, E. Isolauri, Y. Benno and S. Salminen. 2002. Intestinal Bifidobacterium species induce varying cytokine production. J. Allergy Clin. Immunol. 109: 1035–1036.

Hong, H.A., H. Duc le and S.M. Cutting. 2005. The use of bacterial spore formers as probiotics. FEMS Microbiol. Rev. 29: 813–835.

Hsieh, H.H., S.Y. Wang, T.L. Chen, Y.L. Huang and M.J. Chen. 2012. Effects of cow's and goat's milk as fermentation media on the microbial ecology of sugary kefir grains. Int. J. Food Microbiol. 157: 73–81.

Hughes, J. and G. Mellows. 1978. Inhibition of isoleucyltransfer ribonucleic acid synthetase in *Escherichia coli* by pseudomonic acid. Biochem. J. 176: 305–318.

Huys, G., M. Vancanneyt, K. D'Haene, V. Vankerckhoven, H. Goossens and J. Swings. 2006. Accuracy of species identity of commercial bacterial cultures intended for probiotic or nutritional use. Res. Microbiol. 157: 803–810.

Iwasaki, A. and B.L. Kelsall. 2000. Localization of distinct Peyer's patch dendritic cell subsets and their recruitment by chemokines macrophage inflammatory protein (MIP)-3alpha, MIP-3beta, and secondary lymphoid organ chemokine. J. Exp. Med. 191: 1381–1394.

Jackson, M.S., A.R. Bird and A.L. McOrist. 2002. Comparison of two selective media for the detection and enumeration of lactobacilli in human faeces. J. Microbiol. Methods 51: 313–321.

Jensen, H., S. Grimmer, K. Naterstad and L. Axelsson. 2012. *In vitro* testing of commercial and potential probiotic lactic acid bacteria. Int. J. Food Micro. 153: 216–222.

Jian, W., L. Zhu and X. Dong. 2001. New approach to phylogenetic analysis of the genus *Bifidobacterium* based on partial HSP60 gene sequences. Int. J. Syst. Evol. Microbiol. 51: 1633–1638.

Joffre, O., M.A. Nolte, R. Spörri and C. Reis e Sousa. 2009. Inflammatory signals in dendritic cell activation and the induction of adaptive immunity. Immunol. Rev. 227: 234–247.

Junick, J. and M. Blaut. 2012. Quantification of human fecal *Bifidobacterium* species by use of quantitative real-time PCR analysis targeting the *groEL* gene. Appl. Environ. Microbiol. 78: 2613–2622.

Kataoka, K., R. Kibe, T. Kuwahara, M. Hagiwara, H. Arimochi, T. Iwasaki, Y. Benno and Y. Ohnishi. 2007. Modifying effects of fermented brown rice on fecal microbiota in rats. Anaerobe 13: 220–227.

Kawasaki, S., N. Nagasaku, T. Mimura, H. Katashima, S. Ijyuin, T. Satoh and Y. Niimura. 2007. Effect of $CO_2$ on colony development by Bifidobacterium species. Appl. Environ. Microbiol. 73: 7796–7798.

Kawasaki, S., T. Mimura, T. Satoh, K. Takeda and Y. Niimura. 2006. Response of the microaerophilic *Bifidobacterium* species, *B. boum* and *B. thermophilum*, to oxygen. Appl. Environ. Microbiol. 72: 6854–6858.

Kekkonen, R.A., R. Holma, K. Hatakka, T. Suomalainen, T. Poussa, H. Adlercreutz and R. Korpela. 2011. A probiotic mixture including galactooligosaccharides decreases fecal β-glucosidase activity but does not affect serum enterolactone concentration in men during a two-week intervention. J. Nutr. 141: 870–876.

Kingma, S.D.K., N. Li, F. Sun, R.B. Valladares, J. Neu and G.L. Lorca. 2011. *Lactobacillus johnsonii* N6.2 stimulates the innate immune response through Toll-like receptor 9 in Caco-2 cells and increases intestinal crypt Paneth cell number in biobreeding diabetes-prone rats. J. Nutr. 141: 1023–1028.

Kirjavainen, P.V., E.M. Tuomola, R.G. Crittenden, A.C. Ouwehand, D.W. Harty, L.F. Morris, H. Rautelin, M.J. Playne, D.C. Donohue and S.J. Salminen. 1999. *In vitro* adhesion and platelet aggregation properties of bacteremia-associated lactobacilli. Infect. Immun. 67: 2653–2655.

Klose, V., R. Bruckbeck, S. Henikl, G. Schatzmayr and A.P. Loibner. 2010. Identification and antimicrobial susceptibility of porcine bacteria that inhibit the growth of *Brachyspira hyodysenteriae in vitro*. J. Appl. Microbiol. 108: 1271–1280.

Kumura, H., Y. Tanoue, M. Tsukahara, T. Tanaka and K. Shimazaki. 2004. Screening of dairy yeast strains for probiotic applications. J. Dairy Sci. 87: 4050–4056.

Lahtinen, S.J., R.J. Boyle, A. Margolles, R. Frias and M. Gueimonde. 2009. Safety assessment of probiotics. pp. 1193–1225. *In*: D. Charalampopoulos and R.A. Rastall (eds.). Prebiotics and Probiotics Science and Technology. Springer-Verlag, Berlin/Heidelberg, Germany.

Latvala, S., T.E. Pietila, V. Veckman, R.A. Kekkonen, S. Tynkkynen, R. Korpela and I. Julkunen. 2008. Potentially probiotic bacteria induce efficient maturation but differential cytokine production in human monocyte-derived dendritic cells. World J. Gastroenterol. 14: 5570–5583.

Lim, H.J., S.Y. Kim and W.K. Lee. 2004. Isolation of cholesterol-lowering lactic acid bacteria from human intestine for probiotic use. J. Vet. Sci. 5: 391–395.

Lundström, H.S. and J. Björkroth. 2007. Lactic acid bacteria in marinades used for modified atmosphere packaged broiler chicken meat products. J. Food Prot. 70: 766–770.

Luoto, R., E. Isolauri and L. Lehtonen. 2010. Safety of *Lactobacillus* GG probiotic in infants with very low birth weight: twelve years of experience. Clin. Infect. Dis. 50: 1327–1328.

López, P., I. González-Rodríguez, B. Sánchez, P. Ruas-Madiedo, A. Suárez, A. Margolles and M. Gueimonde. 2012. Interaction of *Bifidobacterium bifidum* LMG13195 with HT29 cells influences regulatory-T-cell-associated chemokine receptor expression. Appl. Environ. Microbiol. 78: 2850–2857.

López, P., I. González-Rodríguez, M. Gueimonde, A. Margolles and A. Suárez. 2011. Immune Response to *Bifidobacterium bifidum* Strains Support Treg/Th17 Plasticity. PLoS ONE 6: e24776.

López, P., M. Gueimonde, A. Margolles and A. Suárez. 2010. Distinct *Bifidobacterium* strains drive different immune responses *in vitro*. Int. J. Food Microbiol. 138: 157–165.

Mack, D.R., S. Ahrne, L. Hyde, S. Wei and M. A. Hollingsworth. 2003. Extracellular MUC3 mucin secretion follows adherence of *Lactobacillus* strains to intestinal epithelial cells *in vitro*. Gut. 52: 827–833.

Makino, H., A. Kushiro, E. Ishikawa, D. Muylaert, H. Kubota, T. Sakai, K. Oishi, R. Martin, K. Ben Amor, R. Oozeer, J. Knol and R. Tanaka. 2011. Transmission of intestinal *Bifidobacterium longum* subsp. *longum* strains from mother to infant, determined by multilocus sequencing typing and amplified fragment length polymorphism. Appl. Environ. Microbiol. 77: 6788–6793.

Masco, L., C. Crockaert, K. van Hoorde, J. Swings and G. Huys. 2007. *In vitro* assessment of the gastrointestinal transit tolerance of taxonomic reference strains from human origin and probiotic product isolated of *Bifidobacterium*. J. Dairy Sci. 90: 3572–3578.

Meile, L., W. Ludwig, U. Rueger, C. Gut, P. Kaufmann, G. Dasen, S. Wenger and M. Teuber. 1997. *Bifidobacterium lactis* sp. nov., a moderately oxygen tolerant species isolated from fermented milk. Syst. Appl. Microbiol. 20: 57–64.

Mikkelsen, L.L., C. Bendixen, M. Jakobsen and B.B. Jensen. 2003. Enumeration of bifidobacteria in gastrointestinal samples from piglets. Appl. Environ. Microbiol. 69: 654–658.

Mitsuoka, T., T. Sega and S. Yamamoto. 1965. Eine verbesserte Methodik der qualitativen und quantitativen Analyse der Darmflora von Menschen und Tieren. Zentralbl. Bakteriol. Mikrobiol. Hyg. 1 Abt Orig. A 195: 455–469.

Miyake, T., K. Watanabe, T. Watanabe and H. Oyaizu. 1998. Phylogenetic analysis of the genus *Bifidobacterium* and related genera based on 16S rDNA sequences. Microbiol. Immunol. 42: 661–667.

Mohamadzadeh, M., S. Olson, W.V. Kalina, G. Ruthel, G.L. Demmin, K.L. Warfield, S. Bavari and T.R. Klaenhammer. 2005. Lactobacilli activate human dendritic cells that skew T cells toward T helper 1 polarization. Proc. Natl. Acad. Sci. USA 102: 2880–2885.

Mullié, C., M.B. Romond and D. Izard. 2006. Establishment and follow-up of bifidobacterial species in the gut of healthy bottle-fed infants of 1–4 months age. Folia Microbiol. 51: 473–477.

Muñoz-Quezada, S., E. Chenoll, J. María Vieites, S. Genovés, J. Maldonado, M. Bermúdez-Brito, C. Gomez-Llorente, E. Matencio, M. José Bernal, F. Romero, A. Suárez, D. Ramón and A. Gil. 2013. Isolation, identification and characterisation of three novel probiotic strains (*Lactobacillus paracasei* CNCM I-4034, *Bifidobacterium breve* CNCM I-4035 and *Lactobacillus rhamnosus* CNCM I-4036) from the faeces of exclusively breast-fed infants. Br. J. Nutr. 109(Suppl. 2): S51–62.

Nagata, K., S. Inatsu, M. Tanaka, H. Sato, T. Kouya, M. Taniguchi and Y. Fukuda. 2010. The bifidogenic growth stimulator inhibits the growth and respiration of *Helicobacter pylori*. Helicobacter 15: 422–429.

Naser, S.M., F.L. Thompson, B. Hoste, D. Gevers, P. Dawyndt, M. Vancanneyt and J. Swings. 2005. Application of multilocus sequence analysis (MLSA) for rapid identifi cation of Enterococcus species based on *rpoA* and *pheS* genes. Microbiology 151: 2141–2150.

Naser, S.M., P. Dawyndt, B. Hoste, D. Gevers, K. Vandemeulebroecke, I. Cleenwerck, M. Vancanneyt and J. Swings. 2007. Identification of lactobacilli by *pheS* and *rpoA* gene sequence analyses. Int. J. Syst. Evol. Microbiol. 57: 2777–2789.

Nebra, Y. and A.R. Blanch. 1999. A new selective medium for *Bifidobacterium* spp. Appl. Environ. Microbiol. 65: 5173–5176.

Niers, L.E., M.O. Hoekstra, H.M. Timmerman, N.O. van Uden, P.M. de Graaf, H.H. Smits, J.L. Kimpen and G.T. Rijkers. 2007. Selection of probiotic bacteria for prevention of allergic diseases: immunomodulation of neonatal dendritic cells. Clin. Exp. Immunol. 149: 344–352.

Ouwehand, A.C., S. Tolkko and S. Salminen. 2001. The effect of digestive enzymes on the adhesion of probiotic bacteria *in vitro*. J. Food Sci. 66: 856–859.

Ouwehand, A.C., S. Salminen and E. Isolauri. 2002a. Probiotics: an overview of beneficial effects. Antonie Van Leeuwenhoek. 82: 279–289.

Ouwehand, A.C., S. Salminen, S. Tölkkö, P. Roberts, J. Ovaska and E. Salminen. 2002b. Resected human colonic tissue: new model for characterizing adhesion of lactic acid bacteria. Clin. Diag. Lab. Immunol. 9: 184–186.

Ouwehand, A.C., S. Salminen, P.J. Roberts, J. Ovaska and E. Salminen. 2003. Disease-dependent adhesion of lactic acid bacteria to the human intestinal mucosa. Clin. Diagn. Lab. Immunol. 10: 643–646.

Ouwehand, A., M. Saxelin and S. Salminen. 2004. Assessment of potential risk factors and related properties of clinical, faecal and dairy Bifidobacterium isolates. Biosci. Microflora 23: 37–42.

Pacher, B. and W. Kneifel. 1996. Development of a culture medium for the detection and enumeration of Bifidobacteria in fermented milk products. Int. Dairy J. 6: 43– 64.

Paolillo, R., C. Romano Carratelli, S. Sorrentino, N. Mazzola and A. Rizzo. 2009. Immunomodulatory effects of *Lactobacillus plantarum* on human colon cancer cells. Int. Immunopharmacol. 9: 1265–1271.

Pennacchia, C., G. Blaiotta, O. Pepe and F. Villani. 2008. Isolation of *Saccharomyces cerevisiae* strains from different food matrices and their preliminary selection for a potential use as probiotics. J. Appl. Microbiol. 105: 1919–1928.

Rada, V. 1997. Detection of *Bifidobacterium* species by enzymatic methods and antimicrobial susceptibility testing. Biotechnol. Tech. 11: 909–912.

Rada, V. and J. Petr. 2000. A new selective medium for the isolation of glucose non-fermenting bifi dobacteria from hen caeca. J. Microbiol. Methods 43: 127–132.

Rada, V., E. Vlková, J. Nevoral and I. Trojanová. 2006. Comparison of bacterial flora and enzymatic activity in faeces of infants and calves. FEMS Microbiol Lett. 258: 25–28.

Rajkowska, K. and A. Kunicka-Styczyńska. 2010. Probiotic properties of yeasts isolated from chicken feces and kefirs. Pol. J. Microbiol. 59: 257–263.

Ramiah, K., C.A. van Reenen and L.M. Dicks. 2007. Expression of the mucus adhesion genes Mub and MapA, adhesion-like factor EF-Tu and bacteriocin gene *plaA* of *Lactobacillus plantarum* 423, monitored with real-time PCR. Int. J. Food Microbiol. 116: 405–409.

Rechman, A., F.A. Heinsen, M.E. Koenen, K. Venema, H. Knecht, S. Hellmig, S. Schreiber and S.J. Ott. 2012. Effects of probiotics and antibiotics on the intestinal homeostasis in a computer controlled model of the large intestine. BMC Microbiol. 12: 47.

Reissbrodt, R., W.P. Hammes, F. dal Bello, R. Prager, A. Fruth, K. Hantke, A. Rakin, M. Starcic-Erjavec and P.H. Williams. 2009. Inhibition of growth of Shiga toxin-producing *Escherichia coli* by nonpathogenic *Escherichia coli*. FEMS Microbiol. Lett. 290: 62–69.

Rescigno, M., M. Urbano, B. Valsazina, M. Francoloni, G. Rotta, R. Bonasio, F. Granucci, J.P. Kraehenbuhl and P. Ricciardi-Castagnoli. 2001. Dendritic cells express tight junction proteins and penetrate gut epithelial monolayers to sample bacteria. Nat. Immunol. 2: 361–367.

Riedel, C.U., F. Foata, D.R. Goldstein, S. Blum and B.J. Eikmanns. 2006. Interaction of bifidobacteria with Caco-2 cells-adhesion and impact on expression profiles. Int. J. Food Microbiol. 110: 62–68.

Rivera-Espinoza, Y. and Y. Gallardo-Navarro. 2010. Non-dairy probiotic products. Food Microbiol. 27: 1–11.

Rougé, C., H. Piloquet, M.J. Butel, B. Berger, F. Rochat, L. Ferraris, C. Des Robert, A. Legrand, M.F. de la Cochetière, J.M. N'Guyen, M. Vodovar, M. Voyer, D. Darmaun and J.C. Rozé. 2009. Oral supplementation with probiotics in very-low-birth-weight preterm infants: a randomized, double-blind, placebo-controlled trial. Am. J. Clin. Nutr. 89: 1828–1835.

Scardovi, V. and L.D. Trovatelli. 1974. *Bifidobacterium animalis* (Mitsuoka) comb. nov. and the "*minimum*" and "*subtile*" groups of new bifidobacteria found in sewage. Int. J. Syst. Bacteriol. 24: 21–28.

Schiffrin, E.J., D. Brassart, A.L. Servin, F. Rochat and A. Donnet-Hughes. 1997. Immune modulation of blood leukocytes in humans by lactic acid bacteria: criteria for strain selection. Am. J. Clin. Nutr. 66: 515S–520S.

Schlee, M., J. Harder, B. Köten, E.F. Stange, J. Wehkamp and K. Fellermann. 2008. Probiotic lactobacilli and VSL#3 induce enterocyte beta-defensin 2. Clin. Exp. Immunol. 151: 528–535.

Shima, T., K. Fukushima, H. Setoyama, A. Imaoka, S. Matsumoto, T. Hara, K. Suda and Y. Umesaki. 2008. Differential effects of two probiotic strains with different bacteriological properties on intestinal gene expression, with special reference to indigenous bacteria. FEMS Immunol. Med. Microbiol. 52: 69–77.

Simpson, P.J., G.F. Fitzgerald, C. Stanton and R.P. Ross. 2004a. The evaluation of a mupirocin-based selective medium for the enumeration of bifidobacteria from probiotic animal feed. J. Microbiol. Methods 57: 9–16.

Simpson, P.J., R.P. Ross, G.F. Fitzgerald and C. Stanton. 2004b. *Bifidobacterium psychraerophilum* sp. nov. and *Aeriscardovia aeriphila* gen. nov., sp. nov., isolated from a porcine caecum. Int. J. Syst. Evol. Microbiol. 54: 401–406.

Smits, H.H., A. Engering, D. van der Kleij, E.C. de Jong, K. Schipper, T.M.M. van Capel, B.A.J. Zaat, M. Yazdanbakhsh, E.A. Wierenga, Y. van Kooyk and M.L. Kapsenberg. 2005. Selective probiotic bacteria induce IL-10-producing regulatory T cells *in vitro* by modulating dendritic cell function through dendritic cell-specific intercellular adhesion molecule 3-grabbing nonintegrin. J. Allergy Clin. Immunol. 115: 1260–1267.

Solís, G., C.G. de Los Reyes-Gavilan, N. Fernández, A. Margolles and M. Gueimonde. 2010. Establishment and development of lactic acid bacteria and bifidobacteria microbiota in breast-milk and the infant gut. Anaerobe 16: 307–310.

Sánchez, B., S. Arias, S. Chaignepain, M. Denayrolles, J.M. Schmitter, P. Bressollier and M.C. Urdaci. 2009. Identification of surface proteins involved in the adhesion of a probiotic *Bacillus cereus* strain to mucin and fibronectin. Microbiology. 155: 1708–1716.

Tanaka, R. and M. Mutai. 1980. Improved medium for selective isolation and enumeration of *Bifidobacterium*. Appl. Environ. Microbiol. 40: 866–869.

Thitaram, S.N., G. R. Siragusa and A. Hinton, Jr. 2005. *Bifidobacterium*-selective isolation and enumeration from chicken caeca by a modified oligosaccharide antibiotic-selective agar medium. Lett. Appl. Microbiol. 41: 355–360.

Torriani, S., G.E. Felis and F. Dellaglio. 2001. Differentiation of *Lactobacillus plantarum*, *L. pentosus*, and *L. paraplantarum* by *recA* gene sequence analysis and multiplex PCR assay with *recA* gene-derived primers. Appl. Environ. Microbiol. 67: 3450–3454.

Toure, R., E. Kheadr, C. Lacroix, O. Moroni and I. Fliss. 2003. Production of antibacterial substances by bifidobacterial isolates from infant stool active against Listeria monocytogenes. J. Appl. Microbiol. 95: 1058–1069.

Vankerckhoven, V., G. Huys, M. Vancanneyt, C. Vael, I. Klare, M.B. Romond, J.M. Entenza, P. Moreillon, R.D. Wind, J. Knol, E. Wiertz, B. Pot, E.E. Vaughan, G. Kahlmeter and H. Goossens. 2008a. Biosafety assessment of probiotics used for human consumption: recommendations from the EU-PROSAFE project. Trends Food Sci. Tech. 19: 102–114.

Vankerckhoven, V., G. Huys, M. Vancanneyt, C. Snauwaert, J. Swings, I. Klare, W. Witte, T. Van Autgaerden, S. Chapelle, C. Lammens and H. Goossens. 2008b. Genotypic diversity, antimicrobial resistance, and virulence factors of human isolates and probiotic cultures constituting two intraspecific groups of *Enterococcus faecium* isolates. Appl. Environ. Microbiol. 74: 4247–4255.

Ventura, M., C. Canchaya, A. Del Casale, F. Dellaglio, E. Neviani, G.F. Fitzgerald and D. van Sinderen. 2006. Analysis of bifidobacterial evolution using a multilocus approach. Int. J. Syst. Evol. Microbiol. 56: 2783–2792.

Vesterlund, S., J. Paltta, M. Karp and A.C. Ouwehand. 2005. Measurement of bacterial adhesion—*in vitro* evaluation of different methods. J. Microbiol. Met. 60: 225–233.

Vesterlund, S., V. Vankerckhoven, M. Saxelin, H. Goossens, S. Salminen and A.C. Ouwehand. 2007. Safety assessment of *Lactobacillus* strains: presence of putative risk factors in faecal, blood and probiotic isolates. Int. J. Food Microbiol. 116: 325–331.

Vitali, B., G. Minervini, C.G. Rizzello, E. Spisni, S. Maccaferri, P. Brigidi, M. Gobbetti and R. Di Cagno. 2012a. Novel probiotic candidates for humans isolated from raw fruits and vegetables. Food Microbiol. 31: 116–125.

Vitali, B., M. Ndagijimana, S. Maccaferri, E. Biagi, M.E. Guerzoni and P. Brigidi. 2012b. An *in vitro* evaluation of the effect of probiotics and prebiotics on the metabolic profile of human microbiota. Anaerobe. 18: 386–391.

Ventura, M., D. van Sinderen, G.F. Fitzgerald and R. Zink. 2004. Insights into the taxonomy, genetics and physiology of bifidobacteria. Antonie Van Leeuwenhoek 86: 205–223.


Ventura, M., C. Canchaya, A. Del Casale, F. Dellaglio, E. Neviani, G.F. Fitzgerald and D. van Sinderen. 2006. Analysis of bifidobacterial evolution using a multilocus approach. Int. J. Syst. Evol. Microbiol. 56: 2783–2792.


Vera, A., V. Rigobello and Y. Demarigny. 2009. Comparative study of culture media used for sourdough lactobacilli. Food Microbiol. 26: 728–733.

Watanabe, K., H. Makino, M. Sasamoto, Y. Kudo, J. Fujimoto and S. Demberel. 2009. *Bifidobacterium mongoliense* sp. nov., from airag, a traditional fermented mare's milk product from Mongolia. Int. J. Syst. Evol. Microbiol. 59: 1535–1540.

Wolf, B.W., K.B. Wheeler, D.G. Ataya and K.A. Garleb. 1998. Safety and tolerance of *Lactobacillus reuteri* supplementation to a population infected with the human immunodeficiency virus. Food Chem. Toxicol. 36: 1085–1094.

Zhou, J.S., K.J. Rutherfurd and H.S. Gill. 2005. Inability of probiotic bacterial strains *Lactobacillus rhamnosus* HN001 and *Bifidobacterium lactis* HN019 to induce human platelet aggregation *in vitro*. J. Food Prot. 68: 2459–2464.

Zhu, J. and W.E. Paul. 2008. CD4 T cells: fates, functions, and faults. Blood 112: 1557–1569.

# CHAPTER 2

# A Survey on Established and Novel Strains for Probiotic Applications

*Elisa Salvetti, Sandra Torriani** and *Giovanna E. Felis*

## 1. Introduction

Probiotics are a heterogeneous group of non pathogenic bacteria that are functionally defined by their ability to produce a health benefit to the host when provided in adequate amounts. In recent years, a number of rigorous studies based on double-blind randomized placebo-controlled human intervention trials have been carried out, better supporting the idea that our health can be positively affected by the regular administration of specific probiotics. Indeed, the use of probiotics has recently moved from concept to an actual demonstration of specific benefits by specific microorganisms for specific populations. For this reason, the effects of several microbial strains that are marketed as established probiotics have been increasingly tested in well-conducted clinical studies for the treatment and prevention of selected disease states, both inside and outside the gastrointestinal tract. In the meantime, there is an interest in selecting novel candidate probiotic strains for particular subject groups. In this survey, we focused on both aspects and therefore literature data was collected and analysed (297 papers) for the strains which were applied in clinical trials in the last five years (2008–2013). Information on the identity of the strain, strain identification number and the kind of application were extracted and presented, to show that only a limited number

---

Department of Biotechnology, University of Verona, Italy.
*Postal address*: Strada le Grazie, 15 – Ca' Vignal 2, I-37134 Verona (Italy).
E-mails: elisa.salvetti@univr.it; giovanna.felis@univr.it
* Corresponding author: sandra.torriani@univr.it

*List of abbreviations at the end of the text.*

of species includes the majority of probiotic strains. Also, systematic reviews within the Cochrane Library were analyzed and diseases treated were grouped according to the categories proposed on ClinicalTrials.gov website (http://www.clinicaltrials.gov/), i.e., macro areas following the human organ or system involved. Besides the standard use of alive bacterial cells to be ingested, literature data on health benefits deriving from inactivated bacterial cells have been reported and are summarized. Also, not all the strains proved to be effective for some diseases, confirming that beneficial properties are strain and application dependent. Finally, the perspectives in the field of probiotics brought about genomics are briefly reviewed.

## 2. Established Strains for Probiotic Applications

Regulation on health claims has been introduced in Europe to protect the consumer and to harmonize national legislations on this matter. Therefore, claims associated to functional foods and to probiotics and prebiotics, intended as food supplements, could be used only in presence of evidences of health benefits (van Loveren et al. 2012). In relation to microorganisms, carefully defined health claims could be associated to the physiological functions of the gastrointestinal tract (e.g., contribution to normal bowel function, or improvement in the digestion/absorption of nutrients in specific situations), to effects related to the gut microbiota (e.g., contribution to the defense against pathogens), or to beneficial effects related to the immune system, and other possible claims could be proposed (e.g., reduction of inflammation). However, every possible claim should be linked to scientific evidences of a beneficial physiological or clinical outcome, focusing on the results of controlled human intervention studies after a pre-clinical evaluation of strain/combination of strains properties.

Human intervention studies can be classified in many ways, but blind, randomized controlled intervention trials are generally considered the most important and at the top of the list of possible human studies (randomized non-controlled, controlled non-randomized, observational studies, cohort studies, case-control studies, cross-sectional studies, etc.) (van Loveren et al. 2012).

Strains used in double-blind randomized placebo-controlled clinical studies already present in products on the market or known for more than 10 years or 10 publications from the first citation in the literature are presented in Table 1, which also includes strains recently described but used for applications already known for other strains of the same species. Strains are arranged according to the species they belong to, presented following to their hierarchical taxonomic placement—phylum, order, family (Garrity et al. 2007a, Garrity et al. 2007b) and the infrageneric phylogenetic relationship (Salvetti et al. 2012, Endo et al. 2012). The overview of applications refers to the species rather than to a specific strain. In fact, strains have been tested alone or in combinations, some of them are very well characterized and some are marketed, therefore an exhaustive review strain by strain would be impossible. Notably, known strains belong mainly to two phyla, namely *Firmicutes* and *Actinobacteria*, with the important exceptions of *Escherichia coli* Nissle 1917 and *Saccharomyces boulardii*. *Firmicutes* include genera *Bacillus, Lactobacillus* and *Pediococcus* (order *Lactobacillales*, family *Lactobacillaceae*), *Leuconostoc* (order *Lactobacillales*, family

**Table 1.** Species and strains known for their probiotic properties and relative applications. Strains are ordered according to their hierarchical taxonomic scheme (phylum, class, order, family, genus) and the infrageneric phylogenetic relationships as depicted in Salvetti et al. (2012) for *Lactobacillales* and in Endo et al. (2012) for bifidobacteria.

| Species | Strain designations | Mouth and tooth | Ear nose throat | Respiratory (including cold and influenza) | Digestive system (including *H. pylori* infection, diarrhea anterior resection syndrome, liver diseases, NEC) | Sex organs and urinary tract | Muscle, Bone, and Cartilage Diseases | Skin and connective tissue | Immune system | Metabolic disease/ hormones | Behavior/mental illness | Critical illness | Dietary Supplementation, Safety and Tolerance | Gland and hormone (includes type 2 diabetes mellitus) | nutritional metabolic disease |
|---|---|---|---|---|---|---|---|---|---|---|---|---|---|---|---|
| *Bc. coagulans* | GBI-30, 6086 | | | | x | | x | x | x | | | | | | |
| *Lb. acidophilus* | CL1285, NCIMB 30156, CUL21, CUL60, NCIMB 30157, DDS-1, DSM 24735 (VSL#3), Ecologic641, Florajen, Fortitech, Infloran, Italmex, KB27, KCTC 11906BP, L-92, LA-11, La-14, LA-5, Oxadrop, Probaclac Vaginal, Probiotical, Lafti-L10, NCFM, R0052, 74-2, AD031 | x | x | x | x | x | | x | x | | x | | x | x | x |
| *Lb. crispatus* | CTV-05 | | | | | x | | | | | | | | | |
| *Lb. delbrueckii* subsp. *bulgaricus* | DSM 24734 (VSL#3), Nutricion Medica | | | | x | | | | | | | | | | |
| *Lb. gasseri* | SBT2055 (LG2055), TMC0356, CECT5714, A5 | | | | | | | | x | | | | | | x |
| *Lb. johnsonii* | La1 | | | | x | | | x | | | | | | | |

| | | | | | | | | | | | | | | | |
|---|---|---|---|---|---|---|---|---|---|---|---|---|---|---|---|
| *Lb. casei* | Ecologic641, LBC80R, Italmex, Shirota, DN-114 001 | | x | x | x | | | | x | | | | | | |
| *Lb. paracasei* | DSM13434, CUL08, B21060, IMC 502, CRL431, DSM 24733 (VSL#3), F19, LMG P-17806, Synbiotic2000, ST11, CNCM 1-2116, NCC 2461, LMG P-22043, LPC37 | | x | x | x | | | x | x | | | x | x | | |
| *Lb. rhamnosus* | GG (ATCC 53103), LCR35, LN113, Italmex, GR-1, KCTC 11868BP, LB21 (NCIMB 40564), Antibiophilus, Probaclac Vaginal, HN001, IMC 501, Probiotical, KL53A, Probial LR 04, LPR, E/N, Oxy, Pen, PRSF-L477, and E01-A02-S06 | x | | x | x | x | x | x | x | | | x | x | | x |
| *Lb. plantarum* | HEAL 9 (DSM 15312), LP01 (LMG P-21021), Italmex, KCTC 11867BP (Duolac), CGMCC 1258, PL02, lp-115, FV9, Probial LP 02 (LMG P-21020), LMG P-20606 (2362, Synbiotic2000), 299v (DSM 9843), DSM 24730 (VSL#3), CECT 7315, CECT 7316, SN35N, SN13T, ATCC 20195, LP14 | x | | x | x | x | | | x | | | | | | |
| *Lb. brevis* | CD2, HY7401, Oxadrop | | | | x | x | | | | | | | | | |
| *Lb. coryniformis* | CECT5711 | | | | | | | | x | | | | | | |
| *Lb. fermentum* | VRI-003, LN99, Me-3, CECT5716 | | | x | x | x | | | x | | | | | | |
| *Lb. reuteri* | ATCC 55730, NCIMB 30242, RC-14, DSM 17938, ATCC PTA-5289 | x | | x | x | x | x | x | x | | | | x | | x |
| *Lb. sakei* | KCTC 10755BP | | | | | | | x | x | | | | | | |

*Table 1. contd....*

*Table 1. contd....*

| Species | Strain designations | Mouth and tooth | Ear nose throat | Respiratory (including cold and influenza) | Digestive system (including *H. pylori* infection, diarrhea) anterior resection syndrome, liver diseases, NEC) | Sex organs and urinary tract | Muscle, Bone, and Cartilage Diseases | Skin and connective tissue | Immune system | Metabolic disease / hormones | Behavior/mental illness | Critical illness | Dietary Supplementation, Safety and Tolerance | Gland and hormone (includes type 2 diabetes mellitus) | nutritional metabolic disease |
|---|---|---|---|---|---|---|---|---|---|---|---|---|---|---|---|
| *Lb. salivarius* | CUL61, Ecologic641, LS01 (DSM 22775), PM-A000, Ls-33, FV2 (VSL#3), CECT 5713 | | | | x | x | | x | x | | | | x | | |
| *Pd. acidilactici* | LN23 | | | | | x | | | | | | | | | |
| *Pd. pentosaceus* | LMG P-20608 (Synbiotic2000) | | | | x | | | | | | | | | | |
| *Ln. mesenteroides* | LMG P-20607 (Synbiotic2000) | | | | x | | | | | | | | | | |
| *St. salivarius*† | K12 | | | | | | | | | | | | x | | |
| *St. thermophilus* (*St. salivarius* subsp. *thermophilus*)† | DSM 24731 (VSL#3), Nutricion Medica, Italmex, TH-4, KCTC 11870BP (Duolac), Oxadrop, Probaclac Vaginal, Probiotical, KB19, 510 | | | | x | x | | | | | | | x | | |
| *Lc. lactis* | Ecologic641, W58, A6 | | | | x | | | x | x | | | | | | |

| | | | | | | | | | | | | | | | |
|---|---|---|---|---|---|---|---|---|---|---|---|---|---|---|---|
| *En. faecalis* | FK-23 | | | | | | | | | | | | | | x |
| *En. faecium* | SF68 | | | | | x | | | | | | | | | |
| *Bf. animalis* subsp. *lactis* | DN-173 010, CUL34 (NCIMB 30172), Ecologic641, Bb-12, BS01, UABLA-1, KCTC 11903BP, Bi-07, HN019, Probiotical, BL-04 (ATCC SD5219), W52, AD011, DGCC 420 | | x | x | x | | | x | x | | | | x | x | x |
| *Bf. bifidum* | CUL20 (NCIMB 30153), Ecologic641, BGN4, Fortitech, W23, Infloran, MIMBb75 | | | | x | | | x | x | | | | x | x | x |
| *Bf. breve* | Shirota, BR03 (DSM 16604), BB99 (DSM 13692), M-16V, DSM 24732 (VSL#3) | | x | x | x | | | x | x | | | | x | | |
| *Bf. longum* | BB536, KCTC 11860BP, DSM 24736 (VSL#3), PL03, KB31, R0175, HY8004, 46, BL-88, 2C, *Bf. longum* BL999 (ATCC BAA-999) | | | | x | x | | x | | | x | | x | | |
| *Bf. infantis* (*Bf. longum* subsp. *infantis*)* | Italmex, BB-02, Oxadrop, Probiotical, DSM 24733 (VSL#3) | | | | x | x | | | | | | | x | | |
| *Pr. freudenreichii* subsp. *shermanii* | JS (DSM 7076) | | | | x | | | | x | | | | | | |
| *Es. coli* | Nissle 1917 | | | | x | | | | | | | | | | |
| *Sc. boulardii* | Floratil, CNCM I-1079, Econorm | | | | x | x | | | | | | | | | |

*The species *Bf. infantis* has been moved to *Bf. longum* with the creation of a novel subspecies (Mattarelli et al. 2008); † The species *St. thermophilus* is a subspecies of *St. salivarius* (Farrow and Collins 1984).

*Leuconostocaceae*), *Streptococcus* and *Lactococcus* (order *Lactobacillales*, family *Streptococcaceae*), *Enterococcus* (order *Lactobacillales*, family *Enterococcaceae*), while the *Actinobacteria* group includes *Bifidobacterium* and *Propionibacterium*.

Despite the huge diversity of the genera, species including probiotics strains are a relatively small number. In particular, most of applications are covered by strains belonging to species *Lb. acidophilus* and closely phylogenetically related species, *Lb. casei-paracasei-rhamnosus* (which are phylogenetically closely related as well), *Lb. plantarum*, *Lb. reuteri*, and *Bf. animalis* subsp. *lactis*. On the other hand, applications of *Bc. coagulans*, *E. coli* Nissle 1917 and *Sc. boulardii* are very specific. Those include rheumatoid arthritis for *Bc. coagulans*, digestive system diseases such as ulcerative colitis for *E. coli* Nissle 1917, and diarrhea in infants for *Sc. boulardii* (Mandel et al. 2010, Matthes et al. 2010, Corrêa et al. 2011).

The most frequently used strain of *Lb. acidophilus* is LA-5, which has been applied for allergic diseases, digestive system complications (such as irritable bowel syndrome, ulcerative colitis and antibiotic-associated diarrhea), for treatment of type 2 diabetes mellitus and as a dietary supplementation for gut and oral colonization (Søndergaard et al. 2011, Ejtahed et al. 2012). Interestingly, a strain of *Lb. gasseri*, namely LN40, a phylogenetic relative of *Lb. acidophilus*, has been recently reported to exert positive effects in cases of bacterial vaginosis and vulvovaginal candidiasis. It has been shown that administration of *Lb. gasseri* LN40 together with other probiotic strains after conventional treatment of bacterial vaginosis and/or vulvovaginal candidiasis led to vaginal colonization, fewer recurrences and less malodorous discharge (Ehrström et al. 2010).

The phylogenetic group formed by *Lb. casei-paracasei-rhamnosus* (Salvetti et al. 2012) includes strains with problematic nomenclature, as it has been reported that virtually all the probiotic *Lb. casei* strains belong in fact to *Lb. paracasei* (http://probi.se/files/2009/07/L.-paracasei-8700-2-10.pdf). The most used strains are Shirota, ST-11, F19 and DSM 24733 for digestive system diseases (necrotizing enterocolitis, diarrhea, irritable bowel syndrome, ulcerative colitis and enteropathy), respiratory tract infections, allergic diseases (eczema and allergic rhinitis) and post-operative complications (in particular, anterior resection syndrome and liver regeneration) (e.g., West et al. 2009, Gleeson et al. 2011, Rayes et al. 2012).

Recently, *Lb. paracasei* GMNL-33, administered as a dietary supplement, was shown to reduce the caries-associated salivary populations in adults after two weeks of administration (Chuang et al. 2011), and *Lb. rhamnosus* Lc705 (DSM 7061) has been used in combination with other strains to decrease human bocavirus in the nasopharynx of otitis-prone (Lehtoranta et al. 2012).

The most used strain of *Lb. plantarum* is DSM 24730, which has been applied together with other strains for the treatment of digestive system diseases (such as irritable bowel syndrome, ulcerative colitis and enteropathy), viral diseases (rotavirus-induced diarrhea) and improvement of post-operative complications (anterior resection syndrome) (e.g., Dubey et al. 2008, Montalto et al. 2010, Stephens and Hewitt 2012).

Also, *Lb. reuteri* is a well-known species for probiotic applications thanks mainly to strain ATCC 55730, which has been used as a dietary supplement for oral and gut colonization, and for the treatment of respiratory tract infections, eczema and digestive system diseases (antibiotic-associated diarrhea, ulcerative colitis and *Helicobacter pylori* infections) (Francavilla et al. 2008, Cimperman et al. 2011, Oliva et al. 2012).

As for bifidobacteria, the most applied strains belong to *Bf. animalis* subsp. *lactis*, and in particular strain Bb-12 has been used for treatment of digestive system diseases (such as antibiotic-associated diarrhea and sepsis) and allergic diseases, maintenance of health status, immune system stimulation and improvement of type 2 diabetes mellitus and glucose metabolism (Mohan et al. 2008, de Vrese et al. 2011, Ejtahed et al. 2011).

## 3. Novel Strains for Probiotic Applications

From the above mentioned literature search, it emerged that only three novel strains, belonging to species not previously described to contain beneficial strains, have been reported: *Bacteroides xylanisolvens* DSM 23964, *Lactobacillus helveticus* HY7801, and *Tetragenococcus halophilus* Th221.

***Bacteroides xylanisolvens* DSM 23964.** *Bt. xylanisolvens* is a species described to group xylanolytic, Gram-negative, anaerobic rods isolated from faecal samples in the Phylum *Bacteroidetes* (Chassard et al. 2008). Strain DSM 23964 was shown to lack virulence determinants typical of genus *Bacteroides* to resist the action of gastric enzymes and low-pH conditions (Ulsemer et al. 2012a). It was demonstrated to be safe and was applied in two human studies which confirmed that it is tolerated by healthy humans and could contribute to the maintenance of a healthy condition (Ulsemer et al. 2012b, Ulsemer et al. 2012c). Strain DSM 23964 was proposed as the first natural strain of genus *Bacteroides* which could be proposed as a probiotic.

***Lactobacillus helveticus* HY7801.** First assigned to the species *L. suntoryeus*, it is a Gram-positive bacterium belonging to Family *Lactobacillaceae* in phylum *Firmicutes.* Strain HY7801 has been investigated for its anticolitic effect via the inhibition of TLR-4-linked NF-kappaB activation (Lee et al. 2009a). Also, it was tested in combination with *Lactobacillus brevis* HY7401, and *Bifidobacterium longum* HY8004 in mouse model for colitis with positive effects (Lee et al. 2009b). Finally it was shown to improve health conditions in mice in relation to vulvovaginal candidiasis (Joo et al. 2012). As for humans, in combination with *Bifidobacterium longum* HY8004 and *Lactobacillus brevis* HY7401, it was shown to improve energy homeostasis and liver function in IBS patients (Hong et al. 2011).

***Tetragenococcus halophilus* Th221.** This is a Gram-positive bacterium which belongs to the phylum *Firmicutes* and the family *Enterococcaceae*. The interest in this strain, which was isolated from moromi, a fermented soy sauce composed by soy bean, wheat, water and salt, is derived from its immunomodulatory activity (promotion of T helper type 1 immunity) and it was further investigated in 45 subjects with perennial allergic rhinitis (PAR) within a randomized, double-blind, placebo-controlled study. Significant improvements in terms of disease severity, nasal symptoms, in sneezing and rhinorrhea were detected in subjects who ingested *T. halophilus* Th221 together with a change in total IgE amount in their serum. These data confirmed the efficacy of this strain for PAR improvement and its safety for human consumption as assessed by additional laboratory tests (Nishimura et al. 2009).

Besides these strains, additional novel strains could be considered such as those belonging to known species (reported in Table 1) but which only recently have been used for novel purposes: *Lactobacillus salivarius* WB21, *Enterococcus faecium* IS-27526, *Bifidobacterium breve* CBG-C2 and *Bifidobacterium longum* NCC 3001.

***Lactobacillus salivarius* WB21.** *Lb. salivarius* WB21 is a Gram-positive strain, belonging to phylum *Firmicutes*, which has been involved within a clinical trial for the treatment of periodontitis. Freeze-dried *Lb. salivarius* WB21-containing tablets administration led to the improvement of periodontal clinical parameters and reduction of both salivary lactoferrin and periodontopathic oral bacteria population. These data indicate that this strain could be useful in the improvement/maintenance of oral health status and could contribute to the beneficial effects on periodontal conditions (Shimauchi et al. 2008, Myanagi et al. 2009).

***Enterococcus faecium* IS-27526.** *En. faecium* IS-27526 is a Gram-positive strain, which belongs to phylum *Firmicutes*, family *Enterococcaceae*. This strain was isolated from dadih, a traditional fermented buffalo milk from Indonesia and showed a positive effect on immune system in pre-school children. Within a randomized double-blind placebo-controlled study, Surono and colleagues (2011) observed that the total salivary IgA level and the bodyweight significantly increased in probiotic groups compared to the placebo. The results obtained confirmed that *E. faecium* IS-27526 has significant positive effects on humoral immune response and salivary IgA in underweight pre-school children, and on weight gain of pre-school children.

***Bifidobacterium breve* CBG-C2.** *Bf. breve* CBG-C2 is a Gram-positive bacterium belonging to phylum *Actinobacteria*. This strain was included in a functional yogurt supplemented with a mixture of other probiotic strains for the treatment of metabolic syndrome. Volunteers that consumed the yogurt showed significant beneficial changes in body weight and low-density lipoprotein (LDL)-cholesterol after 8 weeks. The findings of Chang and colleagues supported that regular intake of yogurt supplemented with *Bf. breve* CBG-C2 and other strains improves metabolic syndrome symptoms (Chang et al. 2011).

***Bifidobacterium longum* NCC 3001.** *Bf. longum* NCC 3001 is a Gram-positive bacterium belonging to phylum *Actinobacteria* used in the treatment of critically ill children. A test formula containing a synbiotic blend (composed of *Bf. longum* NCC 3001 in combination with another probiotic strain and prebiotics such as fructooligosaccharides, inulin, and Acacia gum) was administered to children under mechanical ventilation. The enteral formula was safe and well tolerated by children in intensive care units; furthermore it produced an increase in faecal bacterial groups for which beneficial effects were previously reported (Simakachorn et al. 2011).

## 4. Inactivated Microbial cells for Probiotic Applications

Recently, the term "paraprobiotic" was proposed for inactivated microbial cells (intact or broken) or cell fractions that confer a health benefit to the consumer (human or animal) when administered in adequate amounts, orally or topically. Bacterial cells, in fact, interact with the host besides their viability, and microbe/host crosstalk is determined by the recognition of specific bacterial components or products by the host's immune system (Taverniti and Guglielmetti 2011, van Loveren et al. 2012).

The interest in non-viable microorganisms or microbial cell extracts has been enhanced due to the safety problems originating from the large use of live microbial cells. Killed bacteria are harmless thanks to the inactivation treatment and, thus, their utilization potentially decreases shelf-life issues and the risks of microbial transmission and infection for the consumer. It is then clear that the health benefits of paraprobiotics have to be maintained after the inactivation treatment, both quantitatively and qualitatively; once the probiotic properties are confirmed, a strain can be included into the paraprobiotic category without regard to the type of inactivation method (Taverniti and Guglielmetti 2011).

Literature data are available for three strains, which have been administered as non-viable cells and could therefore be indicated as paraprobiotics: *Lactobacillus paracasei* K71, *Bifidobacterium breve* C50 and *Streptococcus thermophilus* 65.

***Lactobacillus paracasei* K71.** This strain was isolated from sake lees, a rice-based Japanese traditional fermented product. The heat-killed strain was shown to down-regulate IgE synthesis both *in vitro* and *in vivo* and, thus, it was considered as a potential safe agent with beneficial effects in adult patients with atopic dermatitis (AD). The randomized, double-blind, placebo-controlled study conducted by Moroi and colleagues (2011) confirmed that heat-killed *L. paracasei* K71 administration to adults with mild to moderate AD caused a statistically significant decrease in the mean skin severity scores. The interaction between microbe and host in patients with AD is found at gut mucosal barrier level: intake of probiotics or paraprobiotics decrease gut mucosa permeability, this alteration induces AD due to the frequent transfer of exogenous antigens. These data support the use of inactivated probiotic lactobacilli, in particular *L. paracasei* K71, in the management of symptomatic AD.

***Streptococcus thermophilus* 065 and *Bifidobacterium breve* C50.** These strains are Gram-positive bacteria, belonging to phylum *Firmicutes* and phylum *Actinobacteria*, respectively. Both strains are human isolates, which were included in a milk infant formula. Once milk fermentation was completed, the formula was heated to inactivate bacteria and then administered to children (0–2 years old) at high risk of atopy. The intake of the non-hydrolyzed infant formula decreased (i) the number of positive skin prick test to cow milk, (ii) the occurrence of digestive and respiratory incidence of potentially allergic adverse events and (iii) the proportion of positive IgE tests against other foods. These beneficial effects may be related to the impact of milk containing inactivated bacteria on intestinal flora and thymus size in healthy infants (Morisset et al. 2010). Several clinical studies were performed on the efficacy of the fermented and heated infant formula related to the reduction of the allergy and acute diarrhea events and, conversely, stimulation of immunity (Campeotto et al. 2011). Based on data obtained, fermented milk and inactivated bacteria could represent a novel approach for the management of hypersensitivity reactions in infants.

## 5. Nonstandard Probiotic Applications

The use of inactivated bacterial cell has been suggested also for atypical applications. This is the case of *Vitreoscilla filiformis* ATCC 15551.

***Vitreoscilla filiformis*** **ATCC 15551.** This strain is a filamentous, Gram-negative, nonpathogenic, nonphotosynthetic, nonfruiting glinding bacterium, which belongs to Phylum *Proteobacteria*. It was isolated from sodic sulphurated thermal springs, traditionally used for the management of dermatological disease due to their local moisturizing properties. Preliminary analyses showed that *V. filiformis* extracts were found to be effective in AD skin symptoms due to its anti-irritant properties and wound healing promotion, suggesting the use of these heat-killed bacteria in skin care preparations (Gueniche et al. 2006). A prospective, randomized, double-blind, placebo-controlled clinical study conducted on children and adults with AD which used a cream supplemented with 5% *V. filiformis* lysate showed amelioration of AD lesions, improvement of score of AD and reduction of pruritus and loss of sleep. Even if the mechanisms of action are not clear yet, the efficacy of the lysate might be related to the combination of perturbation of skin *Staphylococcus aureus* population by the presence of antimicrobial peptides and putative immunomodulatory properties. Based on the data obtained by Gueniche and colleagues (2008), the local application of *V. filiformis* extracts to the skin potentially constitutes a new strategy for the management of inflammatory responses of the skin and, as such, a novel approach for therapies to prevent relapses and stabilize AD skin.

Further studies include the investigation of biomarkers such as immunological parameters in the skin and *in vitro* and *in vivo* analyses to unravel the precise mechanism of action of *V. filiformis* lysate (Gueniche et al. 2008).

Besides the administration of inactivated bacteria which was described above, literature data exists also on the application of live bacteria by spray treatment, which cannot be considered a "standard" probiotic (bacteria are not orally ingested) neither a paraprobiotic (since in this case bacteria are not inactivated) treatment. The strain involved is *Streptococcus sanguinis* 89a.

***Streptococcus sanguinis*** **89a.** Differently from *St. thermophilus*, *St. sanguinis* is included in the viridans streptococci group and it was isolated from the human respiratory tract. Interest in this strain and in viridans streptococci in general was derived from their low presence in the nasopharyngeal flora of otitis-prone children, and, as such, they have been proposed for the treatment of acute otitis and tonsillitis. Skovbjerg and colleagues (2009) reported that spray treatment with *S. sanguinis* 89a in patients with long-standing secretory otitis media (SOM) led to the almost complete restoration of middle ear in a third of them. Although the way the product works is far from being completely understood, it was shown that presence of bacteria in the middle ear is linked to occurrence of inflammation. Activation of innate defense mechanisms and promotion of clearance of bacteria in the middle ear through bacterial spray treatment supports the usefulness and efficacy of this probiotic intake strategy in children with established SOM (Skovbjerg et al. 2009).

## 6. Inefficacy as Revealed by Clinical Trials

Literature data are also available for those strains which when applied proved not to be effective in the condition tested. Such data are particularly useful as they demonstrate that not all the strains from the same species have the same properties but also that sometimes the same strain, when tested in different applications or conditions, might be effective or not effective.

In more detail, data are reported on the inefficacy of *Lb. rhamnosus* Lc706, Lc709, and NIZO 3689, *Lb. casei* DN-114 001, *Lb. plantarum* NIZO 3684 and MF1298, *Bf. breve* BBG-01, *St. thermophilus* F DVS STM 5, and *Bc. coagulans* M/S ESKAG, tested for the treatment of a variety of diseases (hypercholesteromia, airway inflammation, children obstructive pulmonary disease patients—COPD—radiation-induced diarrhea in gynecologic cancer, IBS, and diarrhea) with no success (Giralt et al. 2008, Hatakka et al. 2008, Koning et al. 2010, Perez et al. 2010, Kukkonen et al. 2011, Ligaarden et al. 2010, Dutta et al. 2011, Matsuda et al. 2011).

Interestingly, *Lb. reuteri* DSM 17938, used in many recent studies (Dommels et al. 2009, Coccorullo et al. 2010, Savino et al. 2010, Indrio et al. 2011, Keller et al. 2012a, Keller et al. 2012b, Rojas et al. 2012, Szajewska et al. 2013) showed conflicting results on its efficacy in preventing diarrhea (Wanke et al. 2012, Agustina et al. 2012) and maintenance of oral health status in terms of improvement of gingivitis (Keller et al. 2012c, Iniesta et al. 2012).

Similarly, *Bf. longum* BL999 (ATCC BAA-999), which is used for the treatment of diarrhea and influence on gut microbiota colonization (Chouraqui et al. 2008, Hascoët et al. 2011, Grześkowiak et al. 2012), revealed inconsistent data for the treatment of eczema (Soh et al. 2009, Rautava et al. 2012).

## 7. Future Directions of Probiotic Research: the Impact of Genomics

The increasing availability of entire genome sequences of probiotic strains has allowed the detection of genes underlying health-benefit mechanisms such as those for adaptation to the intestinal niche (i.e., determinants for the resistance to low pH and bile salt or involved in carbohydrate and sugar metabolism), genes involved in the interactions with the host (i.e., cell-surface and mucus-binding proteins and pili) and probiotic factors (effector molecules able to suppress inflammatory responses, antimicrobial compounds for pathogen inhibition, constituents affecting the metabolic behaviour of the indigenous microbiota) (Lebeer et al. 2008, Ventura et al. 2012).

Despite what is already known about the molecular mechanisms by which probiotic bacteria positively impact on host health, they are still far from being completely understood. Furthermore, evolutionary analyses have shown that the adaptation of probiotic bacteria has been driven by the occurrence of DNA duplications and genetic acquisitions to successfully exist and compete in the human gut (e.g., genes involved in sugar metabolism and transport).

The collection of genomic data has led to the development of probiogenomics, which aims at revealing the potential common genetic basis for the health-promoting activities of probiotic bacteria and integrates evolutionary analysis which may have an impact on the molecular basis underlying the health benefits of these strains (Ventura et al. 2009).

Since probiotic features are strain-specific, the sequencing of probiotic genomes is becoming the golden-standard technique for probiotic selection, both for basic research and applied fields. Similar to drugs, whose safety and efficacy are not addressed generically, probiotics have to be evaluated on a strain-by-strain basis.

At present, as seen in Table 1, probiotics are both yeasts and bacteria with the latter including several genera and species, denoting a high heterogeneity underlying the probiotic definition. The use of a collective generic name should be prevented since failure to appreciate the importance of strain specificity for probiotic effects is the most important issue related to the interpretation of data on probiotics. Besides the detection and analysis of traits for which a strain is proposed as a novel probiotic, the deposit of probiotic genome sequences may overcome the current limitation regarding the identification at strain level. The rejection of several probiotic health claims by the European Food Safety Authority due to an inadequate characterization of the microbe led scientists and producers to investigate the genome sequence as the most accurate genetic fingerprint of a given strain, potentially unveiling strain-specific regions which in turn are useful for the development of novel diagnostic methods for strain traceability (Shanahan 2011).

The re-sequencing of the genome of known probiotic strains already used in industrial processes constitutes another approach to assess genome stability and its impact on probiotic functionality. As a matter of fact, the comparative genomic analysis carried out on *Lb. rhamnosus* GG strains isolated from commercial products revealed that genetic rearrangements occurred in dairy product isolates. These events included the deletion of genes, such as SpaCBA pili, that are hypothesized to play a role in probiotic functionality (Sybesma et al. 2013). These data emphasize that genome-based, high-level quality assurance and control measures have to be developed which target the presence and maintenance of genes involved in probiotic functionality, in particular those located in instable genome regions or islands with a relatively high occurrence of mobile genetic elements.

Since pathogenic and probiotic strains can coexist within the same species, the genome sequencing of microorganisms proposed as probiotics should be obtained also in order to assess their safety. Comparative studies elucidated the sequence differences potentially responsible for phenotypic shifts, as revealed for *Escherichia coli* Nissle 1917, a non-pathogenic faecal isolate extensively used as a probiotic agent in the treatment of a range of gastro-intestinal disorders, and virulent *E. coli* strains. It has been shown that both pathogenic and non-pathogenic *E. coli* genomes share highly conserved regions but they also present unique variable regions, which have been acquired by horizontal gene transfer. Differences between probiotic *E. coli* Nissle 1917 and the virulent strains are based on the absence of a number of genes related to genomic islands or prophage regions which role remains unknown but is potentially related to virulence factors (Vejborg et al. 2010). As seen in the latter example, comparative genomics between pathogenic and non-pathogenic strains are supposed to unravel the differences in gene content between safe and non-safe strains and to potentially create a list of pathogenicity or safety markers within a species.

Over the past few years, the application of next-generation sequencing approaches to the study of human-associated microorganisms has shown that the composition of the microbiota within the guts of different individuals is distinct, defining the enterotypes for the first time (Arumugam et al. 2011). Matching host characteristics in health and disease with functional changes and patterns in the microbiota composition is going to be the driver for the next phase of exploration (Jeffrey et al. 2012).

The description of enterotypes and gradients of microbiota composition related to the health status has provided insights into the impact of microbes on the development of the host and the contribution of microbiota to the health maintenance. With these novel findings, the probiotic concept is undergoing transition as knowledge of the gut microbiota becomes translated into the clinic. Analysis of healthy microbiota potentially yields new candidate strains but selection criteria have to be based on demonstrable mechanisms of action, which are disclosed from the analysis of microbe/host and microbe-microbe interaction. Variability among closely related species and, as such, among probiotic strains, in terms of functional properties originates because of the diversity of their signalling with the host, which elicits differential responses (Shanahan et al. 2012).

The exploration of human microbiota composition in health and disease along with the development of molecular interaction models potentially enable the selection of the most suitable probiotic strain for a particular health benefit thanks also to the combination of -omics techniques, such as functional and comparative genomics analysis (Ventura et al. 2009)

Advances in the research of human microbiota and microbiome (the total collection of microorganisms within a community, including the biomolecules produced) and their influence on the host gene expression might allow in the near future the choice of a specific probiotic strain for a particular human genotype along with personalized medicine efforts.

## 8. Non-conventional Abbreviations of Genus

*Bc.*, *Bacillus*; *Bf.*, *Bifidobacterium*; *Bt.*, *Bacteroides*; *Lb.*, *Lactobacillus*; *Lc.*, *Lactococcus*; *Ln.*, *Leuconostoc*; *En.*, *Enterococcus*; *Es.*, *Escherichia*; *Pd.*, *Pediococcus*; *Pr.*, *Propionibacterium*; *Sc.*, *Saccharomyces*; *St.*, *Streptococcus*

## References

http://www.clinicaltrials.gov/.

http://probi.se/files/2009/07/L.-paracasei-8700-2-10.pdf.

Agustina, R., F.J. Kok, O. van de Rest, U. Fahmida, A. Firmansyah, W. Lukito, E.J. Feskens, E.G. van den Heuvel, R. Albers and I.M. Bovee-Oudenhoven. 2012. Randomized trial of probiotics and calcium on diarrhea and respiratory tract infections in Indonesian children. Pediatrics 129: e1155–1164.

Arumugam, M., J. Raes, E. Pelletier, D. Le Paslier, T. Yamada, D.R. Mende, G.R. Fernandes, J. Tap, T. Bruls, J.M. Batto, M. Bertalan, N. Borruel, F. Casellas, L. Fernandez, L. Gautier, T. Hansen, M. Hattori, T. Hayashi, M. Kleerebezem, K. Kurokawa, M. Leclerc, F. Levenez, C. Manichanh, H.B. Nielsen, T. Nielsen, N. Pons, J. Poulain, J. Qin, T. Sicheritz-Ponten, S. Tims, D. Torrents, E. Ugarte, E.G. Zoetendal, J. Wang, F. Guarner, O. Pedersen, W.M. de Vos, S. Brunak, J. Dorè, MetaHIT Consortium, M. Antolín, F. Artiguenave, H.M. Blottiere, M. Almeida, C. Brechot, C. Cara, C. Chervaux, A. Cultrone, C. Delorme, G. Denariaz, R. Dervyn, K.U. Foerstner, C. Friss, M. van de Guchte, E. Guedon, F. Haimet,

W. Huber, J. van Hylckama-Vlieg, A. Jamet, C. Juste, G. Kaci, J. Knol, O. Lakhdari, S. Layec, K. Le Roux, E. Maguin, A. Mérieux, R. Melo Minardi, C. M'rini, J. Muller, R. Oozeer, J. Parkhill, P. Renault, M. Rescigno, N. Sanchez, S. Sunagawa, A. Torrejon, K. Turner, G. Vandemeulebrouck, E. Varela, Y. Winogradsky, G. Zeller, J. Weissenbach, S.D. Ehrlich and P. Bork. 2011. Enterotypes of the human gut microbiome. Nature 473: 174–180.

Campeotto, F., A. Suau, N. Kapel, F. Magne, V. Viallon, L. Ferraris, A.J. Waligora-Dupriet, P. Soulaines, B. Leroux, N. Kalach, C. Dupont and M.J. Butel. 2011. A fermented formula in pre-term infants: clinical tolerance, gut microbiota, down-regulation of faecal calprotectin and up-regulation of faecal secretory IgA. Br. J. Nutr. 22: 1–10.

Chang, B.J., S.U. Park, Y.S. Jang, S.H. Ko, N.M. Joo, S.I. Kim, C.H. Kim and D.K. Chang. 2011. Effect of functional yogurt NY-YP901 in improving the trait of metabolic syndrome. Eur. J. Clin. Nutr. 65: 1250–1255.

Chassard, C., E. Delmas, P.A. Lawson and A. Bernalier-Donadille. 2008. *Bacteroides xylanisolvens* sp. nov., a xylan-degrading bacterium isolated from human faeces. Int. J. Syst. Evol. Microbiol. 58: 1008–1013.

Chouraqui, J.P., D. Grathwohl, J.M. Labaune, J.M. Hascoet, I. de Montgolfier, M. Leclaire, M. Giarre and P. Steenhout. 2008. Assessment of the safety, tolerance, and protective effect against diarrhea of infant formulas containing mixtures of probiotics or probiotics and prebiotics in a randomized controlled trial. Am. J. Clin. Nutr. 87: 1365–1373.

Chuang, L.C. , C.S. Huang, L.W. Ou-Yang and S.Y. Lin. 2011. Probiotic *Lactobacillus paracasei* effect on cariogenic bacterial flora. Clin. Oral Investig. 15: 471–476.

Cimperman, L., G. Bayless, K. Best, A. Diligente, B. Mordarski, M. Oster, M. Smith, F. Vatakis, D. Wiese, A. Steiber and J. Katz. 2011. A randomized, double-blind, placebo-controlled pilot study of *Lactobacillus reuteri* ATCC 55730 for the prevention of antibiotic-associated diarrhea in hospitalized adults. J. Clin. Gastroenterol. 45: 785–789.

Coccorullo, P., C. Strisciuglio, M. Martinelli, E. Miele, L. Greco and A. Staiano. 2010. *Lactobacillus reuteri* (DSM 17938) in infants with functional chronic constipation: a double-blind, randomized, placebo-controlled study. J. Pediatr. 157: 598–602.

Corrêa, N.B., F.J. Penna, F.M. Lima, J.R. Nicoli and L.A. Filho. 2011. Treatment of acute diarrhea with *Saccharomyces boulardii* in infants. J. Pediatr. Gastroenterol. Nutr. 53: 497–501.

de Vrese, M., H. Kristen, P. Rautenberg, C. Laue and J. Schrezenmeir. 2011. Probiotic lactobacilli and bifidobacteria in a fermented milk product with added fruit preparation reduce antibiotic associated diarrhea and *Helicobacter pylori* activity. J. Dairy Res. 78: 396–403.

Dommels, Y.E., R.A. Kemperman, Y.E. Zebregs, R.B. Draaisma, A. Jol, D.A. Wolvers, E.E. Vaughan and R. Albers. 2009. Survival of *Lactobacillus reuteri* DSM 17938 and *Lactobacillus rhamnosus* GG in the human gastrointestinal tract with daily consumption of a low-fat probiotic spread. Appl. Environ. Microbiol. 75: 6198–6204.

Dubey, A.P., K. Rajeshwari, A. Chakravarty and G. Famularo. 2008. Use of VSL[sharp]3 in the treatment of rotavirus diarrhea in children: preliminary results. J. Clin. Gastroenterol. 42: S126–129.

Dutta, P., U. Mitra, S. Dutta, K. Rajendran, T.K. Saha and M.K. Chatterjee. 2011. Randomised controlled clinical trial of *Lactobacillus sporogenes* (*Bacillus coagulans*), used as probiotic in clinical practice, on acute watery diarrhoea in children. Trop. Med. Int. Health. 16: 555–561.

Ehrström, S., K. Daroczy, E. Rylander, C. Samuelsson, U. Johannesson, B. Anzén and C. Påhlson. 2010. Lactic acid bacteria colonization and clinical outcome after probiotic supplementation in conventionally treated bacterial vaginosis and vulvovaginal candidiasis. Microbes Infect. 12: 691–699.

Ejtahed, H.S., J. Mohtadi-Nia, A. Homayouni-Rad, M. Niafar, M. Asghari-Jafarabadi, V. Mofid and A. Akbarian-Moghari. 2011. Effect of probiotic yogurt containing *Lactobacillus acidophilus* and *Bifidobacterium lactis* on lipid profile in individuals with type 2 diabetes mellitus. J. Dairy Sci. 94: 3288–3294.

Ejtahed, H.S., J. Mohtadi-Nia, A. Homayouni-Rad, M. Niafar, M. Asghari-Jafarabadi and V. Mofid. 2012. Probiotic yogurt improves antioxidant status in type 2 diabetic patients. Nutrition 28: 539–543.

Endo, A., Y. Futagawa-Endo, P. Schumann, R. Pukall and L.M. Dicks. 2012. *Bifidobacterium reuteri* sp. nov., *Bifidobacterium callitrichos* sp. nov., *Bifidobacterium saguini* sp. nov., *Bifidobacterium stellenboschense* sp. nov. and *Bifidobacterium biavatii* sp. nov. isolated from faeces of common marmoset (*Callithrixjacchus*) and red-handed tamarin (*Saguinusmidas*). Syst. Appl. Microbiol. 35: 92–97.

Farrow, J.A.E. and M.D. Collins. 1984. DNA base composition, DNA-DNA homology and long-chain fatty acid studies on *Streptococcus thermophilus* and *Streptococcus salivarius*. J. Gen. Microbiol. 130: 357–362.

Francavilla, R., E. Lionetti, S.P. Castellaneta, A.M. Magistà, G. Maurogiovanni, N. Bucci, A. De Canio, F. Indrio, L. Cavallo, E. Ierardi and V.L. Miniello. 2008. Inhibition of *Helicobacter pylori* infection in humans by *Lactobacillus reuteri* ATCC 55730 and effect on eradication therapy: a pilot study. Helicobacter. 13: 127–134.

Garrity, G.M., T.G. Lilburn, J.R. Cole, S.H. Harrison, J. Euzéby and B.J. Tindall. 2007a. Taxonomic outline of the Bacteria and Archaea. Release 7.7 March 6, 2007. Part 9- The Bacteria: Phylum "*Firmicutes*": Class "*Bacilli*". http://www.taxonomicoutline.org/index.php/toba/article/view/186/218.

Garrity, G.M., T.G. Lilburn, J.R. Cole, S.H. Harrison, J. Euzéby and B.J. Tindall. 2007b. Taxonomic outline of the Bacteria and Archaea. Release 7.7 March 6, 2007. Part 9—The Bacteria: Phylum "*Actinobacteria*": Class "*Actinobacteria*". http://www.taxonomicoutline.org/index.php/toba/article/view/186/218.

Giralt, J., J.P. Regadera, R. Verges, J. Romero, I. de la Fuente, A. Biete, J. Villoria, J.M. Cobo and F. Guarner. 2008. Effects of probiotic *Lactobacillus casei* DN-114 001 in prevention of radiation-induced diarrhea: results from multicenter, randomized, placebo-controlled nutritional trial. Int. J. Radiat. Oncol. Biol. Phys. 71: 1213–1219.

Gleeson, M., N.C. Bishop, M. Oliveira and P. Tauler. 2011. Daily probiotic's (*Lactobacillus casei* Shirota) reduction of infection incidence in athletes. Int. J. Sport Nutr. Exerc. Metab. 21: 55–64.

Grześkowiak, Ł., M.M. Grönlund, C. Beckmann, S. Salminen, A. von Berg and E. Isolauri. 2012. The impact of perinatal probiotic intervention on gut microbiota: double-blind placebo-controlled trials in Finland and Germany. Anaerobe 18: 7–13.

Guéniche, A., A. Hennino, C. Goujon, K. Dahel, P. Bastien, R. Martin, R. Jourdain and L. Breton. 2006. Improvement of atopic dermatitis skin symptoms by *Vitreoscilla filiformis* bacterial extract. Eur. J. Dermatol. 16: 380–384.

Guéniche, A., B. Knaudt, E. Schuck, T. Volz, P. Bastien, R. Martin, M. Röcken, L. Breton and T. Biedermann. 2008. Effects of nonpathogenic Gram-negative bacterium *Vitreoscilla filiformis* lysate on atopic dermatitis: a prospective, randomized, double-blind, placebo-controlled clinical study. Br. J. Dermatol. 159: 1357–1363.

Hascoët, J.M., C. Hubert, F. Rochat, H. Legagneur, S. Gaga, S. Emady-Azar and P.G. Steenhout. 2011. Effect of formula composition on the development of infant gut microbiota. J. Pediatr. Gastroenterol. Nutr. 52: 756–762.

Hatakka, K., M. Mutanen, R. Holma, M. Saxelin and R. Korpela. 2008. *Lactobacillus rhamnosus* LC705 together with *Propionibacterium freudenreichii* ssp. *shermanii* JS administered in capsules is ineffective in lowering serum lipids. J. Am. Coll. Nutr. 27: 441–447.

Hong, Y.S., K.S. Hong, M.H. Park, Y.T. Ahn, J.H. Lee, C.S. Huh, J. Lee, I.K. Kim, G.S. Hwang and J.S. Kim. 2011. Metabonomic understanding of probiotic effects in humans with irritable bowel syndrome. J. Clin. Gastroenterol. 45: 415–425.

Indrio, F., G. Riezzo, F. Raimondi, M. Bisceglia, A. Filannino, L. Cavallo and R. Francavilla. 2011. *Lactobacillus reuteri* accelerates gastric emptying and improves regurgitation in infants. Eur. J. Clin. Invest. 41: 417–422.

Iniesta, M., D. Herrera, E. Montero, M. Zurbriggen, A.R. Matos, M.J. Marín, M.C. Sánchez-Beltrán, A. Llama-Palacio and M. Sanz. 2012. Probiotic effects of orally administered *Lactobacillus reuteri*-containing tablets on the subgingival and salivary microbiota in patients with gingivitis. A randomized clinical trial. J. Clin. Periodontol. 39: 736–744.

Jeffery, I.B., M.J. Claesson, P.W. O'Toole and F. Shanahan. 2012. Categorization of the gut microbiota: enterotypes or gradients? Nat. Rev. Microbiol. 10: 591–592.

Joo, H.M., K.A. Kim, K.S. Myoung, Y.T. Ahn, J.H. Lee, C.S. Huh, M.J. Han and D.H. Kim. 2012. *Lactobacillus helveticus* HY7801 ameliorates vulvovaginal candidiasis in mice by inhibiting fungal growth and NF-κB activation. Int. Immunopharmacol. 14: 39–46.

Keller, M.K. and S. Twetman. 2012a. Acid production in dental plaque after exposure to probiotic bacteria. BMC Oral Health. 12: 44.

Keller, M.K., A. Bardow, T. Jensdottir, J. Lykkeaa and S. Twetman. 2012b. Effect of chewing gums containing the probiotic bacterium *Lactobacillus reuteri* on oral malodour. Acta Odontol. Scand. 70: 246–250.

Keller, M.K., P. Hasslöf, G. Dahlén, C. Stecksén-Blicks and S. Twetman. 2012c. Probiotic supplements (*Lactobacillus reuteri* DSM 17938 and ATCC PTA 5289) do not affect regrowth of mutants streptococci after full-mouth disinfection with chlorhexidine: a randomized controlled multicenter trial. Caries Res. 46: 140–146.

Koning, C.J., D. Jonkers, H. Smidt, F. Rombouts, H.J. Pennings, E. Wouters E., Stobberingh and R. Stockbrügger. 2010. The effect of a multispecies probiotic on the composition of the faecal microbiota and bowel habits in chronic obstructive pulmonary disease patients treated with antibiotics. Br. J. Nutr. 10: 1452–1460.

Kukkonen, A.K., M. Kuitunen, E. Savilahti, A. Pelkonen, P. Malmberg and M. Mäkelä. 2011. Airway inflammation in probiotic-treated children at 5 years. Pediatr. Allergy Immunol. 22: 249–251.

Lebeer, S., J. Vanderleyden and S.C. De Keersmaecker. 2008. Genes and molecules of lactobacilli supporting probiotic action. Microbiol. Mol. Biol. Rev. 72: 728–764.

Lee, J.H., B. Lee, H.S. Lee, E.A. Bae, H. Lee, Y.T Ahn, K.S. Lim, C.S. Huh and D.H. Kim. 2009a. *Lactobacillus suntoryeus* inhibits pro-inflammatory cytokine expression and TLR-4-linked NF-kappaB activation in experimental colitis. Int. J. Colorectal. 24: 231–237.

Lee, H., Y.T. Ahn, J.H. Lee, C.S. Huh and D.H. Kim. 2009b. Evaluation of anti-colitic effect of lactic acid bacteria in mice by cDNA microarray analysis. Inflammation 32: 379–386.

Lehtoranta, L., M. Söderlund-Venermo, J. Nokso-Koivisto, H. Toivola, K. Blomgren, K. Hatakka, T. Poussa, R. Korpela and A. Pitkäranta. 2012. Human bocavirus in the nasopharynx of otitis-prone children. Int. J. Pediatr. Otorhinolaryngol. 76: 206–211.

Ligaarden, S.C., L. Axelsson, K. Naterstad, S. Lydersen and P.G. Farup. 2010. A candidate probiotic with unfavourable effects in subjects with irritable bowel syndrome: a randomised controlled trial. BMC Gastroenterol. 10: 16.

Mandel, D.R., K. Eichas and J. Holmes. 2010. *Bacillus coagulans*: a viable adjunct therapy for relieving symptoms of rheumatoid arthritis according to a randomized, controlled trial. BMC Complement. Altern. Med. 10: 1.

Matsuda, F., M.I. Chowdhury, A. Saha, T. Asahara, K. Nomoto, A.A. Tarique, T. Ahmed, M. Nishibuchi, A. Cravioto and F. Qadri. 2011. Evaluation of a probiotics, *Bifidobacterium breve* BBG-01, for enhancement of immunogenicity of an oral inactivated cholera vaccine and safety: a randomized, double-blind, placebo-controlled trial in Bangladeshi children under 5 years of age. Vaccine. 28: 1855–1858.

Mattarelli, P., C. Bonaparte, B. Pot and B. Biavati. 2008. Proposal to reclassify the three biotypes of *Bifidobacterium longum* as three subspecies: *Bifidobacterium longum* subsp. *longum* subsp. nov., *Bifidobacterium longum* subsp. *infantis* comb. nov. and *Bifidobacterium longum* subsp. *suis* comb. nov. Int. J. Syst. Evol. Microbiol. 58: 767–772.

Matthes, H., T. Krummenerl, M. Giensch, C. Wolff and J. Schulze. 2010. Clinical trial: probiotic treatment of acute distal ulcerative colitis with rectally administered *Escherichia coli* Nissle 1917 (EcN). BMC Complement. Altern. Med. 10: 13.

Mayanagi, G., M. Kimura, S. Nakaya, H. Hirata, M. Sakamoto, Y. Benno and H. Shimauchi. 2009. Probiotic effects of orally administered *Lactobacillus salivarius* WB21-containing tablets on periodontopathic bacteria: a double-blinded, placebo-controlled, randomized clinical trial. J. Clin. Periodontol. 36: 506–513.

Mohan, R., C. Koebnick, J. Schildt, M. Mueller, M. Radke and M. Blaut. 2008. Effects of *Bifidobacterium lactis* Bb12 supplementation on body weight, fecal pH, acetate, lactate, calprotectin, and IgA in preterm infants. Pediatr. Res. 64: 418–422.

Montalto, M., A. Gallo, V. Curigliano, F. D'Onofrio, L. Santoro, M. Covino, S. Dalvai, A. Gasbarrini and G. Gasbarrini. 2010. Clinical trial: the effects of a probiotic mixture on non-steroidal anti-inflammatory drug enteropathy—a randomized, double-blind, cross-over, placebo-controlled study. Aliment. Pharmacol. Ther. 32: 209–214.

Morisset, M., C. Aubert-Jacquin, P. Soulaines, D.A. Moneret-Vautrin and C. Dupont. 2010. A non-hydrolyzed, fermented milk formula reduces digestive and respiratory events in infants at high risk of allergy. Eur. J. Clin. Nutr. 65: 175–183.

Moroi, M., S. Uchi, K. Nakamura, S. Sato, N. Shimizu, M. Fujii, T. Kumagai, M. Saito, K. Uchiyama, T. Watanabe, H. Yamaguchi, T. Yamamoto, S. Takeuchi and M. Furue. 2011. Beneficial effect of a diet containing heat-killed *Lactobacillus paracasei* K71 on adult type atopic dermatitis. J. Dermatol. 38: 131–139.

Nishimura, I., T. Igarashi, T. Enomoto, Y. Dake, Y. Okuno and A. Obata. 2009. Clinical efficacy of halophilic lactic acid bacterium *Tetragenococcus halophilus* Th221 from soy sauce moromi for perennial allergic rhinitis. Allergol. Int. 58: 179–185.

Oliva, S., G. Di Nardo, F. Ferrari, S. Mallardo, P. Rossi, G. Patrizi, S. Cucchiara and L. Stronati. 2012. Randomised clinical trial: the effectiveness of *Lactobacillus reuteri* ATCC 55730 rectal enema in children with active distal ulcerative colitis. Aliment. Pharmacol. Ther. 35: 327–34.

Pérez, N., J.C. Iannicelli, C. Girard-Bosch, S. González, A. Varea, L. Disalvo, M. Apezteguia, J. Pernas, D. Vicentin and R. Cravero. 2010. Effect of probiotic supplementation on immunoglobulins, isoagglutinins and antibody response in children of low socio-economic status. Eur. J. Nutr. 9: 173–179.

Rautava, S., E. Kainonen, S. Salminen and E. Isolauri. 2012. Maternal probiotic supplementation during pregnancy and breast-feeding reduces the risk of eczema in the infant. J. Allergy Clin. Immunol. 130: 1355–1360.

Rayes, N., T. Pilarski, M. Stockmann, S. Bengmark, P. Neuhaus and D. Seehofer. 2012. Effect of pre- and probiotics on liver regeneration after resection: a randomised, double-blind pilot study. Benef. Microbes. 3: 237–244.

Rojas, M.A., J.M. Lozano, M.X. Rojas, V.A. Rodriguez, M.A. Rondon, J.A. Bastidas, L.A. Perez, C. Rojas, O. Ovalle, J.E. Garcia-Harker, M.E. Tamayo, G.C. Ruiz, A. Ballesteros, M.M. Archila and M. Arevalo. 2012. Prophylactic probiotics to prevent death and nosocomial infection in preterm infants. Pediatrics 130: e1113–1120.

Salvetti, E., S. Torriani and G.E. Felis. 2012. The Genus *Lactobacillus*: A Taxonomic Update. Probiotics & Antimicro. Prot. 4: 217–226.

Savino, F., L. Cordisco, V. Tarasco, E. Palumeri, R. Calabrese, R. Oggero, S. Roos and D. Matteuzzi. 2010. *Lactobacillus reuteri* DSM 17938 in infantile colic: a randomized, double-blind, placebo-controlled trial. Pediatrics 126: e526–533.

Shanahan, F. 2011. Molecular mechanisms of probiotic action: it's all in the strains! Gut. 60: 1026–1027.

Shanahan, F., T.D. Dinan, P. Ross and C. Hill. 2012. Probiotic in transition. Clin. Gastroenterol. Hepatol. 10: 1220–1224.

Shimauchi, H., G. Mayanagi, S. Nakaya, M. Minamibuchi, Y. Ito, K. Yamaki and H. Hirata. 2008. Improvement of periodontal condition by probiotics with *Lactobacillus salivarius* WB21: a randomized, double-blind, placebo-controlled study. J. Clin. Periodontol. 35: 897–905.

Simakachorn, N., R. Bibiloni, P. Yimyaem, Y. Tongpenyai, W. Varavithaya, D. Grathwohl, G. Reuteler, J.C. Maire, S. Blum, P. Steenhout, J. Benyacoub and E.J. Schiffrin. 2011. Tolerance, safety, and effect on the faecal microbiota of an enteral formula supplemented with pre- and probiotics in critically ill children. J. Pediatr. Gastroenterol. 53: 174–181.

Skovbjerg, S., K. Roos, S.E. Holm, E. Grahn Håkansson, F. Nowrouzian, M. Ivarsson, I. Adlerberth and A.E. Wold. 2009. Spray bacteriotherapy decreases middle ear fluid in children with secretory otitis media. Arch. Dis. Child. 94: 92–98.

Soh, S.E., M. Aw, I. Gerez , Y.S. Chong, M. Rauff, Y.P. Ng, H.B. Wong, N. Pai, B.W. Lee and L.P. Shek. 2009. Probiotic supplementation in the first 6 months of life in at risk Asian infants—effects on eczema and atopic sensitization at the age of 1 year. Clin. Exp. Allergy 39: 571–578.

Søndergaard, B., J. Olsson, K. Ohlson, U. Svensson, P. Bytzer and R. Ekesbo. 2011. Effects of probiotic fermented milk on symptoms and intestinal flora in patients with irritable bowel syndrome: a randomized, placebo-controlled trial. Scand. J. Gastroenterol. 46: 663–672.

Stephens, J.H. and P.J. Hewett. 2012. Clinical trial assessing VSL#3 for the treatment of anterior resection syndrome. ANZ. J. Surg. 82: 420–427.

Surono, I.S., F.P. Koestomo, N. Novitasari, F.R. Zakaria, Yulianasari and Koesnandar. 2011. Novel probiotic *Enterococcus faecium* IS-27526 supplementation increased total salivary sIgA level and bodyweight of pre-school children: a pilot study. Anaerobe 17: 496–500.

Sybesma, W., D. Molenaar, W. van IJcken, K. Venema and R. Kort. 2013. Genome instability in *Lactobacillus rhamnosus* GG. Appl. Environ. Microbiol. 79: 2233–2239.

Szajewska, H., E. Gyrczuk and A. Horvath. 2013. *Lactobacillus reuteri* DSM 17938 for the management of infantile colic in breastfed infants: a randomized, double-blind, placebo-controlled trial. J. Pediatr. 162: 257–262.

Taverniti, V. and S. Guglielmetti. 2011. The immunomodulatory properties of probiotic microorganisms beyond their viability (ghost probiotics: proposal of paraprobiotic concept). Genes Nutr. 6: 261–274.

Ulsemer, P., K. Toutounian, J. Schmidt, U. Karsten and S. Goletz. 2012a. Preliminary safety evaluation of a new *Bacteroides xylanisolvens* isolate. Appl. Environ. Microbiol. 78: 528–535.

Ulsemer, P., K. Toutounian, G. Kressel, J. Schmidt, U. Karsten, A. Hahn and S. Goletz. 2012b. Safety and tolerance of *Bacteroides xylanisolvens* DSM 23964 in healthy adults. Benef. Microbes. 3: 99–111.

Ulsemer, P., K. Toutounian, J. Schmidt, J. Leuschner, U. Karsten and S. Goletz. 2012c. Safety assessment of the commensal strain *Bacteroides xylanisolvens* DSM 23964. Regul. Toxicol. Pharmacol. 62: 336–346.
van Loveren, H., Y. Sanz and S. Salminen. 2012. Health Claims in Europe: Probiotics and Prebiotics as Case Examples. Annu. Rev. Food. Sci. Technol. 3: 247–261.
Vejborg, R.M., C. Friis, V. Hancock, M.A. Schembri and P. Klemm. 2010. A virulent parent with probiotic progeny: comparative genomics of *Escherichia coli* strains CFT073, Nissle 1917 and ABU 83972. Mol. Genet. Genomics 283: 469–484.
Ventura, M., S. O'Flaherty, M.J. Claesson, F. Turroni, T.R. Klaenhammer, D. van Sinderen and P.W. O'Toole. 2009. Genome-scale analysis of health-promoting bacteria: probiogenomics. Nat. Rev. Microbiol. 7: 61–71.
Ventura, M., F. Turroni and D. van Sinderen. 2012. Probiogenomics as a tool to obtain genetic insights into adaptation of probiotic bacteria to the human gut. Bioeng. Bugs. 3: 73–79.
Wanke, M. and H. Szajewska. 2012. Lack of an effect of *Lactobacillus reuteri* DSM 17938 in preventing nosocomial diarrhea in children: a randomized, double-blind, placebo-controlled trial. J. Pediatr. 161: 40–43.
West, C.E., M.L. Hammarström and O. Hernell. 2009. Probiotics during weaning reduce the incidence of eczema. Pediatr. Allergy Immunol. 20: 430–437.

# CHAPTER 3

# Probiotic Cell Cultivation

*Mauricio Santos, Elizabeth Tymczyszyn, Marina Golowczyc, Pablo Mobili* and *Andrea Gomez-Zavaglia**

## 1. Introduction

Research on lactic acid bacteria (LAB) and probiotics is nowadays an interdisciplinary field going beyond the fundamental microbiology. The commercial importance of certain microbial metabolites (i.e., lactic acid) has led to the use of these microorganisms as "biological factories" of valuable products. On the other hand, the worldwide ecological conscience about the importance of reducing industrial contamination has stimulated the use of certain effluents as "non-traditional" culture media for LAB and probiotics. This approach has two advantages: the decrease of industrial costs derived from decontamination of effluents and the lower cost of culture media.

Monitoring bacterial growth is another example of the current interdisciplinarity in the research of LAB and probiotics. Fundamental fields like physics, mathematics and statistics were determinant for the current development of environmentally friendly analytical methods allowing *in situ* and in real time monitoring.

In this chapter, an overview of probiotic cell cultivation and monitoring including the use of both traditional and "non-traditional" culture media, the production of commercially valuable products and biomass is reported. Special emphasis was put on the fundamentals and interpretation of the information provided by analytical methods used for monitoring.

## 2. Cultivation of Microorganisms

The formulation of adequate culture media requires prior knowledge of different culture parameters, including optimal temperature and pH, nutrient necessities, redox potential

Center for Research and Development in Food Cryotechnology (CIDCA, CCT- CONICET La Plata) Argentina.
* Corresponding author: angoza@qui.uc.pt

and the necessity of supplementation for the isolation, identification or obtaining of biomass. A minimum culture medium typically contains water, carbon and nitrogen sources, small amounts of inorganic ions (i.e., sulfur, phosphorus, calcium, potassium and magnesium) and buffer capacity (Roy 2003, Hammes and Hertel 2006). Complex media contain, additionally, vitamins, mineral and growth factors to favor the growth of fastidious microorganisms (Reuter 1985).

Based on these definitions, culture media can be selective and/or differential. Utilization of different carbon sources or the presence of antibiotics may promote the development of a certain type of microorganisms and inhibit the development of others (selective media). On the other hand, the presence of dyes that are sensitive to pH or precipitate in the presence of certain metabolites leads to morphological differences in the colonies that may contribute to differentiate species of microorganisms (differential media). Selective and differential culture media are often used for the identification and enumeration of microorganisms in food and biological samples. Even when there is an increased tendency to use techniques like FISH, flow cytometry and real time quantitative PCR, selective and differential plate counts are still widely used as routine controls in the industry, because of their lower cost (Reuter 1985, Roy 2003).

## *2.1 Cultivation of Probiotics*

According to the World Health Organization (WHO) and the Food and Agriculture Organization (FAO), probiotics are "live microorganisms which when administered in adequate amounts confer a health benefit on the host" (WHO/FAO 2002).

Probiotics are used in the formulation of functional food and pharmaceutical products due to their health promoting effect. Most of the probiotic microorganisms are non sporulated gram positive bacilli from the genera *Lactobacillus* and *Bifidobacterium.* Because of their biosynthetic deficiencies, microorganisms of both genera are fastidious and require complex culture media for growth (Reuter 1985, Hammes and Hertel 2006). Gram positive cocci, such as *Streptococcus thermophilus*, *Lactococcus lactis* and *Enterococcus faecium* have also been reported as probiotics (Reuter 1985, Mitsouka 1992, Roy 2003, Franz et al. 2011).

Lactobacilli, streptococci and lactococci hydrolyze lactose to give glucose and galactose, which can be used as an energy source. This lactase activity explains the capacity of these microorganisms to grow in milk and milk related products such as whey and whey permeate.

### *2.1.1 Culture Media for Lactobacillus*

Most of the probiotic strains of the genus *Lactobacillus* belong to the species: *Lb. acidophilus*, *Lb. casei*, *Lb. paracasei*, *Lb. rhamnosus*, *Lb. gasseri*, *Lb. johnsonii*, *Lb. plantarum*, *Lb. fermentum*, *Lb. reuteri*, *Lb. crispatus* and *Lb. brevis* (Mitsouka 1992, Vinderola et al. 2011). An appropriate culture medium for lactobacilli should have peptides as a nitrogen source, sugars (i.e., lactose or glucose) as carbon source, yeast extract as growth factor, magnesium and manganese in optimal concentrations. Yeast extract provides vitamins, amino acids, purines and pyrimidines (Djeghri-Hocine et al. 2010).

Different growth media have been developed for *Lactobacillus*. Among them, the medium reported by de Man et al. (1960) MRS is the most widely used. MRS is a selective medium used for growth, enumeration and isolation of LAB from food and human feces (Shillinger and Holzapfel 2003, Hammes and Hertel 2006). All species of lactobacilli are able to grow in MRS medium, which is composed of peptone and glucose as nitrogen and carbon sources, respectively. Magnesium, manganese and acetate are growth factors and also inhibit the growth of other microorganisms; citrate inhibits the growth of gram negative bacteria. Polysorbate 80 (also called sorbitan monooleate or Tween 80), provides the fatty acids required by lactobacilli to grow (Sharpe and Fryer 1965).

Different modifications make MRS a more selective and/or differential medium. These modifications include a replacement of the carbon source, addition of antibiotics, decrease of pH or changes in the incubation conditions (temperature, time and presence of oxygen). Vinderola and Reinheimer (1999 and 2000) modified MRS by the addition of trehalose and bile salts, making it selective to enumerate *Lb. acidophilus* in yogurts containing also *Bifidobacterium*, *Lb. bulgaricus* and *St. thermophilus*. In the same way, Lankaputhra et al. (1996) replaced glucose by maltose and Dave et al. (1996) used salicin or sorbitol as the only source of carbon, for the selective enumeration of *Lb. acidophilus*. MRS supplemented with bile salts (Lima et al. 2009) or clindamycin (van de Casteele et al. 2006) has been reported as selective medium for *Lb. acidophilus*. Supplementation with lithium chloride and propionic acid (Lima et al. 2009) or incubation at 15ºC (Champagne et al. 1997) makes MRS selective for *Lb. casei* (Champagne et al. 1997).

Before the formulation of MRS, Rogosa described another selective medium for *Lactobacillus* (Rogosa 1951). Rogosa medium is more selective than MRS because of its higher acidity (pH 5.5) and the presence of high concentrations of acetate, which is inhibitory of other LAB (*Streptococcus*, *Leuconostoc* and *Pediococcus*) (Reuter 1985). This medium contains peptone, yeast extract and ammonium as a nitrogen source, glucose as a source of energy, polysorbate, manganese and magnesium. It also has ammonium citrate, sodium acetate, acetic acid and ferrous sulfate that decrease pH to 5.5 and act as inhibitors of *Streptococcus* and other microorganisms.

Other culture media have been reported for lactobacilli: LBS (Mitsuoka 1978), LaS (Lactobacilli-sorbic acid pH 5) (Reuter 1970), RCM (Johns et al. 1978), TPPY (Ghoddusi and Robinson 1996), LSD (Eloy and Lacrosse 1976), LAMVAB (Hartemink et al. 1997), LPSM (Bujalance et al. 2006) and Elliker broth (Elliker et al. 1956). Their selective capacity for different species of lactobacilli (pH, temperature, addition of dyes, antibiotics supplement) has been widely discussed in several reviews (Reuter 1985, Schillinger and Holzapfel 2003, Ashraf and Shah 2011).

### *2.1.2 Culture Media for Bifidobacterium*

Bifidobacteria are normal inhabitants of the gastrointestinal tract (Mitsuoka 1992 and 1978). The most important probiotic species are: *Bb. animalis, Bb. bifidum, Bb. adolescentis, Bb. lactis, Bb. brevis*, *Bb. longum* and *Bb. longum* subsp. *infantis* (formerly *Bb. infantis*) (Matarelli et al. 2008). The best conditions and growth media for the enumeration of *Bifidobacterium* in food were reported by Roy in 2003.

*Bifidobacterium* is able to grow in MRS or RCM (reinforced clostridial medium) (Munoa and Pares 1988) in anaerobic conditions. Supplementation with neomycin, nalidixic acid and lithium chloride makes these media more selective for *Bifidobacterium*. Other bifidobacteria culture media include Columbia Agar supplemented with sodium propionate and LiCl (Roy 2003), NPNL (neomycin, paromomycin, nalidixic acid, lithium chloride-based media) (Teraguchi et al. 1978) and TPY (tripticase, phytone and yeast extract) (Scardovi 1986), all of them allowing the growth of almost all species of *Bifidobacterium*. Ashraf and Shah (2011) overviewed the media allowing differential and selective enumerations of bifidobacteria in yogurts.

Culture media for bifidobacteria contain short chain fatty acids, propionate and butyrate, yeast extract as growth factor, cysteine to reduce redox potential and improve anaerobic conditions and starch (Hartemink et al. 1996). The beneficial effect of adding α-lactoalbumin and β-lactoglobulin was also reported (Ibrahim and Bezkorovainy 1994).

The addition of bifidogenic oligosaccharides as carbon sources makes culture media selective for *Bifidobacterium*. Bifidogenic oligosaccharides include lactulose, galacto-oligosaccharidos (GOS), also known as trans-galactosyl-oligosaccharides (TOS), fructo-oligosaccharides (FOS), isomalto-oligosaccharides, raffinose and soybean oligo-saccharides (Roy 2003). TOS have demonstrated the greatest bifidogenic effects (Tanaka et al. 1983).

Rada and Koc (2000) and Simpson et al. (2004) reported the addition of mupirocin supplement for the selective enumeration of *Bifidobacterium* in milk and probiotic animal feed. The International Organization for Standardization and International Dairy Federation 2008 (ISO/DIS 29981) recommends the combination of bifidogenic TOS with lithium and mupirocin for the selective enumeration of bifidobacteria in milk products.

## 3. Effect of Growth Conditions on the Probiotic Properties and Biomass Production

The composition of growth media and the conditions of fermentation (i.e., pH, temperature, oxygen, etc.) have a great effect on the probiotic properties of lactobacilli and bifidobacteria (Oh et al. 1995, Saarela et al. 2005, Schar-Zammaretti et al. 2005, Deepika et al. 2012, Dong et al. 2014). Various growth media, such as MRS, skim milk, M-17, Elliker's broth, and whey permeates, have been widely used for probiotic growth at laboratory scale. However, for an industrial scale cheaper culture media are often preferred. When a culture medium is to be chosen, industrials consider other factors including costs, ability to produce large number of cells and the harvesting method. In addition, it is important to take into account the growth phase at harvesting, the physiological state and the method used for bacterial preservation (Kelly et al. 2005, Saarela et al. 2009). Moreover, when probiotic's functionality is evaluated, the food matrix and the technological variables shall also be considered (Kankaanpää et al. 2001, Mättö et al. 2006, Ranadheera et al. 2010, Grześkowiak et al. 2013, Páez et al. 2013).

The effect of growing media and preservation methods on viability and functionality of probiotics is summarized below.

**Cells and mucus adhesion:** The adhesion to the intestinal epithelium and mucosa is important for stimulating the immune system and is crucial to transient colonization. Adhesion to intestinal mucosa may promote bacterial persistence in the gastrointestinal tract and their beneficial effects to the host (Forestier et al. 2001, Mercenier et al. 2003). Different authors reported that the expression of adhesins may be altered when microorganisms are grown in different media and/or conditions, thereby stressing the importance of these factors to maintain probiotic properties (Ouwehand 2001, Deepika et al. 2012).

For laboratory investigation, lactobacilli are usually grown on MRS and most of the reported studies were carried out on bacteria grown in this medium. However, a milk-based media with not completely known composition is generally used at an industrial level (Ouwehand 2001, Páez et al. 2013). Therefore, the reported studies carried out in MRS may not be representative for the adhesive capacities of the same strains when grown industrially (Gilliland 1985, Savage 1992, Ouwehand 2001). In these latter conditions, microorganisms may become coated with milk components or enclosed in a coagulated casein matrix. Furthermore, the presence of different nutrients may affect the expression of adhesins. Ouwehand et al. (2001) reported that the adhesive ability of several probiotic lactobacilli to different substrates (human ileostomy glycoproteins, intestinal glycoproteins and polystyrene) is strongly influenced by the growth medium. In this regard, these authors reported that the adhesion of *Lb. reuteri* and *Lb. brevis* is significantly reduced when grown in milk whey, LDM or API-broth in comparison to MRS broth. In contrast, changes in culture media do not affect the adhesion of *Lb. rhamnosus* LC-705 (Ouwehand et al. 2001). The presence of free polyunsaturated fatty acids (PUFA) in the growth medium also inhibits bacterial adhesion to human intestinal mucus *in vitro* (Kankaanpää et al. 2001).

There are other components that favor the adhesion and persistence of probiotics in the gut when present in the culture media. Jonsson et al. (2001) reported that adding mucin to the growth medium triggers mucus-binding activity in different strains of *Lb. reuteri in vitro* due to the production of cell surface proteins with a mucus-binding capacity. In a recent article, Grześkowiak et al. (2013) reported a higher percentage of adhesion to dog mucus when canine probiotic lactobacilli are grown in laboratory rather than in manufacturing conditions.

**Bacteriocin production:** Bacteriocins are compounds (usually small peptides) with antimicrobial activity, produced by probiotic strains. Bacteriocin production may be affected by the growth temperature, pH, composition of the culture medium, growth phase, etc. (Nebra and Blanch 1999, Aasen et al. 2000, Todorov et al. 2000, Todorov and Dicks 2005a, Todorov et al. 2011, 2012, Schirru et al. 2013). The optimal production of bacteriocins requires a specific combination of environmental parameters, changes in the growth conditions being detrimental (Leal-Sánchez et al. 2002). However, the effect of these factors on the production of bacteriocins in complex environments (i.e., food), is very poorly studied. Some authors have suggested that an increase in the production of bacteriocins occurs under unfavorable growth conditions (i.e., low temperature, nutrient limitation, osmotic stress, etc.) (de Vuyst et al. 1996, Papagianni

2012) whereas others reported opposite results (Aasen et al. 2000). In spite of this controversial issue, it seems to be established that metabolic regulation is crucial for bacteriocin production (Aasen et al. 2000).

The growth temperature also influences the production of bacteriocins. Todorov et al. (2011) observed significant differences in *Lb. plantarum* ST16Pa growth and bacteriocin production when the strain is cultured in MRS broth for 24 h. The optimal temperatures for bacteriocin production are 26 and 30°C for 24 h. When grown at 15°C and 37°C, a reduction of bacteriocin production is observed although the bacterial growth is good. Drosinos et al. (2005) reported that bacteriocin production is favored when *Leuc. Mesenteroides* E131 is grown at temperatures close to the optimal.

The initial pH of MRS broth plays an important role in bacteriocin production. Several authors reported that an initial pH of at least 6.0 is required for the optimal growth of different strains of *Lb. plantarum* as well as for bacteriocin production (Daeschel et al. 1990, Kelly et al. 1996, Todorov and Dicks 2005b, Todorov et al. 2010, 2011).

The composition of the culture medium affects the production of bacteriocins. Microorganisms grown in non-traditional media, such as cheese whey, may have lower production than those grown in MRS (Schirru et al. 2013). The addition of tryptone to the culture medium increases the production of bacteriocins in several strains of *Lb. plantarum*, while the addition of glycerol represses bacteriocin production (Todorov et al. 2005c, 2009). In turn, the presence of vitamins does not stimulate bacteriocin production (Todorov et al. 2005c, 2009). Other studies indicate that carbon/nitrogen ratio influences the production of bacteriocins by *Leuc. mesenteroides* L124 and *Lb. curvatus* L442 (Mataragas et al. 2004).

The nature and concentration of the carbon source is critical for bacteriocin production. Audisio et al. (2001) reported a higher production of bacteriocins against *St. pullorum* when *Enterococcus faecium* CRL 1385 is grown in media containing brown sugar as carbon source rather than glucose or a simple carbohydrate complex. Drosinos et al. (2005) reported that the presence of fructose in the culture medium favors bacteriocin production by *Leuc. mesenteroides* E131.

The addition of enzymes like amylase and lipase, as well as the addition of 1% glucose or 1% peptone to MRS broth has a slightly positive effect on bacteriocin production by *Lb. delbrueckii* strains isolated from yoghurt (Devi et al. 2013). Addition of surfactants such as Tween 80 to co-cultures of *Bifidobacterium* spp. and *Listeria monocytogenes* increases the production of bacteriocins as a consequence of an acceleration of the cell growth (Touré et al. 2003).

The synthesis of bacteriocins is associated with cell growth and is dependent on the biomass production. Todorov and Dicks (2005a) reported that optimal levels of plantaricin ST194BZ, the bacteriocin produced by *Lb. plantarum* ST194BZ, are obtained in growth media allowing high biomass production (i.e., MRS). In contrast, other authors observed that the optimal conditions for bacteriocin production do not correspond with those optimal for growth, ascribing the increase in bacteriocin production to relatively low growth rates (Mataragas et al. 2003). Pasteris et al. (2014) observed that the bacteriocin production by *Lc. lactis* subsp. *lactis* CRL 1584 is associated with bacterial growth phase, the highest production occurring at the end of the exponential growth phase.

Fernandez et al. (2013) have recently demonstrated that probiotic strains of *Lc. lactis* subsp. *lactis* biovar diacetylactis, *Lc. lactis* and *Pediococcus acidilactici* are able to produce bacteriocins in simulated gastrointestinal conditions.

**Immunomodulatory activity:** Immunomodulatory activity is one of the beneficial properties of probiotics and can be affected by the growth medium. However, only few papers dealing with this issue have been published hereto. *Lc. lactis* G50 grown in M17 medium supplemented with glucose induces a higher production of IL-12 by J774 cells than when grown in MRS broth (Kimoto-Nira et al. 2008). This effect was ascribed to changes in cellular components or in the structure of the cell surface induced by the growth media.

**Biomass production:** As mentioned in section 2.1 (*Cultivation of probiotics*), probiotic bacteria are nutritionally fastidious microorganisms requiring numerous nutrients to grow properly. In order to optimize biomass production, a careful formulation of growth media is required. To attain this aim, many authors use the so called Response Surface Methodology (RSM). It consists in a statistical methodology that allows the optimization of experimental designs when several variables are to be considered. RSM can be applied for the construction of models and the evaluation of cause-effect relations. Its main advantage is that the relation among several variables can be established even in the presence of interactions among variables (Box and Wilson 1951, Oh et al. 1995, Azaola et al. 1999, Montgomery et al. 2001, Liew et al. 2005, Liu et al. 2010, Wang and Liu 2008, Chang and Liew 2013).

Yeast extract is the best source of nitrogen in culture media due to the presence of high content of nitrogen compounds, purine and pyrimidine bases and vitamins. This results in a high production of lactobacilli and bifidobacteria biomass (Yoo et al. 1997, Azaola et al. 1999, Nancib et al. 2001, Liew et al. 2005, Dong et al. 2014). However, its high cost challenged some authors to evaluate more economical alternatives. Liu et al. (2010) investigated the effect of five alternative nitrogen sources (malt sprout, corn steep liquor, $NH_4Cl$, $NH_4NO_3$ and diamine citrate) on the biomass production of *Lb. plantarum*, finding that corn steep liquor and malt sprout are the most suitable alternatives. Dong et al. (2014) reported that the use of tryptone or the combination of yeast extract and tryptone increase cell density of *Lb. salivarius* more than the single use of yeast extract.

Monosaccharides and disaccharides are better carbon sources than polysaccharides (Dong et al. 2014). Mataragas et al. (2004) reported that the concentration of glucose does not improve the total biomass produced by *Lb. curvatus* L442 and *Leuc. mesenteroides* L124, whereas Liew et al. (2005) showed a slight effect of glucose on the cell number of *Lb. rhamnosus* ATCC 7469. Polak-Berecka et al. (2010) found that using glucose plus sodium pyruvate strongly increases the biomass production of *Lb. rhamnosus* PEN in comparison with other carbon sources like glucose alone or other carbohydrates. Azaola et al. (1999) observed that although glucose is not a limiting factor in biomass production by *Bb. longum* subsp. *infantis*, its absence may result in a significant decrease of biomass production.

As mentioned in section 2.1.1 (Culture media for *Lactobacillus*), the presence of Tween-80 in culture media strongly favors *Lactobacillus* growth (Partanen et al. 2001, Tan et al. 2012, Dong et al. 2014). Most LAB grown in media containing Tween-80 can integrate oleic acid into their membranes and then transform it into cyclopropane fatty acids. Cyclopropane fatty acid acts as a polyunsaturated fatty acid in the sense that it increases membrane fluidity, protecting bacteria from adverse environmental effects (Partanen et al. 2001).

The addition of minerals and other microelements to the culture media is crucial for the propagation of microorganisms and act as enzymes activators. $Mn^{+2}$ is essential for the growth and biomass production of several species of *Lactobacillus* (Fitzpatrick et al. 2001, Groot et al. 2005, Wegkamp et al. 2009, Dong et al. 2014). Other ions, like $Mg^{+2}$, $Cu^{+2}$, $Fe^{+2}$, showed no significant growth-stimulating effect on *Lb. salivarius* BBE 09–18 (Dong et al. 2014).

The biosynthetic deficiencies of lactobacilli and bifidobacteria are related with their incapacity to produce precursors of B vitamins, amino acids, purines, pyrimidines, and unsaturated fatty acid in anaerobic conditions (Fitzpatrick et al. 2001, Partanen et al. 2001, Liu et al. 2010). Therefore, culture media must include these growth factors. Dong et al. (2014) reported that vitamin C, guanine, uracil, and nicotinic acid have no significant stimulating effect on *Lb. salivarius* BBE 09–18 growth. On the contrary, glycine and folic acid inhibit its propagation.

In addition to the importance of using a good growth medium supporting optimal cell production, environmental factors such as pH and agitation rate strongly determine bacterial growth. In this regard, it was reported that the best condition for the growth of *Bb. pseudocatenulatum* G4 in a milk-based medium is obtained at an impeller tip speed of 0.56 m/s and at controlled pH 6.5 (Stephenie et al. 2007). Bevilacqua et al. (2008) reported that the biomass production of *Lb. plantarum* c19 isolated from table olives is maximum at pH 6.0.

## 4. Non-traditional Culture Media

Industrial culturing of LAB pursues different objectives, namely the production of biomass to be used as probiotic, their use as starters to transform food substrates (i.e., milk, meat or vegetables) into fermented products (carried out through the production of acids), the production of enzymes or the synthesis of metabolites such as exopolysaccharides or organic acids as the desired end-product to be later purified.

As stated in section 2.1 (Cultivation of probiotics), LAB require complex media to grow, with adequate supply of carbon and nitrogen sources, inorganic ions, vitamins and even hormones and growth factors. The media used for the growth of LAB at a laboratory scale, such as MRS or TPPY, may not be economically suitable for the production of bacterial cells and/or extracellular metabolites at large scale, especially when the desired products are lower-value high-volume chemicals, such as lactic acid. Therefore, big efforts are made to find alternative cheaper nutrient sources to sustain industrial biomass and metabolites production. In this sense, the use of agro-industrial wastes as substrates for bacterial growth has gained attention since it addresses two problems at once: on one hand it supplies less expensive carbon and nitrogen sources, reducing the costs of production of bacterial mass and/or its metabolites, and on the other hand it allows the utilization of wastes that otherwise should be treated prior to their elimination, thus adding value to wastes.

Several agricultural residues have been investigated as alternative carbon sources for microbial growth: sugar containing materials such as molasses and maple sap (Kotzamanidis et al. 2002, Cochu et al. 2008, Dumbrepatil et al. 2008), soybean vinasse (Karp et al. 2011), bagasses from sugarcane or cassava (John et al. 2005), starchy raw materials from grains, tubers or carrots (Vishnu et al. 2002, Fukushima et al. 2004, Altaf et al. 2007) and ligno-cellulosic plant materials (Wee and Ryu 2009), among others. Some wastes of animal origin such as cheese whey (rich in lactose and proteins), whey permeate or mussel processing wastewater (rich in glycogen) were also used to grow LAB (Pintado et al. 1999, Guerra and Pastrana 2002, Bergmaier et al. 2005).

Most LAB lack the amylolytic and cellulolytic enzymes necessary for the digestion of starchy, glycogenic or cellulosic residues, thus an acid or enzymatic hydrolysis to maltose and glucose is needed prior to the utilization of those wastes as carbon sources. Only those LAB with high amylase and amyloglucosidase activities, such as several strains of *Lb. plantarum*, *Lb. fermentum*, *Lb. amylovorus* and *Lb. amylophilus*, are able to simultaneously hydrolyze glycogen or starch and ferment the released glucose (Reddy et al. 2008).

Among the bulk-chemicals obtained by fermentation processes, lactic acid is of major importance. Lactic acid is used in the food industry as a preservative, pH regulator and taste-enhancing additive, and in the chemical industry as the starting material for the production of biodegradable plastics based on lactide polymers that could effectively replace those derived from petrol. Such plastics, due to their GRAS (Generally recognized as safe) status, are approved for use in contact with food.

Biological synthesis of lactic acid has several advantages over chemical synthesis. The chemical synthesis always yields the racemic DL-lactic acid mixture, while bacterial conversion of pyruvate to lactate is stereospecific. Therefore, according to the enzyme used by microorganisms for this reaction, pure D- or L-lactic acid are produced. Another advantage of biological synthesis is the lower energy consumption and the lower cost of raw materials (John et al. 2007). Some agro-industrial by-products currently studied as carbon sources for lactic acid production are: potato, cassava, corn and barley starches; corn fiber hydrolyzates, corn cobs and stalks; lignocellulose/hemicellulose hydrolyzates; beet molasses; sugarcane press mud; wheat bran; cheese whey and whey permeate; mussel processing wastes, etc. (John et al. 2007, Reddy et al. 2008, Yadav et al. 2011).

Besides carbon sources, LAB also require peptides, amino acids, vitamins, etc., that should be provided in an inexpensive way to replace the costly yeast extract. Ammonium sulfate supplemented with vitamins showed good results (Nancib et al. 2005), but several agro-industrial wastes were also studied as lower-cost supplements. Grain-derived by-products such as corn steep liquor (Kim et al. 2006), malt-combing nuts (Fitzpatrick et al. 2003), malt sprout (Goksungur and Guvene 1999), wheat bran (Kotzamanidis et al. 2002, Yun et al. 2004, John et al. 2006) and rice bran (Yun et al. 2004), hydrolyzated or treated with proteases, have been used as nitrogen sources in microbiological lactic acid synthesis. With the same goal, yeast autolyzate or spent cells (Gao et al. 2006a, Timbuntam et al. 2006) and wastes from animal origin such as whey protein hydrolyzate (Fitzpatrick and Keefe 2001, Kim et al. 2006), fish and shrimp wastes (Cira et al. 2002, Gao et al. 2006b) have been used.

It is important to take into account that while agro-industrial by-products represent less expensive carbon and nitrogen sources for bacterial growth and consequently for lactic acid synthesis, they also increase the amount of impurities present in the media, generating higher purification costs.

It also should be noted that although the increase of bacterial mass is the main objective in the production of probiotics, it constitutes an undesirable by-product when the production of an extracellular metabolite is the desired goal, since some nutrients that could be used, i.e., for lactic acid or exopolysaccharides synthesis are "wasted" to produce bacterial cells. This has to be taken into account when selecting growth conditions and reactor designs for large scale fermentations.

## 5. Effect of Sub-lethal Stress on Bacterial Growth and Preservation

Industrial standards require that products claiming health benefits contain a minimum of $10^6$ viable cells per gram of product and/or to provide cost effective technologically attainable bacterial concentrations (Lacroix and Yildirim 2007). However, along the industrial dairy processes and also along digestion, probiotics are exposed to different kinds of stress. This exposure negatively affects bacterial viability and fermentative performance. Stress factors include low pH, low or high temperatures, low water activity, oxidative and osmotic stresses, among others. Moreover, most probiotic strains are intestinal isolates, which are strict anaerobes difficult to cultivate. For these reasons, the ability of probiotics to tolerate stress exposure is very important for their successful incorporation into functional foods.

The acid stress results from the bacterial ability to generate large amounts of organic acids, which accumulate in the medium by creating an unfavorable environment for other microorganisms and even for themselves. The thermal stress is associated with temperature changes occurring during the fermentation process. After inoculation, temperatures are generally higher than those optimal for bacterial growth. Therefore, microorganisms must be able to tolerate them without losing their fermentation capacity. The most frequent response to osmotic stress is the accumulation of intracellular solute (by absorption or synthesis). This prevents microbial dehydration (loss of intracellular water) to equilibrate the high external osmolarity (Serrazanetti et al. 2009). In food industry, microorganisms may be exposed to osmotic stress when important amounts of salt (cheese) or sugars (yogurt) (van de Guchte et al. 2002) are added to the product. Oxidative stress is caused by the generation of reactive oxygen species (ROS), such as superoxide anion ($O_2^-$), hydroxyl radical ($OH^{\cdot}$) and hydrogen peroxide ($H_2O_2$) resulting from a partial reduction of oxygen when the rate of generation of these species is higher than that of detoxification (Serrazanetti et al. 2009).

Like other microorganisms, probiotics develop different mechanisms of defense against stress factors that allow them to survive in adverse conditions (van de Guchte 2002, Serrazanetti et al. 2009). When bacteria are exposed to sub-lethal stress, the tolerance to subsequent stresses is enhanced. This exposure results in an adaptation to adverse environments, which is generally associated with the induction of a large number of genes, the synthesis of stress-response proteins and the development of cross-resistance to various types of stress (Girgis et al. 2003).

Based on this knowledge, different approaches have been developed to enhance the long-term survival of probiotics and improve their resistance to environmental stresses occurring during production, storage or digestion. Such strategies are generally related with the incubation under stress conditions (i.e., heat, starving, high concentrations of salt, bile salts, $H_2O_2$ or low pH) (Gilliland and Rich 1990, Desmond et al. 2002). The response of probiotics to sub-lethal stress induces the synthesis of specific stress protecting proteins that improve the bacterial tolerance to further homologous or heterologous stresses (Schmidt and Zink 2000). This capacity is known as "cross-tolerance" and has been used to increase both the performance of fermentative processes and the recovery of microorganisms after storage.

Stress adaptation is a widely extended strategy to enhance the stability of probiotics during industrial processes when strains are not intrinsically resistant. Several papers have been devoted to the molecular bases of stress response in probiotic microorganisms. This fundamental knowledge facilitated the comprehension of such stresses, giving support for the selection of the best strains and industrial conditions in terms of probiotic stability in the final product.

In this regard, it is well known that strains subjected to sub-lethal stress before exposure to harsh industrial conditions are more stable during product manufacture and storage. As the response to sub-lethal stress is strain dependent, a huge amount of articles reporting stress tolerance on lactobacilli and bifidobacteria have been published hereto.

Acidity and heat are some of the main limiting factors affecting strain survival and it has been reported that pre-exposure to stress improves the subsequent survival under acidic conditions (Saarela et al. 2009), and after heat-shock (Ananta and Knorr 2004) and also improves the functionality of probiotics in the gut (Kiliç 2013). Mutagenesis and high pressures have also been used as stress factors.

Most strains of *Lb. plantarum* have shown to be highly tolerant to certain stresses. However, substantial diversity is found for others (Parente et al. 2010). For example, most of the strains respond to heat shock (De Angelis et al. 2004), lactic acid (Pieterse et al. 2005), bile (Bron et al. 2006), oxidative stress (Serrano et al. 2007), low pH and ethanol (Alegría et al. 2004). Sawatari and Yokota (2007) reported that *Lb. plantarum* and *Lb. pentosus* strains can grow up to an external pH of 8.2–8.7. Moreover, the arginine dihydrolase pathway (Arena et al. 1999) and glutamate decarboxylation (Siragusa et al. 2007) can increase acid tolerance of these species in certain environments. Parente et al. (2010) found a high survival upon 1–2 hours exposure of *Lb. plantarum* and *Lb. pentosus* to pH 2.5.

Heat or salt adaptation increase the tolerance of probiotic strains of *Lb. paracasei* to spray drying (Desmond et al. 2001, Corcoran et al. 2006). Sub-lethal stress also enhances the fermentation capacity of *Lb. rhamnosus*, *Lb. brevis* and *Lb. reuteri* (Saarela et al. 2004). *Lb. rhamnosus* GG cells exposed to high hydrostatic pressure are able to resist temperatures that are lethal when cells are not previously exposed to this sub-lethal stress (Ananta and Knorr 2004).

The storage stability of *Lb. rhamnosus* HN001 is substantially increased after heat or osmotic stress during the stationary phase (Prasad et al. 2003). Strains of

*Lb. collinoides* pretreated with heat stress are considerably more tolerant against subsequent acid and ethanol challenges, than the non-adapted control cells (Laplace et al. 1999).

Bifidobacteria are sensitive microorganisms with low capacity to overcome stresses occurring during their production, storage and consumption. Therefore, efforts have been made to improve their survival under stress conditions with the aim of potentiating their technological and probiotic properties. These approaches include incubation under sub-lethal conditions. Cross-tolerance seems to be species dependent (Doleyres and Lacroix 2005).

In *Bb. longum*, it has been reported that a 15 min heat shock at 47ºC enhances thermotolerance 24 to 128 folds. A treatment with salts results in an increased tolerance to freeze-drying and to heat stress. In strains of *Bb. longum* and *Bb. catenulatum*, prolonged incubation at pH 2.0 generates acid resistant strains. In addition, the acid adapted strains show higher resistance to bile salts (1–3%), NaCl (6–10%) and high temperatures (60–70ºC), besides a higher fermentative ability and enzymatic activity (Collado and Sanz 2007).

An adaptation at pH 5.2 for 2 h protects strains of *Bb. brevis* against subsequent exposure to pH 2 to 5, bile (0.2–1.0%), $H_2O_2$ (100–1000 ppm) and during storage at different temperatures (Maus and Ingham 2003). It has also been reported that for certain strains of *Bb. longum* and *Bb. lactis*, sub-lethal $H_2O_2$ treatments might help increase cell resistance to oxidative damage during production and storage of probiotic-containing foods (Oberg et al. 2011). In strains of *Bb. adolescentis*, salt pretreatment results in an increased tolerance to freeze-thawing cycles or lethal heat stress (Schmidt and Zink 2000).

The survival of *St. thermophilus*, *St. salivarius* and *St. macedonicus* after exposure to acid, oxidative, heat and osmotic stresses was addressed by Parente et al. (2010). The bacterial response to acid and heat stresses is ascribed to changes in the intracellular protein patterns (Zotta et al. 2008).

## 6. Fermentation Technologies for Large Scale Production of Probiotics: Technical Aspects

The increasing commercial interest in functional foods requires adequate processes for industrial production. As mentioned in the previous sections, metabolic products negatively influence growth rate and productivity (Luedeking and Piret 2000). Moreover, the high sensitivity of probiotics of human origin to environmental and industrial processes represents a challenge for the large-scale cultivation of high concentrations of probiotic bacteria.

Conventional batch fermentation with suspended cells is the process almost exclusively used for the industrial production of probiotics. However, other less conventional fermentation technologies, like continuous cultures and immobilized cell systems, could be useful to enhance the performance of industrial fermentations.

**Batch fermentation:** In this process, microorganisms are inoculated in a reactor containing culture medium, and fermentation is operated until the required cell density or optimal product concentrations are achieved. Once the fermentation is finished, the bioreactor content is discharged and further prepared for a new fermentation. The bioreactor is only productive during cell growth or product production. Besides minor adjustments of pH or foam control, small additions of essential precursors or continuous supply of air in the case of aerobic fermentation, no additional material is either added or removed during cell growth.

**Continuous fermentation:** In this case, fresh medium is continuously added to the bioreactor and depleted medium is simultaneously removed at identical rate. After a transient period in which the conditions inside the bioreactor change in time, a steady state is achieved. The steady state is characterized by constant conditions, both inside the bioreactor and at the bioreactor outlet (Dunn et al. 2003). This results in a controlled physiological state of cultures that can be manipulated by environmental parameters, like medium composition and dilution rate. The main advantage of using continuous cultures is the removal of secondary growth effects, usually present in batch cultures, which may mask subtle physiological changes. In addition, the constant conditions of continuous cultures in steady state allow the production of microorganisms with constant physiology.

**Membrane bioreactors:** In a membrane system with continuous feeding of fresh medium, cells are retained in the bioreactor by an ultra-filtration or microfiltration membrane whereas small molecules diffuse through the pores of the membrane according to their size. Hence, cells are concentrated and inhibitory metabolic products are eluted in the permeate. In membrane bioreactors, cells are subjected to different stresses, such as low nutrient concentration, oxygen, osmotic and mechanical stresses that could affect sensitive microorganisms, but that might also lead to cross-tolerance for other stresses.

**Immobilized cell technology:** In real systems, microorganisms generally occur as highly organized communities enclosed in a self-produced polymeric matrix (named biofilm) that adheres to an inert or living surface (Costerton et al. 1999). The formation of biofilms is the main strategy for bacterial survival. These natural biofilms also occur in the gut, where microorganisms grow on the gut mucosa, attached to food particles. Biofilms have several advantages over free-cell fermentations: high cell densities, reuse of biocatalysts, improved resistance to contamination and bacteriophage attack, enhancement of plasmid stability, prevention from washing-out during continuous cultures, and physical and chemical protection of cells (Lacroix et al. 2005). Microbial immobilization and growth in porous solid supports during incubation in rich media result in the formation of a high cell density region. Several studies have shown that immobilized and released bacteria exhibit changes in growth, morphology and physiology compared with cells produced in conventional free-cell cultures.

It has been suggested that the combination of cell immobilization and continuous cultures might be used to efficiently produce probiotic bacteria with enhanced tolerance to environmental stresses and improved technological and functional characteristics (Lacroix et al. 2005).

## 7. Monitoring Probiotic Growth

Both industrial and laboratory scale production of probiotics require reliable methods to monitor bacterial growth. These methods can be classified into direct and indirect. Direct methods are based on determinations of the number of living cells (plate count techniques) or the number of particles (microscopic techniques and particle counters) throughout growth kinetics. Indirect methods are based on determinations of culture parameters related with the evolution of bacterial number. The selection of a monitoring method depends on the objective and on the characteristics of each microorganism.

### *7.1 Classical Methods: Absorbance, Dry-cell Weight, Biomass, Plate Counts, Acid Production*

Registration of optical density is the most used method to monitor bacterial growth in real time (Dalgaard et al. 1994, Begot et al. 1996, Presser et al. 1998). Dry-cell weight is also used to determine bacterial growth and is preferred when microorganisms occur as aggregates. Although increase of optical density and dry-cell weight indicate bacterial proliferation, they do not provide information about bacterial physiology because damaged, dead and viable cells contribute altogether to the experimental measures. In addition, determination of dry-cell is time consuming, thereby, results are available too late to take decisions related with the fermentation processes. Moreover, it is inaccurate for low biomass concentrations (early stages of fermentation).

Monitoring biomass was also used to follow cell concentration online. This technique relies on the measurement of the capacitance generated in an intact cell when passing through an electrical field. Although there is a good linear correlation between optical density and capacitance for high biomass concentrations, problems occur when biomass concentrations are low or when the ionic strength of the medium is high. In addition, it is important to point out that all these techniques assume that a microbial population is homogeneous with respect to its physiological state, which is obviously not true. Determining pH is a simple and feasible method to monitor bacterial growth. However, among heterofermentative microorganisms it is not possible to discriminate the contribution of each acid to the decrease of pH.

Plate counts allow overcoming the problem of overestimation in bacterial growth. In spite of that, monitoring microbial growth through plate count is non-practical at an industrial level. Moreover, autoaggregative strains are difficult to count on plates.

### *7.2 Determination of Enzymatic Activity: The Case of Fructose-6-phosphate Phosphoketolase Activity to Monitor Bifidobacteria Growth*

Fructose-6-phosphate phosphoketolase (F6PPK) is the characteristic key enzyme of the 'bifid-shunt' (de Vries and Stouthamer 1967) and is a useful taxonomic tool for the identification of the genus *Bifidobacterium*. F6PPK catalyzes the cleavage of fructose-6-phosphate into acetylphosphate and erythrose-4-phosphate. The acyl part of the acid anhydride further reacts with hydroxylamine to give hydroxamic acid (Lipmann and

Tuttle 1945). Hydroxamic acid forms a brightly purplish complex with trivalent iron, which can be used for qualitative identification of phosphoketolase activity.

As autoaggregation is a frequent feature among bifidobacteria, monitoring strains of this genus through the F6PPK activity offers good advantage over viable plate counts and absorbance measurements (Bibiloni et al. 2000 and 2001).

### 7.3 Determination of Adenosine Triphosphate (ATP)

Determination of ATP is based on the emission of light by luciferin luciferase of firefly (*Photimus pyralis*) in the presence of $O_2$ and ATP. Considering that ATP is an integral part of the bacterial metabolism, the concentration of ATP provides reliable information about the activity of probiotics, having a good correlation with acid production (Cardwell and Sasso 1985, Nordlund et al. 1980, Griffiths 1993).

ATP determination systems are adapted to be used as online systems. This is very important in the case of large volume fermentations because it allows real-time measurements and reduces the risk of contamination.

### 7.4 Real Time Polymerase Chain Reaction (RT-PCR)

In conventional PCR, the amplified DNA product can be observed on agarose gels after the reaction has finished, that is, after a given number of amplification cycles. In contrast, in RT-PCR the amplification products are determined throughout the reaction, which allows quantification after each cycle. The addition of fluorescent probes yielding an increase of fluorescence proportional to DNA product concentration enables the detection of PCR products in real time. This way, the change in fluorescence over time is used to calculate the amount of DNA, allowing quantification of the number of DNA copies produced after each amplification cycle (RT-qPCR). DNA-binding dyes and fluorescently labeled sequence-specific primers are used as fluorescent probes. The specificity of primers allows determining different microorganisms co-existing in the same environment.

RT-qPCR results are often presented as gene copies per milliliter or gram of sample. Considering that each cell contains only one copy of the target gene, gene copies and cell numbers are the same. As the reactions are carried out in closed tubes and information can be evaluated without gel electrophoresis, both risks of contamination and time of analysis (within days) are considerably reduced (Cindy et al. 2009). As a cultivation independent method, it has been used to accurately detect time-demanding microorganisms, like *Oenococcus oeni* (because the sensitivity is $\sim 10^3$ cells/mL) (Solieri and Giudici 2010, Bouix and Ghorbal 2013, Salma et al. 2013). It has also been used for the determination of different probiotics in complex environments (Fujimoto et al. 2008, Dommels et al. 2009, Lahtinen et al. 2009, Reimann et al. 2010, Schmidt et al. 2010, Simões et al. 2010, Angelakis et al. 2011, Fujimoto et al. 2011, Stapleton et al. 2011, Desfossés-Foucault et al. 2012, Filteau et al. 2013).

Even when the use of RT-qPCR for quantification of key microorganisms is continuously growing in probiotic industries, its main limitation is that there are not standard operating protocols for sample collection, storage, preservation, DNA extraction, and nucleic acid targets. In addition, RT-qPCR requires sample preparation and the cost of reagents is relatively high, which is not trivial at an industrial level.

## *7.5 Flow Cytometry*

Traditional microbiological techniques usually fail to monitor simultaneously the evolution of viable, sub-lethally damaged and dead microorganisms. Flow cytometry is a good example of analytical technique decoupled from post sampling growth that allows the evaluation of microorganisms in different physiological states.

Flow cytometry is a rapid technique for an optical analysis of individual cells in suspension labeled with fluorescent probes. Cell suspensions are introduced into the flow cell with flowing sheet fluid. A light source (generally coming from an assortment of lasers) illuminates cells as they pass individually through a beam of light. The light scattering reflects the cell size and structure, and the fluorescence measurements allow determining the cellular content of any constituent labeled with the probe (Walberg et al. 1997). This way, flow cytometry combines the advantage of determining the behavior of a single cell with the capability of measuring a large number of cells in a short time (Hewittl and Nebe-Von-Caron 2001).

Multiparametric flow cytometry allows an assessment to cell physiological state. This approach allows the simultaneous discrimination of functional subpopulations in microbial populations on the bases of different properties (i.e., reproductive activity, metabolic activity, membrane integrity, etc.) (Díaz et al. 2010). Two fluorescent dyes are often used to determine viable, dead and permeabilized cells (sub-lethally injured or damaged cells): 5-(6)-carboxyfluorescein diacetate (cFDA) and propidium iodide (PI). cFDA passively diffuses into cells and its acetate groups are cleaved by intracellular esterases to yield highly fluorescent carboxyfluorescein succinimidyl ester. PI only penetrates bacterial cells when membranes are damaged. Therefore, it penetrates dead or damaged cells to label DNA. Therefore, the use of both cFDA and PI allow distinguishing different populations of cells: a) viable cells cFDA+, PI-, b) dead cells cFDA-, PI+, c) damaged cells PI+ and cFDA+. Hence, this approach allows the resolution of individual cell's physiologic states beyond cultivability (Fuller et al. 2000, Hewitt and Nebe-Caron 2001).

Among LAB and probiotics, flow cytometry is a useful tool to monitor cell damage through preservation processes or after different stress or harmful treatments like wine making (Ueckert et al. 1997, Bunthof et al. 1999, 2001, 2002, Papadimitriou et al. 2006). As microorganisms exposed to stress conditions may be non-culturable, the use of fluorescent dyes like PI and cFDA allows determining their persistence in probiotic products (Bunthof et al. 2002).

Beer production, wine and cider fermentations have also been monitored using flow cytometry (Dinsdale et al. 1995, Lloyd et al. 1996, Hutter 2002, Boyd et al. 2003, Bouchez et al. 2004, Muller et al. 2004, Herrero et al. 2006, Chulp et al. 2007). This analytical method is particularly useful to monitor *Oenococcus oeni* during malolactic fermentation, as results can be obtained in 20 min vs. 10 days required for plate counts based methods (Bouix and Ghorbal 2013).

Another interesting use of flow cytometry is the monitoring of bacteriophage infections based on the lower density of the cell walls of infected cells. As determinations can be made in real time, this approach increases the chance of successful intervention in the fermentation process (Michelsen et al. 2007).

Advances in flow cytometry technology and computational resources led to the development of specific automated systems for direct and on-line determinations throughout fermentations (Zhao et al. 1998, Abu-Absi et al. 2003, Kacmar et al. 2004), which is especially valuable at an industrial level. Considering that certain non-culturable strains of LAB remain metabolically active (Ganesan et al. 2007), preserve their probiotic properties (Rault et al. 2008) and contribute to fermentation processes, the relevance of flow cytometry as analytical tool becomes evident (Díaz et al. 2010). Moreover, due to its high sensitivity and short times of analysis (less than 10000 cells can be counted in less than a minute), flow cytometry becomes a valuable analytical tool that allows overcoming enumeration problems derived from low concentrations of cells (Malacrino et al. 2001, Chaney et al. 2006).

## 7.6 Spectroscopic Methods

### 7.6.1 Dielectric Spectroscopy

Dielectric spectroscopy devices determine conductance and capacitance of microbial metabolites (weakly charged organic molecules) in the culture medium. The most common indicators of cell measurement are the change in capacitance ($\Delta C$) or the relative permittivity ($\varepsilon$), which can be calculated from the experimental determinations of capacitance and the physical constants of the probes. Dielectric spectroscopy has large industrial applications, being the most used method in breweries. It has been used to monitor the growth of *Lb. casei* (Arnoux et al. 2005, Kiviharju et al. 2008).

Impedance is widely used to enumerate bacteria in foods (Nieuwenhof and Hoolwerf 1988, Walker et al. 2005). This approach is particularly useful because it allows overcoming the formation of clumps, a common feature among bifidobacteria, and also other probiotic strains (Mobili et al. 2009).

### 7.6.2 Nuclear Magnetic Resonance Spectroscopy

Nuclear Magnetic Resonance Spectroscopy (NMR) is a powerful analytical technique that allows the characterization of complex mixtures both qualitativelly and quantitativelly. It has an enormous advantage over other analytical techniques: determinations into living cells can be carried out without disturbing the cellular organization. This is particularly useful for the study of bacterial kinetics in real time under physiological conditions (Ramos et al. 2002).

Among LAB, $^{1}H$ and $^{13}C$ NMR have been used to simultaneously monitor the production of alcohols (ethanol, methanol), sugars (fructose, glucose), glycerol, organic acids (malic, tartaric, succinic, acetic and lactic acid) during fermentation (Košir and Kidri 2002, Ercolini et al. 2011, Bouteille et al. 2013). NMR offers various advantages over other methods and can be extended to the on-line monitoring of fermentation of sugars (Gillou and Tellier 1988, Tellier et al. 1989, Ramanjooloo et al. 2009). Different authors used NMR to monitor fermentation in fruit juices and to investigate changes in the concentrations of alcohols, organic acids, sugars and amino acids in port wines (Zhang et al. 1997, Nilsson et al. 2004, Clark et al. 2006).

### *7.6.3 Vibrational Spectroscopy based Methods*

The use of vibrational spectroscopic methods in Food Microbiology has been widely expanded in the last decade as they are able to provide information in a non-destructive and reliable way and at a single-cell level with minimal preparation of samples (Harz et al. 2009). These methods are particularly useful because of their feasibility for monitoring biological processes in real time and without the necessity of exogenous chemical reagents, which saves costs, efforts and time. In addition, the simplicity to interface the modern spectroscopic equipments to almost any computer or collection data system makes vibrational spectroscopic methods very useful for automatic control, process optimization, or quality assessment in real time. Hence, it is not surprising there is an increasing interest of enterprises (i.e., agricultural, chemical, pharmaceutical, textile and many others) in using them for the monitoring of different industrial processes, including fermentations (Naumann et al. 1991, McClure 2003, Cervera et al. 2009). *Lag*, *log* and stationary phases of bacterial growth are characterized by different biochemical reactions related with the synthesis of cellular components necessary for cell growth (Al-Qadiriet al. 2008). These biochemical reactions are translated in chemical changes that can be followed using spectroscopic methods, particularly in the near infrared region (800–2500 nm) (Al-Qadiri et al. 2008, Harz et al. 2009, Palchaudhuri et al. 2011).

Spectral features in NIR, arising from combinations and overtones of the fundamental vibrations associated with C-H, O-H and N-H bonds provide a rich source of information, composed of broad and overlapping bands (Martens and Naes 1989, Esbensen 2005).

NIR spectra can be registered:

- *Off-line*: Samples are collected at a given time and analyzed later, generally in a different place (laboratory).
- *At-line*: Samples are collected and analyzed immediately after being collected (rapid off-line).
- *On-line*: The process is analyzed directly and without manual sample manipulation. Two types of on-line monitoring can be considered:
  a) *In situ* or *in-line*: A fiber-optic probe is immersed into the fermentation broth, carrying the spectral information from the sample to the spectrometer, thus being useful in the production line.
  b) *Ex situ*: The analysis device is physically outside the fermentor. Spectral registration usually involves either a flow-through cell or a fiber optic probe placed on the glass wall of the container.

NIR has been used to detect and monitor microorganisms in food-related applications through detection of biomass, glucose, lactose, exopolysaccharides or lactic acid (Vaccari et al. 1993, Rodriguez-Saona et al. 2001, Sivakesava et al. 2001, Arnold et al. 2002a, Cimander et al. 2002, Macedo et al. 2002, Tamburini et al. 2003, Tosi et al. 2003, Navrátil et al. 2004, Saranwong and Kawano 2008a,b, Cámara-Martos et al. 2012, Feng et al. 2013).

### *7.6.4 Fluorescence Methods for Monitoring Bacterial Growth*

Fluorescence spectroscopy has also been used to monitor fermentations. The method is based on excitation with UV light at one wavelength and measuring the emission of culture components at another wavelength (Kiviharju et al. 2008).

Viable microorganisms contain a variety of intracellular biomolecules that have specific excitation and emission wavelength spectra characterizing their intrinsic fluorescence and that can be used for monitoring fermentations. A good example is the evolution of NAD(P)H concentration (Olsson and Nielsen 1997, Lindemann et al. 1999). If the intracellular concentration of NAD(P)H is constant, the fluorescence of the culture can be correlated with the biomass concentration. However, good correlations between culture fluorescence and biomass concentration are only found for defined culture media and in conditions where the cellular composition do not experience any change (Olsson and Nielsen 1997). Major problems result from medium components that absorb light at excitation wavelength or at the emission wavelength and from other possible fluorophores, like penicillin (Olsson and Nielsen 1997). Some interference problems can be overcome using multiple excitation-fluorescence determinations (2D fluorescence). This strategy enables detection of different fluorophores (i.e., tryptophan, pyridoxine, FAD, FMN, NAD(P)H, and riboflavin) that can be correlated with biomass (Olsson and Nielsen 1997).

## ***7.7 Chemometrics***

The huge amount of information contained in the experimental spectra requires a careful analysis that is generally carried out using chemometric techniques. The great development of computational resources and chemometrics in the last 15 years allowed a wide use of spectroscopic based methods in the fermentation field, making possible the dynamic monitoring of industrial processes *in situ*, in environments consisting of complex chemical mixtures (Bonano et al. 1992, Tosi et al. 2003, Vaidyanathan et al. 2003, Small 2006, Cervera et al. 2009).

Briefly, chemometric methods allow the observation of spectral evolution and also to explore the relationship between spectra and concentration of a given analyte. Multivariate analysis techniques, such as principal component analysis (PCA) or partial least squares (PLS) are common tools to interpret spectroscopic data.

PCA is a mathematical procedure that uses orthogonal transformation to convert multivariate data (the spectra) into a set of values of linearly uncorrelated variables (named principal components). This transformation is defined in such a way that the first principal component has the largest possible variance and each succeeding component in turn has the highest variance possible under the constraint that it be orthogonal to the preceding components. This has the advantage of eliminating multicollinearity when using the PCA results in an analysis of dependence (i.e., regression analysis) (Hall and Pollard 1992).

For a quantitative analysis, a variety of chemometric algorithms, including multiple linear least-squares (MLR), PLS, and principle component regression (PCR) analyses, can be used to define spectroscopic models (Brimmer and Hall 1993, Williams and Norris 2001). In MLR, linear relationships between the absorption in discrete wavelenghts of the spectra and information provided by reference analytical chemistry methods are defined. In PLS, a broad range of wavelengths (50–250 nm or more) containing simultaneously spectral factors related to analyte concentrations are considered. Interfering absorption bands, scattering differences and shifts in band positions are compensated for the definition of a reliable calibration model.

As MLR methods are used to investigate simple spectra, arising from simple chemical systems, they are intrinsically less sensitive to subtle matrix variations (Williams and Norris 2001).

PLS or PCR methods are usually employed for the analysis of more complex chemical systems (i.e., quantification of minor constituents or when analyte bands are highly overlapped by other matrix components).

The integrity and applicability of the derived calibration model is totally dependent on the set of data used to create the model. The most critical aspect to be considered for the definition of calibration models is to ensure that the models are really representative of the variations to be encountered in future samples (Arnold et al. 2002b). Therefore, independent sets of samples collected from at least one independent experiment (an independent fermentation for example) shall be used to validate the performance of the calibration model.

## 8. Conclusions

Cultivation of probiotics at laboratory and industrial scale is, nowadays, an issue that can be afforded from different perspectives. Prior knowledge about bacterial metabolism, resistance and/or sensitivity to stress and persistence of probiotic properties after growth is important to define media for cell cultivation. Considering of all these factors leads to adopt the best cultivation strategy to allow the maximum yield for the purpose pursued.

The development of powerful analytical methods, together with the continuously increasing computational capacities has contributed to facilitate the monitoring of bacterial growth *in situ* and in real time.

In this chapter, we gave an insight on the advances in probiotic cultivation. Efforts were made to put into relevance the contribution of different disciplines into this issue, thus transmitting the idea that probiotic cultivation goes beyond the traditional microbiology.

## 9. Acknowledgements

This work has been funded by the Argentinean Agency for the Promotion of Science and Technology (ANPCyT) [Projects PICT(2008)/145, PICT(2010)/0589, PICT(2010)/2361 and PICT(2011)/0226], the Argentinean National Research Council (CONICET) (Project PIP2012-2014114-201101-00024). EET, MG, PM and AGZ are members of the Research Career, CONICET. MS is fellow of CONICET.

## References

Aasen, I.M., T. Møretrø, T. Katla, L. Axelsson and I. Storrø. 2000. Influence of complex nutrients, temperature and pH on bacteriocin production by *Lactobacillus sakei* CCUG 42687. Appl. Microbiol. Biot. 53: 159–166.

Abu-Absi, N.R., A. Zamamiri, J.A. Kacmar, S.J. Balogh and F. Srienc. 2003. Automated flow cytometry for acquisition of time dependent population data. Cytometry 51: 87–96.

Al-Qadiri, H.M., N.I. Alami, M.A. Al-Holy, M. Lin, A.G. Cavinato and B.A. Rasco. 2008. J. Rapid Meth. Aut. Mic. 16: 73–89.

Alegría, G.E., I. López, J.I. Ruiz, J. Sáenz, E. Fernández, M. Zarazaga, M. Dizy, C. Torres and F. Ruiz-Larrea. 2004. High tolerance of wild *Lactobacillus plantarum* and *Oenococcus oeni* strains to lyophilisation and stress environmental conditions of acid pH and ethanol. FEMS Microbiol. Lett. 230: 53–61.

Altaf, M., M. Venkateshwar, M. Srijana and G. Reddy. 2007. An economic approach for L-(+) lactic acid fermentation by *Lactobacillus amylophilus* GV6 using inexpensive carbon and nitrogen sources. J. Appl. Microbiol. 103: 372–380.

Angelakis, E., H.M. Million and D. Raoult. 2011. Rapid and accurate bacterial identification in probiotics and yoghurts by MALDI-TOF mass spectrometry. J. Food Sci. 76: M568–M572.

Ananta, E. and D. Knorr. 2004. Evidence on the role of protein biosynthesis in the induction of heat tolerance of *Lactobacillus rhamnosus* GG by pressure pre-treatment. Int. J. Food Microbiol. 96: 307–313.

Arena, M.E., F.M. Saguir and M.C. Manca de Nadra. 1999. Arginine dihydrolase pathway in *Lactobacillus plantarum* from orange. Int. J. Food Microbiol. 47: 203–209.

Arnold, S.A.L., R. Gaensakoo, L.M. Harvey and B. McNeil. 2002a. Use of at-line and *in situ* near infrared spectroscopy to monitor biomass in an industrial fed-batch *Escherichia coli* process. Biotechnol. Bioeng. 80: 405–413.

Arnold, S.A., L.M. Harvey, B. McNeil and J.W. Hall. 2002b. Employing near-infrared spectroscopic methods of analysis for fermentation monitoring and control part 1, Method Development. BioPharm International 16: 26–34.

Arnoux, A.S., L. Preziosi-Belloy, G. Esteban, P. Teissier and C. Ghommidh. 2005. Lactic acid bacteria biomass monitoring in highly conductive media by permittivity measurements. Biotechnol. Lett. 27: 1551–1557.

Ashraf, R. and N.P. Shah. 2011. Selective and differential enumerations of *Lactobacillus delbrueckii* subsp. *bulgaricus*, *Streptococcus thermophilus*, *Lactobacillus acidophilus*, *Lactobacillus casei* and *Bifidobacterium* spp. in yoghurt. A review. Int. J. Food Microbiol. 149: 194–208.

Audisio, M.C., G. Olivera and M.C. Apella. 2001. Effect of different complex carbon sources on growth and bacteriocin synthesis of *Enterococcus faecium*. Int. J. Food Microbiol. 63: 235–241.

Azaola, A., P. Bustamante, S. Huerta, G. Saucedo, R. González, C. Ramos and S. Saval. 1999. Use of surface response methodology to describe biomass production of *Bifidobacterium infantis* in complex media. Biotechnol. Tech. 13: 93–95.

Begot, C., I. Desnier, J.D. Daudin, J.C. Labadie and A. Lebert. 1996. Recommendations for calculating growth parameters by optical density measurements. J. Microbiol. Meth. 25: 225–232.

Bergmaier, D., C.P. Champagne and C. Lacroix. 2005. Growth and exopolysaccharide production during free and immobilized cell chemostat culture of *Lactobacillus rhamnosus* RW-9595M. J. Appl. Microbiol. 98: 272–284.

Bevilacqua, A., M.R. Corbo, M. Mastromatteo and M. Sinigaglia. 2008. Combined effects of pH, yeast extract, carbohydrates and di-ammonium hydrogen citrate on the biomass production and acidifying ability of a probiotic *Lactobacillus plantarum* strain, isolated from table olives, in a batch system. 2008. World J. Microbiol. Biotechnol. 24: 1721–1729.
Bibiloni, R., P.F. Pérez and G.L. De Antoni. 2000. An enzymatic-colorimetric assay for the quantification of *Bifidobacterium*. J. Food Prot. 63: 322–326.
Bibiloni, R., A. Gómez-Zavaglia and G.L. De Antoni. 2001. Enzyme-based most probable number method for the enumeration of *Bifidobacterium* in daily products. J. Food Prot. 64: 2001–2006.
Bonano, A.S., J.M. Olinger and P.R. Griffiths. 1992. pp. 19–28. *In*: K.I. Hildrum (ed.). Near Infra-Red Spectroscopy, Horwood, Chichester, UK.
Bouchez, J.C., M. Cornu, M. Danzart, J.Y. Leveau, F. Duchiron and M. Bouix. 2004. Physiological significance of the cytometric distribution of fluorescent yeast after viability staining. Biotechnol. Bioeng. 86: 520–530.
Bouix, M. and S. Ghorbal. 2013. Rapid enumeration of *Oenococcus oeni* during malolactic fermentation by flow cytometry. J. Appl. Microbiol. 114: 1075–1081.
Bouteille, R., M. Gaudet, B. Lecanu and H. This. 2013. Monitoring lactic acid production during milk fermentation by *in situ* quantitative proton nuclear magnetic resonance spectroscopy. J. Dairy Sci. 96: 2071–2080.
Box, G.E.P. and K.B. Wilson. 1951. On the experimental attainment of optimum conditions. J. R. Stat. Soc. B 13: 1–45.
Boyd, A.R., T.S. Gunasekera, P.V. Attfield, K. Simic, S.F. Vincent and D.C. Veal. 2003. A flow-cytometric method for determination of yeast viability and cell number in a brewery. FEMS Yeast Res. 3: 11–16.
Brimmer, P.J. and J.W. Hall. 1993. Determination of nutrient levels in a bioprocess using near-infrared spectroscopy. Appl. Spectrosc. 38: 155–162.
Bron, P.A., D. Molenaar, W.M. de Vos and M. Kleerebezem. 2006. DNA micro-array-based identification of bile-responsive genes in *Lactobacillus plantarum*. J. Appl. Microbiol. 100: 728–738.
Bujalance, C., M. Jimenez-Valera, E. Moreno and A. Ruiz-Bravo. 2006. A selective differential medium for *Lactobacillus plantarum*. J. Microbiol. Meth. 66: 572–575.
Bunthof, C.J., S. Braak, P. Breeuwer, F.M. Rombouts and T. Abee. 1999. Rapid fluorescence assessment of the viability of stressed *Lactococcus lactis*. Appl. Environ. Microbiol. 65: 3681–3689.
Bunthof, C.J., S. Schalkwijk, W. Meijer, T. Abbe and J. Hugenholtz. 2001. Fluorescent method for monitoring cheese starter permeabilization and lysis. Appl. Environ. Microbiol. 67: 4264–4271.
Bunthof, C.J. and T. Abee. 2002. Development of a flow cytometric method to analyze subpopulations of bacteria in probiotic products and dairy starters. Appl. Environ. Microbiol. 68: 2934–2942.
Cámara-Martos, F., G. Zurera-Cosano, R. Moreno-Rojas, R.M. García-Gimeno and F. Pérez-Rodríguez. 2012. Identification and quantification of lactic acid bacteria in a water-based matrix with near-infrared spectroscopy and multivariate regression modeling. Food Anal. Meth. 5: 19–28.
Cardwell, J.T. and Y. Sasso. 1985. The relation of adenosine tri-phosphate activity to titratable activity in three dairy cultures. J. Dairy Sci. 68: 66.
Champagne, C.E., D. Roy and A. Lafond. 1997. Selective enumeration of *Lactobacillus casei* in yoghurt-type fermented milks based on a 15°C incubation temperature. Biotech. Techniques 11: 567–569.
Chang, C.P. and S.L. Liew. 2013. Growth medium optimization for biomass production of a probiotic bacterium, *Lactobacillus rhamnosus* ATCC 7469. J. Food Biochem. 37: 536–543.
Chaney, D., S. Rodríguez, K. Fugelsang and R. Thornton. 2006. Managing high-density commercial scale wine fermentations. J. Appl. Microbiol. 100: 689–698.
Chulp, P.H., T. Wang, E.G. Lee and G.G. Stewart. 2007. Assessment of the physiological status of yeast during high- and low-gravity wort fermentations determined by flow cytometry. Tech. Q. MBAA Commun. 44: 286–295.
Cimander, C., M. Carlsson and C.F. Mandenius. 2002. Sensor fusion for on-line monitoring of yoghurt fermentation. J. Biotechnol. 99: 237–248.
Cindy, J. Smith and A.M. Osborn. 2009. Advantages and limitations of quantitative PCR(Q-PCR)-based approaches in microbial ecology. FEMS Microbiol. Ecol. 67: 6–20.
Cira, L.A., S.G. Huerta, G.M. Hall and K. Shirai. 2002. Pilot scale lactic acid fermentation of shrimp waste for chitin recovery. Process. Biochem. 37: 1359–1366.
Clark, S., N.W. Barnett, M. Adams, I.B. Cook, G.A. Dyson and G. Johnston. 2006. Monitoring a commercial fermentation with proton nuclear magnetic resonance spectroscopy with the aid of chemometrics. Anal. Chim. Acta 563: 338–345.

Cervera, A.E., N. Petersen, A.E. Lantz, A. Larsen and K.V. Gernaey. 2009. Application of near-infrared spectroscopy for monitoring and control of cell culture and fermentation. Biotechnol. Prog. 25: 1561–1581.

Cochu, A., D. Fourmier, A. Halasz and J. Hawari. 2008. Maple sap as a rich medium to grow probiotic lactobacilli and to produce lactic acid. Lett. Appl. Microbiol. 47: 500–507.

Collado, M.C. and Y. Sanz. 2007. Induction of acid resistance in *Bifidobacterium*: a mechanism for improving desirable traits of potentially probiotic strains. J. Appl. Microbiol. 103: 1147–1157.

Corcoran, B.M., R.P. Ross, G.F. Fitzgerald, P. Dockery and C. Stanton. 2006. Enhanced survival of GroESL-overproducing *Lactobacillus paracasei* NFBC 338 under stressful conditions induced by drying. Appl. Environ. Microbiol. 72: 5104–5107.

Costerton, J.W., P.S. Stewart and E.P. Greenberg. 1999. Bacterial biofilms: a common cause of persistent infections. Science 284: 1318–1322.

Daeschel, M.A., M.C. McKeney and L.C. McDonald. 1990. Bacteriocidal activity of *Lactobacillus plantarum* C-11. Food Microbiol. 7: 91–98.

Dalgaard, P., T. Ross, L. Kamperman, K. Neumeyer and T.A. McMeekin. 1994. Estimation of bacterial growth rates from turbidimetric and viable count data. Int. J. Food Microbial. 23: 391–404.

Dave, R.I. and N.P. Shah. 1996. Evaluation of media for selective enumeration of *Streptococcus thermophilus*, *Lactobacillus delbrueckii* ssp. *bulgaricus*, *Lactobacillus acidophilus* and *Bifidobacteria*. J. Dairy Sci. 79: 1529–1536.

De Angelis, M., R. Di Cagno, C. Huet, C. Crecchio, P.F. Fox and M. Gobbetti. 2004. Heat shock response in *Lactobacillus plantarum*. Appl. Environ. Microbiol. 70: 1336–1346.

de Man, J.D., M. Rogosa and M.E. Sharpe. 1960. A medium for the cultivation of lactobacilli. J. Appl. Bact. 23: 130–135.

de Vries, W. and A.H. Stouthamer. 1967. Pathway of glucose fermentation in relation to the taxonomy of bifidobacteria. J. Bacteriol. 93: 574–576.

de Vuyst, L., R. Callewaert and K. Crabbé. 1996. Primary metabolite kinetics of bacteriocin biosynthesis by *Lactobacillus amylovorus* and evidence for stimulation of bacteriocin production under unfavourable growth conditions. Microbiology 142: 817–827.

Deepika, G., E. Karunakaran, C.R. Hurley, C.A. Biggs and D. Charalampopoulos. 2012. Influence of fermentation conditions on the surface properties and adhesion of *Lactobacillus rhamnosus* GG. Microb. Cell Fact. 11: 116–125.

Desfossés-Foucault, E., V. Dussault-Lepage, C. Le Boucher, P. Savard, G. LaPointe and D. Roy. 2012. Assessment of probiotic viability during cheddar cheese manufacture and ripening using propidium monoazide-PCR quantification. Front. Microbiol. 3: 350–361.

Desmond, C., C. Stanton, G.F. Fitzgerald, K. Collins and R.P. Ross. 2001. Environmental adaptation of probiotic lactobacilli towards improvement of performance during spray drying. Int. Dairy J. 11: 801–808.

Desmond, C., C. Stanton, G.F. Fitzgerald, K. Collins and R.P. Ross. 2002. Environmental adaptation of probiotic lactobacilli towards improvement of performance during spray drying. Int. Dairy J. 12: 183–190.

Devi, M., L.J. Rebecca and S. Sumathy. 2013. Characterization of bacteriocin produced by *Lactobacillus delbreukii* isolated from yoghurt. Res. J. Pharm. Biol. Chem. Sci. 4: 1429–1434.

Djeghri-Hocine, B., M. Boukhemis and A. Amrane. 2010. Formulation and evaluation of a selective medium for lactic acid bacteria-validation on some dairy products. J. Agric. Biol. Sci. 5: 148–153.

Díaz, M., M. Herrero, L.A. García and C. Quirós. 2010. Application of flow cytometry to industrial microbial bioprocesses. Biochem. Eng. J. 48: 385–407.

Dinsdale, M.G. and D. Lloyd. 1995. Yeast viability during cider fermentation: two approaches to the measurement of membrane potential. J. Inst. Brew. 101: 453–458.

Doleyres, Y. and C. Lacroix. 2005. Technologies with free and immobilised cells for probiotic bifidobacteria production and protection. Int. Dairy J. 15: 973–988.

Dommels, Y.E.M., R.A. Kemperman, Y.E.M.P. Zebregs, R.B. Draaisma, A. Jol, D.A.W. Wolvers, E.E. Vaughan and R. Albers. 2009. Survival of *Lactobacillus reuteri* DSM 17938 and *Lactobacillus rhamnosus* GG in the human gastrointestinal tract with daily consumption of a low-fat probiotic spread. Appl. Environ. Microbiol. 75: 6198–6204.

Dong, Z., L. Gu, J. Zhang, M. Wang, G. Du, J. Chen and H. Li. 2014. Optimisation for high cell density cultivation of *Lactobacillus salivarius* BBE 09-18 with response surface methodology. Int. Dairy J. 34: 230–236.

Drosinos, E.H., M. Mataragas, P. Nasis, M. Galiotou and J. Metaxopoulos. 2005. Growth and bacteriocin production kinetics of *Leuconostoc mesenteroides* E131. J. Appl. Microbiol. 99: 1314–1323.

Dumbrepatil, A., M. Adsul, S. Chaudhari, J. Khire and D. Gokhale. 2008. Utilization of molasses sugar for lactic acid production by *Lactobacillus delbrueckii* subsp. *delbrueckii* mutant Uc-3 in batch fermentation. Appl. Environ. Microbiol. 74: 333–335.

Dunn, J., E. Heinzle, J. Ingham and J.E. Prenosil. 2003. Biological reaction engineering: Second edition. Weinheim, Wiley-Vch Verlag GmbH & Co. KGaA.

Elliker, P.R., A.W. Anderson and G. Hannesson. 1956. An agar culture medium for lactic acid streptococci and lactobacilli. J. Dairy Sci. 39: 1611–1612.

Eloy, C. and R. Lacrosse. 1976. Composition d'un milieu de culture destine a effectuer le denombrement des microorganismes thermophiles du yoghurt. Bull. Rech. Agron. Gemblou 11: 83–86.

Ercolini, D., I. Ferrocino, A. Nasi, M. Ndagijimana, P. Vernocchi, A. La Storia, L. Laghi, G. Mauriello, M. E. Guerzoni and F. Villani. 2011. Monitoring of microbial metabolites and bacterial diversity in beef stored underdifferent packaging conditions. Appl. Environ. Microbiol. 77: 7372–7481.

Esbensen, K.H. 2005. *In*: Multivariate Data Analysis—In Practice, CAMO Process AS, Esbjerg, Denmark, 5th Edition.

Feng, Y.-Z., G. ElMasry, D.-W. Sun, A.G.M. Scannell, D. Walsh and N. Morcy. 2013. Near-infrared hyperspectral imaging and partial least squares regression for rapid and reagentless determination of Enterobacteriaceae on chicken fillets. Food Chem. 138: 1829–1836.

Fernandez, B., C. Le Lay, J. Jean and I. Fliss. 2013. Growth, acid production and bacteriocin production by probiotic candidates under simulated colonic conditions. J. Appl. Microbiol. 114: 877–885.

Filteau, M., S. Matamoros, P. Savard and D. Roy. 2013. Molecular monitoring of fecal microbiota in healthy adults following probiotic yogurt intake. Pharma Nutrition. 1: 123–129.

Fitzpatrick, J.J., M. Ahrens and S. Smith. 2001. Effect of manganese on *Lactobacillus casei* fermentation to produce lactic acid from whey permeate. Process Biochem. 36: 671–675.

Fitzpatrick, J.J. and U.O. Keeffe. 2001. Influence on whey protein hydrolyste addition to whey permeate batch fermentations for producing lactic acid. Proc. Biochem. 37: 183–186.

Fitzpatrick, J.J., C. Murphy, F.M. Mota and T. Pauli. 2003. Impurity and cost considerations for nutrient supplementation of whey permeate fermentations to produce lactic acid for biodegradable plastics. Int. Dairy J. 13: 575–580.

Forestier, C., C. De Champs, C. Vatouxand and B. Joly. 2001. Probiotic activities of *Lactobacillus casei rhamnosus*: *in vitro* adherence to intestinal cells and antimicrobial properties. Res. Microbiol. 152: 167–173.

Franz, C.M., M. Huch, H. Abriouel, W. Holzapfel and A. Gálvez. 2011. Enterococci as probiotics and their implications in food safety. Int. J. Food Microbiol. 151: 125–140.

Fujimoto, J., T. Matsuki, M. Sasamoto, Y. Tomii and K. Watanabe. 2008. Identification and quantification of *Lactobacillus casei* strain Shirota in human feces with strain-specific primers derived from randomly amplified polymorphic DNA. Int. J. Food Microbiol. 126: 210–215.

Fujimoto, J., K. Tanigawa, Y. Kudo, H. Makino and K. Watanabe. 2011. Identification and quantification of viable *Bifidobacterium breve* strain Yakult in human faeces by using strain-specific primers and propidium monoazide. J. Appl. Microbiol. 110: 209–217.

Fukushima, K., K. Sogo, S. Miura and Y. Kimura. 2004. Production of D-lactic acid by bacterial fermentation of rice starch. Macromol. Biosci. 4: 1021–1027.

Fuller, M.E., S.H. Streger, R.K. Rothmel, B.J. Mailloux, J.A. Hall, T.C. Onstott, J.K. Fredrickson, D.L. Balkwill and M.F. DeFlaun. 2000. Development of a vital fluorescent staining method for monitoring bacterial transport in subsurface environments. Appl. Environ. Microbiol. 66: 4486–4496.

Ganesan, B., M.R. Stuart and B.C. Weimer. 2007. Carbohydrate starvation causes a metabolically active but nonculturable state in *Lactococcus lactis*. Appl. Environ. Microbiol. 73: 2498–2512.

Gao, M., M. Hirata, E. Toorisaka and T. Hano. 2006a. Study on acid hydrolysis of spent cells for lactic acid fermentation. Biochem. Eng. J. 28: 87–91.

Gao, M., M. Hirata, E. Toorisaka and T. Hano. 2006b. Acid-hydrolysis of fish wastes for lactic acid fermentation. Bioresour. Technol. 97: 2414–2420.

Ghoddusi, H.B. and R.K. Robinson. 1996. Enumeration of starters cultures in fermented milks. J. Dairy Res. 63: 151–158.

Gilliland, S.E. and C.N. Rich. 1990. Stability during frozen and subsequent refrigerated storage of *Lactobacillus acidophilus* grown at different pH. J. Dairy Sci. 73: 1187–1192.

Girgis, H.S., J. Smith, J.B. Luchansky and T.R. Klaenhammer. 2003. Stress adaptation of lactic acid bacteria. pp. 159–211. *In*: A.E. Yousef and V.K. Juneja (eds.). Microbial stress adaptation and food safety. Boca Raton: CRC Press.

Griffiths, M.W. 1993. Applications of bioluminescence in the dairy industry. J. Dairy Sci. 76: 3118–3125.

Groot, M.N., E. Klaassens, W.M. de Vos, J. Delcour, P. Hols and M. Kleerebezem. 2005. Genome-based *in silico* detection of putative manganese transport systems in *Lactobacillus plantarum* and their genetic analysis. Microbiology 151: 1229–1238.

Grześkowiak, L., A. Endo, M.C. Collado, L.J. Pelliniemi, S. Beasley and S. Salminen. 2013. The effect of growth media and physical treatments on the adhesion properties of canine probiotics. J. Appl. Microbiol. 115: 539–545.

Gilliland, S.E. 1985. Bacterial starter cultures for food. CRC Press Inc., Boca Raton, FL, USA.

Goksungur, Y. and U. Guvenc. 1999. Batch and continuous production of lactic acid from beet molasses by immobilized *Lactobacillus delbrueckii* IFO 3202. J. Chem. Technol. Biotechnol. 74: 131–136.

Guerra, N.P. and L. Pastrana. 2002. Nisin and pediocin production on mussel-processing waste supplemented with glucose and five nitrogen sources. Lett. Appl. Microbiol. 34: 114–118.

Guillou, M. and C. Tellier. 1988. Determination of ethanol in alcoholic beverages by low-resolution pulsed nuclear magnetic resonance. Anal. Chem. 60: 2182–2185.

Hall, J.W. and A. Pollard. 1992. Near-Infrared spectrophotometry: A new dimension in clinical chemistry. Clin. Chem. 38: 1623–1631.

Hammes, W.P. and Ch. Hertel. 2006. The genera *Lactobacillus* and *Carnobacterium*. Prokaryotes 4: 320–403.

Hartemink, R., B.J. Kok, G.H. Weenk and E.M. Rombouts. 1996. Raffinose-*Bifidobacterium* (RB) agar, a new selective medium for bifidobacteria. J. Microbiol. Methods 27: 33–43.

Hartemink, R., V.R. Domenech and F.M. Rombouts. 1997. LAMVAB, a new selective medium for the isolation of lactobacilli from faeces. J. Microbiol. Meth. 29: 77–84.

Harz, M., P. Rösch and J. Popp. Vibrational Spectroscopy. A powerful tool for the rapid identification of microbial cells at the single-cell level. Cytometry 75A: 104–113.

Herrero, M., C. Quiros, L.A. Garcia and M. Diaz. 2006. Use of flow cytometry to follow the physiological states of microorganisms in cider fermentation processes. Appl. Environ. Microbiol. 72: 6725–6733.

Hewitt, C.J. and G. Nebe-von Caron. 2001. An industrial application of multiparameter flow cytometry: Assessment of cell physiological state and its application to the study of microbial fermentations. Cytometry 44: 179–187.

Hutter, K.J. 2002. Flow cytometry: a new tool for direct control of fermentation processes. J. Inst. Brew. 108: 48–51.

Ibrahim, S. and A. Bezkorovainy. 1994. Growth-promoting factors for *Bifidobacterium longum*. J. Food Sci. 59: 189–191.

International Organization for Standardization and International Dairy Federation. 2008. IDF 220. Draft International Standard (ISO/DIS 29981). Milk products—Enumeration of presumptive bifidobacteria. Colony count technique at 37°C.

John, R.P., K.M. Nampoothiri, A.S. Nair and A. Pandey. 2005. L(+)-Lactic acid production using *Lactobacillus casei* in solid-state fermentation. Biotechnol. Lett. 27: 1685–1688.

John, R.P., K.M. Nampoothiri and A. Pandey. 2006. Simultaneous saccharification and fermentation of cassava bagasse for L (+) lactic acid production using lactobacilli. Appl. Biochem. Biotechnol. 134: 263–272.

John, R.P., K.M. Nampoothiri and A. Pandey. 2007. Fermentative production of lactic acid from biomass: an overview on process developments and future perspectives. Appl. Microbiol. Biotechnol. 74: 524–534.

Johns, F.E., J.F. Gordon and N. Shapton. 1978. The separation from yogurt cultures of lactobacilli and streptococci using reinforced clostridial agar at pH 5.5 and plate count agar incorporating milk. J. Soc. Dairy Technol. 31: 209–212.

Jonsson, H., E. Ström and S. Roos. 2001. Addition of mucin to the growth medium triggers mucus-binding activity in different strains of *Lactobacillus reuteri in vitro*. FEMS Microbiol. Lett. 204: 19–22.

Kacmar, J., A. Zamamiri, R. Carlson, N.R. Abu-Absi and F. Srienc. 2004. Single-cell variability in growing Saccharomyces cerevisiae cell populations measured with automated flow cytometry. J. Biotechnol. 109: 239–254.

Kankaanpää, P.E., S.J. Salminen, E. Isolauri and Y.K. Lee. 2001. The influence of polyunsaturated fatty acids on probiotic growth and adhesion. FEMS Microbiol. Lett. 194: 149–153.

Karp, S.G., A.H. Igashiyama, P.F. Siqueira, J.C. Carvalho, L.P.S. Vandenberghe, V. Thomaz-Soccol, J. Coral, J.-L. Tholozan, A. Pandey and C.R. Soccol. 2011. Application of the biorefinery concept to produce L-lactic acid from the soybean vinasse at laboratory and pilot scale. Bioresour. Technol. 102: 1765–1772.

Kelly, W.J., R.V. Asmundson and C.M. Huang. 1996. Characterization of plantaricin KW30, a bacteriocin produced by *Lactobacillus plantarum*. J. Appl. Bacteriol. 81: 657–662.

Kelly, P., P.B. Maguire, M. Bennett, D.J. Fitzgerald, R.J. Edwards, B. Thiede, A. Treumann, J.K. Collins, G.C. O'Sullivan, F. Shanahan and C. Dunne. 2005. Correlation of probiotic *Lactobacillus salivarius* growth phase with its cell wall associated proteome. FEMS Microbiol. Lett. 252: 153–159.

Kiliç, G.B. 2013. Highlights in probiotic research. Chapter 10. Intech Publishers. Croatia.

Kim, H.-O., Y.-J. Wee, J.-N. Kim, J.-S. Yun and H.-W. Ryu. 2006. Production of lactic acid from cheese whey by batch and repeated batch cultures of *Lactobacillus* sp. RKY2. Appl. Biochem. Biotechnol. 131: 694–704.

Kimoto-Nira, H., C. Suzuki, M. Kobayashi and K. Mizumachi. 2008. Different growth media alter the induction of interleukin 12 by a *Lactococcus lactis* strain. J. Food Protect 10: 1960–2160.

Kiviharju, K., K. Salonen, U. Moilanen and T. Eerikäinen. 2008. Biomass measurement online: the performance of *in situ* measurements and software sensors. J. Ind. Microbiol. Biotechnol. 35: 657–665.

Košir, I.J. and J. Kidri. 2002. Use of modern nuclear magnetic resonance spectroscopy in wine analysis: determination of minor compounds. Anal. Chim. Acta 458: 77–84.

Kotzamanidis, C., T. Roukas and G. Skaracis. 2002. Optimization of lactic acid production from beet molasses by *Lactobacillus delbrueckii* NCIMB 8130. World J. Microbiol. Biotechnol. 18: 441–448.

Lacroix, C., F. Grattepanche, Y. Doleyres and D. Bergmaier. 2005. Immobilized cell technologies for dairy industry. Applications of cell immobilisation biotechnology. Springer-Verlag; Focus on Biotechnology 8B: 295–319.

Lacroix, C. and S. Yildirim. 2007. Fermentation technologies for the production of probiotics with high viability and functionality. Curr. Opin. Biotech. 18: 176–183.

Lahtinen, S.J., L. Tammela, J. Korpela, R. Parhiala, H. Ahokoski, H. Mykkänen and S.J. Salminen. 2009. Probiotics modulate the *Bifidobacterium* microbiota of elderly nursing home residents. Age. 31: 59–66.

Lankaputhra, W.E.V. and N.P. Shah. 1996. A simple method for selective enumeration of *Lactobacillus acidophilus* in yogurt supplemented with *L. acidophilus* and *Bifidobacterium* spp. Milchwissenschaft 51: 446–450.

Laplace, J.M., N. Sauvageot, A. Hartke and Y. Auffray. 1999. Characterization of *Lactobacillus collinoides* response to heat, acid and ethanol treatments. Appl. Microbiol. Biotechnol. 51: 659–663.

Leal-Sánchez, M.V., R. Jiménez-Díaz, A. Maldonado-Barragán, A. Garrido-Fernández and J.L. Ruiz-Barba. 2002. Optimization of bacteriocin production by batch fermentation of *Lactobacillus plantarum* LPCO10. Appl. Environ. Microbiol. 68: 4465–4471.

Liew, S., A. Ariff, A. Raha and Y. Ho. 2005. Optimization of medium composition for the production of a probiotic microorganism, *Lactobacillus rhamnosus*, using response surface methodology. Int. J. Food Microbiol. 102: 137–142.

Lima, K.G.d.C., M.F. Kruger, J. Behrens, M.T. Destro, M. Landgraf and B.D.G.M. Franco. 2009. Evaluation of culture media for enumeration of *Lactobacillus acidophilus*, *Lactobacillus casei* and *Bifidobacterium animalis* in the presence of *Lactobacillus delbrueckii* subsp. *bulgaricus* and *Streptococcus thermophilus*. Lebensm-Wiss. Technol. 42: 491–495.

Lindemann, C., S. Marose, T. Scheper, H.O. Nielsen and K.F. Reardon. 1999. Fluorescence techniques for bioprocess monitoring. pp. 1238–1244. *In*: M.C. Flickinger and S.W. Drew (eds.). Encyclopedia of Bioprocess Technology: Fermentation, Biocatalysis, and Bioseparation. Wiley, USA.

Lipmann, F. and L.C. Tuttle. 1945. A specific micromethod for the determination of acyl phosphates. J. Biol. Chem. 159: 21–28.

Lloyd, D., C.A. Moran, M.T.E. Suller and M.G. Dinsdale. 1996. Flow cytometric monitoring of rhodamine 123 and a cyanine dye uptake by yeast during cider fermentation. J. Inst. Brew. 102: 251–259.

Liu, B., M. Yang, B. Qi, X. Chen, Z. Su and Y. Wan. 2010. Optimizing L-(+)-lactic acid production by thermophile *Lactobacillus plantarum* As. 1.3 using alternative nitrogen sources with response surface method. Biochem. Eng. J. 52: 212–219.

Luedeking, R. and E.L. Piret. 2000. A kinetic study of the lactic acid fermentation batch process at controlled pH. Biotechnol. Bioeng. 67: 393–400.

Macedo, M.G., M.-F. Laporte and C. Lacroix. 2002. Quantification of exopolysaccharide, lactic acid, and lactose concentrations in culture broth by near-infrared spectroscopy. J. Agric. Food Chem. 50: 1774–1779.
Malacrino, P., G. Zapparoli, S. Torriani and F. Dellaglio. 2001. Rapid detection of viable yeasts and bacteria in wine by flow cytometry. J. Microbiol. Meth. 45: 127–134.
Martens, H. and T. Næs. 1989. *In*: Methods for calibration, in Multivariate Calibration, Wiley, Chichester, England. Chapter 3, pp. 97.
Mataragas, M., J. Metaxopoulos, M. Galiotou and E.H. Drosinos. 2003. Influence of pH and temperature on growth and bacteriocin production by *Leuconostoc mesenteroides* L124 and *Lactobacillus curvatus* L442. Meat Sci. 64: 265–271.
Mataragas, M., E.H. Drosinos, E. Tsakalidou and J. Metaxopoulos. 2004. Influence of nutrients on growth and bacteriocin production by *Leuconostoc mesenteroides* L124 and *Lactobacillus curvatus* L442. A. Van Leeuw. 85: 191–198.
Mattarelli, P., C. Bonaparte, B. Pot and B. Biavati. 2008. Proposal to reclassify the three biotypes of *Bifidobacterium longum* as three subspecies: *Bifidobacterium longum* subsp. *longum* subsp. nov., *Bifidobacterium longum* subsp. *infantis* comb. nov. and *Bifidobacterium longum* subsp. *suis* comb. nov. Int. J. Syst. Evol. Micr. 58: 767–772.
Mättö, J., H. Alakomi, A. Vaari, I. Virkajärvi and M. Saarela. 2006. Influence of processing conditions on *Bifidobacterium animalis* subsp. *lactis* functionality with a special focus on acid tolerance and factors affecting it. Int. Dairy J 16: 1029–1037.
Maus, J.E. and S.C. Ingham. 2003. Employment of stressful conditions during culture production to enhance subsequent cool and acid-tolerance of bifidobacteria. J. Appl. Microbiol. 95: 146–154.
McClure, F.W. 2003. 204 years of near infrared technology: 1800–2003. J. Near Infrared Spectrosc. 11: 487–518.
Mercenier, A., S. Pavan and B. Pot. 2003. Probiotics as biotherapeutic agents: present knowledge and future prospects. Curr. Pharma. Design. 9: 175–191.
Michelsen, O., A. Cuesta-Dominguez, B. Albrechtsen and P.R. Jensen. 2007. Detection of bacteriophage-infected cells of *Lactococcus lactis* using flow cytometry. Appl. Environ. Microbiol. 73: 7575–7581.
Mitsuoka, T. 1978. Intestinal bacteria and health. Harcourt Brace Jovanovich Japan, Tokyo, pp. 31.
Mitsuoka, T. 1992. The human gastrointestinal tract. pp. 69–114. *In*: B.J.B. Wood (ed.). The Lactic Acid Bacteria, vol. 1. Elsevier, London.
Mobili, M., A. Londero, T.M.R. Maria, M.E.S. Eusébio, G.L. De Antoni, R. Fausto and A. Gómez-Zavaglia. 2009. Characterization of S-layer proteins of *Lactobacillus* by FTIR spectroscopy and differential scanning calorimetry. Vibrat. Spectrosc. 50: 68–77.
Montgomery, D.C., G.C. Runger and N.F. Hubele. 2001. Engineering Statistics, John Wiley and Sons, Inc., NY.
Muller, S. and A. Losche. 2004. Population profiles of a commercial yeast strain in the course of brewing. J. Food Eng. 63: 375–381.
Munoa, F.J. and R. Pares. 1988. Selective medium for isolation and enumeration of *Bifidobacterium* spp. Appl. Environ. Microbiol. 54: 1715–1718.
Nancib, N., A. Nancib, A. Boudjelal, C. Benslimane, F. Blanchard and J. Boudrant. 2001. The effect of supplementation by different nitrogen sources on the production of lactic acid from date juice by *Lactobacillus casei* subsp. *rhamnosus*. Bioresour. Technol. 78: 149–153.
Nancib, A., N. Nancib, D. Meziane-Cherif, A. Boudenbir, M. Fick and J. Boudrant. 2005. Joint effect of nitrogen sources and B vitamin supplementation of date juice on lactic acid production by *Lactobacillus casei* subsp. *rhamnosus*. Biores. Technol. 96: 63–67.
Naumann, D., D. Helm and H. Labischinski. 1991. Microbiological characterizations by FT-IR spectroscopy. Nature 351: 81–82.
Navrátil, M., C. Cimander and C.-F. Mandenius. 2004. On-line multisensor monitoring of yogurt and filmjölk fermentations on production scale. J. Agric. Food Chem. 52: 415–420.
Nebra, Y. and A.R. Blanch. 1999. A new selective medium for *Bifidobacterium* spp. Appl. Environ. Microbiol. 65: 5173–5176.
Niewenhof, F.F.J. and J.D. Hoolwerf. 1988. Suitability of BactoScan for the estimation of the bacteriological quality of raw milk. Milchwissenschaft 43: 577–586.
Nilsson, M., I.F. Duarte, C. Almeida, I. Delgadillo, B.J. Goodfellow, A.M. Gil and G.A. Morris. 2004. High-resolution NMR and diffusion-ordered spectroscopy of port wine. J. Agricult. Food Chem. 52: 3736–3743.

Nordlund, J., V. Merilainen and V. Andcrssen. 1980. Bioluminescenstiuampninmojligheter inom mjolkhushallningen. Nod. Mejeriindustri 75.
Oberg, T.S., J.L. Steele, S.C. Ingham, V.V. Smeianov, E.P. Briczinski, A. Abdalla and J.R. Broadbent. 2011. Intrinsic and inducible resistance to hydrogen peroxide in *Bifidobacterium* species. J. Ind. Microbiol. Biot. 38: 1947–1953.
Oh, S., S. Rheem, J. Sim, S. Kim and Y. Baek. 1995. Optimizing conditions for the growth of *Lactobacillus casei* YIT 9018 in tryptone-yeast extract-glucose medium by using response surface methodology. Appl. Environ. Microbiol. 61: 3809–3814.
Olsson, L. and J. Nielsen. 1997. On-line and *in situ* monitoring of biomass in submerged cultivations. Trends Biotechnol. 15: 517–522.
Ouwehand, A.C., E.M. Tuomola, S. Tölkkö and S. Salminen. 2001. Assessment of adhesion properties of novel probiotic strains to human intestinal mucus. Int. J. Food Microbiol. 64: 119–126.
Páez, R., L.L. Lavari, L.G. Audero, L.A. Cuatrin, L.N. Zaritzky, J. Reinheimer and G. Vinderola. 2013. Study of the effects of spray-drying on the functionality of probiotic lactobacilli. Int. J. Dairy Technol. 66: 155–161.
Palchaudhuri, S., S.J. Rehse, K. Hamasha, T. Syed, E. Kurtovic, E. Kurtovic and E.J. Stenger. 2011. Raman spectroscopy of xylitol uptake and metabolism in gram-positive and gram-negative bacteria. Appl. Environ. Microbiol. 77: 131–137.
Papadimitriou, K., H. Pratsinis, G. Nebe-von Caron, D. Kletsas and E. Tsakalidou. 2006. Rapid assessment of the physiological status of *Streptococcus macedonicus* by flow cytometry and fluorescence probes. Int. J. Food Microbiol. 111: 197–205.
Papagianni, M. 2012. Effects of dissolved oxygen and pH levels on weissellin A production by *Weissella paramesenteroides* DX in fermentation. Bioprocess Biosyst. Eng. 35: 1035–1041.
Parente, E., F. Ciocia, A. Ricciardi, T. Zotta, G.E. Felis and S. Torriani. 2010. Diversity of stress tolerance in *Lactobacillus plantarum*, *Lactobacillus pentosus* and *Lactobacillus paraplantarum*: A multivariate screening study. Int. J. Food Microbiol. 144: 270–279.
Partanen, L., N. Marttinen and T. Alatossava. 2001. Fats and fatty acids as growth factors for *Lactobacillus delbrueckii.* Syst. Appl. Microbiol. 24: 500–506.
Pasteris, S.E., E. Vera Pingitore, C.E. Ale and M.E.F. Nader-Macías. 2014. Characterization of a bacteriocin produced by *Lactococcus lactis* subsp. *lactis* CRL 1584 isolated from a lithobates catesbeianus hatchery. World J. Microbiol. Biotechnol. 30: 1053–1062.
Pintado, J., J.P. Guyot and M. Raimbault. 1999. Lactic acid production from mussel processing wastes with an amylolytic bacterial strain. Enzyme Microb. Technol. 24: 590–598.
Pieterse, B., R.J. Leer, F.H.J. Schuren and M.J. van der Werf. 2005. Unravelling the multiple effects of lactic acid stress on *Lactobacillus plantarum* by transcription profiling. Microbiology 151: 388–3894.
Polak-Berecka, M., A. Waśko, M. Kordowska-Wiate, M. Podleśny, Z. Targoński and A. Kubik-Komar. 2010. Optimization of medium composition for enhancing growth of *Lactobacillus rhamnosus* PEN using response surface methodology. Polish J. Microbiol. 59: 113–118.
Prasad, J., P. Mc Jarrow and P. Gopal. 2003. Heat and osmotic stress responses of probiotic *Lactobacillus rhamnosus* HN001 (DR20) in relation to viability after drying. Appl. Environ. Microbiol. 69: 917–925.
Presser, K.A., T. Ross and D.A. Ratkowsky. 1998. Modelling the growth limits (growth/no growth interface) of *Escherichia coli* as a function of temperature, pH, lactic acid concentration, and water activity. Appl. Environ. Microbiol. 64: 1773–1779.
Rada, V. and J. Koc. 2000. The use of mupirocin for selective enumeration of bifidobacteria in fermented products. Milchwissenschaft 55: 65–67.
Ramos, A., A.R. Neves and H. Santos. 2002. Metabolism of lactic acid bacteria studied by nuclear magnetic resonance. Anton. Leeuw. Int. J. G. 82: 249–261.
Ramanjooloo, A., A. Bhaw-Luximon, D. Jhurry and F. Cadet. 2009. Quantitative assessment of lactic acid produced by biofermentation of cane sugar juice. Spectrosc. Lett. 42: 296–304.
Ranadheera, R.D.C.S., S.K. Baines and M.C. Adams. 2010. Importance of food in probiotic efficacy. Food Res. Int. 43: 1–7.
Rault, A., M. Bouix and C. Beal. 2008. Dynamic analysis of *Lactobacillus delbrueckii* subsp. *bulgaricus* CFL1 physiological characteristics during fermentation. Appl. Microbiol. Biotechnol. 81: 559–570.
Reimann, S., F. Grattepanche, E. Rezzonico and C. Lacroix. 2010. Development of a real-time RT-PCR method for enumeration of viable *Bifidobacterium longum* cells in different morphologies. Food Microbiol. 27: 236–242.

Reuter, G. 1970. Laktobazillen und eng verwandte Mikroorganismen in Fleisch und Fleischwaren. 2. Mitteilung: Die Charakterisierung der isolierten Laktobazillenstämme. Fleischwirtschaft 50: 954–962.

Reuter, G. 1985. Elective and selective media for lactic acid bacteria. Int. J. Food Microbiol. 2: 55–68.

Reddy, G., M. Altaf, B.J. Naveena, M. Venkateshwar and E. Vijay Kumar. 2008. Amylolytic bacterial lactic acid fermentation—a review. Biotechnol. Adv. 26: 22–34.

Rodriguez-Saona, L., F. Khambaty, F. Fry and E. Calvey. 2001. Rapid detection and identification of bacterial strains by Fourier transform near infrared spectroscopy. J. Agric. Food Chem. 49: 574–579.

Rogosa, M., J.A. Mitchell and R.E. Wisemann. 1951. A selective medium for the isolation and enumeration of oral and fecal lactobacilli. J. Bacteriol. 62: 132–133.

Roy, D. 2003. Media for the detection and enumeration of bifidobacteria in food products. pp. 147–160. *In*: J.E.L. Corry, G.D.W. Curtis and R.M. Baird (eds.). Handbook of Culture Media for Food Microbiology. Elsevier Science B.V.

Saarela, M.H., M. Rantala, K. Hallamaa, L. Nohynek, I. Virkajärvi and J. Mättö. 2004. Stationary-phase acid and heat treatments for improvement of the viability of probiotic lactobacilli and bifidobacteria. J. Appl. Microbiol. 96: 1205–1214.

Saarela, M., I. Virkajarvi, H.L. Alakomi, T. Mattila-Sandholm, A. Vaari, T. Suomalainen and J. Matto. 2005. Influence of fermentation time, cryoprotectant and neutralization of cell concentrate on freeze-drying survival, storage stability, and acid and bile exposure of *Bifidobacterium animalis* ssp. *lactis* cells produced without milk-based ingredients. J. Appl. Microbiol. 99: 1330–1339.

Saarela, M.H., H.L. Alakomi, A. Puhakka and J. Matto. 2009. Effect of the fermentation pH on the storage stability of *Lactobacillus rhamnosus* preparations and suitability of *in vitro* analyses of cell physiological functions to predict it. J. Appl. Microbiol. 106: 1204–1212.

Salma, M., S. Rousseaux, A. Sequeira-Le Grand and H. Alexandre. 2013. Cytofluorometric detection of wine lactic acid bacteria: application of malolactic fermentation to the monitoring. Int. J. Microbiol. Biotechnol. 40: 63–73.

Saranwong, S. and S. Kawano. 2008a. System design for non-destructive near infrared analyses of chemical components and total aerobic bacteria count of raw milk. J. Near Infrared Spec. 16: 389–398.

Saranwong, S. and S. Kawano. 2008b. Interpretation of near infrared calibration structure for determining the total aerobic bacteria count in raw milk: interaction between bacterial metabolites and water absorptions. J. Near Infrared Spec. 16: 497–504.

Savage, D.C. 1992. Growth phase, cellular hydrophobicity, and adhesion *in vitro* of lactobacilli colonizing the keratinizing gastric epithelium in the mouse. Appl. Environ. Microbiol. 58: 1992–1995.

Sawatari, Y. and A. Yokota. 2007. Diversity and mechanisms of alkali tolerance in lactobacilli. Appl. Environ. Microbiol. 73: 3909–3915.

Scardovi, V. 1986. Bifidobacterium. pp. 1418. *In*: P.H. Sneath, N.S. Mair, M.E. Sharpe and J.G. Holt (eds.). Bergey's manual of systematic bacteriology. 9th Ed. Volume 2. Williams and Wilkins Publishers, Baltimore, MD, USA.

Schar-Zammaretti, P., M.L. Dillmann, N. D'Amico, M. Affolter and J. Ubbink. 2005. Influence of fermentation medium composition on physicochemical surface properties of *Lactobacillus acidophilus*. Appl. Environ. Microbiol. 71: 8165–8173.

Schillinger, U. and W.H. Holzapfel. 2003. Culture media for lactic acid bacteria. pp. 127–140. *In*: J.E.L. Corry, G.D.W. Curtis and R.M. Baird (eds.). Handbook of Culture Media for Food Microbiology. Elsevier Science B.V.

Schirru, S., L. Favaro, N.P. Mangia, M. Basaglia, S. Casella, R. Comunian, F. Fancello, B.D.G.M. Franco, R.P.S. Oliveira and S.D. Todorov. 2014. Comparison of bacteriocins production from *Enterococcus faecium* strains in cheese whey and optimised commercial MRS medium. Ann. Microbiol. 64: 321–331.

Schmidt, G. and R. Zink. 2000. Basic features of the stress response in three species of bifidobacteria: *B. longum, B. adolescentis*, and *B. breve*. Int. J. Food Microbiol. 55: 41–45.

Schmidt, M.T., A.K. Olejnik-Schmidt, K. Myszka, M. Borkowska and W. Grajek. 2010. Evaluation of quantitative PCR measurement of bacterial colonization of epithelial cells. Pol. J. Microbiol. 59: 89–93.

Serrano, L.M., D. Molenaar, M. Wels, B. Teusink, P.A. Bron, W.M. de Vos and E.J. Smid. 2007. Thioredoxin reductase is a key factor in the oxidative stress response of *Lactobacillus plantarum* WCFS1. Microbial Cell Factories 6: 29.

Serrazanetti, D., M.E. Guerzoni, A. Corsetti and R. Vogel. 2009. Metabolic impact and potential exploitation of the stress reactions in lactobacilli. Food Microbiol. 26: 700–711.

Sharpe, M.E. and T.F. Fryer. 1965. Media for lactic acid bacteria. Lab. Practice 14: 697–701.

Simões, C., H.L. Alakomi, J. Maukonen and M. Saarela. 2010. Expression of clpL1 and clpL2 genes in *Lactobacillus rhamnosus* VTT E-97800 after exposure to acid and heat stress treatments or to freeze-drying. Beneficial microbes. 1: 253–257.

Simpson, P.J., G.F. Fitzgerald, C. Stanton and R.P. Ross. 2004. The evaluation of a mupirocin-based selective medium for the enumeration of bifidobacteria from probiotics animal feed. J. Microbiol. Meth. 57: 9–16.

Siragusa, S., M. De Angelis, R. Di Cagno, C.G. Rizzello, R. Coda and M. Gobbetti. 2007. Synthesis of γ-aminobutyric acid by lactic acid bacteria isolated from a variety of Italian cheeses. Appl. Environ. Microbiol. 73: 7283–7290.

Sivakesava, S., J. Irudayaraj and D. Ali. 2001. Simultaneous determination of multiple components in lactic acid fermentation using FT-MIR, NIR, and FT-Raman spectroscopic techniques. Proc. Biochem. 37: 371–378.

Small, G.W. 2006. Chemometrics and near-infrared spectroscopy: avoiding the pitfalls. Trends Anal. Chem. 25: 1057–1066.

Solieri, L. and P. Giudici. 2010. PCR assay for strain-specific detection of *Oenococcus oeni* during wine malolactic fermentation. Appl. Environ. Microbiol. 76: 7765–7774.

Stapleton, A.E., M. Au-Yeung, T.M. Hooton, D.N. Fredricks, P.L. Roberts, C.A. Czaja, Y. Yarova-Yarovaya, T. Fiedler, M. Cox and W.E. Stamm. 2011. Randomized, placebo-controlled pPhase 2 trial of a *Lactobacillus crispatus* probiotic given intravaginally for prevention of recurrent urinary tract infection. Clin. Infect. Dis. 52: 1212–1217.

Stephenie, W., B.M. Kabeir, M. Shuhaimi, M. Rosfarizan and A.M. Yazid. 2007. Influence of pH and impeller tip speed on the cultivation of *Bifidobacterium pseudocatenulatum* G4 in a milk-based medium. Biotechnol. Bioprocess Eng. 12: 475–483.

Tamburini, E., G. Vaccari, S. Tosi and A. Trilli. 2003. Near-infrared spectroscopy: a tool for monitoring submerged fermentation processes using an immersion optical-fiber probe. Appl. Spectrosc. 57: 132–138.

Tan, W.S., M.F. Budinich, R. Ward, J.R. Broadbent and J.L. Steele. 2012. Optimal growth of *Lactobacillus casei* in a Cheddar cheese ripening model system requires exogenous fatty acids. J. Dairy Sci. 95: 1680–1689.

Tanaka, R., H. Takayama, M. Morotomi, T. Kurishima, S. Ueyama, K. Matsumoto, A. Kuroda and M. Mutai. 1983. Effects and administration of TOS and *Bifidobacterium breve* 4006 on the human fecal flora. Bifidobacterial Microflora 2: 17–24.

Tellier, C., M. Guillou-Charpin, P. Grenier and D. Le Botlan. 1989. Monitoring alcoholic fermentation by low-resolution pulsed nuclear magnetic resonance. J. Agricult. Food Chem. 37: 988–991.

Teraguchi, S., M. Uehara, K. Ogasa and T. Mitsuoka. 1978. Enumeration of bifidobacteria in dairy products. Jpn. J. Bacteriol. 33: 753–761.

Timbuntam, W., K. Sriroth and Y. Tokiwa. 2006. Lactic acid production from sugar-cane juice by a newly isolated *Lactobacillus* sp. Biotechnol. Lett. 28: 811–814.

Todorov, S., B. Gotcheva, X. Dousset, B. Onno and I. Ivanova. 2000. Influence of growth medium on bacteriocin production in *Lactobacillus plantarum* ST31 Biotechnology and Biotechnological Equipment 1: 50–55.

Todorov, S.D. and L.M.T. Dicks. 2005a. Effect of growth medium on bacteriocin production by *Lactobacillus plantarum* ST194BZ, a strain isolated from boza. Food Technol. Biotechnol. 43: 165–173.

Todorov, S.D. and L.M.T. Dicks. 2005b. *Lactobacillus plantarum* isolated from molasses produces bacteriocins active against gram-negative bacteria. Enzyme Microb. Tech. 36: 318–326.

Todorov, S.D. and L.M.T. Dicks. 2005c. Production of bacteriocin ST33LD, produced by *Leuconostoc mesenteroides* subsp. *mesenteroides*, as recorded in the presence of different medium components. World J. Microb. Biot. 21: 1585–1590.

Todorov, S.D. and L.M.T. Dicks. 2009. Effect of modified MRS medium on production and purification of antimicrobial peptide ST4SA produced by *Enterococcus mundtii.* Anaerobe 15: 65–73.

Todorov, S.D., P. Ho, M. Vaz-Velho and L.M. Dicks. 2010. Characterization of bacteriocins produced by two strains of *Lactobacillus plantarum* isolated from Beloura and Chouriço, traditional pork products from Portugal. Meat Sci. 84: 334–343.

Todorov, S.D., H. Prévost, M. Lebois, X. Dousset, J.G. LeBlanc, B.D.G.M. Franco. 2011. Bacteriocinogenic *Lactobacillus plantarum* ST16Pa isolated from papaya (Carica papaya)—From isolation to application: Characterization of a bacteriocin. Food Res. Int. 44: 1351–1363.

Todorov, S.D., L. Favaro, P. Gibbs and M. Vaz-Velho. 2012. *Enterococcus faecium* isolated from Lombo, a Portuguese traditional meat product: characterisation of antibacterial compounds and factors affecting bacteriocin production. Benef. Microbes 3: 319–330.

Tosi, S., M. Rossi, F. Tamburini, G. Vaccari, A. Amaretti and D. Matteuzzi. 2003. Assessment of in-line near infrared spectroscopy for continuous monitoring of fermentation processes. Biotechnol. Prog. 19: 1816–1821.

Touré, R., E. Kheadr, C. Lacroix, O. Moroni and I. Fliss. 2003. Production of antibacterial substances by bifidobacterial isolates from infant stool active against *Listeria monocytogenes*. J. Appl. Microbiol. 5: 1058–1069.

Ueckert, J.E., G. Nebe von-Caron, A.P. Bos and P.F. Ter Steeg. 1997. Flow cytometric analysis of *Lactobacillus plantarum* to monitor lag times, cell division and injury. Lett. Appl. Microbiol. 25: 295–299.

Vaccari, G., E. Dosi, A.L. Campi, A. González-Vera, R. Matteuzzi and G. Mantovani. 1993. A near-infraroed spectroscopy technique for the control of fermentation processes: An application to lactic acid fermentation. Biotechnol. Bioeng. 43: 913–917.

Vaidyanathan, S., S. White, L.M. Harvey and B. McNeil. 2003. Influence of morphology on the near-infrared spectra of mycelial biomass and its implications in bioprocess monitoring. Biotechnol. Bioeng. 82: 715–724.

van de Casteele, S., T. van Heuverzwijn, T. Ruyssen, P. van Assche, J. Swings and G. Huys. 2006. Evaluation of culture media for selective enumeration of probiotic strains of lactobacilli and bifidobacteria in combination with yoghurt or cheese starters. Int. Dairy J. 16: 1470–1476.

van de Guchte, M., P. Serror, C. Chervaux, T. Smokvina, E. Ehrlich and E. Maguin. 2002. Stress responses in lactic acid bacteria. Anton. Leeuw. Int. J. G. 82: 187–216.

Vinderola, C.G. and J.A. Reinheimer. 1999. Culture media for the enumeration of *Bifidobacterium bifidum* and *Lactobacillus acidophilus* in the presence of yoghurt bacteria. Int. Dairy J. 9: 497–505.

Vinderola, C.G. and J.A. Reinheimer. 2000 Eumeration of *Lactobacillus casei* in the presence of *L. acidophilus*, bifidobacteria and lactic starter bacteria in fermented dairy products. Int. Dairy J. 10: 271–275.

Vinderola, C.G., A.M. Binetti, P. Burns and J.A. Reinheimar. 2011. Cell viability and functionality of probiotic bacteria in dairy products. Frontiers in Microbiology. Article 70. 2: 1–6.

Vishnu, C., G. Seenayya and G. Reddy. 2002. Direct fermentation of various pure and crude starchy substrates to L(+) lactic acid using *Lactobacillus amylophilus* GV6. World J. Microbiol. Biotechnol. 18: 429–433.

Walberg, M., P. Gaustad and H.B. Steen. 1997. Rapid discrimination of bacterial species with different ampicillin susceptibility levels by means of flow cytometry. Cytometry 29: 267–272.

Walker, K., N. Ripandelli and S. Flint. 2005. Rapid enumeration of *Bifidobacterium lactis* in milk powders using impedance. Int. Dairy J. 15: 183–188.

Wang, Z.-W. and X.-L. Liu. 2008. Medium optimization for antifungal active substances production from a newly isolated *Paenibacillus* sp. using response surface methodology. Bioresource Technol. 99: 8245–8251.

Wee, Y.-J. and H.-W. Ryu. 2009. Lactic acid production by *Lactobacillus* sp. RKY2 in a cell-recycle continuous fermentation using lignocellulosic hydrolyzates as inexpensive raw materials. Bioresour. Technol. 100: 4262–4270.

Wegkamp, A., W.M. De Vos and E.J. Smid. 2009. Folate overproduction in *Lactobacillus plantarum* WCFS1 causes methotrexate resistance. FEMS Microbiol. Lett. 297: 261–265.

WHO/FAO. 2002. Guidelines for the evaluation of probiotics in food. London. Available at: ftp://ftp.fao.org/Es/esn/food/wgreport2.pdf.

Williams, P. and K. Norris. 2001. Near-infrared technology in the agriculture and food industries. 2nd ed. American Association of Cereal Chemists, St. Paul, MN, 2001.

Yadav, A.K., A.B. Chaudhari and R.M. Kothari. 2011. Bioconversion of renewable resources into lactic acid: an industrial view. Crit. Rev. Biotechnol. 31: 1–19.

Yun, J.S., Y.J. Wee, J.N. Kim and H.W. Ryu. 2004. Fermentative production of DL-lactic acid from amylase treated rice and wheat bran's hydrolyzate by a novel lactic bacterium, *Lactobacillus* sp. Biotechnol. Lett. 18: 1613–1616.

Yoo, I.K., H.N. Chang, E.G. Lee, Y.K. Chang and S.H. Moon. 1997. Effect of B vitamin supplementation on lactic acid production by *Lactobacillus casei*. J. Ferment. Bioeng. 84: 172–175.

Zhang, B.L., C. Yunianta, Y.-L. Vallet and M.L. Martin. 1997. Natural abundance isotopic fractionation in the fermentation reaction: influence of the fermentation medium. Bioorg. Chem. 25: 117–129.
Zhao, R., A. Natajaran and F. Srienc. 1998. A flow injection flow cytometry system for on-line monitoring of bioreactors. Biotechnol. Bioeng. 62: 609–617.
Zotta, T., A. Ricciardi, F. Ciocia, R. Rossano and E. Parente. 2008. Diversity of stress responses in dairy thermophilic streptococci. Int. J. Food Microbiol. 124: 34–42.

# CHAPTER 4

# Probiotics as Cell Factories for Bioactive Ingredients: Focus on Microbial Polysaccharides and Health Beneficial Effects

*L.E.E. London,*[1,2] *R.P. Ross,*[1] *G.F. Fitzgerald,*[1,3] *Fergus Shanahan,*[1,3] *Noel M. Caplice*[1,4] and *C. Stanton*[1,2,*]

## 1. Introduction

Polysaccharides are sugar polymers commonly found in nature, where they are distributed among plants, fungi, yeasts and bacteria. In nature, polysaccharides have multifaceted biological functions such as cellular communication (glycosaminoglycans), storage of energy (starch and glycogen), cell wall architecture (cellulose and chitin), and biological defence. Polysaccharides are macromolecules varying in molecular weight from hundreds to sometimes several millions of Daltons and are either composed of one type of monosaccharide (homopolymer) or two or more different types of monosaccharide (heteropolymer) and can be substituted by non-carbohydrate compounds. Bacterial polysaccharides are constituents of cell walls, and form part of lipopolysaccharides (LPS), capsular polysaccharides (CPS), and are used as storage units, or secreted into the extracellular environment as exopolysaccharides (EPS).

EPS have been described for many years for their structural diversity, exhibiting a variety of physico-chemical and biological properties. Bacterial EPS are diverse polysaccharides synthesised by bacteria using distinct pathways. The physiological functions of EPS include biological defence against phagocytosis, phage attack,

[1] Alimentary Pharmabiotic Centre, Cork, Ireland.
[2] Teagasc, Food Research Centre, Moorepark, Fermoy, Co. Cork, Ireland.
[3] University College Cork, Ireland.
[4] Centre for Research in Vascular Biology.
* Corresponding author: Catherine.stanton@teagasc.ie

antibiotics, toxic metal ions and to counteract physical stresses such as desiccation and osmotic stress. Nevertheless, the physiological functions of bacterial EPS are still largely unknown and only few bacterial EPS have been exploited for industrial applications (Badel et al. 2011). Polysaccharides, such as dextrans from *Leuconostoc mesenteroides*, gellan from *Sphingomonas paucimobilis,* xanthan from *Xanthomonas campestris* and curdlan from *Agrobacterium biovar* 1 (identified as *Alcaligenes faecali* var. *myxogenes* at the time of discovery) are examples that currently have commercial food and non-food applications (De Vuyst et al. 2001, Sutherland 1988). Lactic acid bacteria (LAB) are well known EPS producers and bacterial EPS have been proven to be effective, non-toxic macromolecules. For example, *L. mesenteroides* excretes dextran which is commercially exploited in the food industry (Monsan et al. 2001). However, the low amount of bacterial EPS produced by most LAB is a barrier to commercial exploitation. Most LAB are food-grade having GRAS status and their EPS production is regarded as safe and thus, they can be utilized in food products. Bacterial EPS have been claimed to have a wide range of health benefits, including cardioprotective properties such as hypocholesterolemic, antioxidant and immunomodulatory activities, in addition to prebiotic and antitumor activities.

## 2. Sources of Polysaccharide-producing Bacteria

Many bacterial species are described as EPS producers, including *Lactobacillus, Leuconostoc, Pediococcus, Lactococcus, Weissella, Streptococcus* and *Bifidobacterium*. Approximately 50 species of *Lactobacillus* have been found to date to produce EPS and among the best known are *Lb. plantarum*, *Lb. johnsonii*, *Lb. rhamnosus*, *Lb. kefiranofaciens*, *Lb. delbrueckii* ssp. *bulgaricus*, *Lb. acidophilus*, *Lb. reuteri*, *Lb. fermentum*, *Lb. pentosus*, *Lb. brevis* and *Lb. helveticus*. Also commonly described for EPS production are *L. mesenteroides*, *L. dextranicium*, *Pediococcus damnosus*, *P. parvulus*, *P. pentosaceus*, *Lactococcus lactis* ssp., *Streptococcus thermophilus*, *Bifidobacterium longum* and *B. animalis*. EPS producing LAB have been found in diverse ecological niches from dairy products and non-dairy based fermented foods (Deutsch et al. 2008, Van der Meulen et al. 2007), beverages (Ibarburu et al. 2010), vegetables (McKellar et al. 2003), breads (Galle et al. 2011), to the human gastrointestinal tract (GIT) (Prasanna et al. 2012b, Ruas-Madiedo et al. 2007, Salazar et al. 2009). Other environmental niches such as soil samples, aqueous solutions, waste water, sludge samples and the marine ecosystem have also been reported to be home to EPS producing bacteria (Asker et al. 2009, Fitriyanto et al. 2011, Freitas et al. 2011, Gray et al. 1991, Iyer et al. 2004, Kambourova et al. 2009, Li et al. 2008, Liu et al. 2010, Maugeri et al. 2002, Nicolaus et al. 2000, Prasertsan et al. 2006, Ruiz-Ruiz et al. 2011, Sun et al. 2011). It has been reported that the quantity of EPS production by LAB depends on various physico-chemical conditions for bacterial growth such as the fermentation medium and temperature. The quantity of EPS produced by LAB strains is generally lower than 1g/L, but recovery is also dependent on the isolation method (Degeest et al. 2001).

## 3. Screening Methods to Detect Polysaccharide Producing Bacteria

The phenotype associated with EPS producing bacteria is described as ropy, mucoid and slimy. Nevertheless, not all mucoid or slime-producing strains are ropy. Ropy colonies form long strands when extended with an inoculation loop, whereas mucoid and slime-producing colonies are not able to produce long filaments (Dierksen et al. 1997). For example, ropy strains exhibit smoother consistency, higher viscosity and lower syneresis in fermented dairy products compared to non-ropy strains (Rawson and Marshall 1997). Some LAB strains are able to produce two EPS with different compositions (Knoshaug et al. 2000), whereas others are able to express both ropy and mucoid phenotypes, depending on the fermentation conditions (Cerning et al. 1994, Dierksen et al. 1997). A simple screening method for detection of EPS producing bacteria is the visual inspection of ropy colonies, which is useful for the identification of LAB strains with potential texture enhancing properties for use in fermented dairy products (Welman et al. 2003). The use of several different sugars in the fermentation medium improves the opportunity for detection of ropy strains (van Geel-Schutten et al. 1998), as EPS production is influenced by the sugar available in the fermentation medium. However, the particular sugar that leads to enhanced EPS production is strain dependent (Cerning et al. 1994, Looijesteijn et al. 1999). Viscometric screening methods for the detection of ropy strains in liquid culture media can also be used for detection of EPS producing bacteria, but increased biomass can interfere with viscosity (Vaningelgem et al. 2004). The use of staining methods for detection of EPS producers has been reported. Ruthenium red-milk agar which uses the high affinity for acidic surfaces such as bacterial polysaccharides and congo red agar which binds with β-1, 3 and β-1,4 linkages of polysaccharides have been reported as useful protocols (Stingele and Mollet 1995). Fluorescently-labelled lectins and carbohydrate binding proteins have been used for the direct visualisation of EPS production by confocal scanning laser microscopy (Hassan et al. 2002, Purohit et al. 2009). However, staining methods are not able to quantify polysaccharides yields (Bouzar et al. 1997).

Milk and de Man, Rogasa and Sharp medium (MRS) are the most widely used media for EPS production by LAB strains but their complexity interferes with EPS extraction and analysis (Cerning et al. 1994). Consequently, synthetic semi-defined, and minimal media have been designed (Kimmel and Roberts 1998, Torino et al. 2000), which allow rapid separation of EPS, by using ultrafiltration and gel permeation chromatography (Tieking et al. 2003), as compounds of complex media are eliminated. PCR screening methods based on the sequences of genes encoding specific enzymes involved in the biosynthesis of EPS have also been developed (Kralj et al. 2003, Tieking et al. 2003). For example, the highly conserved region of the glucosyltransferase gene encoding the extracellular enzyme that synthesises homopolysaccharide EPS in many LAB has been used to generate primers and subsequently used to identify other LAB with similar genes

## 4. Isolation and Quantification of Bacterial Exopolysaccharides

The choice of the EPS production medium influences EPS yield and is therefore of great importance. However, the development of synthetic semi-defined, and minimal media facilitate EPS analysis (De Vuyst et al. 2001). Thus, the method to be employed for EPS isolation depends on the composition of the fermentation medium. The simplest isolation method for semi-defined fermentation medium is dialysis against water after cell removal followed by lyophilisation (Marshall et al. 1995). For more complex environments, additional purification steps are necessary to eliminate components of the medium. For example, media high in protein content are treated with trichloroacetic acid (TCA), digested with proteases or treated with a combination of TCA and proteases to precipitate proteins followed by centrifugation to remove protein (Ruas-Madiedo et al. 2002a). Ethanol precipitation is performed to concentrate polysaccharides (Ricciardi et al. 2002, Rimada and Abraham 2001). Apart from protein removal and ethanol precipitation, microfiltration, ultrafiltration, diafiltration, ion-exchange chromatography and SDS-PAGE are used to purify the EPS fraction (Bergmaier et al. 2001, Nakajima et al. 1990, Pan and Mei 2010, Staaf et al. 2000, van Casteren et al. 1998, Yang et al. 1999).

The simplest method for quantification of EPS is weight determination (Frengova et al. 2000) but values obtained are strongly influenced by the degree of purity. A colorimetric method described by Dubois et al. 1956 to determine sugar and related compounds is widely used to quantify EPS yields. This method is based on the absorbance at 490 nm of the orange-yellow colour of EPS when treated with phenol and concentrated sulphuric acid. However, this protocol quantifies the total carbohydrate content present in the EPS fraction and is also influenced by the degree of purity. More recently, fluorescent marked lectin probes have been used to quantify polysaccharides attached to bacterial cells (Robitaille et al. 2006). Due to the diversity of fermentation media, and various isolation and quantification methods, it is difficult to establish typical EPS yields for specific EPS-producing strains. An easy-to-use method for quantitative analysis of bacterial EPS in fermented milk products was developed by Enikeev 2012. Samples were treated with 12M hydrochloric acid followed by neutralization with 20% (w/v) sodium hydroxide solution with phenolphthalein. After removing proteins by centrifugation and precipitation of EPS, the precipitated EPS was hydrolysed with 2M sulphuric acid and neutralized with 20% (w/v) sodium hydroxide. The acid treatment resulted in protein and lactose removal, thereby avoiding the need for dialysis.

## 5. Microbial Biosynthesis of Exopolysaccharides

The ability of LAB to biosynthesise EPS was first reported by Pasteur and Orla-Jensen in 1943. EPS are known to be synthesised in one of two ways, i.e., via extracellular or intracellular biosynthesis. The extracellular pathway is independent of central carbon metabolism and is rather simple, which limits the variation of structure (Boels et al. 2001, Tieking et al. 2005). Bacterial polysaccharides produced via the extracellular pathway are mostly referred as homopolysaccharides. Homopolysaccharides such as dextrans, levans, alterans and mutans are synthesized by extracellular glycansucrases, which transfer a monosaccharide from a disaccharide to a growing polysaccharide chain. EPS-producing bacteria resident in the gastrointestinal tract of animals and human and mucosal samples have been described as having the extracellular pathway (Ruas-Madiedo et al. 2007, Tieking et al. 2003). The intracellular pathway for EPS biosynthesis produces both homo- and heteropolysacchrides (Cerning 1990, Cerning et al. 1994, De Vuyst and Degeest 1999, Gamar et al. 1997, van Geel-Schutten et al. 1998). However, only a few LAB have been described that are able to produce both homo- and heteropolysaccharides. It was demonstrated that *O. oeni* has this unique property (Dimopoulou et al. 2012), and depending on the strain and on the growth medium some isolates were able to produce either a homo- or a heteropolysaccharide. Various EPS biosynthetic pathways were identified that are potentially active in *O. oeni*. Dextransucrase, fructansucrase and glucosyltransferase genes that can drive the production of homopolysaccharide were identified in this host as well as two EPS operons containing genes necessary for the production of heteropolysaccharides.

In general, the EPS synthesized by the intracellular biosynthesis pathway are comprised of (ir) regular repeating units and consist of a variety of sugars including glucose, galactose, rhamnose, glucuronic acid, fucose, N-acetylglucosamine and N-acetylgalactosamine. EPS are formed from intermediates of central carbon metabolism and several enzymes and/or proteins are involved in their biosynthesis and secretion. Genes coding for the enzymes and/or proteins required for EPS production are either of plasmid or chromosomal origin in LAB strains. In all cases, genes involved in EPS biosynthesis are organized in gene clusters that show significant conservation in organisation and sequence (Lamothe et al. 2002, Stingele and Mollet 1995, vanKranenburg et al. 1997). Hence, the intracellular biosynthetic pathway can be broken down into four separate reaction sequences involving regulation, chain length determination, biosynthesis of repeating units and polymerisation/export. The glycosyltransferases involved in the biosynthesis of repeating units catalyze the transfer of sugar moieties from activated donor molecules to specific acceptor molecules, thereby forming glycosidic bonds (Campbell et al. 1997) resulting in large variation of the EPS formed. Therefore, variations in EPS structure are due to different glycosyltransferase activities.

A number of physiological functions have been described for bacterial EPS. For example, it has been reported to confer a protective function on producing strains against toxic and/or limiting environments such acid stress, starvation survival and the presence of nisin and other antagonists (De Vuyst et al. 2001, Durlu-Ozkaya et al. 2007, Kim et al. 2000a, Looijesteijn et al. 2001). Moderate protection against phage infection was demonstrated by EPS-producing *Lactococcus* ssp. and *Lb. delbrueckii* ssp. (Durlu-Ozkaya et al. 2007, Forde and Fitzgerald 2003, Looijesteijn et al. 2001). Another putative role of EPS is their use as a carbon source by the producing strain, as some LAB strains express enzymes capable of degrading their own EPS and the quantity than can be recovered decreases during prolonged incubation (Badel et al. 2011). Moreover, an adhesive role for bacterial EPS has also been reported (Lebeer et al. 2007).

### *5.1 Bacterial Homopolysacharides*

More than fifty *Lactobacillus* species are reported to produce homopolysaccharides, synthesized via the extracellular biosynthetic pathway (Kralj et al. 2004, Tieking et al. 2005, Tieking et al. 2003). Other LAB strains that have been described as being capable of extracellular biosynthesis of homopolysaccharides are *L. mesenteroides* (Bounaix et al. 2010). The homopolysaccharides are composed of one type of monosaccharide, primarily glucose or fructose (Cerning 1990, De Vuyst et al. 2001, Tieking et al. 2005). Depending on the glycosyl donor, the glycansucrases are either glucosyltransferases (*gtf*) or fructosyltransferases (*ftf*) (Monsan et al. 2001) and the synthesized homopolysaccharides can be clustered into: α-D-glucans, β-D-glucans, β-D-fructans and others (Ruas-Madiedo et al. 2002a). Examples of enzymes produced by glucosyltransferase encoding genes are dextransucrase, mutansucrase and reuteransucrase which synthesize glucans with different structures due to different enzyme activities. For example, *L. mesenteroides* produces an α-1, 6 linkage glucan, *S. mutans* produces an α-1,3 linkage glucan (Wen and Burne 2002). *Lb. reuteri* produces α-1,4 linked glucan (Kralj et al. 2004) and *P. damnosus* 2.6 produces a β-1,3 linked glucan (DuenasChasco et al. 1997). Examples of enzymes produced by fructosyltransferase encoding genes are levansucrase and inulosucrase, which synthesize fructans. Structural differences are described, such as the β-1, 6 linkage present in levans (Kang et al. 1999), while the β-2, 1 linkage is found in inulin (van Hijum et al. 2002). The yield of homopolysaccharides synthesized by the extracellular pathway is reported as high as 1g/L as there is no other energy required than the biosynthesis of the enzyme (De Vuyst et al. 2001, Tieking et al. 2005, Tieking et al. 2003).

### *5.2 Bacterial Heteropolysaccharide*

The quantity of heteropolysaccharides produced by LAB strains is reported to be lower than 1g/L (De Vuyst et al. 2001) and is rather complex. The intracellular biosynthesis of heteropolysaccharides involves the synthesis of EPS from intracellular sugar nucleotides of central carbon metabolism in the cytoplasm. However, glycosyltransferases are the key enzymes involved in heteropolysaccharide biosynthesis, but the production of sugar nucleotides also requires non-EPS-specific enzymes. Hence, complex media are required to maximise heteropolysaccharide production, as they are able to deliver nutrients for both cell growth and polysaccharide formation (De Vuyst and Degeest 1999, De Vuyst et al. 1998). Many different types of heteropolysaccharides are secreted by LAB strains with respect to sugar composition and molecular mass (Cerning 1990, De Vuyst and Degeest 1999). Table 1 provides an overview in the variability of heteropolysaccharides produced by LAB strains. Instability of heteropolysaccharide production was demonstrated by Lamothe et al. (2002) and Laws et al. (2001) to be due to spontaneous loss of plasmids and the plasmid location of the EPS operon which includes genes for several enzymes and/or proteins involved in their biosynthesis and secretion (Bourgoin et al. 1999, Gancel and Novel 1994, Stingele et al. 1996). In strains that do not harbour plasmids encoding EPS operons, genetic instability may be due to mobile genetic elements such as IS and/or to a generalized genomic instability including

**Table 1.** Classification of heteropolysaccharides produced by lactic acid bacteria with respect to sugar composition.

| Structure of repeating unit | Strain | Reference |
|---|---|---|
| Trisaccharide | *Lactobacillus* ssp. G-77 | (Duenas-Chasco et al. 1998) |
| Tetrasaccharide | *Streptococcus thermophilus* | (Marshall et al. 2001a) |
| | *Lactobacillus paracasei* | (Robijn et al. 1996b) |
| Pentasaccharide | *Lactococcus lactis* | (van Casteren et al. 2000a) |
| | *Lactobacillus rhamnosus* | (Vanhaverbeke et al. 1998) |
| | *Lactobacillus delbrueckii* ssp. *bulgaricus* | (Faber et al. 2001a) |
| | *Lactobacillus acidophilus* | (Robijn et al. 1996a) |
| Hexasaccharide | *Lactobacillus helveticus* | (Stingele et al. 1997) |
| | *Streptococcus thermophilus* | (Faber et al. 2001b) |
| Heptasaccharide | *Lactobacillus helveticus* | (Staaf et al. 2000) |
| | *Streptococcus thermophilus* | (Marshall et al. 2001a) |
| | *Lactobacillus delbrueckii* ssp. *bulgaricus* | (Marshall et al. 2001b) |
| | *Lactococcus lactis* | (van Casteren et al. 2000b) |
| Octasaccharide | *Streptococcus thermophilus* | (Low et al. 1998) |
| Nonasaccharide | *Lactobacillus rhamnosus* | (Gorska-Fraczek et al. 2011) |

DNA deletions and rearrangements. Mobile IS elements bordering the EPS operon have been shown to be responsible for the instability of heteropolysaccharide expression (Bourgoin et al. 1999) in *Lb. delbrueckii* subsp. *bulgaricus* and *S. thermophilus* due to spontaneous rearrangement of genes.

## 6. Microbial Beta Glucan

Beta glucans are carbohydrates consisting of linked glucose molecules, which are present in cereals, mushrooms, fungi, yeast, seaweeds and have been recently discovered in bacteria. Examples of beta glucans of different origin are listed in Table 2. Depending on the origin, beta glucans display differences in macromolecular structure, with yeast and fungal beta-glucans for example, consisting of 1,3 β-linked glycopyranosyl residues with small numbers of 1,6 β-linked branches, while cereal beta glucans consist of 1,3 β- and 1,4 β-linked glycopyranosyl residues, and bacterial beta glucans consist of 1,3 β-linked glycopyranosyl residues. Variation in solubility, molecular mass, tertiary structure, degree-of-branching, polymer charge and solution

**Table 2.** Representative beta glucan producing bacteria.

| Name | Source | Reference |
|---|---|---|
| Curdlan | *Agrobacterium* sp., *Alcaligens* sp. | (Nakata et al. 1998) |
| Zymosan | *Saccharomyces cerevisiae* | (Au et al. 1994) |
| Laminaran | *Laminaria digitata* | (Read et al. 1996) |
| Grifolan | *Grifola frondosa* | (Okazaki et al. 1995) |
| Lentinan | *Lentinus eeodes* | (Jong and Birmingham 1993) |

confirmation are characteristics known to influence the physiological and health-promoting effects of beta glucans. A relatively small number of bacterial strains have been reported to produce beta glucans. Curdlan, the first bacterial beta glucan to be described (Harada 1965), is produced by *Alcaligenes faecalis* var. *myxogenes* 10C3 and *Agrobacterium* spp. (Kim et al. 2000a, Kim et al. 2003). Curdlan is an insoluble beta glucan which has been assessed for safety *in vitro* and *in vivo* and is approved for use in certain countries as a food additive (Spicer et al. 1999). The industrial use of curdlan has been widely demonstrated. For example, it is used as a gelling material to improve textural qualities, water-holding capacity and thermal stability of various foods (Jezequel 1998). Sulphated curdlan has been reported to exhibit antiviral (including anti-AIDS virus) activity *in vitro* (Jagodzinski et al. 1994, Mikio 1995). Furthermore, curdlan has been demonstrated to be an auxiliary treatment for severe malaria (Evans et al. 1998), while immunomodulatory activity for curdlan has also been reported (Kim et al. 2003).

Exclusively cider and wine spoiling LAB, including members of the species *Pediococcus* (DuenasChasco et al. 1997), *Lactobacillus* (Duenas-Chasco et al. 1998) and *Oenococcus* (Dols-Lafargue et al. 2008, Lbarburu et al. 2007) have been reported to produce 2-branched 1,3-β-D-glucan. Beta glucan producing pediococci remain difficult to remove from wine because of their natural resistance to traditional wine stabilizing treatment (lysozyme). A study by Coulon et al. (2012) demonstrated that bacterial beta glucan protects the bacterial cell against anti-bacterial agents such as lysozyme by forming a protective barrier around the cell. Kearney et al. (2011) incorporated recombinant beta glucan producing *Lactobacillus paracasei* NFBC 338 into yoghurt as a adjunct culture during product manufacture, which resulted in improved texture qualities and reduced whey separation of the yoghurt, compared to yoghurt made with a non-beta glucan producing isogenic adjunct control strain.

A number of health-promoting properties have been described for microbial beta glucan. For example, beta glucan produced by *P. parvulus* 2.6 and *P. parvulus* CUPV 22 was reported to contribute to bacterial adhesion to human intestinal epithelial cells *in vitro* (de Palencia et al. 2009, Gaizka et al. 2010). In a randomized, double blind study, a fermented, ropy, oat-based product in combination with beta glucan producing *P. damnosus* 2.6 decreased total plasma cholesterol concentrations (Martensson et al. 2005) in healthy human subjects. Subjects received 600 ml per day of this fermented product in combination with beta glucan producing *P. damnosus* 2.6 for 8 weeks, which resulted in increased counts of faecal *Bifidobacterium* spp.

Changes in short chain fatty acid (SCFA) formation in caecum, distal colon and faeces of rats have also been associated with ingestion of fermented oat-based-food in combination with *P. damnosus* 2.6 (Lambo-Fodje et al. 2006). The immunomodulatory properties of *P. parvulus* CUPV 22 and *Lb. suebicus* CUPV 221 have been demonstrated *in vitro* (Gaizka et al. 2010). This effect was evaluated by determining their ability to modulate cytokine production by pro-inflammatory macrophages and anti-inflammatory macrophages in human peripheral blood mononuclear cells. Both bacterial strains triggered production of IL-6, IL-8 and TNF-α in pro-inflammatory macrophages and *Lb. suebicus* CUPV221 induced IL-10 production in both pro-inflammatory and anti-inflammatory macrophages. *Propionibacterium freudenreichii* and *Paenibacillus polymyxa* have also been reported to produce beta glucan (Nordmark et al. 2005). The former synthesises a capsule beta glucan (Deutsch et al. 2008) while the beta glucan produced by the latter has been demonstrated as an immunomodulatory agent in RAW264.7 murine macrophages *in vitro* (Chang et al. 2010). Isolated beta glucan (300 μg/ml) induced the phosphorylation of mitogen-activated protein kinases ERK1/2, JNK/SAPK and p38-MAPKs in cytoplasm of macrophages, which in turn activated transcriptional factors NFκB and AP1.

## 7. Health Beneficial Effects of Bacterial Exopolysaccharides

A range of health beneficial properties have been associated with microbial EPS including hypocholesterolemic, immunomodulatory antimicrobial, anti-adhesive, antitumoral, and antioxidant activities (Table 3), as discussed below. EPS-producing LAB strains have been reported to exhibit anti-adhesive and anti-microbial effects *in vitro* and some LAB are able to counteract the toxic effect of bacterial toxin and enteropathogens, thus having a potential benefit to the host. For example, released-EPS produced by *Lactobacillus acidophilus* A4 was shown to significantly decrease biofilm formation of several pathogenic bacteria such as *Salmonella enteritidis*, *Salmonella typhimurium* KCCM 11806, *Yersinia enterocolitica*, *P. aeruginosa* KCCM 11321, *Listeria monocytogenes* ScottA, and *Bacillus cereus* when tested on both polystyrene and PVC surfaces (Kim et al. 2009). EPS synthesized by *Lb. reuteri* was shown to decrease the ability of enterotoxigenic *Escherichia coli* (ETEC) to bind to porcine erythrocytes (Wang et al. 2010). EPS has been shown to aid bacteria in their ability to adhere to intestinal mucus layers and to colonize the intestinal wall (Gaizka et al.

**Table 3.** Representative exopolysaccharide-producing lactic acid bacteria strains for potential health benefits.

| Health beneficial potential | Microbial EPS-producing strain | Reference |
|---|---|---|
| Phage protection | *Lactococcus* ssp.<br>*Lb. delbruecki* ssp. | (Durlu-Ozkaya et al. 2007)<br>(Fitzgerald and Forde 2003) |
| Protection of pathogen adhesion | *Lactobacillus acidophilus*<br>*Weissella* spp.<br>*Lactobacillus suebicus Pediococcus parvulus* | (Kim et al. 2009)<br>(Schwab et al. 2010)<br>(Paloma et al. 2010) |
| Reduction/prevention of gastritis | *Streptococcus thermophilus* | (de Valdez et al. 2010, de Valdez et al. 2009) |
| Antioxidant agent | *Streptococcus* spp.<br>Bif. *animalis Lactococcuslactis* subsp. *lactis* | (Kanmani et al. 2011)<br>(Xu et al. 2011)<br>(Pan and Mei 2010) |
| Immunomodulation agent | *Streptococcus* spp. *Lactococcus* spp. *Lactobacillus* ssp.<br>*Bifidobacter* spp. | (Rodriguez et al. 2009)<br>(Nakajima et al. 1995, Kitazawa et al. 1996)<br>(Lamontagne et al. 2010)<br>(Hu et al. 2010) |
| Antitumor | *Lactobacillus acidophilus*<br>*Streptococcus thermophilus*<br>*Lactobacillus delbrueckii* subsp.<br>*Lactococcus lactis* ssp. | (Kim et al. 2010)<br>(Hassan et al. 2008) |
| Prebiotic | *Bifidobacterium* spp. | (Salazar et al. 2011) |

2010) and bacterial adhesion to intestinal mucus layer has been reported to reduce chronic gastritis in an animal model (Rodriguez et al. 2009).

Antioxidants primarily protect cells from free radical damage and can thereby reduce the risk of diseases associated with aging, including heart disease, cancer, arthritis, diabetes, Alzheimers and Parkinsons disease. EPS from *Streptococcus* spp. and *B. animalis* have been demonstrated to exhibit antioxidant potential by inhibiting hydroxyl and superoxide anion radicals (Kanmani et al. 2011, Xu et al. 2011). Purified EPS from *Lactococcus lactis* subsp. *lactis* 12 inhibited malondialdehyde formation, an indicator of lipid oxidation, and has been reported to increase antioxidant enzyme activity in Kunming mice in a dose depending manner (Pan and Mei 2010).

It was reported that EPS isolated from *Lactococcus* spp. exhibited immunomodulatory activity when administered intraperitoneally to BALB/c mice (Nakajima et al. 1995). Furthermore, oral administration of fermented milk containing EPS-producing *Streptococcus* spp. to BALB/c mice with induced chronic gastritis (Rodriguez et al. 2009) activated the immune system *in vivo*. Isolated EPS of *Lactococcus* spp., *Lactobacillus* ssp. and *Bifidobacterium* spp. induced inflammatory markers INF-γ, IL-1-α and IL-10 *in vitro* in C57BL/6 spleen antigen presenting macrophages, and in human and mouse cultured immunocompetent cells (Bleau et al. 2010, Chabot et al. 2001, Kitazawa et al. 1996). *In vivo* studies involving oral administration of purified EPS isolated from *Lactobacillus kefiranofaciens* to BALB/c mice resulted in an increased cytokine production by cells of lamina propria and thereby induced a gut mucosal response (Vinderola et al. 2007, Vinderola et al. 2006). The reduction of chronic gastritis and colonic inflammation resulting from ingestion of isolated EPS from *Streptococcus* spp. and other EPS-producing LAB strains was shown by immunosuppression of IFNγ producing cells, an increase of IL10 producing cells, and myeloperoxidase activity (Rodriguez et al. 2010, Rodriguez et al. 2009, Sengul et al. 2006). Oral administration of kefiran promoted immunoglobulin secretion, immunomodulation and altered blood pressure and serum components in rats and mice (Maeda et al. 2004a, Maeda et al. 2004b, Vinderola et al. 2006). The *in vitro* ability of EPS-producing *Bifidobacterium* strains to modify the proliferation of epithelial intestinal cells lines Caco2 and HT29, and the stimulation response and the cytokine production pattern of peripheral blood mononuclear cells elicited by purified EPS fractions was studied by Lopez et al. (2012). The study demonstrated a correlation between biological function and the physico-chemical characteristics of some EPS. About half of the tested EPS-producing *Bifidobacterium* strains showed significantly higher adhesion percentage to both cell lines than the reference strain *B.animalis* subsp. *lactis* Bb12. Furthermore, EPS purified fractions at concentration of 1 μg/ml were able to stimulate the proliferation of the human mononuclear cells. The EPS-producing *Lb. paraplantarum* BGCG11 showed an anti-inflammatory profile *in vitro* whereas the non-EPS derivative strain induced a higher pro-inflammatory response (Nikolic et al. 2012). In addition, when peripheral blood mononuclear cells were stimulated with increasing concentrations of 1, 10 and 100μg/ml of purified EPS, the cytokine profile was similar to that obtained with the EPS-producing *Lactobacillus* with a significant increase of IL-10 and IL-1β production.

Bacterial polysaccharides have been reported to exhibit antitumor activity; however the potency of activity depends on the branching structure (Yoon et al. 2004). Cell-bound EPS isolated from *Lb. acidophilus* 606 promoted autophagy-associated cancer cell death in HT-29 colon cancer cells *in vitro* (Kim et al. 2010). Orally administrated fermented milk containing either EPS-producing *Lb. delbrueckii* ssp. *bulgaricus* or EPS-producing *S. thermophilus* resulted in a reduction of colon cancer in male Fisher rats and exhibited a chemo-preventive effect *in vivo* (Purohit et al. 2009). Kefiran, an EPS produced by *Lactobacillus* spp. increased T-cell activity and thereby decreased tumor growth in mice (Zubillaga et al. 2001).

### 7.1 Microbial Exopolysaccharides and Heart Health

Cardiovascular diseases (CVD), the term used to describe heart, stroke and blood vessel disease constitute a group of disorders of the heart and blood vessels including atherosclerosis and hyperlipidemia. CVD remains the leading cause of death and disabilities globally, responsible for the deaths of approx. 17.3 million people worldwide in 2008 (FAO/WHO 2010) and it is estimated that by 2030, CVD will claim the lives of almost 25 million people annually. Modern dietary patterns in many countries such as high fat diets, rich in saturated fat have led to increased numbers of people who suffer from CVD (Califf 2009). The morbidity and mortality associated with CVD can be significantly reduced by addressing risk factors include hypertension, abnormal blood lipid levels, high saturated fat intakes, obesity and diabetes. Elevated plasma cholesterol is a validated risk factor for CVD and possibly some types of stroke. A 10% reduction in serum cholesterol in men aged 40 years has been reported to result in a 50% reduction in heart disease within 5 years; the same serum cholesterol reduction for men aged 70 years can result in an average 20% reduction in heart disease occurrence in the next 5 years. A number of cardioprotective compounds are present in the diet including soluble dietary fibers composed of polysaccharides which have been shown to decrease risk factors for heart diseases (Table 4).

**Table 4.** Representative bacterial polysaccharides for the reduction of heart health risk factors.

| Risk factor | Bacterial polysaccharide/strain | Reference |
|---|---|---|
| Hypertension | polysaccharide-glycopeptide complex (*Lb. casei*)<br>EPS (*Lb. casei* LC2W) | (Sawada et al. 1990)<br>(Ai et al. 2008) |
| Obesity | Bacterial polysaccharide<br>EPS (*Zymomonas mobilis*) | (Delattre et al. 2012)<br>(Kang et al. 2006, Kido et al. 2003) |
| Obesity induced Type 2 diabetes | Levan (*Bacillus licheniformis*)<br>Bacterial polysaccharide | (Dahech et al. 2011)<br>(Xiu et al. 2010) |
| Hypercholesterolemia | EPS (*Lb. delbrueckii* subsp. *bulgaricus*)<br>EPS (*Lactococcuslactis* subsp. *cremoris* SBT 0495)<br>EPS (*Lb. kefiranofaciens*)<br>EPS (*Lb. mucosae*) | (Tok and Aslim 2010)<br>(Nakajima et al. 1992)<br>(Maeda et al. 2004a)<br>(London et al. 2014) |

### 7.2 Microbial Exopolysaccharides and Hypertension

Few *in vivo* studies have reported the benefits of bacterial polysaccharide for hypertension reduction. Oral administration of a polysaccharide-glycopeptide complex from *Lb. casei* at doses of 1 mg/kg body weight to spontaneously hypertensive and renal hypertensive rats resulted in significantly decreased systolic blood pressure and lasted for 24 h (Sawada et al. 1990). Oral administration of a high molecular weight EPS isolated from *Lb. casei* LC2W at 15 mg/kg body weight daily for 7 days significantly decreased systolic blood pressure without a significant effect on the heart rate in spontaneously hypertensive rats but the blood pressure-lowering effect disappeared 24 h after the administration (Ai et al. 2008).

### 7.3 Microbial Expolysaccharides and Obesity/Type 2 Diabetes

Although body mass and fat regulation result from multiple factors, the American Dietetic Association has recommended that daily dietary intake of polysaccharide (20g fiber per day) has beneficial effects, in terms of weight reduction. A few studies have reported anti-obesity effects of dietary microbially derived polysaccharides. Beta-1, 3-glucan isolated from *Agrobacterium* sp. ATCC 31750 was tested for lipid storage and adipocyte differentiation on pre-adipocytes 3T3-L1 cells *in vitro* (Delattre et al. 2012). The bacterial EPS induced a 26% reduction in the amount of intracellular lipids when it was added at the concentration of 0.5% on the adipocytes culture. It was shown for cells treated with beta-1, 3-glucan that very few contained fatty acids and those containing lipid droplets were smaller in size than those present in untreated cells. Consequently, a significant decrease in the differentiation of pre-adipocytes into mature adipocytes was reported. Kang et al. (2006) studied the suppression of lipogenesis in the liver of high fat fed obese rats, by levan, a bacterial polysaccharide isolated from *Zymomonas mobilis.* Dietary levan supplementation was found to suppress body fat development and adipocyte hypertrophy, resulting in decreased leptin production and secretion. Levan supplementation at a range of 1–10% of body weight markedly suppressed the hyperinsulinemia induced by the high-fat diet and thereby suppressed hepatic lipogenesis and excess body fat development accompanied by decreased serum free fatty acid.

Type 2 diabetes mellitus (T2DM) is a metabolic disorder that is characterized by high blood glucose in the context of insulin resistance and relative insulin deficiency (Charpentier et al. 2003), with obesity contributing to the development of the condition. Oral administration of levan isolated from *Bacillus licheniformis* resulted in an anti-hypoglycemic effect in alloxan induced diabetic Wistar rats (Dahech et al. 2011). The diabetic effect of alloxan is due to excess production of reactive oxygen species leading to toxicity in pancreatic cells which reduces the synthesis and the release of insulin.

### 7.4 Microbial Exopolysaccharide and Hypercholesterolemia

Hypocholesterolemia and abnormal serum lipid concentrations increase the risk of heart disease. Cholesterol homeostasis is governed by the balance between intestinal absorption and elimination as well as total body synthesis. The relationship between EPS production and cholesterol removal rates of 5 *Lb. delbrueckii* subsp. *bulgaricus* strains isolated from home-made yoghurt was studied by Tok and Aslim (2010). Cholesterol removal was studied in fermentation medium supplemented with 0, 1, 2, and 3 mg/ml concentration of oxgall as a bile source and 10 mg/ml cholesterol. It was reported that EPS-producing *Lb. delbrueckii* subsp. *bulgaricus* B3 reduced cholesterol by 40% in the fermentation medium containing 1mg/ml oxgall. The EPS-producing *Lb. delbrueckii* subsp. *bulgaricus* ATCC 11842 strain reduced cholesterol by 30% in fermentation medium containing 1mg/ml oxgall. In both cases it was shown that EPS produced by the strains entrapped cholesterol.

EPS isolated from *Lactococcus lactis* subsp. *cremoris* SBT 0495 was reported to enhance the metabolism of serum cholesterol in rats (Nakajima et al. 1992) and another study reported a reduction in total cholesterol in the liver of cholesterol-fed hamsters supplemented with milk kefir (Liu et al. 2006). Maeda et al. (2004c) reported a reduction of total cholesterol in serum and reduced low-density lipoprotein (LDL)-cholesterol concentrations in the liver of animals fed kefiran, an EPS produced by *Lb. kefiranofaciens*. The intake of kefiran, the EPS produced by microorganisms present in kefir grains also greatly reduced the size of atherosclerotic lesions in cholesterol-fed rabbits (Uchida et al. 2010). Martensson et al. (2005) reported the blood lipid lowering effects in human subjects in a randomized, double blind study of a fermented, ropy, oat-based product in combination with EPS-producing *P. damnosus* 2.6, in which healthy subjects received 600 ml of the product per day for 8 weeks. It was demonstrated that bacterial EPS bound cholic acid *in vitro* (Lee et al. 2007, Pigeon et al. 2002). London et al. (2014) reported that dietary administration of an EPS producing *Lb. mucosae* ($10^9$ CFU/d/animal) for 12 weeks led to reduction of plasma cholesterol by ~33%. Taken together, these data demonstrate that microbial EPS has the potential to reduce risk factors for cardiovascular disease.

## 8. Prebiotic Properties of Bacterial Exopolysaccharides

The term prebiotic was originally defined as a "non-digestible food ingredient that beneficially affects the host by selectively stimulating the growth and/or activity of one or a limited number of bacteria in the colon, and thus improves host health" (Gibson and Roberfroid 1995). A more recent definition stated that "a prebiotic is a selectively fermented ingredient that allows specific changes; both in the composition and/or activity in the gastrointestinal microbiota that confers benefits upon host wellbeing and health" (Gibson et al. 2004). The gut microbiota, in particular *Bifidobacterium* spp. and *Lactobacillus* spp. metabolise prebiotics and improve immunity against pathogens (Gibson and Roberfroid 1995). Hongpattarakere et al. (2012) reported a prebiotic effect associated with bacterial EPS *in vitro*. In this study, bacterial polysaccharides were incorporated during the fermentation of a mixed culture in a continuous colon model system and increased *Bifidobacterium* and *Lactobacillus* growth was observed. Another study demonstrated a change of the *Bifidobacterium* population, whereas *B. longum* were reduced by 1 log but total *Bifidobacterium* numbers were significantly higher in rats fed a bacterial polysaccharide (Salazar et al. 2011). In a human study, Martensson et al. 2005 demonstrated that a combination of native and microbial beta glucan increased the bifidobacteria population.

## 9. Techno-functional Applications of Bacterial Polysaccharides

Industrially important bacterial polysaccharides include dextrans, xanthan gum, alternans, gellans, and xanthan (Sutherland 1988), which have a range of applications in the pharmaceutical and food industries. For example, dextrans are used in the manufacture of gel filtration products, as blood volume extenders and blood flow improvers. They play an important role in paper and metal-plating processes, in enhanced oil recovery, as food syrup-stabilizers and dough improvers (De Vuyst et al. 2001). Xanthan gum is an established food additive and is the major bacterial polysaccharide widely used by the food industry (Evans et al. 1998). Alternan is a low-viscosity bulking agent and extender in foods and cosmetics (De Vuyst et al. 2001).

In the food industry, bacterial polysaccharides have important roles as natural biothickeners and gelling agents (Girard and Schaffer-Lequart 2007, Piermaria et al. 2008). It has been demonstrated that most microbial polysaccharides show higher water solubility compared to plant gums, which is another useful characteristic in various food applications (Ganesh Kumar et al. 2004). Bacterial polysaccharides have been demonstrated to interact with proteins, ions and other compounds in the food matrix and they have stable rheological and emulsifying properties (Bergmaier et al. 2001, Hassan 2008, Katina et al. 2009, Rimada and Abraham 2006, Wang et al. 2010), which limits syneresis in dairy products. Generally, bacterial polysaccharides have been described for their indirect role on product rheology and can enhance texture, "mouthfeel", taste perception and stability of dairy products (Cerning 1990, Folkenberg et al. 2006, Hassan et al. 1995, Lamothe et al. 2002, Laws et al. 2001). In yoghurt, bacterial polysaccharides enhance the structure of the fermented product and in a comparative study between EPS-producers and non-EPS-producers, viscosity values of products manufactured with EPS-producing cultures remained higher after stirring (Amatayakul et al. 2006). Use of bacterial polysaccharides in various food systems avoids the need for addition of animal derived stabilizers. The role of chain stiffness, molecular weight and radius of bacterial polysaccharides are influencing factors for viscosity and rheology (Faber et al. 1998, Tuinier et al. 2001) and different conclusions have been made concerning the presence and concentration of EPS and their influence on flavour. For example, EPS-producing *B. longum* subsp. *infantis* NCIMB 702205, and *B. longum* subsp. *infantis* CCUG 52486 both have potential in the food industry as both produce considerable amount of EPS, together with a higher intrinsic viscosity and apparent viscosity in aqueous solution. However, EPS produced by both strains varies considerably in its microstructure. EPS produced by *B. longum* subsp. *infantis* CCUG 52486 has a porous dense entangled structure with higher emulsification activity, and the EPS of *B. longum* subsp. *infantis* NCIMB 702205 has a higher emulsification activity and smaller emulsion droplets (Prasanna et al. 2012a). Ruas-Madiedo et al. (2002b) reported that the viscosity of stirred fermented milk was strongly related to intrinsic viscosity of EPS producing LAB strains. Guler-Akin et al. (2009) studied the effect of ropy and non-ropy EPS producing LAB strains and reported that ropiness of EPS-producing strains positively influenced the viscosity of reduced fat stirred yoghurt. Bacterial polysaccharides do not exhibit taste *per se* in the final product and flavour discrepancy might be influenced by the performance of the yoghurt starter cultures with the addition of EPS-producers. In cheeses, bacterial polysaccharides influence physico-chemical and sensory properties. The use of EPS-producing strains has been reported to reduce the cheese manufacturing time by improving rennet coagulation properties leading to increased moisture content of cheeses (Awad et al. 2005, Jimenez-Guzman et al. 2009, Rynne et al. 2007), both responsible for the texture improvement and thereby contributing to the reduced energy content of the final product. Additionally, bacterial polysaccharides increase the technological stress tolerance of LAB during food production (De Vuyst et al. 2001, Kim et al. 2000b, Looijesteijn et al. 2001, Stack et al. 2010).

## 10. Conclusions

EPS from LAB exhibit structural diversity via intracellular or extracellular biosynthesis resulting in homopolysaccharides and heteropolysaccharides. The physiological functions of EPS include biological defence against phagocytosis, phage attack, antibiotics, toxic metal ions and the counteracting of physical stresses such as desiccation and osmotic stress. Bacterial EPS have been reported to exhibit beneficial effects on various health-related conditions, heart health in particular, mediating blood pressure lowering activities, and modulation of blood lipids and cholesterol, which are known risk factors for CVD. Additionally, EPS have been shown to inhibit pathogen adhesion in the gastrointestinal tract, reduce and/or prevent gastritis, and they also mediate antioxidant, immunomodulatory, antitumor and prebiotic properties. Furthermore, EPS from LAB strains have been well described for their techno-functional properties and have found applications in dairy and non-dairy products as biothickeners. Additionally, bacterial polysaccharides increase the technological performance of producing strains and subsequently enhance the stress tolerance of LAB strains during food production. Consequently, the application of bacterial EPS in the food industry is advantageous for enhancing both the quality and texture properties of dairy foods, as well as for conferring potential health benefits from their inclusion in health-promoting functional foods.

## 11. Acknowledgments

This work was funded in part by the Alimentary Pharmabiotic Centre, funded by Science Foundation Ireland the FibeBiotics EU project supported by the European Community's Seventh Framework Programme [FP7/2007-2013] under Grant Agreement n° 289517, Enterprise Ireland Commercialisation Fund Contract Reference: CF/2013/3030A/B and FIRM under the National Development Plan, 2000–2006. L.E.E. London was in receipt of a Teagasc Walsh Fellowship, 2007–2010.

## References

Ai, L.Z., H. Zhang, B.H. Guo, W. Chen, Z.J. Wu and Y. Wu. 2008. Preparation, partial characterization and bioactivity of exopolysaccharides from *Lactobacillus casei* LC2W. Carbohydrate Polymers 74: 353–357.

Amatayakul, T., A.L. Halmos, F. Sherkat and N.P. Shah. 2006. Physical characteristics of yoghurts made using exopolysaccharide-producing starter cultures and varying casein to whey protein ratios. International Dairy Journal 16: 40–51.

Asker, M.M.S., Y.M. Ahmed and M.F. Ramadan. 2009. Chemical characteristics and antioxidant activity of exopolysaccharide fractions from Microbacterium terregens. Carbohydrate Polymers 77: 563–567.

Au, B.T., T.J. Williams and P.D. Collins. 1994. Zymosan-induced IL-8 release from human neutrophils involves activation via the CD11b/CD18 receptor and endogenous platelet-activating factor as an autocrine modulator. Journal of immunology 152: 5411–5419.

Awad, S., A. Hassan and K. Muthukumarappan. 2005. Application of exopolysaccharide-producing cultures in making reduced fat Cheddar cheese. Viscoelastic properties. Journal of Animal Science 83: 18–18.

Badel, S., T. Bernardi and P. Michaud. 2011. New perspectives for Lactobacilli exopolysaccharides. Biotechnology Advances 29: 54–66.

Bergmaier, D., C. Lacroix, M.G. Macedo and C.P. Champagne. 2001. New method for exopolysaccharide determination in culture broth using stirred ultrafiltration cells. Applied Microbiology and Biotechnology 57: 401–406.

Bleau, C., A. Monges, K. Rashidan, J.P. Laverdure, M. Lacroix, M.R. Van Calsteren, M. Millette, R. Savard and L. Lamontagne 2010. Intermediate chains of exopolysaccharides from *Lactobacillus rhamnosus* RW-9595M increase IL-10 production by macrophages. Journal of Applied Microbiology 108: 666–675.

Boels, I.C., R. van Kranenburg, J. Hugenholtz, M. Kleerebezem and W.M. de Vos. 2001. Sugar catabolism and its impact on the biosynthesis and engineering of exopolysaccharide production in lactic acid bacteria. International Dairy Journal 11: 723–732.

Bounaix, M.S., V. Gabriel, H. Robert, S. Morel, M. Rernaud-Simeon, B. Gabriel and C. Fontagne-Faucher. 2010. Characterization of glucan-producing Leuconostoc strains isolated from sourdough. International Journal of Food Microbiology 144: 1–9.

Bourgoin, F., A. Pluvinet, B. Gintz, B. Decaris and G. Guedon. 1999. Are horizontal transfers involved in the evolution of the Streptococcus thermophilus exopolysaccharide synthesis loci? Gene 233: 151–161.

Bouzar, F., J. Cerning and M. Desmazeaud. 1997. Exopolysaccharide production and texture-promoting abilities of mixed-strain starter cultures in yogurt production. Journal of Dairy Science 80: 2310–2317.

Califf, R.M. 2009. Key issues for global cardiovascular medicine. Lancet 374: 508–510.

Campbell, J.A., G.J. Davies, V. Bulone and B. Henrissat. 1997. A classification of nucleotide-diphospho-sugar glycosyltransferases based on amino acid sequence similarities. Biochemical Journal 326: 929–939.

Cerning, J. 1990. Exocellular Polysaccharides Produced by Lactic-Acid Bacteria. Fems Microbiology Reviews 87: 113–130.

Cerning, J., C.M.G.C. Renard, J.F. Thibault, C. Bouillanne, M. Landon, M. Desmazeaud and L. Topisirovic. 1994. Carbon Source Requirements for Exopolysaccharide Production by Lactobacillus-Casei Cg11 and Partial Structure-Analysis of the Polymer. Applied and Environmental Microbiology 60: 3914–3919.

Chabot, S., H.L. Yu, L. De Leseleuc, D. Cloutier, M.R. Van Calsteren, M. Lessard, D. Roy, M. Lacroix and D. Oth. 2001. Exopolysaccharides from *Lactobacillus rhamnosus* RW-9595M stimulate TNF, IL-6 and IL-12 in human and mouse cultured immunocompetent cells, and IFN-gamma mouse splenocytes. Lait 81: 683–697.

Chang, Z.Q., J.S. Lee, E. Gebru, J.H. Hong, H.K. Jung, W.S. Jo and S.C. Park. 2010. Mechanism of macrophage activation induced by beta-glucan produced from Paenibacillus polymyxa JB115. Biochemical and Biophysical Research Communications 391: 1358–1362.

Charpentier, G., N. Genes, L. Vaur, J. Amar, P. Clerson, J.P. Cambou, P. Gueret and E.D.S. Investigator. 2003. Control of diabetes and cardiovascular risk factors in patients with type 2 diabetes: a nationwide French survey. Diabetes & metabolism 29: 152–158.

Coulon, J., A. Houles, M. Dimopoulou, J. Maupeu and M. Dols-Lafargue. 2012. Lysozyme resistance of the ropy strain Pediococcus parvulus IOEB 8801 is correlated with beta-glucan accumulation around the cell. International Journal of Food Microbiology 159: 25–29.

Dahech, I., K.S. Belghith, K. Hamden, A. Feki, H. Belghith and H. Mejdoub. 2011. Oral administration of levan polysaccharide reduces the alloxan-induced oxidative stress in rats. International Journal of Biological Macromolecules 49: 942–947.

de Palencia, P.F., M.L. Werning, E. Sierra-Filardi, M.T. Duenas, A. Irastorza, A.L. Corbi and P. Lopez. 2009. Probiotic properties of the 2-substituted (1,3)-beta-D-glucan-producing bacterium *Pediococcus parvulus* 2.6. Applied and Environmental Microbiology 75: 4887–4891.

de Valdez, G.F., C. Rodriguez, M. Medici and F. Mozzi. 2010. Therapeutic effect of *Streptococcus thermophilus* CRL 1190-fermented milk on chronic gastritis. World Journal of Gastroenterology 16: 1622–1630.

de Valdez, G.F., C. Rodriguez, M. Medici, A.V. Rodriguez and F. Mozzi. 2009. Prevention of chronic gastritis by fermented milks made with exopolysaccharide-producing Streptococcus thermophilus strains. Journal of Dairy Science 92: 2423–2434.

De Vuyst, L., F. De Vin, F. Vaningelgem and B. Degeest. 2001. Recent developments in the biosynthesis and applications of heteropolysaccharides from lactic acid bacteria. International Dairy Journal 11: 687–707.

De Vuyst, L. and B. Degeest. 1999. Heteropolysaccharides from lactic acid bacteria. Fems Microbiology Reviews 23: 153–177.

De Vuyst, L., F. Vanderveken, S. Van de Ven and B. Degeest. 1998. Production by and isolation of exopolysaccharides from Streptococcus thermophilus grown in a milk medium and evidence for their growth-associated biosynthesis. Journal of Applied Microbiology 84: 1059–1068.

Degeest, B., F. Vaningelgem and L. De Vuyst. 2001. Microbial physiology, fermentation kinetics, and process engineering of heteropolysaccharide production by lactic acid bacteria. International Dairy Journal 11: 747–757.

Delattre, C., L. Chaisemartin, M. Favre-Mercuret, J.Y. Berthon and L. Rios. 2012. Biological effect of beta-(1,3)-polyglucuronic acid sodium salt on lipid storage and adipocytes differentiation. Carbohydrate Polymers 87: 775–783.

Deutsch, S.M., H. Falentin, M. Dols-Lafargue, G. LaPointe and D. Roy. 2008. Capsular exopolysaccharide biosynthesis gene of *Propionibacterium freudenreichii* subsp. *shermanii*. International Journal of Food Microbiology 125: 252–258.

Dierksen, K.P., W.E. Sandine and J.E. Trempy. 1997. Expression of ropy and mucoid phenotypes in *Lactococcus lactis*. Journal of Dairy Science 80: 1528–1536.

Dimopoulou, M., L. Hazo and M. Dols-Lafargue. 2012. Exploration of phenomena contributing to the diversity of *Oenococcus oeni* exopolysaccharides. International Journal of Food Microbiology 153: 114–122.

Dols-Lafargue, M., H.Y. Lee, C. Le Marrec, A. Heyraud, G. Chambat and A. Lonvaud-Funel. 2008. Characterization of gtf, a glucosyltransferase gene in the genomes of *Pediococcus parvulus* and Oenococcus oeni, two bacterial species commonly found in wine. Applied and Environmental Microbiology 74: 4079–4090.

Dubois, M., K.A. Gilles, J.K. Hamilton, P.A. Rebers and F. Smit. 1956. Colorimetric method for the determination of sugars and related substances. Anal. Chem. 28: 350–356.

Duenas-Chasco, M.T., M.A. Rodriguez-Carvajal, P. Tejero-Mateo, J.L. Espartero, A. Irastorza-Iribas and A.M. Gil-Serrano. 1998. Structural analysis of the exopolysaccharides produced by *Lactobacillus* spp. G-77. Carbohydrate research 307: 125–133.

DuenasChasco, H.T., M.A. RodriguezCarvajal, P.T. Mateo, G. FrancoRodriguez, J.L. Espartero, A. IrastorzaIribas and A.M. GilSerrano. 1997. Structural analysis of the exopolysaccharide produced by *Pediococcus damnosus* 2.6. Carbohydrate research 303: 453–458.

Durlu-Ozkaya, F., B. Aslim and M.T. Ozkaya. 2007. Effect of exopolysaccharides (EPSs) produced by *Lactobacillus delbrueckii* subsp. *bulgaricus* strains to bacteriophage and nisin sensitivity of the bacteria. Lwt-Food Science and Technology 40: 564–568.

Enikeev, R. 2012. Development of a new method for determination of exopolysaccharide quantity in fermented milk products and its application in technology of kefir production. Food Chemistry 134: 2437–2441.

Evans, S.G., D. Morrison, Y. Kaneko and I. Havlik. 1998. The effect of curdlan sulphate on development *in vitro* of Plasmodium falciparum. Transactions of the Royal Society of Tropical Medicine and Hygiene 92: 87–89.

Faber, E.J., J.P. Kamerling and J.F.G. Vliegenthart. 2001a. Structure of the extracellular polysaccharide produced by *Lactobacillus delbrueckii* subsp. *bulgaricus* 291. Carbohydrate research 331: 183–194.

Faber, E.J., M.J. van den Haak, J.P. Kamerling and J.F.G. Vliegenthart. 2001b. Structure of the exopolysaccharide produced by *Streptococcus thermophilus* S3. Carbohydrate research 331: 173–182.

Faber, E.J., P. Zoon, J.P. Kamerling and J.F.G. Vliegenthart. 1998. The exopolysaccharides produced by *Streptococcus thermophilus* Rs and Sts have the same repeating unit but differ in viscosity of their milk cultures. Carbohydrate research 310: 269–276.

FAO/WHO. 2010. Interim Summary of Conclusions and dietary recommendations on total fat & fatty acids. From the Joint FAO/WHO Expert Consultation on Fats and Fatty Acids.

Fitriyanto, N.A., M. Nakamura, S. Muto, K. Kato, T. Yabe, T. Iwama, K. Kawai and A. Pertiwiningrum. 2011. Ce(3)+-induced exopolysaccharide production by *Bradyrhizobium* sp. MAFF211645. Journal of bioscience and bioengineering 111: 146–152.

Folkenberg, D.M., P. Dejmek, A. Skriver, H.S. Guldager and R. Ipsen. 2006. Sensory and rheological screening of exopolysaccharide producing strains of bacterial yoghurt cultures. International Dairy Journal 16: 111–118.

Forde, A. and G.F. Fitzgerald. 2003. Molecular organization of exopolysaccharide (EPS) encoding genes on the lactococcal bacteriophage adsorption blocking plasmid, pCI658. Plasmid 49: 130–142.

Freitas, F., V.D. Alves, C.A.V. Torres, M. Cruz, I. Sousa, M.J. Melo, A.M. Ramos and M.A.M. Reis. 2011. Fucose-containing exopolysaccharide produced by the newly isolated Enterobacter strain A47 DSM 23139. Carbohydrate Polymers 83: 159–165.

Frengova, G.I., E.D. Simova, D.M. Beshkova and Z.I. Simov. 2000. Production and monomer composition of exopolysaccharides by yogurt starter cultures. Canadian Journal of Microbiology 46: 1123–1127.

Gaizka, G.I., D.M. Teresa, I. Ana, S.F. Elena, W.M. Laura, L. Paloma, C.A. Luis and F.D. Pilar. 2010. Naturally occurring 2-substituted (1,3)-beta-D-glucan producing *Lactobacillus suebicus* and *Pediococcus parvulus* strains with potential utility in the production of functional foods. Bioresource Technology 101: 9254–9263.

Galle, S., C. Schwab, E.K. Arendt and M.G. Ganzle. 2011. Structural and rheological characterisation of heteropolysaccharides produced by lactic acid bacteria in wheat and sorghum sourdough. Food microbiology 28: 547–553.

Gamar, L., K. Blondeau and J.M. Simonet. 1997. Physiological approach to extracellular polysaccharide production by *Lactobacillus rhamnosus* strain C83. Journal of Applied Microbiology 83: 281–287.

Gancel, F. and G. Novel. 1994. Exopolysaccharide Production by *Streptococcus-Salivarius* ssp. *Thermophilus* Cultures .2. Distinct Modes of Polymer Production and Degradation among Clonal Variants. Journal of Dairy Science 77: 689–695.

Ganesh Kumar, C., H.S. Joo, J.W. Choi, Y.M. Koo and C.S. Chang. 2004. Purification and characterization of an extracellular polysaccharide from haloalkalophilic *Bacillus* sp. I-450. Enzyme and Microbial Technology 34: 673–681.

Gibson, G.R., H.M. Probert, J. Van Loo, R.A. Rastall and M.B. Roberfroid. 2004. Dietary modulation of the human colonic microbiota: updating the concept of prebiotics. Nutrition Research Reviews 17: 259–275.

Gibson, G.R. and M.B. Roberfroid. 1995. Dietary Modulation of the Human Colonic Microbiota—Introducing the Concept of Prebiotics. Journal of Nutrition 125: 1401–1412.

Girard, M. and C. Schaffer-Lequart. 2007. Gelation of skim milk containing anionic exopolysaccharides and recovery of texture after shearing. Food Hydrocolloids 21: 1031–1040.

Gorska-Fraczek, S., C. Sandstrom, L. Kenne, J. Rybka, M. Strus, P. Heczko and A. Gamian. 2011. Structural studies of the exopolysaccharide consisting of a nonasaccharide repeating unit isolated from *Lactobacillus rhamnosus* KL37B. Carbohydrate research 346: 2926–2932.

Gray, J.X., H.J. Zhan, S.B. Levery, L. Battisti, B.G. Rolfe and J.A. Leigh. 1991. Heterologous Exopolysaccharide Production in *Rhizobium* sp. Strain-Ngr234 and Consequences for Nodule Development. Journal of bacteriology 173: 3066–3077.

Guler-Akin, M.B., M.S. Akin and A. Korkmaz. 2009. Influence of different exopolysaccharide-producing strains on the physicochemical, sensory and syneresis characteristics of reduced-fat stirred yoghurt. International Journal of Dairy Technology 62: 422–430.

Harada, T. 1965. Succinoglucan 10C3: a new acidic polysaccharide of *Alcaligenes faecalis* var. *myxogenes*. Archives of biochemistry and biophysics 112: 65–69.

Hassan, A.N. 2008. ADSA Foundation Scholar Award: Possibilities and challenges of exopolysaccharide-producing lactic cultures in dairy foods. Journal of Dairy Science 91: 1282–1298.

Hassan, A.N., J.F. Frank, M.A. Farmer, K.A. Schmidt and S.I. Shalabi. 1995. Observation of encapsulated lactic acid bacteria using confocal scanning laser microscopy. Journal of Dairy Science 78: 2624–2628.

Hassan, A.N., J.F. Frank and K.B. Qvist. 2002. Direct observation of bacterial exopolysaccharides in dairy products using confocal scanning laser microscopy. Journal of Dairy Science 85: 1705–1708.

Hongpattarakere, T., N. Cherntong, S. Wichienchot, S. Kolida and R.A. Rastall. 2012. *In vitro* prebiotic evaluation of exopolysaccharides produced by marine isolated lactic acid bacteria. Carbohydrate Polymers 87: 846–852.

Hu, C.Y., M.H. Wu, T.M. Pan, Y.J. Wu, S.J. Chang and M.S. Chang. 2010. Exopolysaccharide activities from probiotic bifidobacterium: Immunomodulatory effects (on J774A.1 macrophages) and antimicrobial properties. International Journal of Food Microbiology 144: 104–110.

Ibarburu, I., R. Aznar, P. Elizaquivel, N. Garcia-Quintans, P. Lopez, A. Munduate, A. Irastorza and M.T. Duenas. 2010. A real-time PCR assay for detection and quantification of 2-branched (1,3)-beta-D-glucan producing lactic acid bacteria in cider. International Journal of Food Microbiology 143: 26–31.

Iyer, A., K. Mody and B. Jha. 2004. Accumulation of hexavalent chromium by an exopolysaccharide producing marine Enterobacter cloaceae. Marine Pollution Bulletin 49: 974–977.

Jagodzinski, P.P., R. Wiaderkiewicz, G. Kurzawski, M. Kloczewiak, H. Nakashima, E. Hyjek, N. Yamamoto, T. Uryu, Y. Kaneko, M.R. Posner and D. Kozbor. 1994. Mechanism of the Inhibitory Effect of Curdlan Sulfate on Hiv-1 Infection *in-vitro*. Virology 202: 735–745.

Jezequel, V. 1998. Curdlan: A new functional beta-glucan. Cereal Foods World 43: 361–364.

Jimenez-Guzman, J., A. Flores-Najera, A.E. Cruz-Guerrero and M. Garcia-Garibay. 2009. Use of an exopolysaccharide-producing strain of *Streptococcus thermophilus* in the manufacture of Mexican Panela cheese. Lwt-Food Science and Technology 42: 1508–1512.

Jong, S.C. and J.M. Birmingham. 1993. Medicinal and therapeutic value of the shiitake mushroom. Advances in applied microbiology 39: 153–184.

Kambourova, M., R. Mandeva, D. Dimova, A. Poli, B. Nicoaus and G. Tommonaro. 2009. Production and characterization of a microbial glucan, synthesized by Geobacillus tepidamans V264 isolated from Bulgarian hot spring. Carbohydrate Polymers 77: 338–343.

Kang, E.J., S.O. Lee, J.D. Lee and T.H. Lee. 1999. Purification and characterization of a levanbiose-producing levanase from *Pseudomonas* sp. No. 43. Biotechnology and Applied Biochemistry 29: 263–268.

Kang, S.A., K.H. Hong, K.H. Jang, Y.Y. Kim, R. Choue and Y. Lim. 2006. Altered mRNA expression of hepatic lipogenic enzyme and PPAR alpha in rats fed dietary levan from Zymomonas mobilis. Journal of Nutritional Biochemistry 17: 419–426.

Kanmani, P., R.S. Kumar, N. Yuvaraj, K.A. Paari, V. Pattukumar and V. Arul. 2011. Production and purification of a novel exopolysaccharide from lactic acid bacterium *Streptococcus phocae* PI80 and its functional characteristics activity *in vitro*. Bioresource technology 102: 4827–4833.

Katina, K., N.H. Maina, R. Juvonen, L. Flander, L. Johansson, L. Virkki, M. Tenkanen and A. Laitila. 2009. *In situ* production and analysis of Weissella confusa dextran in wheat sourdough. Food microbiology 26: 734–743.

Kearney, N., H.M. Stack, J.T. Tobin, V. Chaurin, M.A. Fenelon, G.F. Fitzgerald, R.P. Ross and C. Stanton. 2011. *Lactobacillus paracasei* NFBC 338 producing recombinant beta-glucan positively influences the functional properties of yoghurt. International Dairy Journal 21: 561–567.

Kido, Y., S. Hiramoto, M. Murao, Y. Horio, T. Miyazaki, T. Kodama and Y. Nakabou. 2003. Epsilon-polylysine inhibits pancreatic lipase activity and suppresses postprandial hypertriacylglyceridemia in rats. The Journal of nutrition 133: 1887–1891.

Kim, B.S., I.D. Jung, J.S. Kim, J. Lee, I.Y. Lee and K.B. Lee. 2000a. Curdlan gels as protein drug delivery vehicles. Biotechnology Letters 22: 1127–1130.

Kim, D.S., S. Thomas and H.S. Fogler. 2000b. Effects of pH and trace minerals on long-term starvation of *Leuconostoc mesenteroides*. Applied and Environmental Microbiology 66: 976–981.

Kim, M.K., K.E. Ryu, W.A. Choi, Y.H. Rhee and I.Y. Lee. 2003. Enhanced production of (1 -> 3)-beta-D-glucan by a mutant strain of Agrobacterium species. Biochemical Engineering Journal 16: 163–168.

Kim, Y., S. Oh and S.H. Kim. 2009. Released exopolysaccharide (r-EPS) produced from probiotic bacteria reduce biofilm formation of enterohemorrhagic *Escherichia coli* O157:H7. Biochemical and Biophysical Research Communications 379: 324–329.

Kim, Y., S. Oh, H.S. Yun, S. Oh and S.H. Kim. 2010. Cell-bound exopolysaccharide from probiotic bacteria induces autophagic cell death of tumour cells. Letters in Applied Microbiology 51: 123–130.

Kimmel, S.A. and R.F. Roberts. 1998. Development of a growth medium suitable for exopolysaccharide production by *Lactobacillus delbrueckii* ssp. *bulgaricus* RR. International Journal of Food Microbiology 40: 87–92.

Kitazawa, H., T. Itoh, Y. Tomioka, M. Mizugaki and T. Yamaguchi. 1996. Induction of IFN-gamma and IL-1 alpha production in macrophages stimulated with phosphopolysaccharide produced by *Lactococcus lactis* ssp. *cremoris*. International Journal of Food Microbiology 31: 99–106.

Knoshaug, E.P., J.A. Ahlgren and J.E. Trempy. 2000. Growth associated exopolysaccharide expression in *Lactococcus lactis* subspecies *cremoris* Ropy352. Journal of Dairy Science 83: 633–640.

Kralj, S., G.H. van Geel-Schutten, M.M.G. Dondorff, S. Kirsanovs, M.J.E.C. van der Maarel and L. Dijkhuizen. 2004. Glucan synthesis in the genus *Lactobacillus*: isolation and characterization of glucansucrase genes, enzymes and glucan products from six different strains. Microbiology-Sgm 150: 3681–3690.

Kralj, S., G.H. van Geel-schutten, M.J.E.C. van der Maarel and L. Dijkhuizen. 2003. Efficient screening methods for glucosyltransferase genes in *Lactobacillus* strains. Biocatalysis and Biotransformation 21: 181–187.

Lambo-Fodje, A.M., R. Oste and M.E.G.L. Nyman. 2006. Short-chain fatty acid formation in the hindgut of rats fed native and fermented oat fibre concentrates. British Journal of Nutrition 96: 47–55.

Lamontagne, L., C. Bleau, A. Monges, K. Rashidan, J.P. Laverdure, M. Lacroix, M.R. Van Calsteren, M. Millette and R. Savard. 2010. Intermediate chains of exopolysaccharides from *Lactobacillus rhamnosus* RW-9595M increase IL-10 production by macrophages. Journal of Applied Microbiology 108: 666–675.

Lamothe, G.T., L. Jolly, B. Mollet and F. Stingele. 2002. Genetic and biochemical characterization of exopolysaccharide biosynthesis by *Lactobacillus delbrueckii* subsp. *bulgaricus*. Archives of Microbiology 178: 218–228.

Laws, A., Y.C. Gu and V. Marshall. 2001. Biosynthesis, characterisation, and design of bacterial exopolysaccharides from lactic acid bacteria. Biotechnology Advances 19: 597–625.

Lbarburu, I., M.A. Soria-Diaz, M.A. Rodriguez-Carvajal, S.E. Velasco, P. Tejero-Mateo, A.M. Gil-Serrano, A. Irastorza and M.T. Duenas. 2007. Growth and exopolysaccharide (EPS) production by *Oenococcus oeni* I4 and structural characterization of their EPSs. Journal of Applied Microbiology 103: 477–486.

Lebeer, S., T.L.A. Verhoeven, M.P. Velez, J. Vanderleyden and S.C.J. De Keersmaecker. 2007. Impact of environmental and genetic factors on Biofilm formation by the Probiotic strain *Lactobacillus rhamnosus* GG. Applied and Environmental Microbiology 73: 6768–6775.

Lee, K.H., S. Yoo, S.H. Baek and H.G. Lee. 2007. Physicochemical and biological characteristics of DEAE-derivatized PS7 biopolymer of Beijerinckia indica. International journal of biological macromolecules 41: 141–145.

Li, W.W., W.Z. Zhou, Y.Z. Zhang, J. Wang and X.B. Zhu. 2008. Flocculation behavior and mechanism of an exopolysaccharide from the deep-sea psychrophilic bacterium *Pseudoalteromonas* sp. SM9913. Bioresource Technology 99: 6893–6899.

Liu, C.H., J. Lu, L.L. Lu, Y.H. Liu, F.S. Wang and M. Xiao. 2010. Isolation, structural characterization and immunological activity of an exopolysaccharide produced by *Bacillus licheniformis* 8-37-0-1. Bioresource Technology 101: 5528–5533.

Liu, J.R., S.Y. Wang, M.J. Chen, H.L. Chen, P.Y. Yueh and C.W. Lin. 2006. Hypocholesterolaemic effects of milk-kefir and soyamilk-kefir in cholesterol-fed hamsters. British Journal of Nutrition 95: 939–946.

London, L.E.E., A.H. Kumar, R. Wall, P.G. Casey, O. O'Sullivan, F. Shanahan, C. Hill, P.D. Cotter, G.F. Fitzgerald, R.P. Ross, N.M. Caplice and C. Stanton. 2014. Exopolysaccharide-producing probiotic lactobacilli reduce serum cholesterol and modify enteric microbiota in ApoE-deficient mice. J. Nutr. 144: 1956–1962. doi:10.3945/jn.114.191627.

Looijesteijn, P.J., I.C. Boels, M. Kleerebezem and J. Hugenholtz. 1999. Regulation of exopolysaccharide production by *Lactococcus lactis* subsp. *cremoris* by the sugar source. Applied and Environmental Microbiology 65: 5003–5008.

Looijesteijn, P.J., L. Trapet, E. de Vries, T. Abee and J. Hugenholtz. 2001. Physiological function of exopolysaccharides produced by Lactococcus lactis. International Journal of Food Microbiology 64: 71–80.

Lopez, P., D.C. Monteserin, M. Gueimonde, C.G. de los Reyes-Gavilan, A. Margolles, A. Suarez and P. Ruas-Madiedo. 2012. Exopolysaccharide-producing Bifidobacterium strains elicit different *in vitro* responses upon interaction with human cells. Food Research International 46: 99–107.

Low, D., J.A. Ahlgren, D. Horne, D.J. McMahon, C.J. Oberg and J.R. Broadbent. 1998. Role of *Streptococcus thermophilus* MR-1C capsular exopolysaccharide in cheese moisture retention. Applied and Environmental Microbiology 64: 2147–2151.

Maeda, H., X. Zhu, K. Omura, S. Suzuki and S. Kitamura. 2004a. Effects of an exopolysaccharide (kefiran) on lipids, blood pressure, blood glucose, and constipation. Biofactors 22: 197–200.

Maeda, H., X. Zhu, S. Suzuki, K. Suzuki and S. Kitamura. 2004b. Structural characterization and biological activities of an exopolysaccharide kefiran produced by *Lactobacillus kefiranofaciens* WT-2B. Journal of Agricultural and Food Chemistry 52: 5533–5538.

Maeda, H., X. Zhu, S. Suzuki, K. Suzuki and S. Kitamura. 2004c. Structural characterization and biological activities of an exopolysaccharide kefiran produced by *Lactobacillus kefiranofaciens* WT-2B(T). Journal of agricultural and food chemistry 52: 5533–5538.

Marshall, V.M., E.N. Cowie and R.S. Moreton. 1995. Analysis and Production of 2 Exopolysaccharides from *Lactococcus-Lactis* subsp. *cremoris* Lc330. Journal of Dairy Research 62: 621–628.

Marshall, V.M., H. Dunn, M. Elvin, N. McLay, Y. Gu and A.P. Laws. 2001a. Structural characterisation of the exopolysaccharide produced by *Streptococcus thermophilus* EU20. Carbohydrate research 331: 413–422.

Marshall, V.M., A.P. Laws, Y. Gu, F. Levander, P. Radstrom, L. De Vuyst, B. Degeest, F. Vaningelgem, H. Dunn and M. Elvin. 2001b. Exopolysaccharide-producing strains of thermophilic lactic acid bacteria cluster into groups according to their EPS structure. Letters in Applied Microbiology 32: 433–437.

Martensson, O., M. Biorklund, A.M. Lambo, M. Duenas-Chasco, A. Irastorza, O. Holst, E. Norin, G. Welling, R. Oste and G. Onning. 2005. Fermented, ropy, oat-based products reduce cholesterol levels and stimulate the bifidobacteria flora in humans. Nutrition Research 25: 429–442.

Maugeri, T.L., C. Gugliandolo, D. Caccamo, A. Panico, L. Lama, A. Gambacorta and B. Nicolaus. 2002. A halophilic thermotolerant Bacillus isolated from a marine hot spring able to produce a new exopolysaccharide. Biotechnology Letters 24: 515–519.

McKellar, R.C., J. van Geest and W. Cui. 2003. Influence of culture and environmental conditions on the composition of exopolysaccharide produced by Agrobacterium radiobacter. Food Hydrocolloids 17: 429–437.

Mikio, K., O. Yoshiro, O. Hirotomo, K. Yoshikazu, I. Hiroshi, M. Hisashi and K. Takeshi. 1995. Water soluble β-(1,3)-glucan derivative and antiviral agent containing the derivative. 07228601.

Monsan, P., S. Bozonnet, C. Albenne, G. Joucla, R.M. Willemot and M. Remaud-Simeon. 2001. Homopolysaccharides from lactic acid bacteria. International Dairy Journal 11: 675–685.

Nakajima, H., Y. Suzuki, H. Kaizu and T. Hirota. 1992. Cholesterol Lowering Activity of Ropy Fermented Milk. Journal of Food Science 57: 1327–1329.

Nakajima, H., T. Toba and S. Toyoda. 1995. Enhancement of Antigen-Specific Antibody-Production by Extracellular Slime Products from Slime-Forming *Lactococcus-Lactis* subspecies *cremoris* Sbt-0495 in Mice. International Journal of Food Microbiology 25: 153–158.

Nakajima, H., S. Toyoda, T. Toba, T. Itoh, T. Mukai, H. Kitazawa and S. Adachi. 1990. A Novel Phosphopolysaccharide from Slime-Forming *Lactococcus-Lactis* subspecies *cremoris* Sbt-0495. Journal of Dairy Science 73: 1472–1477.

Nakata, M., T. Kawaguchi, Y. Kodama and A. Konno. 1998. Characterization of curdlan in aqueous sodium hydroxide. Polymer 39: 1475–1481.

Nicolaus, B., A. Panico, M.C. Manca, L. Lama, A. Gambacorta, T. Maugeri, C. Gugliandolo and D. Caccamo. 2000. A thermophilic Bacillus isolated from an Eolian shallow hydrothermal vent, able to produce exopolysaccharides. Systematic and Applied Microbiology 23: 426–432.

Nikolic, M., P. Lopez, I. Strahinic, A. Suarez, M. Kojic, M. Fernandez-Garcia, L. Topisirovic, N. Golic and P. Ruas-Madiedo. 2012. Characterisation of the exopolysaccharide (EPS)-producing *Lactobacillus paraplantarum* BGCG11 and its non-EPS producing derivative strains as potential probiotics. International Journal of Food Microbiology 158: 155–162.

Nordmark, E.L., Z. Yang, E. Huttunen and G. Widmalm. 2005. Structural studies of the exopolysaccharide produced by *Propionibacterium freudenreichii* ssp. *shermanii* JS. Biomacromolecules 6: 521–523.

Okazaki, M., Y. Adachi, N. Ohno and T. Yadomae. 1995. Structure-Activity Relationship of (1-]3)-Beta-D-Glucans in the Induction of Cytokine Production from Macrophages, *in-vitro*. Biological & pharmaceutical bulletin 18: 1320–1327.

Paloma, L., G.I. Gaizka, D.M. Teresa, I. Ana, S.F. Elena, W.M. Laura, C.A. Luis and F.D. Pilar. 2010. Naturally occurring 2-substituted (1,3)-beta-D-glucan producing *Lactobacillus suebicus* and *Pediococcus parvulus* strains with potential utility in the production of functional foods. Bioresource Technology 101: 9254–9263.

Pan, D.D. and X.M. Mei. 2010. Antioxidant activity of an exopolysaccharide purified from *Lactococcus lactis* subsp. *lactis* 12. Carbohydrate Polymers 80: 908–914.

Piermaria, J.A., M.L. de la Canal and A.G. Abraham. 2008. Gelling properties of kefiran, a food-grade polysaccharide obtained from kefir grain. Food Hydrocolloids 22: 1520–1527.

Pigeon, R.M., E.P. Cuesta and S.E. Gilliland. 2002. Binding of free bile acids by cells of yogurt starter culture bacteria. Journal of Dairy Science 85: 2705–2710.

Prasanna, P.H.P., A. Bell, A.S. Grandison and D. Charalampopoulos. 2012a. Emulsifying, rheological and physicochemical properties of exopolysaccharide produced by *Bifidobacterium longum* subsp. *infantis* CCUG 52486 and *Bifidobacterium infantis* NCIMB 702205. Carbohydrate Polymers 90: 533–540.

Prasanna, P.H.P., A.S. Grandison and D. Charalampopoulos. 2012b. Screening human intestinal Bifidobacterium strains for growth, acidification, EPS production and viscosity potential in low-fat milk. International Dairy Journal 23: 36–44.

Prasertsan, P., W. Dermlim, H. Doelle and J.F. Kennedy. 2006. Screening, characterization and flocculating property of carbohydrate polymer from newly isolated Enterobacter cloacae WD7. Carbohydrate Polymers 66: 289–297.

Purohit, D.H., A.N. Hassan, E. Bhatia, X. Zhang and C. Dwivedi. 2009. Rheological, sensorial, and chemopreventive properties of milk fermented with exopolysaccharide-producing lactic cultures. Journal of Dairy Science 92: 847–856.

Rawson, H.L. and V.M. Marshall. 1997. Effect of 'ropy' strains of *Lactobacillus delbrueckii* ssp. *bulgaricus* and *Streptococcus thermophilus* on rheology of stirred yogurt. International Journal of Food Science and Technology 32: 213–220.

Read, S.M., G. Currie and A. Bacic. 1996. Analysis of the structural heterogeneity of laminarin by electrospray-ionisation-mass spectrometry. Carbohydrate research 281: 187–201.

Ricciardi, A., E. Parente, M.A. Crudele, F. Zanetti, G. Scolari and I. Mannazzu. 2002. Exopolysaccharide production by *Streptococcus thermophilus* SY: production and preliminary characterization of the polymer. Journal of Applied Microbiology 92: 297–306.

Rimada, P.S. and A.G. Abraham. 2001. Polysaccharide production by kefir grains during whey fermentation. Journal of Dairy Research 68: 653–661.

Rimada, P.S. and A.G. Abraham. 2006. Kefiran improves rheological properties of glucono-delta-lactone induced skim milk gels. International Dairy Journal 16: 33–39.

Robijn, G.W., R.G. Gallego, D.J.C. vandenBerg, H. Haas, J.P. Kamerling and J.F.G. Vliegenthart. 1996a. Structural characterization of the exopolysaccharide produced by *Lactobacillus acidophilus* LMG9433. Carbohydrate research 288: 203–218.

Robijn, G.W., H.L.J. Wienk, D.J.C. vandenBerg, H. Haas, J.P. Kamerling and J.F.G. Vliegenthart. 1996b. Structural studies of the exopolysaccharide produced by *Lactobacillus paracasei* 34-1. Carbohydrate research 285: 129–139.

Robitaille, G., S. Moineau, D. St-Gelais, C. Vadeboncoeur and M. Britten. 2006. Detection and quantification of capsular exopolysaccharides from *Streptococcus thermophilus* using lectin probes. Journal of Dairy Science 89: 4156–4162.

Rodriguez, C., M. Medici, F. Mozzi and G.F. de Valdez. 2010. Therapeutic effect of Streptococcus thermophilus CRL 1190-fermented milk on chronic gastritis. World Journal of Gastroenterology 16: 1622–1630.

Rodriguez, C., M. Medici, A.V. Rodriguez, F. Mozzi and G.F.de Valdez. 2009. Prevention of chronic gastritis by fermented milks made with exopolysaccharide-producing *Streptococcus thermophilus* strains. Journal of Dairy Science 92: 2423–2434.

Ruas-Madiedo, J. P., Hugenholtz and P. Zoon. 2002a. An overview of the functionality of exopolysaccharides produced by lactic acid bacteria. International Dairy Journal 12: 163–171.

Ruas-Madiedo, P., J.A. Moreno, N. Salazar, S. Delgado, B. Mayo, A. Margolles and C.G.de los Reyes-Gavilan. 2007. Screening of exopolysaccharide-producing Lactobacillus and *Bifidobacterium* strains isolated from the human intestinal microbiota. Applied and Environmental Microbiology 73: 4385–4388.

Ruas-Madiedo, P., R. Tuinier, M. Kanning and P. Zoon. 2002b. Role of exopolysaccharides produced by *Lactococcus lactis* subsp. *cremoris* on the viscosity of fermented milks. International Dairy Journal 12: 689–695.

Ruiz-Ruiz, C., G.K. Srivastava, D. Carranza, J.A. Mata, I. Llamas, M. Santamaria, E. Quesada and I.J. Molina. 2011. An exopolysaccharide produced by the novel halophilic bacterium Halomonas stenophila strain B100 selectively induces apoptosis in human T leukaemia cells. Applied Microbiology and Biotechnology 89: 345–355.

Rynne, N.M., T.P. Beresford, A.L. Kelly, M.H. Tunick, E.L. Malin and T.P. Guinee. 2007. Effect of exopolysaccharide-producing adjunct starter cultures on the manufacture, composition and yield of half-fat cheddar cheese. Australian Journal of Dairy Technology 62: 12–18.

Salazar, N., A. Binetti, M. Gueimonde, A. Alonso, P. Garrido, C.G. del Rey, C. Gonzalez, P. Ruas-Madiedo and C.G. de los Reyes-Gavilan. 2011. Safety and intestinal microbiota modulation by the exopolysaccharide-producing strains Bifidobacterium animalis IPLA R1 and Bifidobacterium longum IPLA E44 orally administered to Wistar rats. International Journal of Food Microbiology 144: 342–351.

Salazar, N., A. Prieto, J.A. Leal, B. Mayo, J.C. Bada-Gancedo, C.G. de los Reyes-Gavilan and P. Ruas-Madiedo. 2009. Production of exopolysaccharides by Lactobacillus and Bifidobacterium strains of human origin, and metabolic activity of the producing bacteria in milk. Journal of Dairy Science 92: 4158–4168.

Sawada, H., M. Furushiro, K. Hirai, M. Motoike, T. Watanabe and T. Yokokura. 1990. Purification and characterization of an antihypertensive compound from *Lactobacillus casei*. Agricultural and biological chemistry 54: 3211–3219.

Schwab, C., Y. Wang and M.G. Ganzle. 2010. Exopolysaccharide Synthesized by *Lactobacillus reuteri* Decreases the Ability of Enterotoxigenic *Escherichia coli* To Bind to Porcine Erythrocytes. Applied and Environmental Microbiology 76: 4863–4866.

Sengul, N., B. Aslim, G. Ucar, N. Yucel, S. Isik, H. Bozkurt, Z. Sakaogullari and F. Atalay. 2006. Effects of exopolysaccharide-producing probiotic strains on experimental colitis in rats. Diseases of the Colon & Rectum 49: 250–258.

Spicer, E.J.F., E.I. Goldenthal and T. Ikeda. 1999. A toxicological assessment of curdlan. Food and Chemical Toxicology 37: 455–479.

Staaf, M., Z.N. Yang, E. Huttunen and G. Widmalm. 2000. Structural elucidation of the viscous exopolysaccharide produced by *Lactobacillus helveticus* Lb161. Carbohydrate research 326: 113–119.

Stack, H.M., N. Kearney, C. Stanton, G.F. Fitzgerald and R.P. Ross. 2010. Association of beta-glucan endogenous production with increased stress tolerance of intestinal lactobacilli. Applied and Environmental Microbiology 76: 500–507.

Stingele, F., J. Lemoine and J.R. Neeser. 1997. *Lactobacillus helveticus* Lh59 secretes an exopolysaccharide that is identical to the one produced by *Lactobacillus helveticus* TN-4, a presumed spontaneous mutant of *Lactobacillus helveticus* TY1--2. Carbohydrate research 302: 197–202.

Stingele, F. and B. Mollet. 1995. Homologous integration and transposition to identify genes involved in the production of exopolysaccharides in *Streptococcus thermophilus*. Genetics of Streptococci, Enterococci and Lactococci 85: 487–493.

Stingele, F., J.R. Neeser and B. Mollet. 1996. Identification and characterization of the eps (Exopolysaccharide) gene cluster from *Streptococcus thermophilus* Sfi6. Journal of bacteriology 178: 1680–1690.

Sun, S.S., Z.Z. Zhang, Y.Z. Luo, W.Z. Zhong, M. Xiao, W.J. Yi, L. Yu and P.C. Fu. 2011. Exopolysaccharide production by a genetically engineered Enterobacter cloacae strain for microbial enhanced oil recovery. Bioresource Technology 102: 6153–6158.

Sutherland, I.W. 1988. Bacterial Surface Polysaccharides—Structure and Function. International Review of Cytology-a Survey of Cell Biology 113: 187–231.

Tieking, M., S. Kaditzky, R. Valcheva, M. Korakli, R.F. Vogel and M.G. Ganzle. 2005. Extracellular homopolysaccharides and oligosaccharides from intestinal lactobacilli. Journal of Applied Microbiology 99: 692–702.

Tieking, M., M. Korakli, M.A. Ehrmann, M.G. Ganzle and R.F. Vogel. 2003. *In situ* production of exopolysaccharides during sourdough fermentation by cereal and intestinal isolates of lactic acid bacteria. Applied and Environmental Microbiology 69: 945–952.

Tok, E. and B. Aslim. 2010. Cholesterol removal by some lactic acid bacteria that can be used as probiotic. Microbiology and Immunology 54: 257–264.

Torino, M.I., F. Sesma and G.F. de Valdez. 2000. Semi-defined media for the exopolysaccharide (EPS) production by *Lactobacillus helveticus* ATCC 15807 and evaluation of the components interfering with the EPS quantification. Milchwissenschaft-Milk Science International 55: 314–316.

Tuinier, R., W.H.M. van Casteren, P.J. Looijesteijn, H.A. Schols, A.G.J. Voragen and P. Zoon. 2001. Effects of structural modifications on some physical characteristics of exopolysaccharides from *Lactococcus lactis*. Biopolymers 59: 160–166.

Uchida, M., I. Ishii, C. Inoue, Y. Akisato, K. Watanabe, S. Hosoyama, T. Toida, N. Ariyoshi and M. Kitada. 2010. Kefiran reduces atherosclerosis in rabbits fed a high cholesterol diet. Journal of atherosclerosis and thrombosis 17: 980–988.

van Casteren, W.H.M., P. de Waard, C. Dijkema, H.A. Schols and A.G.J. Voragen. 2000a. Structural characterisation and enzymic modification of the exopolysaccharide produced by *Lactococcus lactis* subsp. *cremoris* B891. Carbohydrate research 327: 411–422.

van Casteren, W.H.M., C. Dijkema, H.A. Schols, G. Beldman and A.G.J. Voragen. 1998. Characterisation and modification of the exopolysaccharide produced by *Lactococcus lactis* subsp. *cremoris* B40. Carbohydrate Polymers 37: 123–130.

van Casteren, W.H.M., C. Dijkema, H.A. Schols, G. Beldman and A.G.J. Voragen. 2000b. Structural characterisation and enzymic modification of the exopolysaccharide produced by *Lactococcus lactis* subsp. *cremoris* B39. Carbohydrate research 324: 170–181.

Van der Meulen, R., S. Grosu-Tudor, F. Mozzi, F. Vaningelgem, M. Zamfir, G.F. de Valdez and L. de Vuyst. 2007. Screening of lactic acid bacteria isolates from dairy and cereal products for exopolysaccharide production and genes involved. International Journal of Food Microbiology 118: 250–258.

van Geel-Schutten, G.H., F. Flesch, B. ten Brink, M.R. Smith and L. Dijkhuizen. 1998. Screening and characterization of Lactobacillus strains producing large amounts of exopolysaccharides. Applied Microbiology and Biotechnology 50: 697–703.

van Hijum, S.A.F.T., G.H. van Geel-Schutten, H. Rahaoui, M.J.E.C. van der Maarel and L. Dijkhuizen. 2002. Characterization of a novel fructosyltransferase from *Lactobacillus reuteri* that synthesizes high-molecular-weight inulin and inulin oligosaccharides. Applied and Environmental Microbiology 68: 4390–4398.

Vanhaverbeke, C., C. Bosso, P. Colin-Morel, C. Gey, L. Gamar-Nourani, K. Blondeau, J.M. Simonet and A. Heyraud. 1998. Structure of an extracellular polysaccharide produced by *Lactobacillus rhamnosus* strain C83. Carbohydrate research 314: 211–220.

Vaningelgem, F., M. Zamfir, F. Mozzi, T. Adriany, M. Vancanneyt, J. Swings and L. De Vuyst. 2004. Biodiversity of exopolysaccharides produced by *Streptococcus thermophilus* strains is reflected in their production and their molecular and functional characteristics. Applied and Environmental Microbiology 70: 900–912.

vanKranenburg, R., J.D. Marugg, I.L. vanSwam, N.J. Willem and W.M. deVos. 1997. Molecular characterization of the plasmid-encoded eps gene cluster essential for exopolysaccharide biosynthesis in *Lactococcus lactis*. Molecular Microbiology 24: 387–397.

Vinderola, G., C. Matar, J. Palacios and G. Perdigon. 2007. Mucosal immunomodulation by the non-bacterial fraction of milk fermented by *Lactobacillus helveticus* R389. International Journal of Food Microbiology 115: 180–186.

Vinderola, G., G. Perdigon, J. Duarte, E. Farnworth and C. Matar. 2006. Effects of the oral administration of the exopolysaccharide produced by *Lactobacillus kefiranofaciens* on the gut mucosal immunity. Cytokine 36: 254–260.

Wang, Y., M.G. Ganzle and C. Schwab. 2010. Exopolysaccharide Synthesized by *Lactobacillus reuteri* Decreases the Ability of Enterotoxigenic *Escherichia coli* To Bind to Porcine Erythrocytes. Applied and Environmental Microbiology 76: 4863–4866.

Welman, A.D., I.S. Maddox and R.H. Archer. 2003. Screening and selection of exopolysaccharide-producing strains of *Lactobacillus delbrueckii* subsp. *bulgaricus*. Journal of Applied Microbiology 95: 1200–1206.

Wen, Z.T. and R.A. Burne. 2002. Functional genomics approach to identifying genes required for biofilm development by Streptococcus mutans. Applied and Environmental Microbiology 68: 1196–1203.

Xiu, A.H., Y. Kong, M.Y. Zhou, B. Zhu, S.M. Wang and J.F. Zhang. 2010. The chemical and digestive properties of a soluble glucan from *Agrobacterium* sp. ZX09. Carbohydrate Polymers 82: 623–628.

Xu, R.H., N. Shang and P.L. Li. 2011. *In vitro* and *in vivo* antioxidant activity of exopolysaccharide fractions from Bifidobacterium animalis RH. Anaerobe 17: 226–231.

Yang, Z.N., E. Huttunen, M. Staaf, G. Widmalm and H. Tenhu. 1999. Separation, purification and characterisation of extracellular polysaccharides produced by slime-forming *Lactococcus lactis* ssp. cremoris strains. International Dairy Journal 9: 631–638.

Yoon, E.J., S.H. Yoo, J.H. Cha and H.G. Lee. 2004. Effect of levan's branching structure on antitumor activity. International Journal of Biological Macromolecules 34: 191–194.

Zubillaga, M., R. Weill, E. Postaire, C. Goldman, R. Caro and J. Boccio. 2001. Effect of probiotics and functional foods and their use in different diseases. Nutrition Research 21: 569–579.

# PART II

# Molecular Aspects and Related Techniques

CHAPTER 5

# Identification and Enumeration of Probiotics

*Knut J. Heller** and *Diana Meske*

## 1. Introduction

Probiotic bacteria, or more specifically bacterial strains with probiotic properties, are identified by their specific probiotic properties. The latter are the health effects they exert on the human host. These health effects have to be established in double-blind, placebo-controlled clinical studies. It goes without saying that it is not practicable to identify established probiotic bacteria in food or faeces by their health effects, because this would mean that for every strain to be identified a clinical study would have to be carried out. The situation would change, if the bacterial gene products (or gene product, if there is only one) responsible for the probiotic effects, their states of activity and their cellular concentrations necessary to exert the probiotic effects were known. Then, identification of the bacterial species in conjunction with the correctly expressed active gene products would be sufficient for identification of a probiotic strain. However, as long as a clear correlation between gene products of a specific bacterial strain on one hand and manifestation of a health effect on the other hand has not been established, surrogate assays are needed. Such surrogate assays aim at identifying probiotic bacteria by strain-specific methods. This is in line with the statement made by FAO/WHO (2006): "*Strain typing has to be performed with a reproducible genetic method or using a unique phenotypic trait.*"

However, one has to bear in mind that upon isolation of a new potentially probiotic strain correct assignment to a species or sub-species is necessary. With this assignment, the genetic background is defined the probiotic strain is embedded in. Thus, all information available on the species or sub-species, whether or not spoilage

Department of Microbiology and Biotechnology, Max Rubner-Institut (Federal Research Institute of Nutrition and Food), Hermann-Weigmann-Str. 1, D-24103 Kiel, Germany.
E-mails: knut.heller@mri.bund.de; diana.meske @mri.bund.de
* Corresponding author

or pathogenic organisms are members of that species or sub-species, can be taken into account to judge upon the safety of the isolated microorganism (Pot et al. 1997).

In this chapter we will present an overview about current methods used for identification and enumeration of bacteria with established probiotic properties applied in food. To do so, we will focus on two aspects: (i) identification and enumeration in food products, where the probiotic bacteria constitute a major part of the microbiota present; and (ii) identification and enumeration in faeces, where the probiotic bacteria represent a minority within the fecal microbiota. However, we will start this chapter with a short description of the two major groups of probiotics applied in food and especially in dairy products, i.e., lactobacilli and bifidobacteria, and some general remarks on taxonomy and identification of bacteria.

## 2. Taxonomy and Typing of Lactobacilli and Bifidobacteria

Taxonomy forms the frame in which identification positions unknown isolates. The basis for taxonomic grouping is the species. In contrast to eukaryotes, where species definition follows a clear biological concept—i.e., the species is a population, whose members are capable of interbreeding and producing fertile offspring—species definition in prokaryotes is rather arbitrary. In modern phylogeny, the species is defined by DNA-DNA hybridization: all strains with more than 70% DNA-DNA relatedness are considered to be members of the same species, provided the phenotypic characteristics agree with this definition. Therefore, modern methods of identification follow polyphasic approaches, integrating phylogenetic and phenotypic characteristics (Trüper and Schleifer 2006).

The two major groups of probiotics applied in food and especially in dairy products are lactobacilli and bifidobacteria. They are members of the two phyla *Firmicutes* and *Actinobacteria* (Fig. 1), respectively, which together form the group of Gram-positive bacteria.

Members of the genus *Lactobacillus* together with those of other closely related genera of the family *Lactobacillaceae* (like, e.g., *Lactococcus*, *Pediococcus*, *Streptococcus*, etc.) are often addressed as lactic acid bacteria (LAB). Due to the large phylogenetic distance of this group to the genus *Bifidobacterium* it becomes obvious that the latter should not be included in the LAB group although one of the major fermentation products is lactose.

| | **Domain:** *Bacteria* | |
|---|---|---|
| *Actinobacteria* (high GC) | **Phylum** | *Firmicutes* (low GC) |
| *Actinobacteria* | **Class** | *Bacilli* |
| *Bifidobacteriales* | **Order** | *Lactobacillales* |
| *Bifidobacteriaceae* | **Family** | *Lactobacillaceae* |
| *Bifidobacterium* | **Genus** | *Lactobacillus* |
| e.g., *Bifidobacterium longum* | **Species** | e.g., *Lactobacillus casei* |
| e.g., *Bifidobacterium longum* BB536 | **Strain** | e.g., *Lactobacillus casei* Shirota |

**Fig. 1.** Bifidobacteria and lactobacilli and their taxonomic position. The presentation is based on information taken from web site: http://www.bacterio.cict.fr/classifphyla.html [Euzéby 1997].

Today, one of the most important approaches for taxonomic grouping of bacteria is the comparative sequencing of 16S rDNA genes (Woese et al. 1975). However, as demanded by polyphasic taxonomy, this approach should be complemented by morphological and/or biochemical analyses. In Table 1 some of the most important and frequently tested phenotypes of the two genera *Lactobacillus* and *Bifidobacterium* are listed.

**Table 1.** Some phenotypic properties of the two genera *Lactobacillus* and *Bifidobacterium*.

| **Phenotype** | ***Lactobacillus*** | ***Bifidobacterium*** |
|---|---|---|
| Morphology | rod-shaped | branched rods |
| Gram stain | positive | positive |
| Motility | variable | non-motile |
| Growth conditions | anaerobic, facultatively aerobic | anaerobic |
| Predominant metabolites of hexose consumption | homolactic: lactic acid; heterolactic: lactic acid:ethanol:$CO_2$ = 1:1:1 | lactic acid:acetic acid = 2:3 |
| Growth temperature | 15–45°C | 37°C |
| GC-content of DNA | low | high |

Lactobacilli are rod-shaped, motile or non-motile, Gram-positive bacteria with anaerobic metabolism. Hexoses are metabolized either by homo- or by heterolactic fermentation, thereby producing as end products either solely lactic acid or lactic acid, ethanol and carbon dioxide in equimolar amounts (Table 1). While homolactic lactobacilli prefer hexoses over pentoses, heterolactic prefer pentoses. The latter is due to the higher energy yield from pentoses (2 net ATPs) than from hexoses (1 net ATP). When oxygen is present, some lactobacilli are able to regenerate $NAD^+$ by NADH oxidase, using oxygen as electron acceptor (Sakamoto and Komagata 1996). In that case, the energy yield of hexose consumption may double from 1 net ATP to two net ATPs: the intermediate acetyl-phosphate can be converted to acetic acid to yield one more ATP rather than to ethanol (produced to regenerate $NAD^+$ under anaerobic conditions). In addition, the end products of homolactic or heterolactic hexose fermentation shift to non-equimolar amounts of lactate and acetate or lactic acid, acetic acid, ethanol and carbon dioxide, respectively (Sakamoto and Komagata 1996). Lactobacilli live in many different habitats. Potentially probiotic strains have mostly been isolated from the intestinal tract of healthy humans or from fermented foods (Heller 2009).

Bifidobacteria are Gram-positive, non-motile bacteria, shaped as branched, irregular rods. They are strictly anaerobic. Hexoses are metabolized by the fructose-6-phosphate shunt, to yield from one mole hexose 1.5 moles of acetate and 1 mole of lactate; gas is not produced (Biavati and Mattarelli 2006). Fructose-6-phosphate phosphoketolase therefore is an important phenotypic property for identification of bifidobacteria (Martin et al. 2009). The typical habitat of bifidobacteria is the intestinal tract. Potentially probiotic strains have been isolated from the intestinal tract of healthy humans, with the exception of *Bifidobacterium animalis* subsp. *lactis*, which has been isolated from fermented dairy product.

For further characterization or identification of lactobacilli and bifidobacteria, phenotypic testing, e.g., growth on different carbohydrates or activities of different enzymes, may be applied, as described for lactobacilli (Hammes and Hertel 2006) and bifidobacteria (Biavati and Mattarelli 2006). Usually such investigations are carried out using commercially available test systems like, e.g., API, Analytical Profile Index (bioMérieux SA, Marcy l'Etoile, France) or Biolog Microbial ID System (Biolog Inc., Hayward, CA, USA).

For the genera *Lactobacillus* and *Bifidobacterium* 185 and 48 species, respectively, are listed in the Bacterial Nomenclature Up-to-Date database of DSMZ (DSMZ 2012). Of these, only few species have been investigated so far for potential application as probiotics: *Lactobacillus* species investigated were *acidophilus, casei, crispatus, fermentum, gasseri, johnsonii, paracasei, plantarum, reuteri, rhamnosus, salivarius*, and *Bifidobacterium* species were *adolescentis, animalis, bifidum, breve, infantis, longum, pseudocatenulatum* (Heller 2009, Kneifel and Domigk 2009). However, when it comes down to commercial application of probiotics, the numbers of species applied are considerably lower: *L. acidophilus, casei, johnsonii, reuteri, rhamnosus*, and *B. animalis, breve, longum* (de Vrese and Schrezenmeir 2009).

## 3. Identification and Enumeration of Probiotics in Foods

The probiotic concept originated from the ideas of E. Metchnikoff (1907), who thought that longevity of the Caucasian population was a consequence of intense consumption of fermented dairy products and the lactic acid bacteria therein. Therefore, probiotics are mostly applied in foods of dairy origin, with yoghurt or yoghurt-like products prevailing. In such products, probiotics are present in considerable amounts: usually around $10^8$ and no fewer than $10^6$ colony forming units (cfu) per gram are applied (BGVV 1999). Furthermore, microbial diversity in those products is rather simple (e.g., in yoghurt it consists of *Streptococcus thermophilus* and *Lactobacillus delbrueckii* subsp. *bulgaricus*) or even non-existent, when milk-based probiotic products like Yakult are considered, which contain just the probiotic strain. Therefore, isolation of probiotics for identification and enumeration is usually not a problem and can be done by bacteriological methods. Under such conditions, MRS agar pH 6.2 (de Man et al. 1960) is recommended for lactobacilli. The same may hold true for analyzing "yoghurt mild". "Yoghurt mild" is produced with the aid of a starter culture consisting of *Streptococcus thermophilus* and a thermophilic *Lactobacillus*. Thus, when the probiotic strain is such a thermophilic *Lactobacillus*, like, e.g., *L. acidophilus, L. johnsonii, L. helveticus*, etc., it may be applied as part of the starter culture and may thus be the only *Lactobacillus* present in the product. However, when analyzing a probiotic strain in traditional yoghurt, which is manufactured with the defined starter culture consisting of *S. thermophilus* and *L. delbrueckii* subs. *bulgaricus*, or when analyzing dairy products manufactured with the aid of complex starter cultures containing several strains of different lactic acid bacteria, the selectivity of the medium may be increased by lowering the pH to 5.7 (ISO 1998) or by applying selective SL agar as described by Rogosa et al. (1953). For enumeration, serial dilutions of samples are spread on the agar plates. After anaerobic incubation at 30°C for up to five days, colony-forming units (cfu) are counted and the original concentration of lactobacilli in the sample is calculated. If different types of colonies can be distinguished, which could, for example, be the case when several different lactobacilli are present in the product, they have to be differentiated for species identification, e.g., by molecular methods as described by Singh et al. (2009) and as discussed in the last chapter. Thereafter, counts of cfu for specific lactobacilli have to be adjusted accordingly.

For enumeration of bifidobacteria in the absence of accompanying micro flora, TPY agar (Scardovi 1986) is recommended. Plates are incubated anaerobically at 37°C for 3–4 days. Serial dilutions are used for determination of cfu as described above for lactobacilli. However, in the presence of accompanying micro flora, AMC agar (Arroyo et al. 1995), supplemented with Tween 80 (Engel 2002) is recommended as selective medium. Again, as already stated for lactobacilli, bifidobacteria counts in the original sample are determined by counting cfu and eventually correcting the values obtained by subtracting the cfu of accompanying microorganisms.

The method of choice for species identification is 16S rDNA sequence analysis. DNA is extracted either directly from colonies or from sedimented cells of liquid cultures inoculated with purified, single colonies. The genomic region encoding the 16S RNA gene is then amplified by PCR (Ventura et al. 2001) with the aid of the primer pairs which may be genus-specific, as is the case with bifidobacteria, or higher taxa-specific, as is the case for lactobacilli (Table 2). Sequences obtained are compared to publicly available nucleotide sequence databases by means of BLAST analysis (Altschul et al. 1990) or in a more refined way by the ARB package (Ludwig et al. 2004).

Determining cfu is usually considered the reference method for enumeration. However, one has to bear in mind that this method does not count individual cells but the number of aggregates that contain at least one viable cell. Thus, the number of individual cells is usually underestimated, e.g., for chain-forming organisms. Molecular methods like quantitative real-time PCR, which do not count living entities but just DNA or RNA molecules, on the other hand, have the disadvantage of counting the numbers of extractable genomes without differentiating whether those were isolated from living or dead cells. However, the advantage of these methods is that colony formation is not required; instead, DNA may be extracted from the food product directly. For probiotic products, the latter should be no problem since probiotics have to be present in numbers sufficient to cause the health effect claimed for the product. Such numbers are always higher than $10^6$ and usually around $10^8$ per ml or per gram. Furthermore, recently a quantitative PCR technique has been described, which allows to restrict amplifiable DNA to that isolated from intact cells (Desfossés-Foucault et al.

**Table 2.** Primers used for amplification of 16 S rDNA of lactobacilli and bifidobacteria.

| Primer | Location | Sequence 5' ® 3' | Reference |
|---|---|---|---|
| ***Lactobacillus*** | | | |
| fD1 | 8–27 | AGAGTTTGATCCTGGCTCAG | Weisburg et al. 1991 |
| rD1 | 1541–1525 | AAGGAGGTGATCCAGCC | Weisburg et al. 1991 |
| ***Bifidobacterium*** | | | |
| P0 | 6–28 (16S rRNA) | GAGAGTTTGATCCTGGCTCAG | Di Cello et al. 1996 |
| Lm3 | 1432–1412 (16S rRNA) | CGGGTGCTNCCCACTTTCATG | Kaufmann et al. 1997 |

2012). The method relies on addition of propidium monoazide to bacterial suspensions for blocking of DNA not contained within intact cells. After centrifugation of the intact cells and removal of the propidium monoazide containing supernatant, DNA is extracted from the cells. Only this DNA then is accessible for qPCR. Also, quantitative reverse-transcription PCR may be applied (Bej et al. 1991). Bacterial mRNAs have small half lifes of usually a few minutes only. Thus, intact but dead cells, which are no longer active in transcription, are usually depleted off their mRNAs within a few minutes. Therefore, the amount of mRNA is directly proportional to the number of living cells, provided that a transcript is targeted, which is not subject to extensive regulation. Such transcripts are those of house-keeping genes, which only show marginal responses to varying environmental signals (Fey et al. 2004). However, one still has to bear in mind that for enumeration of a probiotic strain, enumerating all members of a species is not sufficient, as long as it is not guaranteed that only one strain of a species is present in the product.

## 4. Identification and Enumeration of Probiotics in Faeces or Intestinal Content

The major challenge for identifying and enumerating probiotics in faeces or intestinal content and the major difference for doing so in probiotic foods is that probiotics constitute only a minority within a microbiota of enormous quantity and biodiversity. When considering the human intestinal microbiotia to consist of $10^{14}$ microbial cells per host organism (Lepage et al. 2013), it becomes clear that the ca. $10^{10}$ cfu of a probiotic strain taken up with a probiotic food represent only about 0.01% of the intestinal microbiota. Identification and enumeration of a probiotic strain directly by determining cfu would only be a realistic option, if the strain expressed a very strong and specific selectable phenotype, e.g., a rare and unusually high antibiotic resistance. It is realistic though to enrich the strain among a population of microbes with similar properties. Then, as already stated for probiotics in food, strain-specific identification of single colonies would be the next essential step. Such identification can be done by strain-specific colony-hybridization (Grunstein and Hogness 1975), where all colonies on a plate can be analyzed at once, or by strain-specific conventional PCR (Singh et al. 2009), which requires analysis of single colonies. Further techniques for strain-specific identification are discussed in the next chapter. Since all these techniques are rather tedious and time consuming, there is a demand for applying techniques which do not rely on counting cfu. Such methods have been developed over the years (Vaughan et al. 1999). One important method is *in situ* fluorescence hybridization (FISH), which allows direct identification of single specific cells within a population of non-related microbes in an environmental sample (Amann et al. 1995, Harmsen et al. 2000). The target for binding of fluorescence-labeled probes usually is rRNA, because it is present in very high copy numbers within the cells. However, since this method involves microscopic inspection of the samples, the cells to be detected should not be present in too low numbers. Otherwise, detection could be very tedious and time-consuming. Quantitative PCR (qPCR) may also be applied (Fey et al. 2004). Again, to increase sensitivity, rRNA may be targeted by the primers applied. However, then this PCR will not allow for strain-specific enumeration.

## 5. Strain-specific Identification

Methods for strain identification are typing methods: they allow demonstrating identity or non-identity of a strain isolated with that of the probiotic strain applied. In Fig. 2, a consensus of the taxonomic resolving powers of typing methods is presented as discussed and compared in several reviews (Vandamme et al. 1996, Salvelkoul et al. 1999, Domig et al. 2003). Of the methods listed in Fig. 2, several like rRNA sequencing, cell wall structure, % G+C, FAME, DNA-DNA hybridization, bacteriophage and bacteriocin typing, and ARDRA discriminate at higher taxa levels only and are thus not suitable for strain identification.

"Phenotype determination by classical methods", including the commercially available typing systems, may in principle be applicable for strain identification. However, for discriminating at this level, they simply require so much time and experimental effort that they should not be considered appropriate.

Broadest range of application is possible for "DNA hybridization probes" and "DNA sequencing". However, many "DNA hybridization probes" have been developed for higher taxonomic levels and only few for strain level (Singh et al. 2009). The reasons for this may be the very limited application for DNA probes at strain level and the extremely intensive and laborious testing for their development. DNA sequencing, on the other hand, bears the potential of becoming the standard method for resolution at strain level. Powerful sequencing techniques have been developed during recent years, which allow sequencing of enormous numbers of nucleotides at low costs. For strain identification, 16S rDNA, due to limited variability of this gene, cannot solely be the target for sequencing. Instead, "Multi Locus Sequencing Typing/Multi Locus Sequence Analysis (MLST/MLSA)" technique (Maiden et al. 1998) is applied. This method is based on sequencing several gene loci, the sequence information of which is catenated, compared by bioinformation techniques like BLAST (Altschul et al. 1990). The comparisons are then used for differentiation and typing purposes. Usually,

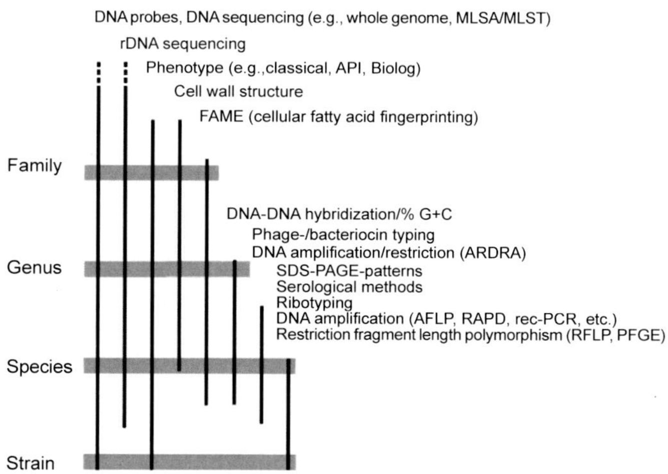


**Fig. 2.** Taxonomic resolution of different typing techniques. The figure represents a consensus of the information presented in Vandamme et al. (1996), Savelkoul et al. (1999), Domig et al. (2003).

house-keeping genes (Hanage et al. 2006) like *atpD*, *gyrB*, *recA*, *rpoB*, *rpoD*, etc. are chosen, which may be combined with the rather conserved 16S and 23S rRNA genes (Bennasar et al. 2010) to allow for inclusion of all phylogenetic data produced so far for species identification. MLSA has been applied for analysing lactic acid bacteria populations in sourdough (De Vuyst and Vancanneyt 2007). A database is available at http://pubmlst.org/.

The so-called "next generation" sequencing techniques, the "454 sequencing" (Margulies et al. 2005) and the "Solexa/Illumina" and "SOLiD" sequencing (Hudson 2008), allow generating raw sequencing data for entire microbial genomes within one or a few days. Thus, whole genome sequencing may eventually become the standard method of strain typing. However, before whole genome sequencing or MLSA/MLST can become a standard typing method, a clear concept is needed for defining the extent of allowable sequence deviation within one and the same strain.

Serology targets the surfaces of micro-organisms. It has been a very successfully method for differentiation of pathogenic bacteria below species level. However, it remains to be determined whether serology is capable of differentiating a probiotic strain from other strains within the same species.

In SDS-PAGE (sodium dodecyl sulfate polyacrylamide gel electrophoresis), total cellular proteins are separated according to their molecular masses. It has been successfully applied in lactic acid bacteria for differentiation at strain level (Pot et al. 1997). This method is powerful when used as an "in-house" method, because it requires growing and harvesting the cells under strictly defined conditions in order to generate reproducible protein patterns. The necessity for strictly defined conditions also applies for DNA-amplification techniques, which are based on conditions allowing mismatches between primers and target sequences. The best example is RAPD (random amplification of polymorphic DNA), where primers with arbitrary sequences are used. To a lesser degree, this also applies for rep-PCR (Repetitive Element PCR), ERIC-PCR (Enterobacterial Repetitive Intergenic Consensus PCR) (de Bruijn 1992), Box-PCR (BOX-A1R-based repetitive extragenic palindromic PCR) (Louws et al. 1994), etc. Small changes in the amplification procedures lead to significant changes in the electrophoretic patterns generated.

AFLP (amplified fragment length polymorphism) (Vos et al. 1995, Salvelkoul et al. 1999) is a PCR-based method of very high discriminating power, allowing for high throughput of samples. The method can be described as follows. At the sticky ends of restriction fragments obtained from genomic DNA by hydrolysis with two different restriction enzymes, primer binding sites are generated by ligation of two different adapters fitting to the sticky ends of the fragments. For subsequent PCR, primers targeting the adapter sequences are applied. The selectivity of the primers can be increased by extending the site targeted by the primers by one or two nucleotides into the restriction fragment. In ribotyping (Snipes et al. 1989), chromosomal DNA isolated from single colonies or pure liquid cultures is hydrolysed by restriction enzymes and separated by agarose gel electrophoresis. In a following Southern blot, an rDNA-specific probe identifies DNA fragments, which comprise regions of the rRNA genes. The largest heterogeneities of DNA regions flanking the rRNA genes result in banding patterns with very high intra-species variation. In IS-RFLP (Leclercq et al. 2007), probes targeting insertion sequences are applied for Southern hybridization.

However, this method depends on knowledge about the presence of specific insertion sequences within the genomes of the strains of interest. Applicability of this method for potential probiotic bacteria has been demonstrated (Turgay et al. 2011). For PFGE (Snell and Wilkins 1986, Briczinski and Roberts 2006), chromosomal DNA is cleaved by a rare cutting restriction enzyme. The very large DNA fragments generated are separated by pulsed-field gel electrophoresis (PFGE), which is capable of separating DNA fragments with a length of several hundred thousand basepairs. Banding patterns obtained by different enzymes are used for fingerprinting.

AFLP, ribotyping, IS-RFLP and PFGE generate reproducible electrophoretic banding patterns. The discriminating powers of some of the typing techniques have been compared. Hänninen et al. (2001) discriminated 8 out of 35 different *Campylobacter jejuni* strains by ribotyping, 10 by PFGE and 10 by AFLP. Tynkkynen et al. (1999) discriminated 7 out of 19 different *Lactobacillus rhamnosus* strains by RAPD, 10 by ribotyping, and 12 by PFGE. Mättö et al. (2004) discriminated 7 different *B. longum* strains by RAPD, 13 by ribotyping, and 14 by PFGE (only 14 of the 18 isolates were analysed by PFGE) out of 18 *Bifidobacterium longum*. Furthermore, for 10 *Bifidobacterium adolescentis* isolates, 6 strains could be differentiated by RAPD, whereas ribotyping and PFGE each differentiated 9 strains. Apparently, PFGE and AFLP are the two methods with the highest powers of discriminating at strain level (Vogel et al. 2004). PFGE has been called the "gold-standard" of strain identification by FAO/WHO (FAO/WHO 2006). For epidemiologic studies, PFGE is used as a standardized effective method for identification of strains in foodborne disease outbreaks by concerted actions of laboratories joined in "Pulse-Net" (Boxrud et al. 2010).

When applying fingerprinting techniques one has to bear in mind that they are just typing techniques: they analyze one characteristic phenotype, which then is used for assigning the strains to groups of identical (or very similar) or to non-identical ones. With the exception of techniques based on DNA sequencing, fingerprinting methods are not applicable for phylogenetic considerations. Strains grouped together by such a method do not form a taxonomic unit but form a group of members which have been assigned to the group on the basis of an arbitrary definition. For identical strains, the definition may either not allow for any deviation—e.g., from a banding pattern in PFGE—or may allow for small deviations at a maximum level of 10%. For PFGE this could mean that one out of 10 bands in the banding pattern may be different and the strains would still be considered identical. When tracing back to the original source of infection microorganisms in outbreaks (Chiou et al. 2001), this definition is frequently applied.

One has to be aware, however, that when assigning microorganisms to one group on the basis of just one fingerprinting method, this group may in fact be heterogeneous. Recently, *Lactobacillus fermentum* strains with identical PFGE patterns have been described, which differed in up to four phenotypes tested: growth temperature, sugar metabolism, ARDRA pattern (Njeru et al. 2010). This raises the question, whether fingerprinting methods are appropriate for assigning microorganisms to a group of microorganisms having in common one or few important functional phenotypes not tested by the fingerprinting method. One may conclude that for probiotic strains typing methods like PFGE are useful only as long as specific tests are not available for those genes or for the activity of those genes that make a strain probiotic. When such information becomes available, typing methods combined with methods testing the activities of genes relevant for probiotic properties will be capable of demonstrating that a functional trait is present within the strain exhibiting the correct genetic background.

## References


Altschul, S.F., W. Gish, W. Miller, E.W. Myers and D.J. Lipman. 1990. Basic logical alignment search tool. J. Mol. Biol. 215: 403–410.

Amann, R.I., W. Ludwig and K.H. Schleifer. 1995. Phylogenetic identification and *in situ* detection of individual microbial cells without cultivation. Microbiol. Rev. 59: 143–169.

Arroyo, L., L.N. Cotton and J.H. Martin. 1995. AMC agar: a composite medium for selective enumeration of *Bifidobacterium longum*. Cultured Dairy products Journal 30: 12–15.

Bej, A.K., M.H. Mahbubani and R.M. Atlas. 1991. Detection of viable Legionella pneumophila in water by polymerase chain reaction and gene probe methods. Appl. Environ. Microbiol. 57: 597–600.

Bennasar, A., M. Mulet, J. Lalucat and E. Garcia-Valdés. 2010. PseudoMLSA: a database for multigenic sequence analysis of Pseudomonas species. BMC Microbiol. 10: 118.

[BGVV]. 1999. Probiotische Mikroorganismenkulturen in Lebensmitteln. Abschlussbericht der Arbeitsgruppe "Probiotische Mikroorganismen in Lebensmitteln" am Bundesinstitut für gesundheitlichenVerbraucherschutz und Veterinärmedizin (BgVV), Berlin. Ernährungs-Umschau 47: 191–195.

Biavati, B. and P. Mattarelli. 2006. The family *Bifidobacteriaceae*. pp. 322–382. *In*: M. Dworkin et al. (eds.). Prokaryotes, Vol. 1, 3rd edition. Springer Science+Business Media LLC, New York, USA.

Boxrud, D., T. Monson, T. Stiles and J. Besser. 2010. The role, challenges, and support of Pulse-Net laboratories in detecting foodborne disease outbreaks. Publ. Health Rep. 125: 57–62.

Briczinski, E.P. and R.F. Roberts. 2006. *Technical Note*: A rapid pulsed-field gel electrophoresis method for analysis of bifidobacteria. J. Dairy Sci. 89: 2424–2427.

Chiou, C.S., W.B. Hsu, H.L. Wei and J.H. Chen. 2001. Molecular epidemiology of a Shigella flexneri outbreak in a mountainous township in Taiwan, Republic of China. J. Clin. Microbiol. 39: 1048–1056.

de Man, J.D., M. Rogosa and M.E. Sharpe. 1960. A medium for the cultivation of lactobacilli. J. Appl. Bact. 23: 130–135.

de Bruijn, F.J. 1992. Use of repetitive (Repetitive Extragenic Palindromic and Enterobacterial Repetitive Intergeneric Consensus) sequences and the polymerase chain reaction to fingerprint the genomes of *Rhizobium meliloti* isolates and other soil bacteria. Appl. Environ. Microbiol. 58: 2180–2187.

Desfossés-Foucault, E., V. Dussault-Lepage, C. Le Boucher, P. Savard, G. Lapointe and D. Roy. 2012. Assessment of Probiotic Viability during Cheddar Cheese Manufacture and Ripening Using Propidium Monoazide-PCR Quantification. Front Microbiol. 3, 350 [Epub ahead of print].

De Vrese, M. and J. Schrezenmeir. 2009. Präventive Bedeutung von probiotischen Joghurts. pp. 174–193. *In*: S.C. Bischoff (ed.). Probiotika, Präbiotika und Synbiotika. Georg Thieme Verlag KG, Stuttgart, Deutschland.

De Vuyst, L. and M. Vancanneyt. 2007. Biodiversity and identification of sourdough lactic acid bacteria. Food Microbiol. 24: 120–127.

Di Cello, F.P. and R. Fani. 1996. A molecular strategy for the study of natural bacterial communities by PCR-based techniques. Minerva Biotec. 8: 126–134.

Domig, K.J., H.K. Mayer and W. Kneifel. 2003. Methods used for the isolation, enumeration, characterisation and identification of *Enterococcus* spp. Int. J. Food Microbiol. 88: 165–188.

[DSMZ] Deutsche Stammsammlung für Mikroorganismen und Zellkulturen. 2012. Bacterial Nomenclature Up-to-Date database. http://old.dsmz.de/microorganisms/bacterial_nomenclature.php?

Engel, G. 2002. Investigations on suitability of substrates for quantitative determination of bifidobacteria. Kieler Milchw. Forschungsber. 54: 21–33 (in German).

Euzéby, J.P. 1997. List of Bacterial Names with Standing in Nomenclature: a folder available on the Internet. Int. J. Syst. Bacteriol (http://www.bacterio.cict.fr Last full update: December 08, 2012) 47: 590–592.

[FAO/WHO] Food and Agriculture Organization of the United Nations/World Health Organization. 2006. Probiotics in food. Health and nutritional properties and guidelines for evaluation. FAO Food and Nutrition Paper 85.

Fey, A., S. Eichler, S. Flavier, R. Christen, M.G. Höfle and C.A. Guzmán. 2004. Establishment of a real-time PCR-based approach for accurate quantification of bacterial RNA targets in water, using Salmonella as a model organism. Appl. Environ. Microbiol. 70: 3618–3623.

Grunstein, M. and D.S. Hogness. 1975. Colony hybridization—a method for isolation of cloned DNAs that contain a specific gene. Proc. Natl. Acad. Sci. U.S.A. 72: 3961–3965.

Hammes, W.P. and C. Hertel. 2006. The genera *Lactobacillus* and *Carnobacterium*. pp. 320–403. *In*: M. Dworkin et al. (eds.). Prokaryotes, Vol. 1, 3rd edition. Springer Science+Business Media LLC, New York, USA.

Hanage, W.P., C. Fraser and B.G. Spratt. 2006. Sequences, sequence clusters and bacterial species. Philos. Trans. R. Soc. Lond. B Biol. Sci. 361: 1917–1927.

Hänninen, M.L., P. Perko-Mäkelä, H. Rautelin, B. Duim and J.A. Wagenaar. 2001. Genomic relatedness within five common Finnish Campylobacter jejuni pulsed-field gel electrophoresis genotypes studied by amplified fragment length polymorphism analysis, ribotyping, and serotyping. Appl. Environ. Microbiol. 67: 1581–1586.

Harmsen, H.J.M., A.C.M. Wildeboer-Veloo, G.C. Raangs, A.A. Wagendorp, N. Klijn, J.G. Bindels and G.W. Welling. 2000. Analysis of intestinal flora development in breast-fed and formula-fed infants by using molecular identification and detection methods. J. Pediatr. Gastr. Nutr. 30: 61–67.

Heller, K.J. 2009. "Pharmakokinetik" und Sicherheit von Probiotika. pp. 88–94. *In*: S.C. Bischoff (ed.). Probiotika, Präbiotika und Synbiotika. Georg Thieme Verlag KG, Stuttgart, Deutschland.

Hudson, M.E. 2008. Sequencing breakthroughs for genomic ecology and evolutionary biology. Mol. Ecol. Resour. 8: 3–17.

ISO 15214. 1998. Microbiology of food and animal feeding stuffs—Horizontal method for the enumeration of mesophilic lactic acid bacteria—Colony-count technique at 30 degrees C.

Kaufmann, P., A. Pfefferkorn, M. Teuber and L. Meile. 1997. Identification and quantification of *Bifidobacterium* species isolated from food with genus-specific 16S rRNA-targeted probes by colony hybridization and PCR. Appl. Environ. Microbiol. 63: 1268–1273.

Kneifel, W. and K.J. Domingk. 2009. Taxonomie von Milchsäurebakterien mit probiotischer Kapazität. pp. 103–117. *In*: S.C. Bischoff (ed.). Probiotika, Präbiotika und Synbiotika. Georg Thieme Verlag KG, Stuttgart, Deutschland.

Leclercq, A.J., G. Torrea, V. Chenal-Francisque and E. Carniel. 2007. 3 IS-RFLP: a powerful tool for geo-graphical clustering of global isolates of Yersinia pestis. Adv. Exp. Med. Biol. 603: 322–326.

Lepage, P., M.C. Leclerc, M. Joosens, S. Mondot, H.M. Blottière, J. Raes, D. Ehrlich and J. Doré. 2013. A metagenomic insight into our gut's microbiome. Gut 62: 146–158.

Louws, F.J., D.W. Fulbright, C. Taylor-Stephens and F.J. de Bruijn. 1994. Specific Genomic Fingerprints of Phytopathogenic *Xanthomonas* and *Pseudomonas* Pathovars and Strains Generated with Repetitive Sequences and PCR. Appl. Environ. Microbiol. 60: 2286–2295.

Ludwig, W., O. Strunk, R. Westram, L. Richter, H. Meier, Yadhukumar, A. Buchner, T. lai, S. Steppi, G. Jobb, W. Forster, I. Brettske, S. Gerber, A.W. Ginhart, O. Gross, S. Grumann, S. Hermann, R. Jost, A. Konig, T. Liss, R. Lussmann, M. May, B. Nonhoff, B. Reichel, R. Strehlow, A. Stamatakis, N. Stuckmann, A. Vilbig, M. Lenke, T. Ludwig, A. Bode and K.-H. Schleifer. 2004. ARB: a software environment for sequence data. Nucl. Acid Res. 32: 1363–1371.

Maiden, M.C.J., J.A. Bygraves, E. Feil, G. Morelli, J.E. Russell, R. Urwin, Q. Zhang, J.J. Zhou, K. Zurth, D.A. Caugant, I.M. Feavers, M. Achtman and B.G. Spratt. 1998. Multilocus sequence typing: A portable approach to the identification of clones within populations of pathogenic microorganisms Proc. Natl. Acad. Sci. U.S.A. 95: 3140–3145.

Margulies, M., M. Egholm, W.E. Altman, S. Attiya, J.S. Bader, L.A. Bemben, J. Berka, M.S. Braverman, Y.-J. Chen, Z. Chen, S.B. Dewell, L. Du, J.M. Fierro, X.V. Gomes, B.C. Godwin, W. He, S. Helgesen, C.H. Ho, G.P. Irzyk, S.C. Jando, M.L.I. Alenquer, T.P. Jarvie, K.B. Jirage, J.-B. Kim, J.R. Knight, J.R. Lanza, J.H. Leamon, S.M. Lefkowitz, M. Lei, J. Li, K.L. Lohman, H. Lu, V.B. Makhijani, K.E. McDade, M.P. McKenna, E.W. Myers, E. Nickerson, J.R. Nobile, R. Plant, B.P. Puc, M.T. Ronan, G.T. Roth, G.J. Sarkis, J.F. Simons, J.W. Simpson, M. Srinivasan, K.R. Tartaro, A. Tomasz, K.A. Vogt, G.A. Volkmer, S.H. Wang, Y. Wang, M.P. Weiner, P. Yu, R.F. Begley and J.M. Rothberg. 2005. Genome sequencing in microfabricated high-density picolitre reactors. Nature 437: 376–380.

Martin, R., E. Jiménez, H. heilig, L. Fernández, M.L. Marin, E.G. Zoetendal and J.M. Rodrígues. 2009. Isolation of bifidobacteria from breast milk and assessment of bifidobacterial population by PCR-denaturing gradient gel electrophoresis and quantitative real-time PCR. Appl. Environ. Microbiol. 75: 965–969.

Mättö, J., E. Malinen, M.-L. Suihko, M. Alander, A. Palva and M. Saarela. 2004. Genetic heterogeneity and functional properties of intestinal bifidobacteria. J. Appl. Microbiol. 97: 459–462.

Metchnikoff, E. 1907. The Prolongation of Life—Optimistic Studies. Heinemann, London, UK.

Njeru, P.N., N. Rösch, D. Ghadimi, A. Geis, W. Bockelmann, M. de Vrese, J. Schrezenmeir and K.J. Heller. 2010. Identification and characterization of lactobacilli isolated from *Kimere*, a spontaneously fermented pearl millet dough from Mbeere, Kenya (East Africa). Beneficial Microbes 1: 243–252.

Pot, B., T. Coenye and K. Kersters. 1997. The taxonomy of microorganisms used as probiotics with special focus on enterococci, lactococci, and lactobacilli. Microecol. Therapy 26: 11–25.

Rogosa, M., R.F. Wiseman, J.A. Mitchell, M.N. Disraely and A.J. Beaman. 1953. Species differentiation of oral lactobacilli from man including descriptions of *Lactobacillus-salivarius* nov-spec. and *Lactobacillus-cellobiosus* nov-spec. J. Bacteriol. 65: 681–699.

Sakamoto, M. and K. Komagata. 1996. Aerobic growth of and activities of NADH Oxidase and NADH Peroxidase in lactic acid bacteria. J. Ferment. Bioengin. 82: 210–216.

Savelkoul, P.H., H.J.M. Aarts, J. de Haas, L. Dijkshoorn, B. Duim, M. Otsen, J.L.W. Rademaker, L. Schouls and J.A. Lenstra. 1999. Amplified-fragment length polymorphism analysis: the state of an art. J. Clin. Microbiol. 37: 3083–91.

Scardovi, V. 1986. Genus *Bifidobacterium* Orla-Jensen 1924, 472 [al.]. pp. 1418–1434. *In*: P.H.A. Sneath, N.S. Mair, M.E. Sharpe and J.G. Holt (eds.). Bergey's Manual of Systematic Bacteriology, Vol. 2, 1st ed. Williams and Wilkins. Baltimore, MD.

Singh, S., P. Goswami, R. Singh and K.J. Heller. 2009. Application of molecular identification tools for *Lactobacillus*, with a focus on discrimination between closely related species: a review. LWT—Food Sci. Technol. 42: 448–457.

Snell, R.G. and R.J. Wilkins. 1986. Separation of chromosomal DNA molecules from *C. albicans* by pulsed field gel electrophoresis. Nucleic Acids Res. 14: 4401–4406.

Snipes, K.P., D.C. Hirsh, R.W. Kasten, L.M. Hansen, D.W. Hird, T.E. Carpenter and R.H. McCapes. 1989. Use of an rRNA probe and restriction endonuclease analysis to fingerprint *Pasteurella multocida* isolated from turkeys and wildlife. J. Clin. Microbiol. 27: 1847–1853.

Trüper, H.G. and K.-H. Schleifer. 2006. Prokaryote characterization and Identification. pp. 58–79. *In*: M. Dworkin et al. (eds.). Prokaryotes, Vol. 1, 3rd edition. Springer Science+Business Media LLC, New York, USA.

Tynkkynen, S., R. Satokari, M. Saarela, T. Mattila-Sandholm and M. Saxelin. 1999. Comparison of ribotyping, randomly amplified polymorphic DNA analysis, and pulsed-field gel electrophoresis in typing of Lactobacillus rhamnosus and L. casei strains. Appl. Environ. Microbiol. 65: 3908–3914.

Turgay, M., S. Irmler, D. Isolini, R. Amrein, M.-T. Fröhlich-Wyder, H. Bertoud, E. Wagner and D. Wechsler. 2011. Biodiversity, dynamics, and characteristics of Propionibacterium freudenreichii in Swiss Emmentaler PDO cheese. Dairy Science & Technology 91: 471–489.

Vandamme, P., B. Pot, M. Gillis, P. De Vos, K. Kersters and J. Swings. 1996. Polyphasic taxonomy, a consensus approach to bacterial systematics. Microbiol. Rev. 60: 407–438.

Vaughan, E.E., B. Mollet and W.M. deVos. 1999. Functionality of probiotics and intestinal bacteria: light in the intestinal tract tunnel. Curr. Opinion Biotechnol. 10: 505–510.

Ventura, M., R. Reniero and R. Zink. 2001. Specific Identification and Targeted Characterization of *Bifidobacterium lactis* from Different Environmental Isolates by a Combined Multiplex-PCR Approach. Appl. Environ. Microbiol. 67: 2760–2765.

Vogel, B., V. Fussing, B. Ojeniyi, L. Gram and P. Ahrens. 2004. High-resolution genotyping of Listeria monocytogenes by fluorescent amplified fragment length polymorphism analysis compared to pulsed-field gel electrophoresis, random amplified polymorphic DNA analysis, ribotyping, and PCR-restriction fragment length polymorphism analysis. J. Food Prot. 67: 1656–1665.

Vos, P., R. Hogers, M. Bleeker, M. Reijans, T. van de Lee, M. Hornes, A. Friters, J. Pot, J. Paleman, M. Kuiper and M. Zabeau. 1995. AFLP—a new technique for DNA-fingerprinting. Nucl. Acids Res. 23: 4407–4414.

Weisburg, W.G., S.M. Barns, D.A. Pelletier and D.J. Lane. 1991. 16S ribosomal DNA amplification for phylogenetic studies. J. Bacteriol. 173: 697–703.

Woese, C.R., G.E. Fox, L. Zablen, T. Uchida, L. Bonen, K. Pechmann, B.J. Lewis and D. Stahl. 1975. Conservation of primary structure in 16S ribosomal RNA. Nature 254: 83–96.

CHAPTER 6

# Investigation of Probiotic Functionalities by Proteomics

*Maria Fiorella Mazzeo* and *Rosa Anna Siciliano**

## 1. Introduction

The last decades experienced a significant scientific revolution prompted by the introduction of "genomics", a new discipline that shifted the focus of research from the study of a single gene to the overall set of genes of a living organism and their inter-relationships. The release of complete genomic sequences, from bacteria to human, provided the scientific community with massive and fundamental information but, at the same time, highlighted that knowing when, how and where proteins are expressed is mandatory to truly understand the way a biological system works in optimal or adverse conditions. This paved the way to the onset of the "-omic sciences" (transcriptomics, proteomics, metabolomics and so on), i.e., the study of RNA, proteins and metabolites in a holistic view. On the other hand, it was soon evident that the integration of knowledge provided by different -omic sciences in a systems biology perspective, could be crucial to decipher functionality and dynamics of a biological system as a single entity or a part of complex ecosystems.

The term 'proteome' was coined about twenty years ago by Marc Wilkins to describe the entire set of proteins encoded by the genome of organisms, cells or tissues at a given time (Wilkins et al. 1996). The ambitious goal of proteomics is to depict a global, dynamic and integrated view of cell physiology that includes protein identification, analysis of post-translational modifications (PTMs) and changes in expression level, in time and in space intracellular distributions, stabilities and turnover rates, as well as protein-protein interactions.

Centro di Spettrometria di Massa Proteomica e Biomolecolare, Istituto di Scienze dell'Alimentazione, CNR, via Roma 64, 83100 Avellino, Italy.

E-mails: fmazzeo@isa.cnr.it; rsiciliano@isa.cnr.it

* Corresponding author

Undoubtedly, two major events allowed the development of proteomics: the growing number of completely sequenced genomes of living organisms and the availability of two-dimensional electrophoresis methods (2-DE) (Klose 1975, O'Farrell 1975). However, since the beginning of the proteomic era, the development and application of proteomics strictly relied on the capability of analytical resources and technologies to perform high-throughput proteomic studies. In fact, proteome analysis is quite challenging as proteins in a cell have different physicochemical features, subcellular localizations, abundance levels (spanning 10 orders of magnitude), exist in different forms due to PTMs, and have alternative splicing or protease action. Moreover, proteome is highly dynamic, being affected by environmental stimuli and pathological conditions, among other factors. The introduction of novel mass spectrometric instruments could be regarded as the turning point in proteomics, providing the appropriate analytical tools for large-scale protein identification and, later on, protein quantification. In the last years, technical implementations drastically improved analytical performances in terms of sensitivity, mass accuracy and resolution, and data acquisition speed (now allowing the identification and quantification up to 5,000 proteins in a single experiment). At the same time, sophisticated computational bioinformatics provided *ad hoc* tools to process complex mass spectrometric datasets.

Proteomics has represented a revolutionary event in the microbiology field too, as Professor Frederick C. Neidhart clearly stated: "(before the proteomics advent) the study of cell physiology was largely reductionistic as the living cell was taken apart and studied biochemically even if the powerful marriage of biochemistry and genetic analysis led to notable triumphs, culminating in the field of molecular biology". "(Proteomics allows) to understand completely the workings of a cell observing the integrated behaviour of the entire panoply of gene regulatory devices" (Neidhart 2011). The main issues addressed by microbial proteomics could be summarized as follows: (i) identification of differentially expressed proteins to shed light on metabolic pathways, molecular processes and regulatory networks activated or deactivated in cell adaptation to various environmental stimuli (growth or stress conditions, starvation, etc.); (ii) analysis of PTMs (glycosylation, phosphorylation, and so on) that play pivotal roles in the regulation of protein functions and biological processes; (iii) elucidation of host-pathogen interactions underlying infection dynamics, including the study of virulence factors and surface proteins; (iv) characterization of microbial communities lifestyle in complex ecosystems (including biofilms), termed metaproteomics. For extended read on different applications of proteomics in microbiology please refer to the following reviews (Hecker et al. 2008, Di Cagno et al. 2011, Graham et al. 2011, Malmström et al. 2011, Chao and Hansmeier 2012, Munday et al. 2012, Otto et al. 2012).

In the microbial world, Lactic Acid Bacteria (LAB) hold a role of great importance for their widespread use as starters in food fermentation, contributing to the organoleptic properties, and for the probiotic traits of certain LAB strains, that beneficially affect human and animal health (FAO/WHO 2006). Probiotics promote digestion and uptake of dietary nutrients, strengthen intestinal barrier function, modulate immune response and enhance antagonism towards pathogens. Nowadays, the use of probiotics in clinical applications and in the production of functional foods has significantly increased, leading to the growing need to clarify their mechanisms of action at a molecular level. In this contest, -omic sciences have greatly contributed to widen knowledge, and proteomics has been thoroughly applied to unravel the molecular basis of probiotic functionalities and their health promoting effects as the leading actors of these mechanisms are proteins.

This chapter focuses on the role of proteomics in the field of probiotics, with main emphasis on the most recent advancements. Analytical strategies routinely used in proteomics are briefly reviewed and a glossary of terms is reported in Table 1, while for theoretical and technical details of proteomics and mass spectrometry (MS), readers should refer to recently published reviews (Aebersold and Mann 2003, Yates 2004, Walther and Mann 2010, May et al. 2011, Angel et al. 2012, Nikolov et al. 2012, Roepstorff 2012, Altelaar et al. 2013).

**Table 1.** Glossary of proteomic methodologies.

| Methodology | Abbreviation | Definition | Main References |
|---|---|---|---|
| Two-Dimensional Electrophoresis | 2-DE | Powerful analytical technique used for separation of complex protein mixtures. Proteins are separated in first dimension according to their isoelectric point using IEF and in second dimension according to their molecular weight using SDS-PAGE. Protein spots are visualized on 2-DE maps by Coomassie, silver or fluorescent staining. | Klose 1975<br>O'Farrell 1975 |
| DIfference Gel Electrophoresis | DIGE | Variant of gel electrophoresis that allows to avoid inter-gel variations and improve quantitative analysis. Up to three different protein samples are labelled with spectrally different fluorescent dyes, mixed prior to 2-DE and simultaneously analyzed on the same gel. | Unlü et al. 1997 |
| Gel-based proteomic approaches | | Proteomic strategies that rely on 1 or 2-DE for proteome separation, followed by in-gel protein digestion and MS analysis of the obtained peptide mixtures. | |
| Matrix Assisted Laser Desorption Ionization-Time of Flight-Mass Spectrometry | MALDI-TOF-MS | Rapid and simple MS technique that allows the analysis of complex peptide mixtures with high resolution in m/z measures (m/z error < 20 ppm) and sensitivity (up to attomolar level) and does not require sample purification prior to the analysis. It is generally applied to the identification of 2-DE separated proteins. | Karas and Hillenkamp 1988 |
| Peptide Mass Fingerprint | PMF | Proteomic strategy based on MALDI-TOF-MS analysis of peptide mixtures obtained from the in-gel digestion of 2-DE separated proteins. MALDI-TOF-MS data are used to identify proteins by means of searches in specific protein sequence databases using dedicated bioinformatic tools. | Henzel et al. 1993<br>Pappin et al. 1993 |
| Liquid Chromatography coupled to tandem mass spectrometry | LC-MS/MS | A versatile technique that allows the on line MS analysis of peptides separated by reverse-phase liquid chromatography. In a LC-MS/MS experiment, a dataset is produced alternating the acquisition of a mass spectrum containing precursor ions generated by peptides eluted | |

*Table 1. contd....*

*Table 1. contd.*

| Methodology | Abbreviation | Definition | Main References |
|---|---|---|---|
| | | along the chromatographic run, and MS/MS spectra obtained by the fragmentation of these ions. The entire dataset is used to identify proteins by means of database searches using dedicated bioinformatic tools. | |
| Two-or Multi-Dimensional Liquid Chromatography | 2D-LC nD-LC | Methods that combine more than one chromatographic step prior to MS analyses. A combination of strong cation or anion exchange chromatography as first separation dimension and reverse phase chromatography as second separation dimension is usually applied, thus dramatically improving resolution. This approach is mandatory in shotgun proteomics where very complex peptide mixtures have to be analyzed. | Yates et al. 1999 |
| Shotgun proteomics | | Proteomic gel-free approach in which the whole proteome extracted from a biological system is submitted to enzymatic digestion (usually with trypsin) and the obtained complex peptide mixture is analyzed by 2D-LC/nD-LC-MS/MS. The obtained dataset is used to identify proteins by means of database searches using dedicated bioinformatic tools. | |
| Stable Isotope Labelling with Amino acids in Cell culture | SILAC | Method for relative protein quantification based on metabolic protein labelling and MS analysis. Proteomes extracted from cells in specific physiological/pathological states are labelled *in vivo* during cell growth introducing an isotopically labelled amino acid ($^{13}C$, $^{15}N$) in the culture medium. Proteomes from unlabelled and labelled cells are mixed and analyzed by gel-based or shotgun proteomic approaches. Relative quantification is achieved comparing the intensities of ions generated by the same labelled and unlabelled peptide in mass spectra. | Ong et al. 2002 |
| Isotope-Coded Affinity Tag | ICAT | Method for relative protein quantification based on chemical protein labelling and MS analysis. Proteomes that have to be compared are labelled via a chemical reaction using a polyfunctional reagent, synthetized in two forms, light (8 hydrogen atoms) or heavy (8 deuterium atoms), that links protein cysteine residues. Labelled proteomes are mixed and analyzed by shotgun proteomics. Relative quantification is achieved comparing the intensities of ions generated by the same labelled and unlabelled peptide in mass spectra. | Gygi et al. 1999 |

*Table 1. contd....*

*Table 1. contd.*

| Methodology | Abbreviation | Definition | Main References |
|---|---|---|---|
| Isobaric Tags for Relative and Absolute Quantitation | iTRAQ | Method for relative/absolute protein quantification based on chemical peptide labelling and MS analysis. Peptide mixtures, obtained from the hydrolysis of proteomes that have to be compared, are labelled with isobaric reagents targeting the peptide N-terminus and the epsilon-amino group of lysine residues, mixed and analyzed by LC-MS/MS. Relative quantification is performed comparing the intensities of specific fragment ions in the MS/MS spectra. iTRAQ allows multiplexed analysis of up to eight different samples. | Ross et al. 2004 |
| Absolute QUAntification | AQUA | Method for absolute protein quantification based on the use of synthetic stable-isotope-labelled peptides (proteotypic peptides representative of proteins of interest) and MS analysis. Proteotypic peptides of known concentration are spiked into peptide or protein samples to determine the absolute amount of the natural counterpart by comparing the intensity of ions generated by the corresponding labelled and unlabelled peptides in mass spectra. This is a so-called targeted method and requires preliminary information on chemical-physical features of the target protein to design the AQUA experiment. | Gerber et al. 2003 |
| Artificial gene encoding a concatenation of tryptic peptides | QconCAT | QconCAT and PSAQ are variants of the AQUA methodology. QconCAT uses as standard for protein quantification synthetic metabolically isotope-labelled proteins in which more standard peptides are concatenated. | Beynon et al. 2005 |
| Protein Standard Absolute Quantification | PSAQ | PSAQ uses full-length isotope-labelled recombinant proteins as standard for protein quantification. | Brun et al. 2007 |
| Label-free methodologies | | Methods for quantitative analyses based exclusively on LC-MS/MS data that do not require any metabolic or chemical labelling step. Quantitative analysis is performed assuming a correlation between peak areas (or signal intensities) of precursor ions originated by peptides in LC-MS/MS experiments and the abundance of protein generating those peptides. Alternatively, those methods take into account the number of MS/MS spectra generated by peptides assigned to a protein (spectral counting methods). | Grossmann et al. 2010<br>Lundgren et al. 2010<br>Neilson et al. 2011 |

## 2. Mass Spectrometry Methodologies in Proteomics

### 2.1 The Classical Approach: 2-DE and Protein Identification by Mass Spectrometry

Two-dimensional electrophoresis (2-DE) is a powerful separation technique that allows resolving and visualizing complex protein mixtures extracted from biological systems. In classical 2-DE experiments, proteins are separated according to their isoelectric point in the first dimension by isoelectric focusing (IEF) and, orthogonally, according to their molecular mass by sodium dodecyl sulphate—polyacrylamide gel electrophoresis (SDS-PAGE) in the second dimension which finally produces 2-DE proteomic maps. The maps obtained from samples extracted from biological systems under different physiological and/or pathological states are then compared using specific imaging analysis software in order to detect spots showing quantitative differences in their intensities, likely originated from differentially expressed proteins (Fig. 1 Flow chart of the two main strategies applied in proteomics). Proteins contained in spots of interest are submitted to in gel-digestion (commonly using trypsin) and identified by analyzing the obtained peptide mixtures by MS. The most simple and rapid approach, termed peptide mass fingerprint (PMF), relies on the analysis of peptide mixtures by matrix assisted laser desorption ionization-time of flight-mass spectrometry (MALDI-TOF-MS). MALDI-TOF-MS data (i.e., lists of mass/charge

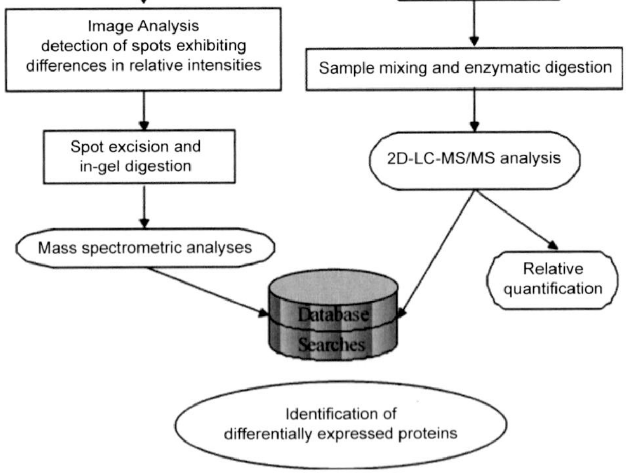


**Fig. 1.** Flow chart of the two main strategies applied in proteomics.
*This step could be ruled out performing relative quantification by label-free methods.

(m/z) values extrapolated from mass spectra) are used to identify proteins by querying specific protein sequence databases (NCBInr, Uniprot, etc.) with search engines such as Mascot (www.matrixscience.com, Eng et al. 1994) or Protein Prospector (http://prospector.ucsf.edu, Clauser et al. 1999). Experimental m/z values are compared with theoretical m/z values obtained from the virtual digestion of all the proteins contained in the database and proteins are identified based on the number of matched peptides taking into account a specific mass accuracy. A score value above an arbitrary confidence threshold indicates reliable identifications (Henzel et al. 1993, Pappin et al. 1993).

Alternatively, peptide mixtures are analyzed by liquid chromatography coupled to tandem mass spectrometry (LC-MS/MS). In a MS/MS experiment, precursor ions originated from peptides chromatographically separated, are selected and fragmented into a collision cell. Typically, two complementary ion series are formed by the cleavage of amide bonds along the polypeptide chain, named *b* and *y*, which contain the N and C terminus of the peptide respectively. The dataset that includes both MS (m/z value of the precursor ions) and MS/MS data (fragmentation mass spectra) produced during the entire analysis is used for protein identification by means of database searches using various bioinformatic tools (Eng et al. 1994, Clauser et al. 1999). This approach also allows the identification of proteins from organisms whose genomes are not completely sequenced yet.

One of the most notable improvements in 2-DE was the development of 2-D difference gel electrophoresis (2D-DIGE) (Unlü et al. 1997, Lilley and Friedman 2004) in which two different samples are labelled with different fluorescent dyes and, together with a labelled standard, analyzed on a single 2-DE gel, thus avoiding problems of inter gel variability and improving quantitative analysis.

The major drawbacks of the classical approach regard the limits of 2-DE in the analysis of hydrophobic proteins (membrane proteins), low abundance proteins, or samples that contain proteins with concentrations in a large dynamic range (Wilkins et al. 1998, Corthals et al. 2000, Rabilloud et al. 2010). Notwithstanding its limits the classical approach remains the method of choice in many proteomics studies mainly in bacterial proteomics (Hecker et al. 2008, Curreem et al. 2012).

### 2.2 Gel-free Proteomic Approaches

In the last ten years, technological advancements in mass spectrometric instruments (introduction of hybrid mass spectrometers, more accurate and high resolution mass analyzers, nanoelectrospray ionization sources, etc.) (Yates 2004, Domon and Aebersold 2006, Nilsson et al. 2010, Thelen and Miernyk 2012) and improvements in LC systems (nanoHPLC, UPLC, etc.) (Holcapek et al. 2012) opened the way for the development of gel-free proteomic approaches, also named shotgun proteomics. In such approaches the entire proteome of a biological system is enzymatically digested and the resulting peptide mixture directly analyzed by nanoLC-MS/MS, thus overcoming the 2-DE limits. The obtained dataset is used to identify all the proteins of the proteome by means of database searches using dedicated bioinformatic tools. As the obtained peptide mixtures are very complex (thousands of peptides), their analysis requires the application of two- or multi-dimensional LC separation methods, i.e., a combination of strong cation or anion exchange chromatography or high-pH reversed phase chromatography as first separation dimension, followed by reverse phase chromatography as the second separation dimension (multi-dimensional protein identification technology (MudPIT)) (Yates et al. 1999, Washburn et al. 2001, Fournier et al. 2007, Di Palma et al. 2012) (Fig. 1). In order to better detect and characterize PTMs, strategies based on affinity chromatography have been developed that allow the selective enrichment of proteins or peptides modified with specific functional groups, such as immobilized metal affinity chromatography (IMAC) or hydrophilic interaction chromatography (HILIC), specifically designed for phosphoproteome analysis (Thingholm et al. 2009, Zhao and Jensen 2009, Medvedev et al. 2012).

The introduction of shotgun proteomics highlighted the need of novel relative and absolute protein quantification strategies and led to the development of quantitative mass spectrometry-based methodologies (Ong and Mann 2005, Domon and Aebersold 2010, Cox and Mann 2011, Bantscheff et al. 2012). Briefly, two different methodological approaches have been introduced: (i) metabolic or chemical protein/peptide stable-isotope labelling strategies; (ii) label-free quantification methods. Stable-isotope labelling strategies for peptide and protein quantification rely on the assumption that, in the same sample, labelled peptides and proteins exhibited physico-chemical features, including ionization features, very similar to those of their natural counterparts. Therefore, relative and absolute quantification could be achieved by comparing the intensity of ions generated by each couple of labelled and unlabelled peptides. The most used metabolic method is stable isotope labelling with amino acids in cell culture (SILAC), in which protein labelling occurs during cell growth and division introducing isotopically labelled arginine and lysine ($^{13}C$, $^{15}N$) in the culture medium (Ong et al. 2002, Ong 2012). Alternatively, chemical labels may be introduced in proteins or peptides after proteome extraction via reactions that involve specific functional groups. The pioneering method was the isotope-coded affinity tag (ICAT) in which protein cysteine residues were targeted (Gygi et al. 1999). More recently, labelling methods based on isobaric reagents targeting the peptide N-terminus and the epsilon-amino group of lysine residues have been developed (iTRAQ) allowing multiplexed analysis of eight different biological samples (Ross et al. 2004).

Mass spectrometric methods that allow the absolute protein/peptide quantification have also been developed. Synthetic stable-isotope-labelled peptide, termed AQUA (Gerber et al. 2003) or protein standards, termed QconCAT (Beynon et al. 2005) or PSAQ (Brun et al. 2007) of known concentration are spiked into a peptide/protein sample prior to the MS analysis to determine the absolute amount of the natural counterpart.

More recently, label-free methodologies have been introduced to achieve peptide/protein quantification avoiding the use of any metabolic or chemical labelling. Two different approaches were developed: (i) methods based on the correlation between peak areas (or signal intensities) of precursor ions originated by peptides in LC-MS/MS experiments and the abundance of proteins generating those peptides; (ii) methods of spectral counting that take into account the number of MS/MS spectra generated by peptides assigned to a protein in an LC-MS/MS experiment. Drawbacks, capabilities and future prospective of label free quantification have been extensively reviewed (Grossmann et al. 2010, Lundgren et al. 2010, Neilson et al. 2011).

## 3. Proteomics for Studying Molecular Mechanisms of Probiotic Action

### *3.1 Adaptation Mechanisms to Gastro-Intestinal Tract Environment*

In order to exert their biological functions, probiotics should be able to sense the gastro-intestinal tract (GIT) environment, to adhere to human intestinal mucosa and to compete with other microorganisms for nutrients, these being essential features for efficient gut colonization (Fig. 2) and also traits used as criteria in the selection of potential probiotic strains. It is well-known that the two major challenges probiotics have to overcome through GIT passage are exposure to acid pH (≈2) in the stomach and to bile in the small intestine (from 40 to 1 mM). Bile, composed of bile acids (weak organic acids) and bile salts (taurine or glycine conjugates that are deconjugated by bile salt hydrolases (BSHs) in the bacterial cytoplasm), causes intracellular acidification. Proteomics was successfully applied to investigate the molecular basis of probiotic adaptation and response to acid pH and bile exposure, pointing out that different probiotic species adopt similar strategies to face with adverse environmental conditions including: (i) activation of folding machinery and molecular chaperones (overexpression of Clp

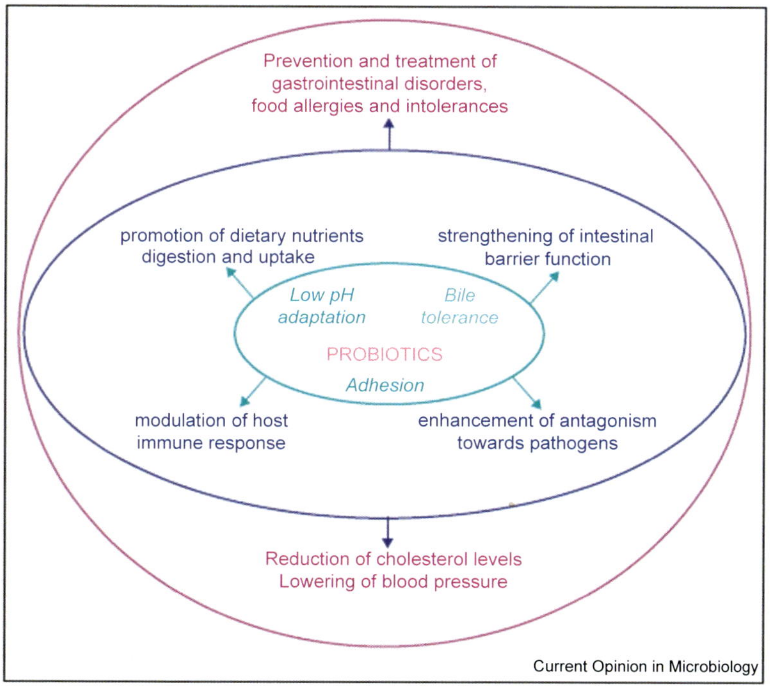


**Fig. 2.** Beneficial effects exerted by probiotics on human health. Reproduced from Siciliano and Mazzeo 2012. Curr. Opin. Microbiol. 15: 1–7 with permission from Elsevier, copyright 2013.

family proteins, GroES, GroEL, DnaK and DnaJ) to counteract protein misfolding, denaturation and aggregation; (ii) changes in the expression level of proteins involved in carbohydrate metabolism, mainly glycolysis, to enhance energy production; (iii) increased activity of proton translocating ATPase ($F_0$-$F_1$-ATPase), an enzyme responsible for mantaining cytoplasmic pH homeostasis in anaerobic microorganisms facilitating proton extrusion (Fig. 3) (Siciliano and Mazzeo 2012). Furthermore, the expression level of several proteins involved in transcription and translation (such as RNA polymerases, elongation factors, ribosomal proteins), nucleotide and amino acid biosynthesis and related to DNA-repair were also influenced by bile and low pH exposure (Sanchez et al. 2008a, Ruiz et al. 2011).

Proteomics highlighted peculiar response mechanisms to bile exposure that were adopted by different species such as the overexpression of ABC-type multidrug transporters, the action of bile efflux pumps for bile salts extrusion (Ctr and BetA in *Bifidobacterium*) (Sanchez et al. 2008a, Ruiz et al. 2009, Koskenniemi et al. 2011,

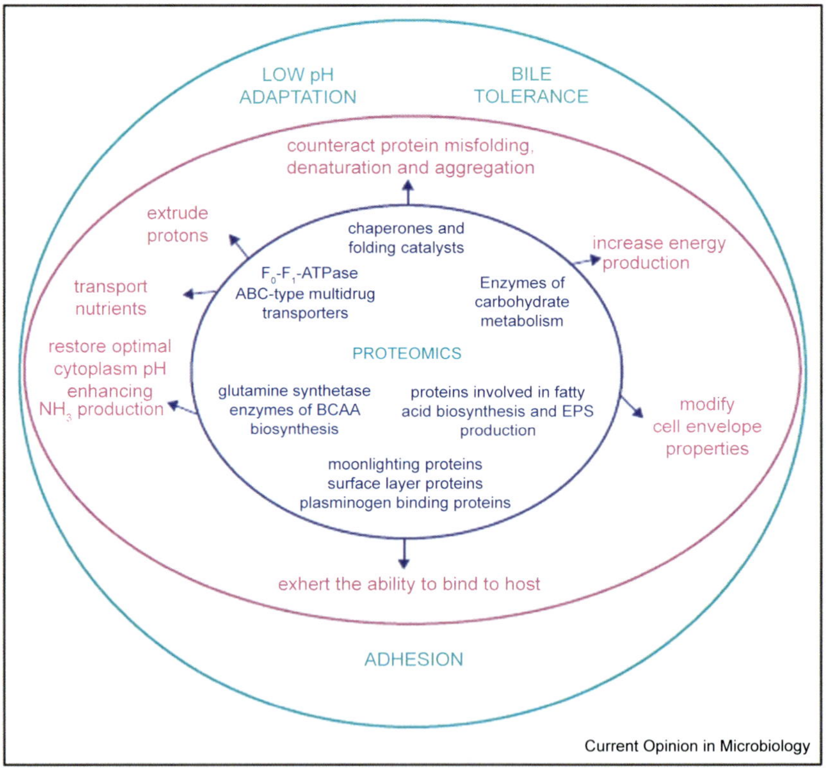


**Fig. 3.** Molecular mechanisms underlying the adaptation and adhesion of probiotics to GIT, as investigated by proteomics. Reproduced from Siciliano and Mazzeo 2012. Curr. Opin. Microbiol. 15: 1–7 with permission from Elsevier, copyright 2013.

Hamon et al. 2011, 2012) and the activation of the bifid shunt, a unique hexose metabolism of *Bifidobacterium* that occurs via the phosphoketolase pathway and leads to an enhanced acetic acid and ATP production (Sanchez et al. 2005, Sanchez et al. 2007a, Ruiz et al. 2011). Bile exposure also triggered response mechanisms involving modifications in cell-envelope and variations in exopolysaccharides (EPS) biosynthesis in *Bifidobacteria* and *Lactobacilli* (Ruiz et al. 2007, Sanchez et al. 2007a, Burns et al. 2010, Hamon et al. 2011, Koskenniemi et al. 2011). In *Bifidobacterium*, proteomic evidence, paralleled with an increased production of EPS in bile tolerant strains, supported the hypothesis that a protective coating of EPS may allow the bacterium to better withstand stomach pH and bile salts (Ruas-Madiedo et al. 2009, Ruiz et al. 2011). On the contrary, in *L. rhamnosus* GG, the repression of genes and underexpression of proteins involved in EPS biosynthesis could be related to a reduction of EPS layer thickness in order to promote bacterial adhesion to gut. The overexpression of proteins participating in lysinylation of membrane lipids and in D-alanyl ester substitution of lipoteichoic acids suggested changes of cell surface charge as a potential defence mechanism to bile exposure (Koskenniemi et al. 2011). Moreover, in *L. rhamnosus* and *L. casei*, bile exposure triggered the overexpression of Nag A and Nag B, proteins involved in the catabolism of N-acetyl-glucosamine, in the early steps of cell-wall biosynthesis (Wu et al. 2010, Koskenniemi et al. 2011, Hamon et al. 2012). The overexpression of these proteins was also induced in *L. casei* Zhang by acid pH exposure (Wu et al. 2011). A more comprehensive proteomic study focused on response mechanisms to acid stress in *L. casei* Zhang, performed integrating 2D-DIGE and iTRAQ methodologies, confirmed common response mechanisms and showed an increased synthesis of specific proteins (MurA and MurG) that participate in peptidoglycan biosynthesis (Wu et al. 2012). In addition, probiotics adopt different metabolic strategies to strengthen cell membrane and reduce its fluidity, either by deactivating the biosynthesis of long-chain saturated fatty acid and/or by increasing the synthesis of cyclic fatty acids (Sanchez et al. 2007a, Koskenniemi et al. 2011, Ruiz et al. 2011). These findings were recently confirmed by both proteomic and transcriptomic analyses in *L. casei* BL 23, which showed a reducted rate of fatty acid biosynthesis induced by bile stress (Alcántara and Zúñiga 2012).

Interestingly, the role of BSHs in *Bifidobacteria* as response mechanism to bile is still controversial. These proteins are responsible for the deconjugation of bile salts and should play a pivotal role in bacterial cell detoxification, as ionized bile acids can be excreted by specific transporters. Although synthesis and activity of BSHs increased in a bile-resistant derivative of *B. animalis* subsp. *lactis*, there is no clear evidence that their expression level was regulated by bile exposure in *Bifidobacteria* (Sanchez et al. 2007a, Ruiz et al. 2011). On the contrary, transcriptomic and proteomic data suggested that, in *Lactobacillus*, the changes of BSHs activity as response to bile salts might be species-specific. In fact, a BSH encoding gene was strongly induced immediately upon bile exposure in *L. rhamnosus* GG (Koskenniemi et al. 2011), while BSH expression was not modified in *L. casei* Zhang (Wu et al. 2010) and it was down regulated in *L. plantarum* (Hamon et al. 2011). Most of those findings were also confirmed by an *in vivo* proteomic study performed on a model for rabbit intestinal culture of *B. longum* (Yuan et al. 2008). A recent study aimed at examining bile tolerance of *L. johnsonii*

PF01 using iTRAQ shotgun proteomics, highlighted the overexpression of three BSHs with different bile salt specificity, suggesting the possibility that *L. johnsonii* was able to hydrolyze all types of bile salts. Notably, this proteomic study led to the identification of a higher number of differentially expressed proteins (215 proteins), compared with gel-based approaches, thus providing a better overview of the complex network of mechanisms adopted by *L. johnsonii* to withstand bile exposure (Lee et al. 2013). Finally, it cannot be ruled out that cells exposed to bile also undergo an oxidative damage, that triggers the expression of proteins responsible in maintaining redox balance, such as thioredoxin, or involved in redox reactions (Sanchez et al. 2007a, Lee et al. 2008a, Wu et al. 2010, Hamon et al. 2011).

Overexpression of glutamine synthetase and enzymes involved in the biosynthesis of branched-chain amino acids (BCAA) has been reported in *B. longum* as response to acid pH. These findings could be related to the restoration of cytoplasm optimal pH, as the biosynthesis of BCCA leads to ammonia production by glutamine synthetase (Sanchez et al. 2007b, Ruiz et al. 2011). However, in *L. rhamnosus* GG, acid stress induced a down regulation of glutamine synthetase and, more importantly, a significant phosphorylation of different proteins mainly involved in carbohydrate metabolism (Koponen et al. 2012). In *Lactobacillus*, proteomic analyses suggested that a similar intracellular pH control could be achieved by activating the arginine deiminase (ADI) pathway, which is absent in probiotic *Bifidobacteria* described so far (De Angelis and Gobbetti 2004, Lee et al. 2008b, Lee and Pi 2010).

### 3.2 Adhesion Mechanisms to the Host Mucosa

Adhesion of probiotics to the intestinal host mucosa is a key feature for the transient colonization of GIT, making probiotic bacteria able to exert their beneficial health effects. Different bacterial species exhibit different colonization strategies and adhesion processes, involving proteinaceous compounds as well as teichoic and lipoteicoic acids, peptidoglycans and EPS. A pivotal role in host-probiotic interactions is played by extracellular proteins that include cell-envelope and secreted proteins having the ability to interact with the mucus layer of the GIT or with extracellular matrix proteins (fibronectin, collagen and laminin) as well as to bind to plasminogen (Plg) (Sanchez et al. 2008b,c, 2010). Proteomics, properly integrated by *in vitro* tests, has significantly contributed to identify proteins directly involved in adhesion processes (Fig. 3) (Beck et al. 2011, Siciliano and Mazzeo 2012). Studies in this field have been also prompted by the development of innovative and specific strategies for membrane and cell-wall proteomics (Solis and Cordwell 2011).

In the last years, proteomics has remarkably helped to assess the presence of cytoplasmic housekeeping proteins in extracellular proteome, acting as adhesion promoting factors. Such proteins, defined as moonlighting proteins, display different seemingly unrelated functions in different cell locations (Jeffery 2003). Most of them lack any extracytoplasmic sorting sequence and although first insights on mechanisms of secretion and cell anchoring have been recently reported, they still remain to be fully understood (Yang et al. 2011). This class of proteins includes several enzymes of the glycolytic pathway (such as enolase (ENO), glyceraldehyde-3-phosphate dehydrogenase (GAPDH), phosphoglycerate mutase (PGM), triosephosphate isomerase), ribosomal proteins, molecular chaperones (such as DnaK and GroEL), translational factors (such as elongation factor-Tu (EF-Tu)) (Sanchez et al. 2008b,c, 2010, Beck et al. 2011). A plasminogen/fibronectin/mucin binding activity has been assessed for GAPDH, ENO, EF-Tu, GroEL and DnaK (Granato et al. 2004, Bergonzelli et al. 2006, Antikainen et al. 2007, Siciliano et al. 2008, Castaldo et al. 2009, Sanchez et al. 2010, Vastano et al. 2013). As these proteins were also involved in the adhesion of pathogens to GIT, the mechanisms used by probiotics to attach to intestinal mucosa mimic those of pathogens, and may thus explain the exclusion of enteropathogenic bacteria by competing for binding sites (Lebeer et al. 2010). In *L. plantarum,* the overexpression of several proteins, including EF-Tu, GAPDH, GroEL, DnaK, GroES and PGM in the cell wall proteoma of the highly adhesive strain *L. plantarum* WHE 92, compared to those of the less adhesive *L. plantarum* 299v and *L. plantarum* CECT 4185, indicated a strict relation between composition of cell-wall proteome and adhesion properties to mucin. In *L. plantarum* WHE 92, an increased expression level of conjugated bile acid hydrolase was observed, suggesting a role of these proteins in the mucin adhesion process (Izquierdo et al. 2009). A proteomic map of surface associated proteins of *L. plantarum* 299v, that included 29 proteins, was also reported by Beck et al. (Beck et al. 2009). Many of these proteins were also identified in the surface-associated proteome of *L. rhamnosus* GG and could have a role in adhesion process (Deepika et al. 2012). In this bacterium, an extensive proteomic study also suggested a pivotal role of a sortase and the pilin elements SpaC and SpaA, involved in the assembly of a pilus structure, in intestinal mucus binding (Savijoki et al. 2011). Interestingly, the first phosphoproteomic study of *L. rhamnosus* GG led to the characterization of phosphorylated forms of proteins participating in central cellular pathways (mainly glycolysis) and of moonlighting proteins (Koponen et al. 2012). As phosphorylation is also implicated in infection and adhesion of pathogens (Jers et al. 2008), these findings bring to light the urgent need for detailed structural studies aimed at clarifying the role of PTMs in modulating the multifunctional biological properties and subcellular localization of moonlighting proteins.

Differential proteomics also demonstrated the involvement of the surface layer protein A (SlpA) of *L. acidiphilus* in adhesion to Caco-2 cells (Ashida et al. 2011), further confirming results obtained from a study on mutant strains (Buck et al. 2005).

It is worth mentioning that the probiotic recruitment of Plg and its subsequent conversion to plasmin plays a central role in host colonization processes. In this way, probiotics acquire a surface-associated and host-derived proteolytic activity useful for facilitating degradation of extracellular matrices and migration across physical and molecular barriers. A functional cell wall proteomic study of *B. longum* demonstrated, for the first time, a potential Plg-binding activity for BSHs and glutamine synthetase, as well as for DnaK, ENO, and PGM, already identified as human Plg receptors in different microorganisms, including pathogens (Candela et al. 2007). A thorough investigation of the secretome and membrane proteome of *B. animalis* subsp. *lactis* BB-12 led to the identification of several factors involved in bacteria-host interactions (Gilad et al. 2011, 2012) thus integrating data previously reported for *B. longum* (Sanchez et al. 2008b,c, Ruiz et al. 2009). Among them, proteins BIF_00998 and BIF_00999 showed high homology with proteins known to participate in fimbriae formation (long fibrous structures which serve as adhesive anchors to facilitate attachment to host cells) and/

or in collagen binding. Furthermore, fibronectin-binding protein BIF_01178 was potentially involved in fibronectin binding and/or fimbriae formation (Gilad et al. 2011, 2012).

It should be underlined that several proteins involved in the adhesion mechanisms were also implicated in acid pH and bile adaptation (Ruiz et al. 2011). For instance, bile exposure caused the upregulation of Plg receptors such as DnaK, ENO and PGM in *B. animalis* subsp. *lactis* BI07 (Candela et al. 2010). Similarly, bile promoted changes in the cell envelope proteome of *B. longum*, enhancing the expression level of glutamine synthetase, EF-Tu, ENO and GAPDH (Ruiz et al. 2009). These findings confirmed the potential role of bile as a key signal triggering the probiotic-host crosstalk and then the colonization process.

### 3.3 Molecular Mechanisms of Probiotic Health Effects

Prophylactic and/or therapeutic effects of probiotics in several gastrointestinal disorders such as inflammatory bowel diseases, antibiotic associated-diarrhoea, neonatal necrotizing enterocolitis, irritable bowel syndrome, *Helicobacter pylori* infection, etc., as well as food allergies and intolerances, have been well-documented to date. Moreover, reduction of cholesterol levels and lowering of blood pressure have also been associated to probiotic consumption (Ooi and Liong 2010, Yan and Polk 2010). However the molecular basis of their modes of action and identity of effector molecules (peptidoglycan, teichoic acid, cell surface polysaccharides, extracellular proteins) underlying the strain-specific clinical effects of probiotics remain to be fully understood.

Proteomics, suitably supported by *in vitro* and *in vivo* analyses, has contributed to identify and structurally characterize extracellular proteins mediating probiotic immunomodulatory mechanisms. The analysis of the extracellular proteome of *B. animalis* BB12 led to the identification of six proteins with potential immunogenic features. Sequence alignment showed that BIF_00825 and BIF_01398 shared similar regions near the C-terminus with p40, the surface antigen protein of *L. rhamnosus* GG, while a cell wall-associated hydrolase (BIF_01914) was homologous to the cell wall-associated glycoside hydrolase of *L. rhamnosus* GG p75 (NLP/P60 protein). Both p40 and p75 were able to stimulate the activation of Akt (protein kinase B protein family) and inhibit TNF-induced apoptosis in intestinal epithelial cells (Yan et al. 2007, Gilad et al. 2011). Recently p75, renamed Msp1, was shown to be O-glycosylated. Glycosylation did not appear to be crucial for Akt activation even if a possible role of the glycan chains could not be ruled out. However, glycan components might act to shield host cell receptors, possibly reducing the capacity of Msp1 to interact with intestinal epithelial cells, thus modulating the extent of Akt signalling (Lebeer et al. 2012). Although glycosylation of surface proteins in probiotics has been so far much less documented compared to pathogens, it holds important potential as it can mediate very specific interactions in microbe-host signalling, especially with the host innate immune system (Lebeer et al. 2010).

In *L. johnsonii* NCC533, moonlighting proteins EF-Tu and GroEL were able to stimulate interleukin-8 secretion in macrophages and HT29 cells in a monocyte ($CD14^+$) mediated proinflammatory response (Granato et al. 2004, Bergonzelli et al. 2006), thus suggesting that immunomodulation could be included in the multitasking functions of moonlighting proteins.

Recently, the overexpression of GroES and GroEL, cell-wall associated moonlighting proteins of *L. acidophilus,* induced by the pH decrease in culture media, was suggested to affect the release of particular cytokines from splenocytes, with an increase of IL-12, IFN-γ and IL-10, and a decrease of IL-4 (Kuwana and Yamamoto 2012).

The S-layer protein SlpA of *L. acidophilus* has been identified as the first probiotic bacterial dendritic cells (DCs)—specific ICAM-3-grabbing nonintegrin (DC-SIGN) ligand, functionally involved in the modulation of DCs and T cells functions. In fact a knockout mutant of *L. acidophilus* NCFM lacking SlpA showed a reduced capacity to bind to DCs leading to an enhanced proinflammatory cytokine production (Konstantinov et al. 2008). The amount of SlpA in different *L. acidophilus* strains was related to adhesion ability to Caco-2 cells as well as to the release of IL-12 from DCs (Ashida et al. 2011). SlpA was also involved in immunostimulatory activity of *L. helveticus* M92 and immune protection against *Salmonella thyphimurium* infection (Beganovic et al. 2011).

Communication between intestinal bacteria and DCs might be also mediated by immunomodulatory peptides, enclosed in extracellular proteins and secreted by commensal bacteria. In fact, *L. plantarum* secreted a bioactive peptide (STp, homologous to gi|28270057 from *L. plantarum* WCFS1) resistant to intestinal proteolysis, found in the human colonic microenvironment and capable of modulating phenotype and function of human blood-enriched DCs (Bernardo et al. 2012).

Recently, it has been demonstrated that specific soluble factors produced by probiotics, called postbiotics, elicit host immunomodulatory response. These findings open the way for a new generation of nutraceuticals or functional foods, that could become a safer alternative for clinical applications, especially in chronic inflammatory conditions like inflammatory bowel disease, where somministration of probiotics seems to be less effective (Tsilingiri and Rescigno 2013).

Although scientific evidence suggested that the administration of probiotics was effective in improving lipid profiles, including the reduction of serum/plasma total cholesterol, LDL-cholesterol and triglycerides or increment of HDL-cholesterol, metabolic strategies underlying these beneficial effects remain still unclear (Ooi and Liong 2010). Kim and co-workers assessed that changes in the expression level of specific proteins associated with stress responses, translation, and metabolic processes in *L. acidophilus* were related to cholesterol reduction (Kim et al. 2008). A proteomic and genetic study has recently found that the catabolite control protein A (CcpA) could be a novel factor involved in the *L. acidophilus* A4 ability to lower serum cholesterol, mainly affecting the expression level of membrane associated proteins (Lee et al. 2010).

The design of *ad hoc* proteomic studies will be of paramount importance to open new perspectives in the molecular basis of probiotic beneficial effects.

## 4. Proteomics for Studying Human Microbiota Dynamics

Microorganisms in nature are not isolated, they coexist in complex communities specific to each environment and are constituted by thousands of different species. The behaviour of microorganisms cultivated in laboratory could not necessary resemble their lifestyle in natural habitat, being affected by various factors such as competition for nutrients, microbial heterogeneity, peculiar environmental conditions, etc. Furthermore, many microorganisms could not be cultivated using standard techniques (Amann et al. 1995), so their analysis requires cultivation-independent methods such as RNA and DNA sequencing technologies. These evidence and the development of high-throughput shotgun genomic methods, led to the advent of metagenomics, i.e., the analysis of complex microbial communities by direct extraction and cloning of DNA from their natural environment and, therefore, to the release of multiple metagenome sequences or, simply, metagenomes. As for proteomics, metagenomics paved the way to metaproteomics, intended as the large-scale characterization of the entire protein complement of microbiota in environmental samples at a given point in time (Wilmes and Bond 2004).

Human gut is a very complex ecosystem including over a thousand bacterial species (Qin et al. 2010), and approaches integrating metagenomics and metaproteomics hold promise to successfully investigate molecular mechanisms and metabolic processes through which microbes impact on human health (Aires and Butel 2011, Hettich et al. 2012, Siggins et al. 2012). As matter of fact, gut microbiota participates in the digestion process but, most importantly, it has a critical function in the development and response of immune system, while alterations in its composition have been associated to different gastrointestinal diseases (Sekirov et al. 2010). Probiotics tightly interact with gut microbiota, having a role in modulating its composition and/or activity, so that they have the potential to counteract such alterations (Gueimonde and Collado 2012).

The first pioneering metaproteomic work was aimed at characterizing variations in the infant faecal microbiota of two individuals over a five months period. Although several differences in the proteomic maps were detected, it was possible to identify just one protein, having high homology with *B. infantis* and *B. longum* transaldolase (Klaassens et al. 2007). Interestingly, transaldolase was also the first moonlighting protein detected in the extracellular proteome of *B. longum* (Sánchez et al. 2008c) and an important role in mucin adhesion and aggregation process has been recently reported for an extracellular transaldolase from *B. bifidum* (Gonzalez-Rodriguez et al. 2012).

The availability of high-throughput mass spectrometric methodologies and new suitable bioinformatic tools (including the built up of *ad hoc* databases containing human metagenomes (Gill et al. 2006, Qin et al. 2010) and fully sequenced genomes of known gut inhabitants), as well as the set up of protocols for sample preparation, prompted the development of metagenomic and metaproteomic integrated studies in order to decipher gut metaproteome (Cantarel et al. 2011). Up to now, metaproteomics led to the identification and functional classification of thousand proteins of human faecal samples and provided direct evidence on: (i) dominant microbial populations in gut ecosystem; (ii) key functional activities of this microbial community; (iii) human proteins involved in the host-microbiota interactions; (iv) subject-specificity and stability over one-year period of human gut microbiota (Verberkmoes et al. 2009,

Rooijers et al. 2011, Kolmeder et al. 2012). Substantial inter-individual variation and significant anatomic region-related (biogeographic) features in colon have been assessed by metaproteomic profile of human mucosal ecosystem (Li et al. 2011). A thorough metaproteomic investigation carried out on rats showed differences in composition and functional activities among bacterial communities inhabiting the mucus, those associated with the contents of selected intestinal regions and those of faecal samples and highlighted a biogeographical distribution of microbes along the digestive tract, thus leading to conclude that the analysis of faeces samples could reflect the functions of gut microbiota only to a minor extent (Haange et al. 2012).

More recently the integrated metaproteomics/metagenomics approach was applied in the exploitation of dysbiosis of gut microbiota due to gastrointestinal diseases, shedding light on functional differences at the gene and protein level and changes in bacterial community related to Ileum Crohn's disease (ICD). Results highlighted that general processes (including carbohydrate transport and metabolism, energy production and conversion, amino acid/lipid/nucleotide transport and metabolism, defence mechanisms) were deficient in ICD patients (Erickson et al. 2012). The bacterial community composition of mucosal-luminal interface, as analysed by metaproteomics and metagenomics, was also shown to be related to different types of diseases (Crohn's disease and ulcerative colitis) (Presley et al. 2012).

As the first metaproteomic studies undeniably demonstrate the innovative strength of integrated metagenomics and metaproteomics in the analysis of microbial communities, the near future will then witness a boost in studies aimed at investigating gut microbiota dynamics and its alterations due to diseases. This knowledge could conduct to new criteria for the selection of novel probiotic strains exhibiting specific ability in maintaining microbiota homeostasis.

## 5. Conclusions

Proteomics has significantly contributed to decipher the molecular basis of probiotic functionality, providing a global picture of proteins involved in adhesion and adaptation to GIT environment. As many of these proteins were actually implicated in both processes, proteomics further assessed the role of bile as a host signal triggering changes crucial for probiotic activities (Ruiz et al. 2009, Candela et al. 2010).

Furthermore integrated metaproteomic and metagenomic studies are emerging as new powerful tools in exploiting the physiological or pathological dynamics of the gut microbial community.

However, the understanding of how probiotics exert beneficial effects and their species-specific mechanisms of action, notably downstream in the gastrointestinal mucosa, is still open to investigation. The identification of protein patterns closely linked to specific probiotic properties could prompt the development of new and faster *in vitro* assays for the assessment of health effects, currently revealed only through biological tests or clinical trials. This could lead, in the near future, to the definition of molecular biomarkers for the identification and selection of innovative probiotic strains with predictable and improved functionality.

## 6. Acknowledgments

The authors are grateful to Dr. Borja Sanchez for critically reading the manuscript.

## References

Aebersold, R. and M. Mann. 2003. Mass spectrometry-based proteomics. Nature 422: 198–207.

Aires, J. and M.J. Butel. 2011. Proteomics, human gut microbiota and probiotics. Expert Rev. Proteomics 8: 279–288.

Alcántara, C. and M. Zúñiga. 2012. Proteomic and transcriptomic analysis of the response to bile stress of *Lactobacillus casei* BL23. Microbiology 158: 1206–1218.

Altelaar, A.F., J. Munoz and A.J. Heck. 2013. Next-generation proteomics: towards an integrative view of proteome dynamics. Nat. Rev. Genet. 14: 35–48.

Amann, R.I., W. Ludwig and K.H. Schleifer. 1995. Phylogenetic identification and *in situ* detection of individual microbial cells without cultivation. Microbiol. Rev. 59: 143–169.

Angel, T.E., U.K. Aryal, S.M. Hengel, E.S. Baker, R.T. Kelly, E.W. Robinson and R.D. Smith. 2012. Mass spectrometry-based proteomics: existing capabilities and future directions. Chem. Soc. Rev. 41: 3912–3928.

Antikainen, J., V. Kuparinen, K. Lähteenmäki and K.T. Korhonen. 2007. pH-dependent association of enolase and glyceraldehyde-3-phosphate dehydrogenase of *Lactobacillus crispatus* with the cell wall and lipoteichoic acids. J. Bacteriol. 189: 4539–4543.

Ashida, N., S. Yanagihara, T. Shinoda and N. Yamamoto. 2011. Characterization of adhesive molecule with affinity to Caco-2 cells in *Lactobacillus acidophilus* by proteome analysis. J. Biosci. Bioeng. 112: 333–337.

Bantscheff, M., S. Lemeer, M.M. Savitski and B. Kuster. 2012. Quantitative mass spectrometry in proteomics: critical review update from 2007 to the present. Anal. Bioanal. Chem. 404: 939–965.

Beck, H.C., S.M. Madsen, J. Glenting, J. Petersen, H. Israelsen, M.R. Nørrelykke, M. Antonsson and A.M. Hansen. 2009. Proteomic analysis of cell surface-associated proteins from probiotic *Lactobacillus plantarum*. FEMS Microbiol. Lett. 297: 61–66.

Beck, H.C., S. Feddersen and J. Petersen. 2011. Application of probiotic proteomics in enteric cytoprotection. pp. 155–168. *In*: J.J. Malago, J.F.J.G. Koninkx and R. Marinsek-Logar (eds.). Probiotic Bacteria and Enteric Infections. Springer Science + Business Media B.V., Dordrecht, Netherlands.

Beganović, J., J. Frece, B. Kos, A. Leboš Pavunc, K. Habjanič and J. Šušković. 2011. Functionality of the S-layer protein from the probiotic strain *Lactobacillus helveticus* M92. Antonie Van Leeuwenhoek 100: 43–53.

Bergonzelli, G.E., D. Granato, R.D. Pridmore, L.F. Marvin-Guy, D. Donnicola and I.E. Corthésy-Theulaz. 2006. GroEL of *Lactobacillus johnsonii* La1 (NCC 533) is cell surface associated: potential role in interactions with the host and the gastric pathogen *Helicobacter pylori*. Infect. Immun. 74: 425–434.

Bernardo, D., B. Sánchez, H.O. Al-Hassi, E.R. Mann, M.C. Urdaci, S.C. Knight and A. Margolles. 2012. Microbiota/host crosstalk biomarkers: regulatory response of human intestinal dendritic cells exposed to *Lactobacillus* extracellular encrypted peptide. PLoS One 7: e36262.

Beynon, R.J., M.K. Doherty, J.M. Pratt and S.J. Gaskell. 2005. Multiplexed absolute quantification in proteomics using artificial QCAT proteins of concatenated signature peptides. Nat. Methods 2: 587–589.

Brun, V., A. Dupuis, A. Adrait, M. Marcellin, D. Thomas, M. Court, F. Vandenesch and J. Garin. 2007. Isotope-labelled protein standards: toward absolute quantitative proteomics. Mol. Cell Proteomics 6: 2139–2149.

Buck, B.L., E. Altermann, T. Svingerud and T.R. Klaenhammer. 2005. Functional analysis of putative adhesion factors in *Lactobacillus acidophilus* NCFM. Appl. Environ. Microbiol. 71: 8344–8351.

Burns, P., B. Sánchez, G. Vinderola, P. Ruas-Madiedo, L. Ruiz, A. Margolles, J. Reinheimer and C.G. de los Reyes-Gavilán. 2010. Inside the adaptation process of *Lactobacillus delbrueckii* subsp. *lactis* to bile. Int. J. Food Microbiol. 142: 132–141.

Candela, M., S. Bergmann, M. Vici, B. Vitali, S. Turroni, B.J. Eikmanns, S. Hammerschmidt and P. Brigidi. 2007. Binding of human plasminogen to *Bifidobacterium*. J. Bacteriol. 189: 5929–5936.

Candela, M., M. Centanni, J. Fiori, E. Biagi, S. Turroni, C. Orrico, S. Bergmann, S. Hammerschmidt and P. Brigidi. 2010. DnaK from *Bifidobacterium animalis* subsp. *lactis* is a surface-exposed human plasminogen receptor upregulated in response to bile salts. Microbiology 156: 1609–1618.

Cantarel, B.L., A.R. Erickson, N.C. VerBerkmoes, B.K. Erickson, P.A. Carey, C. Pan, M. Shah, E.F. Mongodin, J.K. Jansson, C.M. Fraser-Liggett and R.L. Hettich. 2011. Strategies for metagenomic-guided whole-community proteomics of complex microbial environments. PLoS One 6: e27173.

Castaldo, C., V. Vastano, R.A. Siciliano, M. Candela, M. Vici, L. Muscariello, R. Marasco and M. Sacco. 2009. Surface displaced alfa-enolase of *Lactobacillus plantarum* is a fibronectin binding protein. Microb. Cell Fact. 8: 14.

Chao, T.C. and N. Hansmeier. 2012. The current state of microbial proteomics: where we are and where we want to go. Proteomics 12: 638–650.

Clauser, K.R., P.R. Baker and A.L. Burlingame. 1999. Role of accurate mass measurement (+/– 10 ppm) in protein identification strategies employing MS or MS/MS and database searching. Anal. Chem. 71: 2871–2882.

Corthals, G.L., V.C. Wasinger, D.F. Hochstrasser and J.C. Sanchez. 2000. The dynamic range of protein expression: a challenge for proteomic research. Electrophoresis 21: 1104–1115.

Cox, J. and M. Mann. 2011. Quantitative, high-resolution proteomics for data-driven systems biology. Annu. Rev. Biochem. 80: 273–299.

Curreem, S.O.T., R.M. Watt, S.K.P. Lau and P.C.Y. Woo. 2012. Two-dimensional gel electrophoresis in bacterial proteomics. Protein Cell 3: 346–363.

De Angelis, M. and M. Gobbetti. 2004. Environmental stress responses in *Lactobacillus*: a review. Proteomics 4: 106–122.

Deepika, G., E. Karunakaran, C.R. Hurley, C.A. Biggs and D. Charalampopoulos. 2012. Influence of fermentation conditions on the surface properties and adhesion of *Lactobacillus rhamnosus* GG. Microb. Cell Fact. 11: 116.

Di Cagno, R., M. De Angelis, M. Calasso and M. Gobbetti. 2011. Proteomics of the bacterial cross-talk by quorum sensing. J. Proteomics 74: 19–34.

Di Palma, S., M.L. Hennrich, A.J. Heck and S. Mohammed. 2012. Recent advances in peptide separation by multidimensional liquid chromatography for proteome analysis. J. Proteomics 75: 3791–3813.

Domon, B. and R. Aebersold. 2006. Mass spectrometry and protein analysis. Science 312: 212–217.

Domon, B. and R. Aebersold. 2010. Options and considerations when selecting a quantitative proteomics strategy. Nat. Biotechnol. 28: 710–721.

Eng, J.K., A.L. McCormack and J.R. III Yates. 1994. An approach to correlate tandem mass spectral data of peptides with amino acid sequences in a protein database. J. Am. Soc. Mass Spectrom. 5: 976–989.

Erickson, A.R., B.L. Cantarel, R. Lamendella, Y. Darzi, E.F. Mongodin, C. Pan, M. Shah, J. Halfvarson, C. Tysk, B. Henrissat, J. Raes, N.C. Verberkmoes, C.M. Fraser, R.L. Hettich and J.K. Jansson. 2012. Metagenomics/metaproteomics reveals human host-microbiota signatures of Crohn's disease. PLoS One 7: e49138.

FAO/WHO. 2006. Probiotics in food. Health and nutritional properties and guidelines for evaluation. FAO food and nutrition paper no. 85, ISBN 92-5-105513-0.

Fournier, M.L., J.M. Gilmore, S.A. Martin-Brown and M.P. Washburn. 2007. Multidimensional separations-based shotgun proteomics. Chem. Rev. 107: 3654–3686.

Gerber, S.A., J. Rush, O. Stemman, M.W. Kirschner and S.P. Gygi. 2003. Absolute quantification of proteins and phosphoproteins from cell lysates by tandem MS. Proc. Natl. Acad. Sci. U.S.A. 100: 6940–6945.

Gilad, O., B. Svensson, A.H. Viborg, B. Stuer-Lauridsen and S. Jacobsen. 2011. The extracellular proteome of *Bifidobacterium animalis* subsp. *lactis* BB-12 reveals proteins with putative roles in probiotic effects. Proteomics 11: 2503–2514.

Gilad, O., K. Hjernø, E.C. Osterlund, A. Margolles, B. Svensson, B. Stuer-Lauridsen, A.L. Møller and S. Jacobsen. 2012. Insights into physiological traits of *Bifidobacterium animalis* subsp. *lactis* BB-12 through membrane proteome analysis. J. Proteomics 75: 1190–1200.

Gill, S.R., M. Pop, R.T. Deboy, P.B. Eckburg, P.J. Turnbaugh, B.S. Samuel, J.I. Gordon, D.A. Relman, C.M. Fraser-Liggett and K.E. Nelson. 2006. Metagenomic analysis of the human distal gut microbiome. Science 312: 1355–1359.

González-Rodríguez, I., B. Sánchez, L. Ruiz, F. Turroni, M. Ventura, P. Ruas-Madiedo, M. Gueimonde and A. Margolles. 2012. Role of extracellular transaldolase from *Bifidobacterium bifidum* in mucin adhesion and aggregation. Appl. Environ. Microbiol. 78: 3992–3998.

Graham, C., G. McMullan and R.L. Graham. 2011. Proteomics in the microbial sciences. Bioeng. Bugs. 2: 17–30.

Granato, D., G.E. Bergonzelli, R.D. Pridmore, L. Marvin, M. Rouvet and I.E. Corthésy-Theulaz. 2004. Cell surface-associated elongation factor Tu mediates the attachment of *Lactobacillus johnsonii* NCC533 (La1) to human intestinal cells and mucins. Infect. Immun. 72: 2160–2169.

Grossmann, J., B. Roschitzki, C. Panse, C. Fortes, S. Barkow-Oesterreicher, D. Rutishauser and R. Schlapbach. 2010. Implementation and evaluation of relative and absolute quantification in shotgun proteomics with label-free methods. J. Proteomics 73: 1740–1746.

Gueimonde, M. and M.C. Collado. 2012. Metagenomics and probiotics. Clin. Microbiol. Infect. 18: 32–34.

Gygi, S.P., B. Rist, S.A. Gerber, F. Turecek, M.H. Gelb and R. Aebersold. 1999. Quantitative analysis of complex protein mixtures using isotope-coded affinity tags. Nat. Biotechnol. 17: 994–999.

Haange, S.B., A. Oberbach, N. Schlichting, F. Hugenholtz, H. Smidt, M. von Bergen, H. Till and J. Seifert. 2012. Metaproteome analysis and molecular genetics of rat intestinal microbiota reveals section and localization resolved species distribution and enzymatic functionalities. J. Proteome Res. 11: 5406–5417.

Hamon, E., P. Horvatovich, E. Izquierdo, F. Bringel, E. Marchioni, D. Aoudé-Werner and S. Ennahar. 2011. Comparative proteomic analysis of *Lactobacillus plantarum* for the identification of key proteins in bile tolerance. BMC Microbiol. 11: 63.

Hamon, E., P. Horvatovich, M. Bisch, F. Bringel, E. Marchioni, D. Aoudé-Werner and S. Ennahar. 2012. Investigation of biomarkers of bile tolerance in *Lactobacillus casei* using comparative proteomics. J. Proteome Res. 11: 109–118.

Hecker, M., H. Antelmann, K. Büttner and J. Bernhardt. 2008. Gel-based proteomics of Gram-positive bacteria: A powerful tool to address physiological questions. Proteomics 8: 4958–4975.

Henzel, W.J., T.M. Billeci, J.T. Stults, S.C. Wong, C. Grimley and C. Watanabe. 1993. Identifying proteins from two-dimensional gels by molecular mass searching of peptide fragments in protein sequence databases. Proc. Natl. Acad. Sci. U.S.A. 90: 5011–5015.

Hettich, R.L., R. Sharma, K. Chourey and R.J. Giannone. 2012. Microbial metaproteomics: identifying the repertoire of proteins that microorganisms use to compete and cooperate in complex environmental communities. Curr. Opin. Microbiol. 15: 373–380.

Holčapek, M., R. Jirásko and M. Lísa. 2012. Recent developments in liquid chromatography-mass spectrometry and related techniques. J. Chromatogr. A. 1259: 3–15.

Izquierdo, E., P. Horvatovich, E. Marchioni, D. Aoude-Werner, Y. Sanz and S. Ennahar. 2009. 2-DE and MS analysis of key proteins in the adhesion of *Lactobacillus plantarum*, a first step toward early selection of probiotics based on bacterial biomarkers. Electrophoresis 30: 949–956.

Jeffery, C.J. 2003. Moonlighting proteins: old proteins learning new tricks. Trends Genet. 19: 415–417.

Jers, C., B. Soufi, C. Grangeasse, J. Deutscher and I. Mijakovic. 2008. Phosphoproteomics in bacteria: towards a systemic understanding of bacterial phosphorylation networks. Expert Rev. Proteomics 5: 619–627.

Karas, M. and F. Hillenkamp. 1988. Laser desorption ionization of proteins with molecular masses exceeding 10,000 daltons. Anal. Chem. 60: 2299–2301.

Kim, Y., J.Y. Whang, K.Y. Whang, S. Oh and S.H. Kim. 2008. Characterization of the cholesterol-reducing activity in a cell-free supernatant of *Lactobacillus acidophilus* ATCC 43121. Biosci. Biotechnol. Biochem. 72: 1483–1490.

Klaassens, E.S., W.M. de Vos and E.E. Vaughan. 2007. Metaproteomics approach to study the functionality of the microbiota in the human infant gastrointestinal tract. Appl. Environ. Microbiol. 73: 1388–1392.

Klose, J. 1975. Protein mapping by combined isoelectric focusing and electrophoresis of mouse tissues. A novel approach to testing for induced point mutations in mammals. Humangenetik 26: 231–243.

Kolmeder, C.A., M. de Been, J. Nikkilä, I. Ritamo, J. Mättö, L. Valmu, J. Salojärvi, A. Palva, A. Salonen and W.M. de Vos. 2012. Comparative metaproteomics and diversity analysis of human intestinal microbiota testifies for its temporal stability and expression of core functions. PLoS One 7: e29913.

Konstantinov, S.R., H. Smidt, W.M. de Vos, S.C. Bruijns, S.K. Singh, F. Valence, D. Molle, S. Lortal, E. Altermann, T.R. Klaenhammer and Y. van Kooyk. 2008. S layer protein A of *Lactobacillus acidophilus* NCFM regulates immature dendritic cell and T cell functions. Proc. Natl. Acad. Sci. U.S.A. 105: 19474–19479.

Koponen, J., K. Laakso, K. Koskenniemi, M. Kankainen, K. Savijoki, T.A. Nyman, W.M. de Vos, S. Tynkkynen, N. Kalkkinen and P. Varmanen. 2012. Effect of acid stress on protein expression and phosphorylation in *Lactobacillus rhamnosus* GG. J. Proteomics 75: 1357–1374.

Koskenniemi, K., K. Laakso, J. Koponen, M. Kankainen, D. Greco, P. Auvinen, K. Savijoki, T.A. Nyman, A. Surakka, T. Salusjärvi, W.M. de Vos, S. Tynkkynen, N. Kalkkinen and P. Varmanen. 2011. Proteomics and transcriptomics characterization of bile stress response in probiotic *Lactobacillus rhamnosus* GG. Mol. Cell Proteomics 10: M110.002741.

Kuwana, R. and N. Yamamoto. 2012. Increases in GroES and GroEL from *Lactobacillus acidophilus* L-92 in response to a decrease in medium pH, and changes in cytokine release from splenocytes: Transcriptome and proteome analyses. J. Biosci. Bioeng. 114: 9–16.

Lebeer, S., J. Vanderleyden and S.C. De Keersmaecker. 2010. Host interactions of probiotic bacterial surface molecules: comparison with commensals and pathogens. Nat. Rev. Microbiol. 8: 171–184.

Lebeer, S., I.J. Claes, C.I. Balog, G. Schoofs, T.L. Verhoeven, K. Nys, I. von Ossowski, W.M. de Vos, H.L. Tytgat, P. Agostinis, A. Palva, E.J. Van Damme, A.M. Deelder, S.C. De Keersmaecker, M. Wuhrer and J. Vanderleyden. 2012. The major secreted protein Msp1/p75 is O-glycosylated in *Lactobacillus rhamnosus* GG. Microb. Cell Fact 11: 15.

Lee, K., H.G. Lee and Y.J. Choi. 2008a. Proteomic analysis of the effect of bile salts on the intestinal and probiotic bacterium *Lactobacillus reuteri*. J. Biotechnol. 137: 14–19.

Lee, K., H.G. Lee, K. Pi and Y.J. Choi. 2008b. The effect of low pH on protein expression by the probiotic bacterium *Lactobacillus reuteri*. Proteomics 8: 1624–1630.

Lee, K. and K. Pi. 2010. Effect of transient acid stress on the proteome of intestinal probiotic *Lactobacillus reuteri*. Biochemistry (Mosc) 75: 460–465.

Lee, J., Y. Kim, H.S. Yun, J.G. Kim, S. Oh and S.H. Kim. 2010. Genetic and proteomic analysis of factors affecting serum cholesterol reduction by *Lactobacillus acidophilus* A4. Appl. Environ. Microbiol. 76: 4829–4835.

Lee, J.Y., E.A. Pajarillo, M.J. Kim, J.P. Chae and D.K. Kang. 2013. Proteomic and transcriptional analysis of *Lactobacillus johnsonii* PF01 during bile salt exposure by iTRAQ shotgun proteomics and quantitative RT-PCR. J. Proteome Res. 12: 432–443.

Li, X., J. LeBlanc, A. Truong, R. Vuthoori, S.S. Chen, J.L. Lustgarten, B. Roth, J. Allard, A. Ippoliti, L.L. Presley, J. Borneman, W.L. Bigbee, V. Gopalakrishnan, T.G. Graeber, D. Elashoff, J. Braun and L. Goodglick. 2011. A metaproteomic approach to study human-microbial ecosystems at the mucosal luminal interface. PLoS One 6: e26542.

Lilley, K.S. and D.B. Friedman. 2004. All about DIGE: Quantification technology for differential-display 2D-gel proteomics. Expert Rev. Proteomics 1: 401–409.

Lundgren, D.H., S.I. Hwang, L. Wu and D.K. Han. 2010. Role of spectral counting in quantitative proteomics. Expert Rev. Proteomics 7: 39–53.

Malmström, L., J. Malmström and R. Aebersold. 2011. Quantitative proteomics of microbes: Principles and applications to virulence. Proteomics 11: 2947–2956.

May, C., F. Brosseron, P. Chartowski, C. Schumbrutzki, B. Schoenebeck and K. Marcus. 2011. Instruments and methods in proteomics. Methods Mol. Biol. 696: 3–26.

Medvedev, A., A. Kopylov, O. Buneeva, V. Zgoda and A. Archakov. 2012. Affinity-based proteomic profiling: Problems and achievements. Proteomics 12: 621–637.

Munday, D.C., R. Surtees, E. Emmott, B.K. Dove, P. Digard, J.N. Barr, A. Whitehouse, D. Matthews and J.A. Hiscox. 2012. Using SILAC and quantitative proteomics to investigate the interactions between viral and host proteomes. Proteomics 12: 666–672.

Neidhart, F.C. 2011. How microbial proteomics got started. Proteomics 11: 2943–2946.

Neilson, K.A., N.A. Ali, S. Muralidharan, M. Mirzaei, M. Mariani, G. Assadourian, A. Lee, S.C. van Sluyter and P.A. Haynes. 2011. Less label, more free: approaches in label-free quantitative mass spectrometry. Proteomics 11: 535–553.

Nikolov, M., C. Schmidt and H. Urlaub. 2012. Quantitative mass spectrometry-based proteomics: an overview. Methods Mol. Biol. 893: 85–100.

Nilsson, T., M. Mann, R. Aebersold, J.R. III Yates, A. Bairock and J.J.M. Bergeron. 2010. Mass spectrometry in high-throughput proteomics: ready for the big time. Nat. Methods 7: 681–685.

O'Farrell, P.H. 1975. High resolution two-dimensional electrophoresis of proteins. J. Biol. Chem. 250: 4007–4021.

Ong, S.E., B. Blagoev, I. Kratchmarova, D.B. Kristensen, H. Steen, A. Pandey and M. Mann. 2002. Stable isotope labeling by amino acids in cell culture, SILAC, as a simple and accurate approach to expression proteomics. Mol. Cell Proteomics 1: 376–386.

Ong, S.E. and M. Mann. 2005. Mass spectrometry-based proteomics turns quantitative. Nat. Chem. Biol. 1: 252–262.

Ong, S.E. 2012. The expanding field of SILAC. Anal. Bioanal. Chem. 404: 967–976.

Ooi, L.G. and M.T. Liong. 2010. Cholesterol-lowering effects of probiotics and prebiotics: a review of *in vivo* and *in vitro* findings. Int. J. Mol. Sci. 11: 2499–2522.

Otto, A., J. Bernhardt, M. Hecker and D. Becher. 2012. Global relative and absolute quantitation in microbial proteomics. Curr. Opin. Microbiol. 15: 364–372.

Pappin, D.J., P. Hojrup and A.J. Bleasby. 1993. Rapid identification of proteins by peptide-mass fingerprinting. Curr. Biol. 3: 327–332.

Presley, L.L., J. Ye, X. Li, J. Leblanc, Z. Zhang, P.M. Ruegger, J. Allard, D. McGovern, A. Ippoliti, B. Roth, X. Cui, D.R. Jeske, D. Elashoff, L. Goodglick, J. Braun and J. Borneman. 2012. Host-microbe relationships in inflammatory bowel disease detected by bacterial and metaproteomic analysis of the mucosal-luminal interface. Inflamm. Bowel Dis. 18: 409–417.

Qin, J., R. Li, J. Raes, M. Arumugam, K.S. Burgdorf, C. Manichanh, T. Nielsen, N. Pons, F. Levenez, T. Yamada, D.R. Mende, J. Li, J. Xu, S. Li, D. Li, J. Cao, B. Wang, H. Liang, H. Zheng, Y. Xie, J. Tap, P. Lepage, M. Bertalan, J.M. Batto, T. Hansen, D. Le Paslier, A. Linneberg, H.B. Nielsen, E. Pelletier, P. Renault, T. Sicheritz-Ponten, K. Turner, H. Zhu, C. Yu, S. Li, M. Jian, Y. Zhou, Y. Li, X. Zhang, S. Li, N. Qin, H. Yang, J. Wang, S. Brunak, J. Doré, F. Guarner, K. Kristiansen, O. Pedersen, J. Parkhill, J. Weissenbach, MetaHIT Consortium, P. Bork, S.D. Ehrlich and J. Wang. 2010. A human gut microbial gene catalogue established by metagenomic sequencing. Nature 464: 59–65.

Rabilloud, T., M. Chevallet, S. Luche and C. Lelong. 2010. Two-dimensional gel electrophoresis in proteomics: past, present and future. J. Proteomics 73: 2064–2077.

Roepstorff, P. 2012. Mass spectrometry based proteomics, background, status and future needs. Protein Cell 3: 641–647.

Rooijers, K., C. Kolmeder, C. Juste, J. Doré, M. de Been, S. Boeren, P. Galan, C. Beauvallet, W.M. de Vos and P.J. Schaap. 2011. An iterative workflow for mining the human intestinal metaproteome. BMC Genomics 12: 6.

Ross, P.L., Y.N. Huang, J.N. Marchese, B. Williamson, K. Parker, S. Hattan, N. Khainovski, S. Pillai, S. Dey, S. Daniels, S. Purkayastha, P. Juhasz, S. Martin, M. Bartlet-Jones, F. He, A. Jacobson and D.J. Pappin. 2004. Multiplexed protein quantitation in *Saccharomyces cerevisiae* using amine-reactive isobaric tagging reagents. Mol. Cell Proteomics 3: 1154–1169.

Ruas-Madiedo, P., M. Gueimonde, F. Arigoni, C.G. de los Reyes-Gavilán and A. Margolles. 2009. Bile affects the synthesis of exopolysaccharides by *Bifidobacterium animalis*. Appl. Environ. Microbiol. 75: 1204–1207.

Ruiz, L., B. Sánchez, P. Ruas-Madiedo, C.G. de Los Reyes-Gavilán and A. Margolles. 2007. Cell envelope changes in *Bifidobacterium animalis* ssp. *lactis* as a response to bile. FEMS Microbiol. Lett. 274: 316–322.

Ruiz, L., Y. Couté, B. Sánchez, C.G. de los Reyes-Gavilán, J.C. Sanchez and A. Margolles. 2009. The cell-envelope proteome of *Bifidobacterium longum* in an *in vitro* bile environment. Microbiology 155: 957–967.

Ruiz, L., P. Ruas-Madiedo, M. Gueimonde, C.G. de Los Reyes-Gavilán, A. Margolles and B. Sánchez. 2011. How do bifidobacteria counteract environmental challenges? Mechanisms involved and physiological consequences. Genes Nutr. 6: 307–318.

Sánchez, B., M.C. Champomier-Vergès, P. Anglade, F. Baraige, C.G. de Los Reyes-Gavilán, A. Margolles and M. Zagorec. 2005. Proteomic analysis of global changes in protein expression during bile salt exposure of *Bifidobacterium longum* NCIMB 8809. J. Bacteriol. 187: 5799–5808.

Sánchez, B., M.C. Champomier-Vergès, B. Stuer-Lauridsen, P. Ruas-Madiedo, P. Anglade, F. Baraige, C.G. de los Reyes-Gavilán, E. Johansen, M. Zagorec and A. Margolles. 2007a. Adaptation and response of *Bifidobacterium animalis* subsp. *lactis* to bile: a proteomic and physiological approach. Appl. Environ. Microbiol. 73: 6757–6767.

Sánchez, B., M.C. Champomier-Vergès, C.M. Collado, P. Anglade, F. Baraige, Y. Sanz, C.G. de los Reyes-Gavilán, A. Margolles and M. Zagorec. 2007b. Low-pH adaptation and the acid tolerance response of *Bifidobacterium longum* biotype *longum*. Appl. Environ. Microbiol. 73: 6450–6459.

Sanchez, B., L. Ruiz, C.G. de los Reyes-Gavilan and A. Margolles. 2008a. Proteomics of stress response in *Bifidobacterium*. Front. Biosci. 13: 6905–6919.

Sánchez, B., P. Bressollier and M.C. Urdaci. 2008b. Exported proteins in probiotic bacteria: adhesion to intestinal surfaces, host immunomodulation and molecular cross-talking with the host. FEMS Immunol. Med. Microbiol. 54: 1–17.

Sánchez, B., M.C. Champomier-Vergès, P. Anglade, F. Baraige, C.G. de los Reyes-Gavilan, A. Margolles and M. Zagorec. 2008c. A preliminary analysis of *Bifidobacterium longum* exported proteins by two-dimensional electrophoresis. J. Mol. Microbiol. Biotechnol. 14: 74–79.

Sanchez, B., M. Urdaci and A. Margolles. 2010. Extracellular proteins secreted by probiotic bacteria as mediators of effects that promote mucosal-bacteria interactions. Microbiology 156: 3232–3242.

Savijoki, K., N. Lietzén, M. Kankainen, T. Alatossava, K. Koskenniemi, P. Varmanen and T.A. Nyman. 2011. Comparative proteome cataloging of *Lactobacillus rhamnosus* strains GG and Lc705. J. Proteome Res. 10: 3460–3473.

Sekirov, I., S.L. Russell, L.C. Antunes and B.B. Finlay. 2010. Gut microbiota in health and disease. Physiol. Rev. 90: 859–904.

Siciliano, R.A., G. Cacace, M.F. Mazzeo, L. Morelli, M. Elli, M. Rossi and A. Malorni. 2008. Proteomic investigation of the aggregation phenomenon in *Lactobacillus crispatus*. Biochim. Biophys. Acta 1784: 335–342.

Siciliano, R.A. and M.F. Mazzeo. 2012. Molecular mechanisms of probiotic action: a proteomic perspective. Curr. Opin. Microbiol. 15: 390–396.

Siggins, A., E. Gunnigle and F. Abram. 2012. Exploring mixed microbial community functioning: recent advances in metaproteomics. FEMS Microbiol. Ecol. 80: 265–280.

Solis, N. and S.J. Cordwell. 2011. Current methodologies for proteomics of bacterial surface-exposed and cell envelope proteins. Proteomics 11: 3169–3189.

Thelen, J.J. and J.A. Miernyk. 2012. The proteomic future: where mass spectrometry should be taking us. Biochem. J. 444: 169–181.

Thingholm, T.E., O.N. Jensen and M.R. Larsen. 2009. Analytical strategies for phosphoproteomics. Proteomics 9: 1451–1468.

Tsilingiri, K. and M. Rescigno. 2013. Postbiotics: what else? Benef. Microbes 4: 101–107.

Unlü, M., M.E. Morgan and J.S. Minden. 1997. Difference gel electrophoresis: A single gel method for detecting changes in protein extracts. Electrophoresis 18: 2071–2077.

Vastano, V., U. Capri, M. Candela, R.A. Siciliano, L. Russo, M. Renda and M. Sacco. 2013. Identification of binding sites of *Lactobacillus plantarum* enolase involved in the interaction with human plasminogen. Microbiol. Res. 168: 65–72.

Verberkmoes, N.C., A.L. Russell, M. Shah, A. Godzik, M. Rosenquist, J. Halfvarson, M.G. Lefsrud, J. Apajalahti, C. Tysk, R.L. Hettich and J.K. Jansson. 2009. Shotgun metaproteomics of the human distal gut microbiota. ISME J. 3: 179–189.

Walther, T.C. and M. Mann. 2010. Mass spectrometry-based proteomics in cell biology. J. Cell Biol. 190: 491–500.

Washburn, M.P., D. Wolters and J.R. 3rd Yates. 2001. Large-scale analysis of the yeast proteome by multidimensional protein identification technology. Nat. Biotechnol. 19: 242–247.

Wilkins, M.R., C. Pasquali, R.D. Appel, K. Ou, O. Golaz, J.C. Sanchez, J.X. Yan, A.A. Gooley, G. Hughes, I. Humphery-Smith, K.L. Williams and D.F. Hochstrasser. 1996. From proteins to proteomes: large scale protein identification by two-dimensional electrophoresis and amino acid analysis. Biotechnology (NY) 14: 61–65.

Wilkins, M.R., E. Gasteiger, J.C. Sanchez, A. Bairoch and D.F. Hochstrasser. 1998. Two-dimensional gel electrophoresis for proteome projects: the effects of protein hydrophobicity and copy number. Electrophoresis 19: 1501–1505.

Wilmes, P. and P.L. Bond. 2004. The application of two-dimensional polyacrylamide gel electrophoresis and downstream analyses to a mixed community of prokaryotic microorganisms. Environ. Microbiol. 6: 911–920.

Wu, R., Z. Sun, J. Wu, H. Meng and H. Zhang. 2010. Effect of bile salts stress on protein synthesis of *Lactobacillus casei* Zhang revealed by 2-dimensional gel electrophoresis. J. Dairy Sci. 93: 3858–3868.

Wu, R., W. Zhang, T. Sun, J. Wu, X. Yue, H. Meng and H. Zhang. 2011. Proteomic analysis of responses of a new probiotic bacterium *Lactobacillus casei* Zhang to low acid stress. Int. J. Food Microbiol. 147: 181–187.

Wu, C., J. Zhang, W. Chen, M. Wang, G. Du and J. Chen. 2012. A combined physiological and proteomic approach to reveal lactic-acid-induced alterations in *Lactobacillus casei* Zhang and its mutant with enhanced lactic acid tolerance. Appl. Microbiol. Biotechnol. 93: 707–722.

Yan, F., H. Cao, T.L. Cover, R. Whitehead, M.K. Washington and D.B. Polk. 2007. Soluble proteins produced by probiotic bacteria regulate intestinal epithelial cell survival and growth. Gastroenterology 132: 562–575.

Yan, F. and D.B Polk. 2010. Probiotics: progress toward novel therapies for intestinal diseases. Curr. Opin. Gastroenterol. 26: 95–101.
Yang, C.K., H.E. Ewis, X. Zhang, C.D. Lu, H.J. Hu, Y. Pan, A.T. Abdelal and P.C. Tai. 2011. Nonclassical protein secretion by *Bacillus subtilis* in the stationary phase is not due to cell lysis. J. Bacteriol. 193: 5607–5615.
Yates, J.R. 3rd., E. Carmack, L. Hays, A.J. Link and J.K. Eng. 1999. Automated protein identification using microcolumn liquid chromatography-tandem mass spectrometry. Methods Mol. Biol. 112: 553–569.
Yates, J.R. III. 2004. Mass spectral analysis in proteomics. Annu. Rev. Biophys. Biomol. Struct. 33: 297–316.
Yuan, J., B. Wang, Z. Sun, X. Bo, X. Yuan, X. He, H. Zhao, X. Du, F. Wang, Z. Jiang, L. Zhang, L. Jia, Y. Wang, K. Wei, J. Wang, X. Zhang, Y. Sun, L. Huang and M. Zeng. 2008. Analysis of host-inducing proteome changes in *Bifidobacterium longum* NCC2705 grown *in vivo*. J. Proteome Res. 7: 375–385.
Zhao, Y. and O.N. Jensen. 2009. Modification-specific proteomics: Strategies for characterization of post-translational modifications using enrichment techniques. Proteomics 9: 4632–4641.

CHAPTER 7

# Requirements for a Successful Future of Probiotics

*Hörmannsperger Gabriele*

## 1. Introduction

In light of the extensive efforts that are put in technology developments in the field of probiotics, it is of primary importance to be aware of an array of widespread but misleading notions on probiotics that hamper their successful use and endanger their currently high acceptance by consumers. Most importantly, many so-called probiotics do not fulfill the FAO/WHO definition, which states that probiotics are live bacteria that confer a health benefit to the host when ingested in sufficient amounts (FAO/WHO guidelines, http://www.who.int/foodsafety/fsmanagement/en/probiotic_guidelines.pdf). Apart from few exceptions, almost none of the bacterial strains that are advertised and sold as "probiotics" have been proven to have such a beneficial effect on consumer health. The reason for this discrepancy is the common practice to use *in vitro*-determined characteristics, like high acid and bile acid resistance, as simple selection criteria for probiotics, whereas the obligatory proof of physiologically relevant beneficial effects on the host is neglected. The analysis of acid and/or bile acid resistance is valid and important in order to assess the potential of a given microbe to survive the harsh conditions during the gastrointestinal transit; however, gastrointestinal survival is not sufficient to label a given microbe as probiotic. Furthermore, members of some bacterial genera, like *Lactobacillus* and *Bifidobacterium*, are often generally classified as probiotics, again without any scientific proof of a beneficial health effect of the specific strain. Microbes are also very often referred to as probiotics on the sole account of promising *in vitro* functions, like pathogen killing or the modulation of cytokine profiles in cell culture systems. However, this is misleading as long as the hypothesized beneficial effect has not been proven in high quality experimental or human trials in the respective target population. Last but not least, it is important to understand that a

Nutrition and Immunology, ZIEL Institute for Food and Health, Technische Universität München, Gregor-Mendel-Str. 2, 85350 Freising-Weihenstephan, Germany.
E-mail: gabriele.hoermannsperger@tum.de

given probiotic strain is never generally beneficial. Proven probiotic traits of a specific microbe are always restricted to the host function, or disease, for which it has been tested. In consequence, a proven probiotic function of a given microbe cannot simply be extrapolated to additional host functions or diseases. The extensive over-and misuse of the term "probiotic" for microbes without any proven beneficial effects on the host is the underlying reason for a lot of confusion, disappointment and consequently loss of confidence in the probiotic concept in the general public.

Nevertheless, the obvious lack of functional probiotics is in sharp contrast to the promising scientific findings of the last decade that unraveled a pivotal role of the intestinal microbiota (the community of all living microorganisms in the gut) in host health (Clemente et al. 2012, Flint et al. 2012), which in turn strongly support the probiotic concept. Experimental studies and especially the analysis of germfree animals revealed that the intestinal microbiota does not only have a major impact on intestinal health but also on immune functions and the metabolic status of the host (Thompson and Trexler 1971, Wostmann 1983). Due to the enormous complexity and interindividual variability of the intestinal microbiota, which comprises over a thousand different bacterial species in varying amounts and with different functionalities, a "normal" intestinal microbiota cannot easily be defined (Huttenhower and Gevers 2012). Nevertheless, more and more intestinal and extraintestinal diseases like inflammatory bowel diseases (IBD), allergies, autoimmune diseases and diabetes are found to be associated with changes in the intestinal microbiota (Qin et al. 2012, Qin et al. 2010, Greenhill 2013, Ly et al. 2011).

## 2. A Next Generation of Probiotics is Needed

In view of the accumulating evidence for the importance of microbe-host interactions for host health, it is obvious that the identification of true probiotics holds the potential to be of great value for the health of the consumer. However, the identification of true probiotics is a time-consuming, expensive and highly complex process that, from the first *in vitro* steps onwards, needs to take the respective target function (e.g., intestinal barrier function, digestion, absorption, immune function, metabolic function...) or disease (irritable bowel syndrome, allergies, inflammatory bowel diseases, infections ....) into account. It is clear that this process requires deep understanding of the physiology and molecular mechanisms underlying the respective target function or target disease. As reliable biomarkers for these host functions are often lacking, the identification and proper choice of target functions that can be used for initial *in vitro* screenings of probiotics are major challenges.

Taking the huge variety of different bacterial species and their relative abundance in the human intestine into account (Qin et al. 2010), it becomes clear that the strains with the most prominent impact on host health are probably not to be found within the, up to date, most extensively investigated genus *Lactobacillus,* which makes up only a minor portion in the intestinal microbiota of adults (Walter 2008). Therefore, the *in vitro* screenings should be broadened to all kinds of commensal bacterial species, depending on the restrictions due to the intended use of a given probiotic, like for example in Functional Food, as a food supplement or in a pharmaceutical product. Whenever available, the results of human studies that link alterations in the abundance of specific genera or species with certain diseases, like for example the observed reduction of *Faecalibacterium prausnitzii* in Crohn's Disease patients (Sokol et al. 2008), could be used to identify potential probiotic candidate species. After initial *in vitro* screenings, the successful candidate strains need to be tested in experimental pre-clinical setups in order to assess their probiotic potential. Such an elaborate selection process has for example been successfully performed by Kwon et al. (2010), who selected bacterial strains that are able to induce anti-inflammatory immune mechanisms *in vitro*. The most promising bacteria, in this case *L. acidophilus*, *L. casei*, *Lactobacillus reuteri*, *Bifidobacterium bifidium*, and *Streptococcus thermophilus*, were then combined and the effect of this bacterial mixture on the severity of immune mediated diseases like IBD and autoimmune diseases was assessed in appropriate animal models. The results of these pre-clinical studies revealed a significant anti-inflammatory potential of the bacterial mixture in the context of these diseases, which needs to be confirmed in clinical studies in the future. In summary, there is a long way to go from the initial identification of strains with beneficial *in vitro* effects to the clinical confirmation of beneficial effects for the healthy consumer or patients, which in turn sets up the basis for the use of such a strain as probiotic in the respective target population.

## 3. Unraveling of Probiotic Mechanisms is Pivotal for the Targeted use of Probiotics

Apart from the identification of new probiotics, the characterization of the probiotic mechanisms that mediate the observed beneficial effects is of major importance for the effective and targeted use of probiotics in the future. There are several single bacterial strains or bacterial mixtures, such as *L. rhamnosus* GG in the context of diarrhea (Szajewska et al. 2011) and *Escherichia coli* Nissle 1917 or the probiotic mixture VSL#3 in the context of IBD (Kruis et al. 2004, Mimura et al. 2004, Tursi et al. 2010), that have already been shown to be clinically relevant probiotics. However, the protective molecular mechanisms remain to be elucidated. The lack of mechanistic understanding often hampers the targeted use of these probiotics, as it is impossible to predict which patients will or will not respond to the treatment, what is the best time point for the application or which patients might even be endangered by the treatment with a large body of living bacteria (Sanders et al. 2010, Besselink et al. 2008). In addition, due to the lack of mechanistic understanding, it is unclear whether probiotics that have been shown to be protective in the context of a specific disease might also be protective in the context of other dysfunctions or diseases. This situation is in sharp contrast to the application of pharmaceutical agents which are defined structures that are mostly known or even designed to target certain host functions. There is therefore a clear need for the identification of probiotic structure-function relationships in order to enable the targeted clinical use of probiotics.

In theory, probiotics are able to influence host health via two distinct pathways (Fig. 1). On the one hand, probiotic bacteria might exert indirect protective effects on the host via probiotic-microbe interactions. The probiotic bacteria might either positively affect the composition and functionality of the intestinal microbiota or directly prevent the growth and activity of pathogens. The latter effect might be mediated either by direct antagonism, for example via antibacterial molecules, or

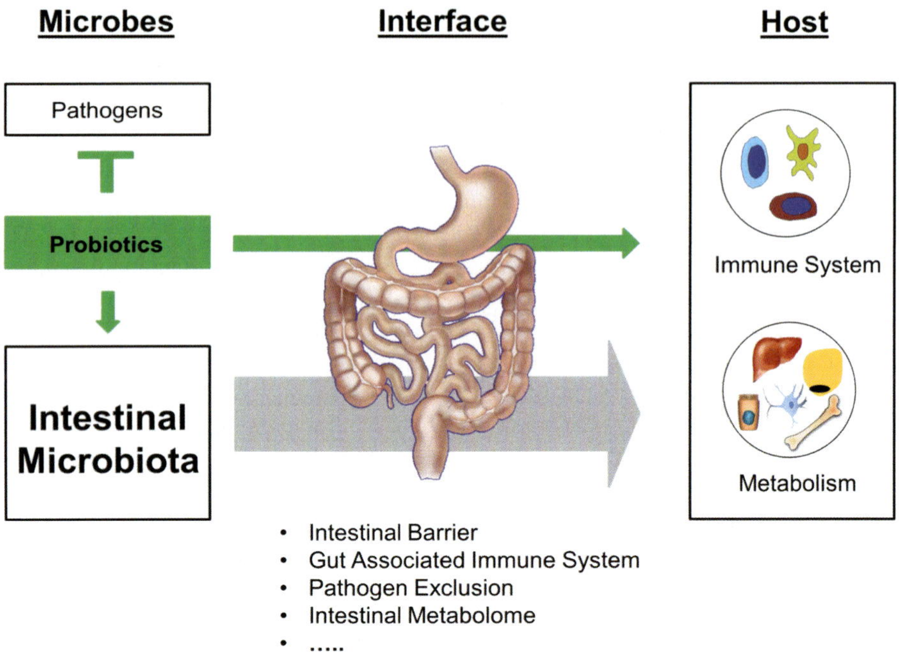


**Fig. 1. Probiotic modulation of host health can be mediated via diverse pathways.** The intestine is a major interface for microbe-host interactions. The intestinal microbiota is known to have a pivotal impact on the intestinal and systemic health of the host. Probiotics are hypothesized to either directly affect host functions (probiotic-host interactions) or to mediate indirect protective effects either via the modulation of the intestinal microbiota or by direct antagonism of pathogens (probiotic-microbe interactions).

by competition for nutrients, trace elements or intestinal attachment sites (pathogen exclusion). On the other hand, beneficial effects of probiotics might be mediated by direct probiotic-host interactions. It is hypothesized that probiotics are able to detoxify harmful compounds as well as to positively affect intestinal barrier functions or immune functions. Concerning the probiotic structures that mediate these microbe-microbe or microbe-host interactions, *in vitro, ex vivo* and experimental studies revealed that secreted as well as bacteria-associated structures like metabolites (Fukuda et al. 2011), proteins (Yan et al. 2011, von Schillde 2012), cell-wall components (Shen et al. 2012) or even bacterial DNA (Jijon et al. 2004) can exert protective effects.

Detailed knowledge of the structure-function relationship that underlies an observed protective effect of a specific probiotic is highly relevant for its appropriate application. The major impact of mechanistic knowledge for the rational use of probiotics can for example be illustrated using the results of a study that revealed *Lactobacillus salivarius* UCC118 to be able to prevent systemic infections of the food-borne pathogen *Listeria monocytogenes* in mice (Corr et al. 2007). Mechanistically, the observed protection was subsequently found to be mediated by the *Lactobacillus salivarius* UCC118-produced bacteriocin Abp118 (bacterial structure) that directly antagonizes the pathogen (function). Firstly, the identification of this protective mechanism (a microbe-microbe interaction) clearly indicates that the protective effect of *Lactobacillus salivarius* UCC118 is restricted to the prevention of infections of pathogens that are sensitive to the bacteriocin Abp118. Secondly, as the protection was found to be due to a direct antimicrobial effect of *Lactobacillus salivarius* UCC118, it is obvious that *Lactobacillus salivarius* UCC118 must be present in sufficient amounts in the intestine at the moment of exposure to *Listeria monocytogenes* in order to prevent infection. As probiotics usually do not colonize the intestinal tract, this result implies that constant uptake of sufficient amounts of *Lactobacillus salivarius* UCC118 by the consumer would be necessary to reduce the risk of *Listeria monocytogenes* infection. Thirdly, on the basis of the identified structure-function relationship one can assume that *Lactobacillus salivarius* UCC118 will not have any therapeutic effect in established systemic *Listeria monocytogenes* infections. In summary, this example clearly demonstrates that the characterization of probiotic structure-function relationships allows the development of appropriate application protocols with respect to timing and dosage of probiotic uptake as well as with respect to the specific indication. The characterization of probiotic structure-function relationships is therefore a prerequisite for the rational design and, in consequence, success of human trials that need to be performed in order to confirm the probiotic potential of a specific microbial strain.

Recent experimental studies in the context of IBD support the hypothesis that not only living bacteria but also isolated bacterial structures are able to confer beneficial effects on the host (Yan et al. 2011, von Schillde et al. 2012). The initial *in vitro* observation that *Lactobacillus rhamnosus* GG (LGG) is able to inhibit cytokine-induced cell death of intestinal epithelial cells was followed by the finding that this barrier-protective effect was mediated by a secreted protein of LGG, p40. Subsequent experimental studies showed that not only the application of LGG, but also the application of isolated encapsulated p40, was able to protect intestinal epithelial cells and to reduce inflammation in IBD models (Yan et al. 2011). Furthermore, *Bacteroides fragilis*, a prominent human commensal, was found to mediate anti-inflammatory effects in experimental IBD models via the immunomodulatory structure Polysaccharide A (PSA). The subsequent application of isolated PSA resulted in significantly reduced intestinal inflammation compared to placebo-treated mice (Round et al. 2010). These studies demonstrate that the characterization of probiotic structure-function relationships opens new ways for the application of isolated, microbe-derived bioactive molecules, so called "postbiotics" (Tsilingiri et al. 2012). One major advantage of the application of postbiotics is thought to be their favorable safety profile, as there is no need for the uptake of billions of living microbes. In addition, these isolated bacterial structures can be applied in a controlled and standardized way, whereas, in the case of the application of living probiotics, the level of the active structure in the intestine is dependent on the number and metabolic activity of the respective strain. However, sometimes, the living probiotic strain may be indispensable in order to ensure appropriate delivery of the active structure at the intestinal target site. In these cases, it remains to be elucidated whether the application of the active probiotic structure, e.g., in pharmaceutical setups, can be optimized by the use of genetically engineered bacterial carrier strains like *Lactococcus lactis*, that show a favorable safety profile and produce high amounts of the respective active structure (Van huynegm and Steidler 2012).

## 4. Conclusion

In summary, it is obvious that probiotics have a high potential for the maintenance of health and the treatment of diseases. However, there is a clear need for the identification of a next generation of probiotics as well as for the characterization of probiotic structure-function relationships in order to enable a targeted, and in turn more successful, application of probiotics or probiotic-derived structures in the future (Fig. 2).

**Future Directions:**

- Functional screenings for candidate probiotics
- Broadened spectrum of candidate bacteria
- Confirmation of beneficial effects in high quality clinical studies
- Characterization of probiotic mechanisms
- Identification of active probiotic structures
- Use of isolated probiotic structures (post-biotics)

**Fig. 2.** Future directions in the field of probiotics.

## References

Besselink, M.G., H.C. van Santvoort, E. Buskens, M.A. Boermeester, H. van Goor, H.M. Timmerman, V.B. Nieuwenhuijs, T.L. Bollen, B. van Ramshorst, B.J. Witteman, C. Rosman, R.J. Ploeg, M.A. Brink, A.F. Schaapherder, C.H. Dejong, P.J. Wahab, C.J. van Laarhoven, E. van der Harst, C.H. van Eijck, M.A. Cuesta, L.M. Akkermans, H.G. Gooszen; Dutch Acute Pancreatitis Study Group. 2008. Probiotic prophylaxis in predicted severe acute pancreatitis: a randomised, double-blind, placebo-controlled trial. Lancet 371: 651–9.

Clemente, J.C., L.K. Ursell, L.W. Parfrey and R. Knight. 2012. The impact of the gut microbiota on human health: an integrative view. Cell 148: 1258–70.

Corr, S.C., Y. Li, C.U. Riedel, P.W. O'Toole, C. Hill and C.G. Gahan. 2007. Bacteriocin production as a mechanism for the antiinfective activity of *Lactobacillus salivarius* UCC118. Proc. Natl. Acad. Sci. USA 104: 7617–21.

FAO/WHO guidelines, http://www.who.int/foodsafety/fsmanagement/en/probiotic_guidelines.pdf.

Flint, H.J., K.P. Scott, P. Louis and S.H. Duncan. 2012. The role of the gut microbiota in nutrition and health. Nature reviews Gastroenterology & Hepatology 9: 577–89.

Fukuda, S., H. Toh, K. Hase, K. Oshima, Y. Nakanishi, K. Yoshimura, T. Tobe, J.M. Clarke, D.L. Topping, T. Suzuki, T.D. Taylor, K. Itoh, J. Kikuchi, H. Morita, M. Hattori and H. Ohno. 2011. Bifidobacteria can protect from enteropathogenic infection through production of acetate. Nature 469: 543–7.

Greenhill, C. 2013. Paediatrics: Differences in the faecal microbiota of children with beta-cell autoimmunity. Nature reviews Endocrinology.

Huttenhower, C. and D. Gevers. 2012. Structure, function and diversity of the healthy human microbiome. Nature 486: 207–14.

Jijon, H., J. Backer, H. Diaz, H. Yeung, D. Thiel, C. McKaigney, C. De Simone and K. Madsen. 2004. DNA from probiotic bacteria modulates murine and human epithelial and immune function. Gastroenterology 126: 1358–73.

Kruis, W., P. Fric, J. Pokrotnieks, M. Lukás, B. Fixa, M. Kascák, M.A. Kamm, J. Weismueller, C. Beglinger, M. Stolte, C. Wolff and J. Schulze. 2004. Maintaining remission of ulcerative colitis with the probiotic *Escherichia coli* Nissle 1917 is as effective as with standard mesalazine. Gut 53: 1617–23.
Kwon, H.K., C.G. Lee, J.S. So, C.S. Chae, J.S. Hwang, A. Sahoo, J.H. Nam, J.H. Rhee, K.C. Hwang and S.H. Im. 2010. Generation of regulatory dendritic cells and CD4+Foxp3+ T cells by probiotics administration suppresses immune disorders. Proc. Natl. Acad. Sci. USA 107: 2159–64.
Ly, N.P., A. Litonjua, D.R. Gold and J.C. Celedon. 2011. Gut microbiota, probiotics, and vitamin D: interrelated exposures influencing allergy, asthma, and obesity? The Journal of Allergy and Clinical Immunology 127: 1087–94; quiz 95–6.
Mimura, T., F. Rizello, U. Helwig, G. Poggioli, S. Schreiber, I.C. Talbot, R.J. Nicholls, P. Gionchetti, M. Campieri and M.A. Kamm. 2004. Once daily high dose probiotic therapy (VSL#3) for maintaining remission in recurrent or refractory pouchitis. Gut 53: 108–14.
Qin, J., Y. Li, Z. Cai, S. Li, J. Zhu, F. Zhang, S. Liang, W. Zhang, Y. Guan, D. Shen, Y. Peng, D. Zhang, Z. Jie, W. Wu, Y. Qin, W. Xue, J. Li, L. Han, D. Lu, P. Wu, Y. Dai, X. Sun, Z. Li, A. Tang, S. Zhong, X. Li, W. Chen, R. Xu, M. Wang, Q. Feng, M. Gong, J. Yu, Y. Zhang, M. Zhang, T. Hansen, G. Sanchez, J. Raes, G. Falony, S. Okuda, M. Almeida, E. LeChatelier, P. Renault, N. Pons, J.M. Batto, Z. Zhang, H. Chen, R. Yang, W. Zheng, S. Li, H. Yang, J. Wang, S.D. Ehrlich, R. Nielsen, O. Pedersen, K. Kristiansen and J. Wang. 2012. A metagenome-wide association study of gut microbiota in type 2 diabetes. Nature 490: 55–60.
Qin, J., R. Li, J. Raes, M. Arumugam, K.S. Burgdorf, C. Manichanh, T. Nielsen, N. Pons, F. Levenez, T. Yamada, D.R. Mende, J. Li, J. Xu, S. Li, D. Li, J. Cao, B. Wang, H. Liang, H. Zheng, Y. Xie, J. Tap, P. Lepage, M. Bertalan, J.M. Batto, T. Hansen, D. Le Paslier, A. Linneberg, H.B. Nielsen, E. Pelletier, P. Renault, T. Sicheritz-Ponten, K. Turner, H. Zhu, C. Yu, S. Li, M. Jian, Y. Zhou, Y. Li, X. Zhang, S. Li, N. Qin, H. Yang, J. Wang, S. Brunak, J. Doré, F. Guarner, K. Kristiansen, O. Pedersen, J. Parkhill, J. Weissenbach, MetaHIT Consortium, P. Bork, S.D. Ehrlich and J. Wang. 2010. A human gut microbial gene catalogue established by metagenomic sequencing. Nature 464: 59–65.
Round, J.L. and S.K. Mazmanian. 2010. Inducible Foxp3+ regulatory T-cell development by a commensal bacterium of the intestinal microbiota. Proc. Natl. Acad. Sci. USA 107: 12204–9.
Sanders, M.E., L.M. Akkermans, D. Haller, C. Hammerman, J. Heimbach, G. Hormannsperger et al. 2010. Safety assessment of probiotics for human use. Gut Microbes 1: 164–85.
Shen, Y., M.L. Torchia, G.W. Lawson, C.L. Karp, J.D. Ashwell and S.K. Mazmanian. 2012. Outer membrane vesicles of a human commensal mediate immune regulation and disease protection. Cell Host & Microbe 12: 509–20.
Sokol, H., B. Pigneur, L. Watterlot, O. Lakhdari, L.G. Bermudez-Humaran, J.J. Gratadoux, S. Blugeon, C. Bridonneau, J.P. Furet, G. Corthier, C. Grangette, N. Vasquez, P. Pochart, G. Trugnan, G. Thomas, H.M. Blottière, J. Doré, P. Marteau, P. Seksik and P. Langella. 2008. *Faecalibacterium prausnitzii* is an anti-inflammatory commensal bacterium identified by gut microbiota analysis of Crohn disease patients. Proc. Natl. Acad. Sci. USA 105: 16731–6.
Szajewska, H., M. Wanke and B. Patro. 2011. Meta-analysis: the effects of *Lactobacillus rhamnosus* GG supplementation for the prevention of healthcare-associated diarrhoea in children. Aliment Pharm. Ther. 34: 1079–87.
Thompson, G.R. and P.C. Trexler. 1971. Gastrointestinal structure and function in germ-free or gnotobiotic animals. Gut 12: 230–5.
Tsilingiri, K. and M. Rescigno. 2012. Postbiotics: what else? Beneficial Microbes 69–75.
Tursi, A., G. Brandimarte, A. Papa, A. Giglio, W. Elisei, G.M. Giorgetti et al. 2010. Treatment of relapsing mild-to-moderate ulcerative colitis with the probiotic VSL#3 as adjunctive to a standard pharmaceutical treatment: a double-blind, randomized, placebo-controlled study. Am. J. Gastroenterol. 105: 2218–27.
Van Huynegem, K. and L. Steidler. 2012. Clinical development of lactocepin: a novel bacterial biologic? Expert review of Clinical Immunology 8: 597–9.
von Schillde, M.A., G. Hormannsperger, M. Weiher, C.A. Alpert, H. Hahne, C. Bauerl, K. van Huynegem, L. Steidler, T. Hrncir, G. Pérez-Martínez, B. Kuster and D. Haller. 2012. Lactocepin secreted by Lactobacillus exerts anti-inflammatory effects by selectively degrading proinflammatory chemokines. Cell Host & Microbe 11: 387–96.
Walter, J. 2008. Ecological role of lactobacilli in the gastrointestinal tract: implications for fundamental and biomedical research. Appl. Environ. Microbiol. 74: 4985–96.
Wostmann, B.S., C. Larkin, A. Moriarty and E. Bruckner-Kardoss. 1983. Dietary intake, energy metabolism, and excretory losses of adult male germfree Wistar rats. Laboratory Animal Science 33: 46–50.
Yan, F., H. Cao, T.L. Cover, M.K. Washington, Y. Shi and L. Liu. 2011. Colon-specific delivery of a probiotic-derived soluble protein ameliorates intestinal inflammation in mice through an EGFR-dependent mechanism. J. Clin. Invest. 121: 2242–53.

## CHAPTER 8

# High-throughput Techniques for Studying of Gut Microbiota

*Clotilde Rousseau*[1,2,*] and *Marie-José Butel*[1]

## 1. Introduction

Of the microbial populations associated with the human body, the largest and most complex is that present in the gastrointestinal tract. The human digestive tract is inhabited by more than 100 trillion bacteria and Archaea, which together make up the gut microbiota (Box 1). Microbial communities, predominantly anaerobes, are distributed all along our digestive tract and progressively increase in number from the jejunum to the colon to reach a maximum density in the distal colon with over $10^{11}$ bacteria per gram of fecal content and are of great diversity (Suau et al. 1999). Thus, our gut resident bacteria outnumber our own human cells by a factor of 10 (Ley et al. 2006). All higher organisms such as other mammals, insects and fish, have their specific microbiota. These microbial communities interact extensively with the host and participate in its development and homeostasis. We can therefore be seen as a 'super-organism' consisting of a combination of microbial and eukaryotic cells. The intestinal microbiota can be considered a real organ as its impact on our lives (O'Hara and Shanahan 2006). Indeed, it makes a major contribution to the physiological processes of the host with (i) metabolic and nutritional functions as digestion of certain components of our food and vitamin synthesis (Tremaroli and Bäckhed 2012), (ii) structural functions as intestine wall development and (iii) protective functions as immune system maturation and protection against bacterial pathogens (Hooper et al. 2012), also, we could not live without each other. Therefore, any imbalance in the

[1] EA 4065 - Ecosystème Intestinal, Probiotiques, Antibiotiques, Faculté des Sciences Pharmaceutiques et Biologiques, Université Paris Descartes, 4 avenue de l'Observatoire, 75270 Paris cedex 06, France. E-mail: marie-jose.butel@parisdescartes.fr

[2] Microbiologie, Hôpital Saint-Louis, Assistance Publique des Hôpitaux de Paris, Paris.

* Corresponding author

**Box 1.** Glossary.

| **Glossary** |
|---|
| **Microbiota** |
| Community of microbes inhabiting a given environment. |
| **Microbiome** |
| The totality of the microorganisms and their genomes in a given environment subject to environmental interactions. The human microbiome corresponds to the collection of microbes associated with the human body and their collective genomes (metagenome). However, microbiome and metagenome are usually interchangeably used. |
| **Core human microbiome** |
| The human core microbiome includes a set of organisms (phylogenetic core) and their genomes (metagenomic core), which are conserved between individuals in the human population. This could support core functions of the microbiota while inter-individual variability could be the basis of our differences of attributes. |
| **Meta-omics** |
| Meta-omics approaches include the characterization of the totality of DNA, mRNA, proteins or metabolites present in a given environment, which respectively correspond to metagenomics, metatranscriptomics, metaproteomics and metabolomics. |
| **Metagenomics** |
| Metagenomics is the study of the total DNA (metagenome) that can be directly recovered from a given environment. The human metagenome encompass the DNA of the host and the microbes. Shotgun or direct sequencing of total microbial DNA and high-throughput sequencing of metagenomic libraries are used to screen for phylogenetic diversity and potential functions of an ecosystem. Functional metagenomics is the screening of a metagenomic library of clones to identify the functions encoded by the metagenomic inserts. |

gut bacterial communities, i.e., dysbiosis, may be associated with alterations of gut functions and lead to diseases. Recently, several diseases such as obesity, diabetes and inflammatory bowel diseases (IBD) have been linked to gut ecosystem imbalance (Tremaroli and Bäckhed 2012). Consequently, high hopes are placed in therapy to modulate the gut microbiota balance and its functions, for example by the use of probiotics.

However, we have at present only fragmentary knowledge of the gut microbiota. Several difficulties are encountered in its study; first, all compartments of the gastrointestinal tract are not easily accessible, second, the gut microbiota presents extreme complexity and diversity which, moreover, differs between individuals, intestinal location and age of the subject (Zoetendal et al. 2008), and third, the majority of the gut bacterial species are not currently cultivated in laboratories (Eckburg et al. 2005, Suau et al. 1999). The study of the gut microbial ecology involves both the analysis of its bacterial composition in abundance and diversity, of the activity of these bacteria, and of their relationships with each other (synergistic or competitive interactions) and the host, which requires the use of different approaches. Recent technological advances based on culture-independent approaches have been made to develop new high-throughput techniques for studying complex microbial communities. These new approaches at the interface between microbial ecology, genomics and post-genomics (Zoetendal et al. 2008) permit to reassess the importance of the biodiversity of the human intestinal microbiota (Eckburg et al. 2005, Suau et al. 1999) and its functional impact on our health.

In this chapter, we will review current knowledge on the composition and diversity of the intestinal microbiota. We will discuss how molecular approaches have provided novel insights toward the phylogenetic and functional characterization of the gut microbiota focusing on the recent developments and applications of high-throughput techniques. In addition, recent insights on the link between the gut microbiota, the human health and the use of probiotics are provided.

## 2. Studying the Gut Microbiota: From Culture to Molecular High-throughput Techniques

During the past decades, major advances in our knowledge of the gut microbiota resulted from development of anaerobic culture techniques, animal models to study relationships between intestinal bacteria and the host, and gnotobiotic approaches involving germfree or defined-microbiota animal models (Savage 2001). Culture-based methods have learned a lot about the intestinal ecosystem; however, they have important limitations. First, they are time consuming because of the huge number of species to identify. Second, many bacteria are not cultivable. Indeed, some bacteria have very specific nutritional requirements difficult or impossible to satisfy *in vitro* and others have complex relationship with the host or dependent metabolic activities of each other make it almost impossible to cultivate. Comparison of bacterial counts of the same sample by culture and microscopic observation confirmed that culture underestimated populations. This 'counting anomaly' indicated that a large proportion of the bacterial species of the dominant intestinal microbiota of healthy humans escaped our attention (Suau et al. 1999). Thus, in the late 1980s and early 1990s, the development of culture-independent molecular tools revolutionized the fields of microbial ecology by providing new access to this 'uncultured majority'.

Total nucleic acids can be obtained directly from any complex microbial ecosystem and used for analysis of all the members of this ecosystem without requiring culturing step. A short segment of the ribosomal RNA called 16S rRNA is largely used as a taxonomic marker for bacterial classification and phylogeny. Many molecular techniques based on the analysis of 16S rRNA gene sequences were successfully developed to study the gut microbiota. However, it is only recently that advances in sequencing technology and development of bioinformatics tools have led to the advent of high-throughput technologies, which have opened up new ways to investigate the $10^{14}$ microorganisms inhabiting the human gut and their functions. Thus were born the global 'omics' approaches or simultaneous analysis of all the dominant members of a given ecosystem (Fig. 1). Metagenomics permit direct access to the entire pool of genes of the gut microbiota, while metatranscriptomics, metaproteomics and metabolomics respectively target total mRNA, proteins and metabolites of the gut microbial communities.

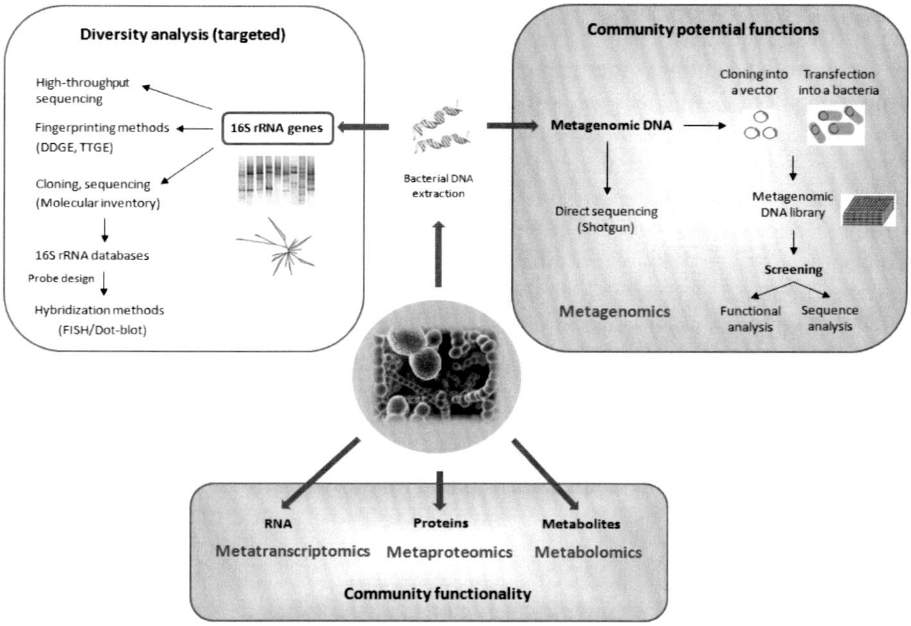


**Fig. 1.** Schematic representation of the various techniques for studying the gut microbiota and its functions.

## 3. 16S rRNA Gene-based Approaches

The 16S rRNA molecule that contains approximately 1,500 bp is present in all prokaryote organisms. Ten regions of the gene have been conserved throughout evolution from which primers or probes called 'universal' can be extracted. Between these conserved regions are nine regions called 'hypervariable' and annotated V1 to V9 whose sequence can be used for the identification and differentiation of all bacteria (Woese et al. 1990). These hypervariable regions offer opportunities for specific priming of a given taxon or a bacterial species. To date, the 16S rRNA gene remains the most knowledgeable marker entered in the databases of nucleotide sequences and *de facto* is the most widely used for bacterial identification.

### *3.1 Molecular Species–OTU—Phylotype*

The exploration of the gut microbiota diversity by comparative sequence analysis of the 16S rRNA gene is based on DNA sequences discrimination according to their phylogenetic relationship at the species level. The 16S rRNA hypervariable regions represent the target for the amplification-based sequencing, which could be for example the V1 to V3 or V4 to V6 regions. Within the same species, the 16S rRNA gene sequence may vary from one member to another, which explains that a molecular species, i.e., phylotype or operational taxonomic unit (OTU), is defined by a certain degree of similarity of the 16S rRNA gene sequence. According to studies, the similarity threshold chosen can vary from 97% to 99% and lead to differences in classification and richness for a given sequence library.

### 3.2 16S rRNA Gene Sequence Databases

To determined bacterial identity from its 16S rRNA gene sequence, databases comprising rRNA sequences of known microorganisms are required. In the past decade, applications using small subunit ribosomal RNA properties have dramatically increased in the area of microbial ecology to explore diversity of bacterial communities in a variety of ecosystems including the human intestinal ecosystem. Currently, over 2.6 million of bacterial and Archaea 16S rRNA sequences are available in DNA databases, which are far greater than for any other gene (http://rdp.cme.msu.edu/; RDP Release 10, Update 31 dated 7 December 2012). Nevertheless, early studies based on comparison of DNA sequences obtained from fecal samples to those of known microorganisms contained in databases revealed that most of these sequences were not listed. Thus, nearly 80% of the bacterial species present in our digestive ecosystem are yet unknown (Eckburg et al. 2005, Hayashi et al. 2002, Suau et al. 1999). Even if 16S rRNA gene sequences of only a small fraction of all bacterial isolates of the gut microbiota has been deposited in DNA databases, this number is growing tremendously. Obviously, the molecular descriptions that can be made of the diversity of the intestinal microbiota are continuously outdated given the steady reporting of novel inhabitants.

### 3.3 16S rRNA-based Techniques and Main Advances

From a given sample, the total bacterial DNA is extracted and purified. 16S rRNA hypervariable regions are then amplified using polymerase chain reaction (PCR). For microbiota diversity analyzes, several approaches targeting 16S rRNA gene sequences have been successfully developed (Fig. 1). They correspond to different resolution levels and the choice depends on the question to be answered; however, they are often complementary. Common methods are rRNA gene library (or molecular inventory) based on cloning and subsequent sequencing of 16S rRNA genes, hybridization approaches as fluorescence *in situ* hybridization (FISH) and microarray, real-time PCR, and DNA fingerprinting approaches such as denaturing gradient gel electrophoresis (DGGE), temperature gradient gel electrophoreses (TGGE) and denaturing high performance liquid chromatography (dHPLC) (Zoetendal et al. 2008). The highest resolution level is obtained with rRNA gene library, which allows estimating the bacterial diversity at the species level by generating phylogenetic trees. A more general view of the microbiota composition in dominant bacterial groups is achieved using fingerprinting approaches. These techniques are most often used to monitor changes in microbial communities or compare microbiota of several samples. These different approaches also differ according to the qualitative or quantitative information they can generate. Qualitative data are provided by fingerprinting techniques, while FISH, microarray and real-time PCR allow quantification of specifically targeted microbes or populations, and by involving FISH, their visualization into the sample. These approaches are relatively low-throughput and none of them gives exhaustive or absolutely accurate information. Nevertheless, their combination brought many novel insights into the phylogenetic diversity of the gut ecosystem. Recently, the development of high-throughput sequencing technologies has allowed the determination of 16S rRNA genes sequences directly from the total DNA extracted and amplified, without the need for cloning. This approach is much more comprehensive and less laborious than the rRNA gene library and has significantly increased the number of bacteria identified in the human gut microbiota.

### 3.3.1 16S rRNA Gene Library

The 16S rRNA gene is amplified from samples with PCR using broad-range bacterial and Archaea primers. PCR products are then cloned and sequenced. To inventory bacteria present in a given environment by comparative sequence analysis, the construction of 16S rRNA gene clone library is required. After library construction, the sequences of the cloned amplicons are determined and compared to sequences deposited in rRNA databases, and followed by phylogenetic analysis. Although this technique gives significant information about identity of uncultured bacteria, the data are not quantitative and the cloning step is not without biases. In addition, labor and expenses incurred by cloning and sequencing of rDNA libraries constitute an important limitation of this method.

Pioneer studies of the gut communities which were restricted to culture-based approaches estimated microbial diversity at little more than 400 species (Savage 1977). DNA library sequencing provided novel insights into the gut microbiota community showing that the biodiversity is more complex than previously described (Eckburg et al. 2005, Suau et al. 1999). Indeed, according to studies, as low as 20–30% of total bacterial species of the gut ecosystem corresponded to known cultivable organisms, indicating that the vast majority of the dominant bacteria of the human gut microbiota have so far eluded scientific description. Furthermore, among the new bacterial species identified by 16S rRNA gene clones sequencing, the low G+C Gram positive phylum (Firmicutes) was the most represented, indicating a specific defect in identification of this group by culture methods. DNA library sequencing also provided new insights into the phylogenetic composition of the gut microbiota. Suau and colleagues studied the fecal microbiota of a man (40 years-old) for the first time using this technique (Suau et al. 1999). Among the 284 clones analyzed, there were 82 distinct OTUs and the vast majority (95%) belonged to two eubacterial divisions, the *Cytophaga-Flavobacterium-Bacteroides* and the Firmicutes, and fell into three main groups; *Bacteroides*, *Clostridium coccoides* (*Clostridium* rRNA subcluster XIVa), and *Clostridium leptum* (*Clostridium* rRNA subcluster IV). This was subsequently confirmed (Eckburg et al. 2005, Tap et al. 2009).

This type of molecular approach has also recently highlighted the existence of a phylogenetic core within the human intestinal microbiota. Tap et al. demonstrated by the analysis of thousands of 16S rDNA sequences from 17 human fecal samples that despite a high individual specificity, 2.1% of the OTUs (36% of the total sequences) are shared among more than 50% of the individuals (Tap et al. 2009). These 66 dominant and prevalent OTUs included members of the genera *Faecalibacterium, Ruminococcus, Eubacterium, Dorea, Bacteroides, Alistipes* and *Bifidobacterium*. These common bacterial species might be implicated in the conserved functions of the intestinal microbiota between individuals.

### *3.3.2 Molecular Fingerprinting Methods*

All molecular fingerprinting methods are PCR-based and generate profiles representing the sequence diversity within the selected ecosystem. Principles and technical procedures, however, vary (Zoetendal et al. 2008). Denaturing gradient gel electrophoresis (DGGE), temperature gradient gel electrophoresis (TGGE) and temporal temperature gradient gel electrophoresis (TTGE) are based on sequence specific melting behavior of 16S rDNA amplicons under denaturing gel electrophoresis. The denaturing gradient can be chemical, urea or formamide, in the case of DGGE or temperature in the cases of TGGE and TTGE. Electrophoretic profiles are real characteristic fingerprints of the microbiota, in which each band corresponds to a dominant species or a group of taxonomically close dominant species and may be identified by sequencing. The use of universal primers allows the characterization of 90 to 99% of the dominant species of the fecal microbiota, while the detection of minority populations can be achieved by using specific bacterial groups or species primers. Another fingerprinting method, denaturing high performance liquid chromatography (dHPLC) was primarily used to detect variations in DNA sequence resulting in different elution times of the DNA fragments on the chromatogram. It has also been used to discriminate between PCR-amplified bacterial 16S rRNA genes thereby, allowing identification of bacteria at the species level and/or biotype level. The method is now adapted for the analysis of complex microbial communities such as the human gut microbiota (Rougé et al. 2010). Regarding other fingerprinting methods, single-strand conformation polymorphism (SSCP) is based on the conformational polymorphism between the single-stranded DNA fragments, and terminal-restriction fragment length polymorphism (T-RFLP) on specific target sites for restriction enzymes. For all, with improvements of statistical softwares, similarity indices can be calculated and cluster analysis of 16S rDNA profiles can be performed. In principle, fingerprinting methods are sensitive enough to separate rDNA molecules differing by a single base pair. These techniques are able to detect bacteria representing up to 1% of the total bacterial populations, i.e., dominant bacteria. They are particularly effective for monitoring shifts in microbial communities over time or during environmental changes in complex samples, and to compare large numbers of samples.

Fingerprinting methods revealed that the prevalence of bacterial groups and species varies along the length of the gut. Density and diversity increase from the stomach to the colon at the same time as the physico-chemical conditions prevailing in the different digestive compartments evolve such as decreasing oxygen gradient. Moreover, studies on mucosal biopsies showed that the fecal microbiota differed from the mucosal microbiota in composition (Lepage et al. 2005). Zoetendal et al. (1998) analyzing by TTGE the microbiota of 16 volunteers revealed that species composition is highly individual specific (Zoetendal et al. 1998). They also demonstrated a remarkable stability over time (Vanhoutte et al. 2004, Zoetendal et al. 1998) and resilience aptitude after a stress such as antibiotics use of the dominant bacterial communities of the gut (De La Cochetière et al. 2005). However, whether the compositional stability reflect a functional stability remains to be investigated. Previous cultivation-based studies of the fecal microbiota revealed that community shifts do occur, especially in infants and elderly people (Mackie et al. 1999). Confirming these findings, fingerprinting approaches also revealed that host genotype and diet affect microbial diversity in the digestive tract, making them essential consideration to describe the gut microbiota and to establish links between microbial populations and host factors or some pathology (Zoetendal et al. 1998). During chronic IBD, such as Crohn's disease and ulcerative colitis, these methods have observed a decrease in the microbial diversity associated with structure instability (Sokol et al. 2006).

### *3.3.3 Hybridization Methods*

Hybridization methods are based on the use of 16S rRNA-targeted probes. They were made possible thanks to the knowledge of specific sequences corresponding to domains (Bacteria probe), phyla, major bacterial groups or species, which have enabled the development of specific probes. Hybridization is performed on DNA or RNA extracts obtained from a sample (dot-blot) or more frequently directly on the sample by fluorescent *in situ* hybridization (FISH). Semi-quantitative (dot-blot) or quantitative information (FISH) on different bacterial populations is obtained (Rigottier-Gois et al. 2003, Sokol et al. 2008). In FISH, the probes are labeled with a fluorochrome and directly target the content cells. The high copy number of 16S rRNA molecules in a bacterial cell allows detection of a single cell. The analysis of the fluorescent signal, cell by cell, permit not only evaluating the morphology of non-cultivable bacteria, but also its spatial distribution *in situ*, particularly useful in the study of tissues or biofilms. At present, analysis has been automated by coupling to image analyzers and combination with flow cytometry (FISH-FC) (Rigottier-Gois et al. 2003, Sokol et al. 2008). Hybridization techniques are relatively rapid and do not require prior enrichment, amplification or gel separation. However, by their nature these methods are dependent on sequence data availability and hence fail to detect novel sequences. Regarding dot-blot, it has the further drawback of using radioactive isotopes and is delicate when working on RNA.

FISH has been mainly used to quantify major groups of bacteria in human feces of healthy subjects. The sensitivity obtained is presently $10^6$ cells per g of feces. It also enabled to identify bacterial subpopulations and to locate them in the digestive tract. Fallani et al. studied the establishment of intestinal populations in 606 European infants and showed that delivery mode and feeding type significantly influenced the microbiota composition, these effects being still evident after weaning (Fallani et al. 2011). Interestingly, they observed that the impact of country of birth was even more pronounced, with dominant bifidobacteria in northern countries and greater early species diversification in southern European countries (Fallani et al. 2010). This technique also provided important information in pathological conditions of the bowel. Sokol et al. demonstrated that a reduction of a major member of Firmicutes, *Faecalibacterium prausnitzii,* exhibiting anti-inflammatory effects was associated with a higher risk of post-operative recurrence of ileal Crohn's disease (Sokol et al. 2008).

### *3.3.4 Phylogenetic Microarray*

DNA microarray (or DNA chip) is a high-throughput derivative of dot-blot technique. It is based on the quantification, usually by fluorescence, of the hybridization of oligonucleotides covalently linked to the glass surface with DNA or RNA present in a sample. In bacteriology, microarray is a commonly used approach, which is principally applied for monitoring gene expression or detection of DNA sequence polymorphisms and mutations in the arrayed genes. Advances in robotics have allowed the miniaturization of blots and thus increasing their number on a limited surface support. However, the use of microarrays in microbial community studies occurred later due to the great difficulty in defining the oligonucleotides that must target the great number of bacteria that may be present. A better understanding of complex microbiota and expanding databases of 16S rRNA were necessary in order to draw microarrays representative in the field. Thus, most of the first environmental chips were confined to a single bacterial family (Castiglioni et al. 2004, Loy et al. 2005). Antibiotic resistance genes were also successfully screened in intestinal microbiota (Call et al. 2003). Miniaturization is the advantage of microarray, allowing the treatment of a very important number of oligonucleotide sequences. Today, chips of over a million oligonucleotides are marketed. However, its cost remains one of the highest in bacterial identification. In addition, although DeSantis et al. (2007) reported a greater microbial diversity using microarray than clone library, DNA chips cannot, by their design, identify and classify unknown bacteria.

Since 2005, DNA microarrays supposed to encompass the entire bacterial domain have emerged (Huyghe et al. 2008, Palmer et al. 2006) and can currently be used for diversity analysis, i.e., phylogenetic microarray. The first studies using phylogenetic microarray to characterize the gut microbiota revealed the usefulness of this approach for the study of the structure and population dynamics in the digestive tract. Palmer et al. followed 14 infants during their first year of life and confirmed that a chaotic microbiota characterized early months of life, while a convergence of the microbiota composition was observed by the end of the first year of life; nonetheless, individual-specific community patterns were still recognizable (Palmer et al. 2007). Another study recently showed that adult subject-specific gut microbiota was stable not only in short-term intervals (<1 year) but also during long periods of time (>10 years) (Rajilić-Stojanović et al. 2012). The data showed that the gut ecosystem contains a core community of permanent colonizers, and that environmentally introduced changes of the microbiota throughout adulthood are primarily affecting the abundance but not the presence of specific microbial species. Phylogenetic microarrays were also applied for disorders studies of the gut microbiota, such as IBD (Kang et al. 2010).

### *3.3.5 Real-time PCR*

The real-time PCR (or qPCR) is the most recent quantitative PCR method. It is based on the monitoring of the amplification reaction over time by means of a fluorescent label capable of emitting a radiation proportional to the amount of PCR product. Number of cycles necessary to obtain detectable amount of PCR product is inversely proportional to the number of nucleic acid molecules initially present. In microbial ecology, qPCR requires special conditions. Indeed, the DNA extracted from complex ecosystems as stool samples may contain PCR inhibitory molecules, which have to be detected prior to performing the analysis to prevent the amplification efficiency varies from a sample to another. In addition, the specificity of the qPCR for a target molecule can be affected in a complex sample because of multiple interactions that may arise.

Nevertheless, qPCR revealed suitable for studying complex bacterial communities and allows quantifying very low concentration of bacterial targets, i.e., specific groups or phylogenetic species, which is difficult using other approaches (Furet et al. 2009). However, the quantification is only relative since 16S rRNA gene copy numbers differ among bacterial genomes, which hampers the extrapolation of the data to bacterial numbers. It can also be coupled to microarray to determine the diversity and relative abundance of groups of low abundance, which has a particular interest in the study of the population dynamics of pathogenic bacteria or probiotics that are usually present in low numbers.

### *3.3.6 Amplification-based Sequencing*

The human gut has been identified as one of the most densely populated ecosystems on Earth. The identification of all the molecular species of this ecosystem therefore require considerable sequencing capabilities. In the 70s, the advent of enzymatic sequencing by the Sanger method has allowed the realization of the first analyze of the human intestinal microbiota using 16S rRNA gene clone library sequencing (Suau et al. 1999). Nevertheless, the information obtained by cloning and subsequent sequencing was dependent on the number of clones sequenced. Other disadvantages are the high cost and the considerable time required due to the isolation of bacterial clones. Since then, many technical obstacles have been removed leading to high-throughput approaches based on the recent advances in sequencing technologies and the development of bioinformatics tools.

In the 2000s, next-generation sequencing has been used to analyze the gut microbiota diversity in amplification-based sequencing approaches using 16S rRNA as phylogenetic marker. Based on total DNA extracted from a sample, 16S rRNA hypervariable regions are amplified and then directly sequenced. This sequencing technology generates several million reads per reaction where manual techniques did not produce a single sequence (Metzker 2010). High-throughput DNA sequencing is thus an effective technique that allows the cataloguing of millions of sequences at a time for a more affordable cost. The sequences are aligned and compared with the sequences of DNA databases to search for homologous regions and affiliation with a phylogenetic group. The degree of diversity or homology is proportional to the genetic distance when plotted onto a phylogenetic tree. Hence, these characteristics allow the identification of bacterial Phyla, Classes, Orders, Families, Genera and Species when compared with 16S rRNA databases (Wang et al. 2007). The longer the sequence the more precise is the analysis, but the analysis time is thereby also extended. The use of tags to identify different samples allows the analysis of several samples at once, eliminating hence the influence of different sequencing efficiency.

However, this strategy also reduces the depth of the analysis by limiting the number of reads per sample. Next-generation sequencers allow the generation of a number of reads per sample not previously possible and this is a major gain in depth analysis. Three high-throughput DNA sequencing-based technologies are currently commonly used for gut microbiota diversity analyze: Roche/454 GS FLX Titanium sequencer, Illumina genome analyser, and Applied Biosystems SOLiD (Medini et al. 2008).

Koren et al. analyzed the diversity and the phylogenetic composition of gut microbial communities using 454 pyrosequencing during pregnancy (Koren et al. 2012). They showed that dramatic changes occur from the first to third trimester, with a wide diversity expansion, a richness reduction and an overall increase in Proteobacteria and Actinobacteria. High-throughput 454 sequencing technology was also used to analyze gut microbiota of obese patients, and revealed that obesity was associated with phylum-level changes in the microbiota and reduced bacterial diversity (Turnbaugh et al. 2009). The same team highlight that the relative proportion of Bacteroidetes was reduced in obese people by comparison with lean people, and that this proportion increased with weight loss (Ley et al. 2006). This work, demonstrating that obesity has a microbial component, open new potential therapeutic ways. Overall, these techniques have contributed enormously to the description of the human intestinal microbiota, both in healthy subjects and in various pathological contexts.

## 4. Community-based Approaches: 'Omics'

Molecular approaches, such as broad-range sequencing of 16S rRNA genes, are presently being used to define shifts in the composition of the gut microbiota during an individual's life or to assess its biodiversity in geographically defined healthy populations and in different disease states. Although they are powerful approaches for describing the microbial diversity, the potential functions of these microbes cannot be extracted from 16S rDNA data. Thus, other approaches were needed to gain insight into the roles of microbes in the human gut ecosystem and their relation to health and diseases.

Functions of individual organisms and their interactions with the host are usually studied using *in vitro* or animal models. Since genome sequences are available, genes involved in these interactions have also been discovered for both pathogens and commensal bacteria. Thus, the impact of colonization by some commensal bacteria, such as *Bacteroides thetaiotaomicron, Bifidobacterium longum* or *Akkermansia muciniphila,* on nutritional and defensive functions of the host intestine has been shown (Derrien et al. 2011, Hooper et al. 2001, Sonnenburg et al. 2006). Similarly, the impact of the host on the microbes has been studied (Marco et al. 2010, Sokol et al. 2008). However, these approaches are reductionist as they mainly focus on individual organisms and cannot be extrapolated to what interactions happen in a complex ecosystem such as the gastrointestinal tract. Community approaches, also called 'omics' approaches, open the way to explore potential functions and activities of microbes without the need of cultivation, thus directly drawing a picture of the gut ecosystem as a 'molecular scanner'. Metagenomics is a DNA-based approach to gain insights into the genetic potential of microbial communities, while the other meta-'omics' approaches focus on activity biomarkers, such as messenger RNA, proteins or metabolites (Fig. 1).

### 4.1 Metagenomics

Metagenomics is defined as simultaneous analysis of the entire genomes of bacterial populations in a given ecosystem, giving insight into the phylogenetic and functional properties of cultivated and non-cultivated bacteria (Box 1). It is based on the direct extraction of total microbial DNA from an environment that circumvents culturing the organisms under study, and cloning of the DNA fragments into a cultured organism (usually *E. coli*) to capture it for study and preservation. This results in a metagenomic library that can be used for sequence-based and function-based analysis. Sequence-based analyses allow to get an overview of the genetic diversity and functional potential of an ecosystem, while function-based analyses allow to screen the library for novel functions of interest, and thus meet the two major questions: 'Who is out there?' and 'What are they doing?' (Lepage et al. 2013).

#### 4.1.1 Sequence-based Analysis

The emergence of next-generation sequencing technologies allowed sequencing the inserts of clones from metagenomic libraries in a rapid and cost-effective manner. The sequence reads are assigned to databases of reference sequences as GreenGene, NCBI non-redundant, Clusters of Orthologous Groups (COG), or Kyoto Encyclopedia of Genes and Genomes (KEGG) databases. The assembly of the sequencing data, which requires the use of specific algorithms, allows the reconstruction of all or part of the genome sequence of interest, and is the crucial step of the process. A taxonomic classification of the reads is calculated and allows describing the phylogenetic composition and diversity of the sample. However, sequence-based metagenomic approaches are primarily used to create a catalogue of the genetic potential at functional level that is present in an ecosystem. Comparison of the reads to either the COG and/or the KEGG classifications, enable identification of major functions and genes that encode enzymes of clinical interest. According to the highest scoring BLAST match to a protein sequence for which the functional role is known, the reads are assigned to enzymes and pathways. Finally, the collection of different datasets can be compared to estimate their distance and similarities.

The first sequence-based studies of the intestinal microbiota focused on the 16S rDNA sequences that are present in two metagenomic libraries constructed from the feces of six healthy individuals and six patients with Crohn's disease (Manichanh et al. 2006, 2008). These studies indicated, as expected from previous works, markedly reduced species diversity belonging to Firmicutes in patients with Crohn's disease compared to healthy subjects. Besides such targeted sequence-based metagenomic approaches, large-scale approaches have also been applied to investigate the genetic potential in the digestive tract ecosystem (Gill et al. 2006, Kurokawa et al. 2007). The first study based on analysis of samples from two healthy volunteers showed that many genes of the gut microbiome correspond to metabolic pathways that are essential for the host, such as biosynthesis of vitamins or metabolism of glycans, amino-acids and xenobiotics (Gill et al. 2006). The second study, which compared the genetic diversity in the colon of 13 healthy volunteers including adults and infants, indicated that human microbiota hosts an important activity of horizontal gene transfers (Kurokawa et al.

2007). In addition, this study demonstrated the chaotic pattern of unweaned infant's gut microbiome, while the adult-type microbiome appeared more complex with a high functional uniformity between individuals compare to infants. Gene families were also found to be commonly enriched in adult-type (237 genes) and infant-type (136 genes) microbiomes suggesting a bacterial adaptation to the intestinal environment.

In 2008, an expert group launched the International Human Microbiome Consortium (IHMC), whose objective is to harmonize practices and facilitate data comparisons among worldwide projects studying the human microbiome. The goal is to achieve an exhaustive analysis of the human microbiome to characterize the relationship between its composition and health and diseases and to use that knowledge to improve diseases prevention and/or treatment (http://www.human-microbiome.org/). Two major studies carried out in the MetaHIT project funded by European Union have been recently published establishing the existence of a functional metagenomic core in the human microbiome (Qin et al. 2010) and 'enterotypes' (Arumugam et al. 2011) (Box 2). Qin et al. established a reference gene catalog of the human gut microbiome based on sequencing 3.3 million microbial genes from fecal samples of 124 European individuals using metagenomics (Qin et al. 2010). This study revealed a microbiota-associated gene set that is approximately 150 times larger than the human host gene repertoire. Indeed, whereas the number of genes in the human genome is currently estimated at 23,000, the intestinal metagenome, or all combined genomes of our gut microbiota, exceeds nine billion. Deep metagenomic sequencing provided the opportunity to reveal the existence of a common set of microbial genes (metagenomic core) between individuals. Indeed, despite large variation in community membership, ~50% of the genes of an individual are shared by at least 50% of individuals (Qin et al. 2010). The same study also estimated at more than 1,000 bacterial species per person and more than 5,000 bacterial species in total the number of bacterial species comprising the gut microbiota. At present, it is considered that this number reflects

**Box 2.** Key issues.

| **Key issues** |
|---|
| • The human gut microbiota is composed of $10^{14}$ bacteria, 10 times the number of eukaryotic human cells, which are distributed in three major phyla: Firmicutes, Bacteroidetes and Actinobacteria. |
| • The human gut microbiota plays essential functions to the health of the host, so that any dysbiosis can lead to disease. |
| • In healthy adults, the gut microbiota presents a relatively stable composition affected by host genetics (individual specificity) and environmental factors such as diet and geographic location. A phylogenetic core of conserved species has been identified within individuals. |
| • Novel insights into the gut microbial diversity and functionality are provided by recently developed high-throughput technologies. |
| • The total number of bacterial genes (metagenome) present in the gut microbiota exceeds 150 times that of the human genome, and presents a huge diversity. |
| • A pool of genes is shared by a majority of individuals (functional metagenomic core) and is likely to be responsible of the conserved functions of the human digestive tract. Enterotypes, or microbiota profiles based on species composition and genes diversity, have been identified within the human population. |
| • Deepening a better knowledge of the healthy and pathological microbiota will lead to identify diagnostic and prognostic biomarkers of diseases, and to develop targeted preventive and therapeutic strategies such as modulation of the microbiota (fecal transplantation, probiotics/prebiotics). |

only a fraction of the actual digestive microbial diversity. Following an effort to characterize the 'average' human intestinal microbiota, Arumugam, Rea et al. observed an organization of intestinal metagenome in assemblies of genes and microbial taxa; they called 'enterotypes', which might respond differently to environmental changes such as diet and drug intake (Arumugam et al. 2011). To date, the individuals studied (n ~1000) were split into three 'enterotypes', each characterized by a particular ecological landscape dominated by either *Prevotella*, *Bacteroides* or *Ruminococcus*/*Methanobrevibacter*. They also noted that abundant molecular functions are not necessarily provided by abundant species, highlighting the importance of a functional analysis to understand microbial communities and to identify potential diagnostic biomarkers. The genes richness (diversity) also differs between the 'enterotypes'. Such quantitative metagenomics require both optimized sequencing for counting of genes that are carried on different genomes with highest accuracy, as each target gene has highly variable sequence, and highest number of sequences for detection of genes that are rare. In addition, important bioinformatics requirements are needed as it relies on comparing huge amounts of data generated within a project in order to identify new biomarkers.

A continuous effort is made to develop high-throughput sequencing technologies to generate an increasing number of sequences and bioinformatics tools for post-metagenomic analysis making it a more efficient reassembly of the genomic fragment that are sequenced. Direct sequencing techniques, also called 'shotgun', are now available by the development of high-speed sequencers and allow direct sequencing of metagenomic DNA of a sample without acid nucleic amplification step or cloning. Nonetheless, a complete coverage of the human gut microbial metagenome is at the moment almost impossible given the enormous abundance and complexity of the microbial populations that comprise the human gut microbiota. However, comparative metagenomics are relevant to identify genes that are specific or enriched in certain health or disease context.

### *4.1.2 Function-based Analysis*

Function-based metagenomics is a powerful approach to discover new bacterial genes that express a function. Functional screening is conditioned by the ability to express the respective genes and to secrete the genes products in the surrogate host that was used for the metagenome library construction. The power of the approach is that it does not require that the genes of interest be recognizable by sequence analysis. Contrarily, the major limitation is that many genes will not be expressed in any particular host bacterium selected for cloning. The diversity of the organism whose DNA has been successfully expressed in *E. coli* is surprisingly high (Gabor et al. 2004). Thus, *E. coli* is the most used host for functional metagenomics; however, other species such as *Lactococcus lactis*, *Streptomyces* or *Bacillus subtilis* can also be used to facilitate the heterologous expression of Gram-positive bacterial DNA.

Several functional metagenomic studies have focused on the carbohydrate active enzymes produced by digestive bacteria involved in the degradation of dietary fiber into metabolizable components. Many new families of glycoside hydrolases have been described, making the human gut metagenome one of the most important sources of carbohydrate active enzymes (Tasse et al. 2010). Other studies have focused on the interactions between the microbiota and the host, especially with intestinal epithelial cells that constitute the first line of contact and are involved in the regulatory of immune and inflammatory processes. Thus, bacterial genes modulating the nuclear factor κB pathway have been identified (Lakhdari et al. 2010). Functional analysis of metagenomic clones have led to assignment of function to numerous 'hypothetical proteins' in the databases. In the future, progress must be made to overcome the barrier to the heterologous gene expression and to efficiently identify rare clones in the libraries.

### 4.2 Other Meta-'omics'

Beyond the widely explored phylogenetic level, the metagenomics open the way to study the functionality of the human gut microbiome. Nevertheless, metagenomics can only describe the genetic potential of microorganisms present in our digestive tract, but does not provide information about their activities and gene expression *in situ.* The next level is thus to assess the actual functions exerted by the microbiota. This goal can be achieved by studying the transcripts, proteins and metabolites of the intestinal microbiota, which opens today under the name of 'microbiomics' and declined in metatranscriptomics, metaproteomics and metabolomics (Fig. 1). Comparison of metagenome and metatranscriptome analyses showed that the most expressed genes as the main activities were not linked to the most represented in the metagenome in a marine ecosystem (Frias-Lopez et al. 2008). In addition, DNA/RNA extraction procedures have an important impact on the results (Ó Cuív et al. 2011). Thus, understanding the role of gut microbial communities requires both standardization of procedures and combination of several meta-'omics' approaches.

Among the recent applications of 'omics' approaches in the study of the gut microbiota, metatranscriptomics was used to analyze fecal microbiota from 10 healthy humans and revealed a uniform functional pattern in healthy individual (Gosalbes et al. 2011). The main functional roles of the gut microbiota were carbohydrate metabolism, energy production and synthesis of cellular components. Metaproteomics and especially metabolomics methods have also been applied to study the gut microbiota. The determination of the fecal metabolome catalogue in healthy individuals, patients with Crohn's disease and patients with ulcerative colitis highlighted differences in metabolome, such as reduced levels of butyrate and acetate and elevated quantity of amino-acids in IBD subjects (Marchesi et al. 2007). Similarly, differences in gut metabolome were identified among healthy subjects and patients with colorectal cancer. Thus, a way that develops recently in the field of metabolomics is the search for metabolites biomarkers of diseases or prediction of their onset. The integration of data provided by all 'omics' approaches will help in understanding the functions of our intestinal microbes and their relation to the host; and particularly whether microbiota alterations associated with certain pathologies are a consequence rather than a cause of these disorders.

## 5. High-throughput Techniques for Studying the Impact of Probiotics on Gut Microbiota

Gut microbiota aberrations related to different human diseases, such as IBD or obesity, have been identified (Tremaroli and Bäckhed 2012). To develop intervention strategies targeted to modulate aberrant microbiota states towards healthy composition and activity, probiotics and the most likely combinations of probiotic strains seem promising. However, one of the major challenges is first to define the composition and activity of the normal microbiota linked to a health status. In this field, difficulties are still encountered as for example microbiota composition can differ according to the geographical location (Mueller et al. 2006). Nevertheless, metagenomic studies showed that the normal microbiota may be better defined at the functional level rather than at the phylogenetic one. In addition, probiotics may have an efficacy either by modulating community structure, which would result in changes in the functions of the ecosystem, or by modifying the metabolic functions without altering the composition of the gut microbiota, or both. Thus, in order to understand the functional shift and restoration of the healthy condition induced by the probiotic, studies should go beyond the taxonomic composition and explore the functions of the gut microbiota (Qin et al. 2010). So far, gut microbiome data available from human intervention studies with probiotics are very limited. Incorporating such data with studies on the host genes response to microbiota modulation and probiotics administration will allow us to understand host–microbe interactions and therefore how the gut microbiota affects our physiology.

High-throughput approaches have been exploited in some studies on the impact of probiotics on the gut microbiota and the interaction with other gut members. In mice and pig, the administration of a bacteriocin (Abp118) producing strain of *Lactobacillus salivarius* showed an impact on members of the Firmicutes phylum using pyrosequencing, which was associated with the bacteriocin production (Riboulet-Bisson et al. 2012). Surprisingly, in both models, *L. salivarius* administration and production of Abp118 had an effect on Gram-negative microorganisms, even though Abp118 is normally not active *in vitro* against this group of microorganisms. New molecular approaches have also been used in combination with metabolomics to study alteration of energy homeostasis using gnotobiotic mice. The effect of dietary supplementation with *Bifidobacterium breve* strains on the gut microbiota composition and fat distribution and composition has been studied by 16S rRNA pyrosequencing and gas-liquid chromatography respectively (Wall et al. 2012). A variable influence on modulating the gut microbial communities and the fatty acid metabolism was observed depending on the *B. breve* strain administered. The impact of probiotic strains on the catabolic activity of *Bacteroides thetaiotaomicron* was studied by others. Indeed, obesity is accompanied by the enrichment of bacteria belonging to the Firmicutes, which purportedly derive energy in the form of short-chain fatty acids, thus contributing to the total energy of the host. In a simplified germ-free mouse model colonized with *B. thetaiotaomicron*, *Bifidobacterium longum* was shown to expand the catabolic activity of *B. thetaiotaomicron* (Sonnenburg et al. 2006). However, another

*Bifidobacterium* species, *B. animalis,* induced no change in carbohydrate usage capabilities. Thus, also in this study, the probiotic impact on gut microbiota members was highly species specific. A similar up-regulation of genes involved in carbohydrate transport and metabolism was reported with *Lactobacillus plantarum* WCFS1 in germ-free mice maintained on high-fat/high-sugar diet (Marco et al. 2009).

Another recent study analyzed the impact of consumption of a probiotic containing a consortium of species (*Bifidobacterium*, *Lactobacillus*, *Lactococcus* and *Streptococcus*) on fecal microbiome and metatranscriptome in human and mice (McNulty et al. 2011). No significant changes were observed in either bacterial species composition or the composition of known enzymes encoding genes. However, interestingly RNA sequence analysis revealed significant changes in expression of microbiome-encoded enzymes involved in numerous metabolic pathways, most prominently those related to carbohydrate metabolism. In humans, these changes were confined to the period of the probiotic consumption. However, all these studies are preliminary and the impact of probiotics on our microbiome may be clear only when our knowledge of this ecosystem has progressed.

## 6. Concluding Remarks and Future Perspectives

Evidence is building that the resident community of microbes of our digestive tract, called the microbiome, plays a major role in health and disease. Molecular methods based on 16S rRNA gene is redefining our vision of this organ, revealing the unsuspected huge diversity of the gut microbiota and its individual-specificity. Recently, advances in sequencing technology and the development of metagenomics and bioinformatics tools have offered new insights into the functional potential of this ecosystem. However, genetic sequenced-based data such as 16S rRNA or metagenomics data do not give information about the functions exerted by the microbial communities in the gut. To elucidate microbiome functions other omics technologies, which are the large-scale studies of mRNA transcripts, proteins and metabolites, have been developed. It is now well established due to the complexity of the gut microbiome that only integrative 'omics' approaches are likely to effectively characterize this ecosystem and to establish links between microbial populations and their functions, host factors and some pathology. Their potential to develop diagnosis or therapy monitoring biomarkers and new therapeutic strategies has spawned many human microbiome projects worldwide. At present, large-scale human cohorts are required to validate the robustness of identified microbiome-health status relationships. A broader representation of the human population, including multiple age groups, wider geographical origin, different diet modes, twins and unrelated individuals and extended follow-up will distinguish immutable characteristics of the microbiota of group effects.

Most studies of the human gut microbiota are performed on fecal samples representing the luminal microbiota that differs from the mucosa-associated microbiota. Indeed, surface-adherent and luminal bacterial populations are distinct and may fulfill different roles within the ecosystem. Indeed, the biofilm-like architecture of the mucosal microbiota, in close contact with the underlying gut epithelium, may facilitate beneficial functions including nutrient exchange and induction of host innate immunity (Sonnenburg et al. 2004). Thus, the mucosa-associated microbiota also needs to be further investigated in relation to health and disease.

Despite their considerable power, molecular approaches have limitations and some critical technical steps are likely to have an influence on the results obtained and avoid comparing data from different studies. Sample conservation, cell lysis, DNA extraction, DNA amplification and cloning procedures can have a major impact on the analysis of phylogenetic and genetic diversity of the gut microbiota. For example, limited sensitivity of broad-range PCR may hinder detection of rare phylotypes and the use of different primers may lead to a different representation of certain bacterial groups (Hill et al. 2010). In addition, some bacteria, especially among Gram-positive are hardly lysed causing difficulty in assessing their presence or quantity. In this regard, it is surprising that in numerous metagenomic studies sequences from bifidobacteria are either absent or present at low levels, whereas using other methods these microorganisms appear to be present at relatively high levels in the same samples. Similarly, in one of the first metagenomic study of the gut microbiome sequences affiliated to Bacteroidetes, one of the dominant microbial groups in the colon, were not detected (Gill et al. 2006). Another point to consider in metagenomic studies is the sequencing depth, as it correlates directly with the cost, as well as the management and analysis of the massive amount of generated data.

It is important to underline that the molecular assessment of the gut bacterial communities is limited to the dominant faction. Therefore, our knowledge about the sub-dominant bacteria, representing less than $10^8$ bacteria per g of feces, is very incomplete and comes mainly from cultivated isolates. In addition, the molecular approach is characterized by its inability to distinguish live from dead organisms contrary to culture. Thus, complementary approaches are needed. Novel cultivation strategies focusing on the particular nutrient intake of certain bacteria according to their metabolic capabilities have allowed the isolation of previously unknown species related to *Verrucomicrobia* and *Clostridium* XIV, a cluster of Gram-positive bacteria (Barcenilla et al. 2000, Zoetendal et al. 2003). A recent study combining culture techniques with metagenomics demonstrated that thousands of isolates from a donor can be clonally archived to create personalized microbiota collections (Goodman et al. 2011). Thus, gnotobiotic animal models colonized by defined microbiota characteristic of host status may be useful tools to determine the impact of diet modulations, probiotic intake, fecal transplant and genetic defects.

Finally, molecular methods and especially 'omics' approaches have considerably increased our knowledge of the human gut microbiome. The next step is to integrate all datasets generated by 'omics' analyzes in a coherent biological model of the human gut taking into account host-microbiome interactions. This would help to characterize the pathogenic process and allow identification of prognostic and diagnosis biomarkers, and new therapeutic targets.

## References

http://rdp.cme.msu.edu/

Arumugam, M., J. Raes, E. Pelletier, D. Le Paslier, T. Yamada, D.R. Mende, G.R. Fernandes, J. Tap, T. Bruls, J.-M. Batto, M. Bertalan, N. Borruel, F. Casellas, L. Fernandez, L. Gautier, T. Hansen, M. Hattori, T. Hayashi, M. Kleerebezem, K. Kurokawa, M. Leclerc, F. Levenez, C. Manichanh, H.B. Nielsen, T. Nielsen, N. Pons, J. Poulain, J. Qin, T. Sicheritz-Ponten, S. Tims, D. Torrents, E. Ugarte, E.G. Zoetendal, J. Wang, F. Guarner, O. Pedersen, W.M. de Vos, S. Brunak, J. Doré, J. Weissenbach, S.D. Ehrlich and P. Bork. 2011. Enterotypes of the human gut microbiome. Nature 473: 174–180.

Barcenilla, A., S.E. Pryde, C. Martin, S.H. Duncan, C.S. Stewart, C. Henderson and H.J. Flint. 2000. Phylogenetic relationships of butyrate-producing bacteria from the human gut. Appl. Environ. Microbiol. 66: 1654–1661.

Call, D.R., M.K. Bakko, M.J. Krug and M.C. Roberts. 2003. Identifying antimicrobial resistance genes with DNA microarrays. Antimicrob. Agents Chemother. 47: 3290–3295.

Castiglioni, B., E. Rizzi, A. Frosini, K. Sivonen, P. Rajaniemi, A. Rantala, M.A. Mugnai, S. Ventura, A. Wilmotte, C. Boutte, S. Grubisic, P. Balthasart, C. Consolandi, R. Bordoni, A. Mezzelani, C. Battaglia and G. De Bellis. 2004. Development of a universal microarray based on the ligation detection reaction and 16S rrna gene polymorphism to target diversity of cyanobacteria. Appl. Environ. Microbiol. 70: 7161–7172.

De La Cochetière, M.F., T. Durand, P. Lepage, A. Bourreille, J.P. Galmiche and J. Doré. 2005. Resilience of the dominant human fecal microbiota upon short-course antibiotic challenge. J. Clin. Microbiol. 43: 5588–5592.

Derrien, M., P. Van Baarlen, G. Hooiveld, E. Norin, M. Müller and W.M. de Vos. 2011. Modulation of Mucosal Immune Response, Tolerance, and Proliferation in Mice Colonized by the Mucin-Degrader Akkermansia muciniphila. Front. Microbiol. 2: 166.

DeSantis, T.Z., E.L. Brodie, J.P. Moberg, I.X. Zubieta, Y.M. Piceno and G.L. Andersen. 2007. High-density universal 16S rRNA microarray analysis reveals broader diversity than typical clone library when sampling the environment. Microb. Ecol. 53: 371–383.

Eckburg, P.B., E.M. Bik, C.N. Bernstein, E. Purdom, L. Dethlefsen, M. Sargent, S.R. Gill, K.E. Nelson and D.A. Relman. 2005. Diversity of the human intestinal microbial flora. Science 308: 1635–1638.

Fallani, M., D. Young, J. Scott, E. Norin, S. Amarri, R. Adam, M. Aguilera, S. Khanna, A. Gil, C.A. Edwards and J. Doré. 2010. Intestinal microbiota of 6-week-old infants across Europe: geographic influence beyond delivery mode, breast-feeding, and antibiotics. J. Pediatr. Gastroenterol. Nutr. 51: 77–84.

Fallani, M., S. Amarri, A. Uusijarvi, R. Adam, S. Khanna, M. Aguilera, A. Gil, J.M. Vieites, E. Norin, D. Young, J.A. Scott, J. Doré and C.A. Edwards. 2011. Determinants of the human infant intestinal microbiota after the introduction of first complementary foods in infant samples from five European centres. Microbiology 157: 1385–1392.

Frias-Lopez, J., Y. Shi, G.W. Tyson, M.L. Coleman, S.C. Schuster, S.W. Chisholm and E.F. Delong. 2008. Microbial community gene expression in ocean surface waters. Proc. Natl. Acad. Sci. U.S.A. 105: 3805–3810.

Furet, J.-P., O. Firmesse, M. Gourmelon, C. Bridonneau, J. Tap, S. Mondot, J. Doré and G. Corthier. 2009. Comparative assessment of human and farm animal faecal microbiota using real-time quantitative PCR. FEMS Microbiol. Ecol. 68: 351–362.

Gabor, E.M., W.B.L. Alkema and D.B. Janssen. 2004. Quantifying the accessibility of the metagenome by random expression cloning techniques. Environ. Microbiol. 6: 879–886.

Gill, S.R., M. Pop, R.T. Deboy, P.B. Eckburg, P.J. Turnbaugh, B.S. Samuel, J.I. Gordon, D.A. Relman, C.M. Fraser-Liggett and K.E. Nelson. 2006. Metagenomic analysis of the human distal gut microbiome. Science 312: 1355–1359.

Goodman, A.L., G. Kallstrom, J.J. Faith, A. Reyes, A. Moore, G. Dantas and J.I. Gordon. 2011. Extensive personal human gut microbiota culture collections characterized and manipulated in gnotobiotic mice. Proc. Natl. Acad. Sci. U.S.A. 108: 6252–6257.

Gosalbes, M.J., A. Durbán, M. Pignatelli, J.J. Abellan, N. Jiménez-Hernández, A.E. Pérez-Cobas, A. Latorre and A. Moya. 2011. Metatranscriptomic approach to analyze the functional human gut microbiota. PLoS ONE 6: e17447.

Hayashi, H., M. Sakamoto and Y. Benno. 2002. Phylogenetic analysis of the human gut microbiota using 16S rDNA clone libraries and strictly anaerobic culture-based methods. Microbiol. Immunol. 46: 535–548.

Hill, J.E., W.M.U. Fernando, G.A. Zello, R.T. Tyler, W.J. Dahl and A.G. Van Kessel. 2010. Improvement of the representation of bifidobacteria in fecal microbiota metagenomic libraries by application of the cpn60 universal primer cocktail. Appl. Environ. Microbiol. 76: 4550–4552.

Hooper, L.V., M.H. Wong, A. Thelin, L. Hansson, P.G. Falk and J.I. Gordon. 2001. Molecular analysis of commensal host-microbial relationships in the intestine. Science 291: 881–884.

Hooper, L.V., D.R. Littman and A.J. Macpherson. 2012. Interactions Between the Microbiota and the Immune System. Science 336: 1268–1273.

Huyghe, A., P. Francois, Y. Charbonnier, M. Tangomo-Bento, E.-J. Bonetti, B.J. Paster, I. Bolivar, D. Baratti-Mayer, D. Pittet and J. Schrenzel. 2008. Novel microarray design strategy to study complex bacterial communities. Appl. Environ. Microbiol. 74: 1876–1885.

Kang, S., S.E. Denman, M. Morrison, Z. Yu, J. Dore, M. Leclerc and C.S. McSweeney. 2010. Dysbiosis of fecal microbiota in Crohn's disease patients as revealed by a custom phylogenetic microarray. Inflamm. Bowel Dis. 16: 2034–2042.

Koren, O., J.K. Goodrich, T.C. Cullender, A. Spor, K. Laitinen, H.K. Bäckhed, A. Gonzalez, J.J. Werner, L.T. Angenent, R. Knight, F. Bäckhed, E. Isolauri, S. Salminen and R.E. Ley. 2012. Host remodeling of the gut microbiome and metabolic changes during pregnancy. Cell 150: 470–480.

Kurokawa, K., T. Itoh, T. Kuwahara, K. Oshima, H. Toh, A. Toyoda, H. Takami, H. Morita, V.K. Sharma, T.P. Srivastava, T.D. Taylor, H. Noguchi, H. Mori, Y. Ogura, D.S. Ehrlich, K. Itoh, T. Takagi, Y. Sakaki, T. Hayashi and M. Hattori. 2007. Comparative metagenomics revealed commonly enriched gene sets in human gut microbiomes. DNA Res. 14: 169–181.

Lakhdari, O., A. Cultrone, J. Tap, K. Gloux, F. Bernard, S.D. Ehrlich, F. Lefèvre, J. Doré and H.M. Blottière. 2010. Functional metagenomics: a high throughput screening method to decipher microbiota-driven NF-κB modulation in the human gut. PLoS ONE 5: e13092.

Lepage, P., P. Seksik, M. Sutren, M.-F. De la Cochetière, R. Jian, P. Marteau and J. Doré. 2005. Biodiversity of the mucosa-associated microbiota is stable along the distal digestive tract in healthy individuals and patients with IBD. Inflamm. Bowel Dis. 11: 473–480.

Lepage, P., M.C. Leclerc, M. Joossens, S. Mondot, H.M. Blottière, J. Raes, D. Ehrlich and J. Doré. 2013. A metagenomic insight into our gut's microbiome. Gut 62: 146–158.

Ley, R.E., P.J. Turnbaugh, S. Klein and J.I. Gordon. 2006. Microbial ecology: human gut microbes associated with obesity. Nature 444: 1022–1023.

Loy, A., C. Schulz, S. Lücker, A. Schöpfer-Wendels, K. Stoecker, C. Baranyi, A. Lehner and M. Wagner. 2005. 16S rRNA gene-based oligonucleotide microarray for environmental monitoring of the betaproteobacterial order "Rhodocyclales". Appl. Environ. Microbiol. 71: 1373–1386.

Mackie, R.I., A. Sghir and H.R. Gaskins. 1999. Developmental microbial ecology of the neonatal gastrointestinal tract. Am. J. Clin. Nutr. 69: 1035S–1045S.

Manichanh, C., L. Rigottier-Gois, E. Bonnaud, K. Gloux, E. Pelletier, L. Frangeul, R. Nalin, C. Jarrin, P. Chardon, P. Marteau, J. Roca and J. Dore. 2006. Reduced diversity of faecal microbiota in Crohn's disease revealed by a metagenomic approach. Gut 55: 205–211.

Manichanh, C., C.E. Chapple, L. Frangeul, K. Gloux, R. Guigo and J. Dore. 2008. A comparison of random sequence reads versus 16S rDNA sequences for estimating the biodiversity of a metagenomic library. Nucleic Acids Res. 36: 5180–5188.

Marchesi, J.R., E. Holmes, F. Khan, S. Kochhar, P. Scanlan, F. Shanahan, I.D. Wilson and Y. Wang. 2007. Rapid and noninvasive metabonomic characterization of inflammatory bowel disease. J. Proteome Res. 6: 546–551.

Marco, M.L., T.H.F. Peters, R.S. Bongers, D. Molenaar, S. van Hemert, J.L. Sonnenburg, J.I. Gordon and M. Kleerebezem. 2009. Lifestyle of *Lactobacillus plantarum* in the mouse caecum. Environ. Microbiol. 11: 2747–2757.

Marco, M.L., M.C. de Vries, M. Wels, D. Molenaar, P. Mangell, S. Ahrne, W.M. de Vos, E.E. Vaughan and M. Kleerebezem. 2010. Convergence in probiotic Lactobacillus gut-adaptive responses in humans and mice. ISME J 4: 1481–1484.

McNulty, N.P., T. Yatsunenko, A. Hsiao, J.J. Faith, B.D. Muegge, A.L. Goodman, B. Henrissat, R. Oozeer, S. Cools-Portier, G. Gobert, C. Chervaux, D. Knights, C.A. Lozupone, R. Knight, A.E. Duncan, J.R. Bain, M.J. Muehlbauer, C.B. Newgard, A.C. Heath and J.I. Gordon. 2011. The Impact of a Consortium of Fermented Milk Strains on the Gut Microbiome of Gnotobiotic Mice and Monozygotic Twins. Sci. Transl. Med. 3: 106ra106.

Medini, D., D. Serruto, J. Parkhill, D.A. Relman, C. Donati, R. Moxon, S. Falkow and R. Rappuoli. 2008. Microbiology in the post-genomic era. Nat. Rev. Microbiol. 6: 419–430.

Metzker, M.L. 2010. Sequencing technologies—the next generation. Nat. Rev. Genet. 11: 31–46.

Mueller, S., K. Saunier, C. Hanisch, E. Norin, L. Alm, T. Midtvedt, A. Cresci, S. Silvi, C. Orpianesi, M.C. Verdenelli, T. Clavel, C. Koebnick, H.-J.F. Zunft, J. Doré and M. Blaut. 2006. Differences in fecal microbiota in different European study populations in relation to age, gender, and country: a cross-sectional study. Appl. Environ. Microbiol. 72: 1027–1033.

Ó Cuív, P., D. Aguirre de Cárcer, M. Jones, E.S. Klaassens, D.L. Worthley, V.L.J. Whitehall, S. Kang, C.S. McSweeney, B.A. Leggett and M. Morrison. 2011. The effects from DNA extraction methods on the evaluation of microbial diversity associated with human colonic tissue. Microb. Ecol. 61: 353–362.

O'Hara, A.M. and F. Shanahan. 2006. The gut flora as a forgotten organ. EMBO Rep. 7: 688–693.

Palmer, C., E.M. Bik, M.B. Eisen, P.B. Eckburg, T.R. Sana, P.K. Wolber, D.A. Relman and P.O. Brown. 2006. Rapid quantitative profiling of complex microbial populations. Nucleic Acids Res. 34: e5.

Palmer, C., E.M. Bik, D.B. DiGiulio, D.A. Relman and P.O. Brown. 2007. Development of the human infant intestinal microbiota. PLoS Biol. 5: e177.

Qin, J., R. Li, J. Raes, M. Arumugam, K.S. Burgdorf, C. Manichanh, T. Nielsen, N. Pons, F. Levenez, T. Yamada, D.R. Mende, J. Li, J. Xu, S. Li, D. Li, J. Cao, B. Wang, H. Liang, H. Zheng, Y. Xie, J. Tap, P. Lepage, M. Bertalan, J.-M. Batto, T. Hansen, D. Le Paslier, A. Linneberg, H.B. Nielsen, E. Pelletier, P. Renault, T. Sicheritz-Ponten, K. Turner, H. Zhu, C. Yu, S. Li, M. Jian, Y. Zhou, Y. Li, X. Zhang, S. Li, N. Qin, H. Yang, J. Wang, S. Brunak, J. Doré, F. Guarner, K. Kristiansen, O. Pedersen, J. Parkhill, J. Weissenbach, P. Bork, S.D. Ehrlich and J. Wang. 2010. A human gut microbial gene catalogue established by metagenomic sequencing. Nature 464: 59–65.

Rajilić-Stojanović, M., H.G.H.J. Heilig, S. Tims, E.G. Zoetendal and W.M. de Vos. 2012. Long-term monitoring of the human intestinal microbiota composition. Environ. Microbiol.

Riboulet-Bisson, E., M.H.J. Sturme, I.B. Jeffery, M.M. O'Donnell, B.A. Neville, B.M. Forde, M.J. Claesson, H. Harris, G.E. Gardiner, P.G. Casey, P.G. Lawlor, P.W. O'Toole and R.P. Ross. 2012. Effect of *Lactobacillus salivarius* bacteriocin Abp118 on the mouse and pig intestinal microbiota. PLoS ONE 7: e31113.

Rigottier-Gois, L., A.G. Bourhis, G. Gramet, V. Rochet and J. Doré. 2003. Fluorescent hybridisation combined with flow cytometry and hybridisation of total RNA to analyse the composition of microbial communities in human faeces using 16S rRNA probes. FEMS Microbiol. Ecol. 43: 237–245.

Rougé, C., O. Goldenberg, L. Ferraris, B. Berger, F. Rochat, A. Legrand, U.B. Göbel, M. Vodovar, M. Voyer, J.-C. Rozé, vD. Darmaun, H. Piloquet, M.-J. Butel and M.-F. de La Cochcetière. 2010. Investigation of the intestinal microbiota in preterm infants using different methods. Anaerobe 16: 362–370.

Savage, D.C. 1977. Microbial ecology of the gastrointestinal tract. Annu. Rev. Microbiol. 31: 107–133.

Savage, D.C. 2001. Microbial biota of the human intestine: a tribute to some pioneering scientists. Curr. Issues Intest. Microbiol. 2: 1–15.

Sokol, H., P. Lepage, P. Seksik, J. Doré and P. Marteau. 2006. Temperature gradient gel electrophoresis of fecal 16S rRNA reveals active *Escherichia coli* in the microbiota of patients with ulcerative colitis. J. Clin. Microbiol. 44: 3172–3177.

Sokol, H., B. Pigneur, L. Watterlot, O. Lakhdari, L.G. Bermúdez-Humarán, J.-J. Gratadoux, S. Blugeon, C. Bridonneau, J.-P. Furet, G. Corthier, C. Grangette, N. Vasquez, P. Pochart, G. Trugnan, G. Thomas, H.M. Blottière, J. Doré, P. Marteau, P. Seksik and P. Langella. 2008. *Faecalibacterium prausnitzii* is an anti-inflammatory commensal bacterium identified by gut microbiota analysis of Crohn disease patients. Proc. Natl. Acad. Sci. U.S.A. 105: 16731–16736.

Sonnenburg, J.L., L.T. Angenent and J.I. Gordon. 2004. Getting a grip on things: how do communities of bacterial symbionts become established in our intestine? Nat. Immunol. 5: 569–573.

Sonnenburg, J.L., C.T.L. Chen and J.I. Gordon. 2006. Genomic and metabolic studies of the impact of probiotics on a model gut symbiont and host. PLoS Biol. 4: e413.

Suau, A., R. Bonnet, M. Sutren, J.J. Godon, G.R. Gibson, M.D. Collins and J. Doré. 1999. Direct analysis of genes encoding 16S rRNA from complex communities reveals many novel molecular species within the human gut. Appl. Environ. Microbiol. 65: 4799–4807.

Tap, J., S. Mondot, F. Levenez, E. Pelletier, C. Caron, J.-P. Furet, E. Ugarte, R. Muñoz-Tamayo, D.L.E. Paslier, R. Nalin, J. Dore and M. Leclerc. 2009. Towards the human intestinal microbiota phylogenetic core. Environ. Microbiol. 11: 2574–2584.

Tasse, L., J. Bercovici, S. Pizzut-Serin, P. Robe, J. Tap, C. Klopp, B.L. Cantarel, P.M. Coutinho, B. Henrissat, M. Leclerc, J. Doré, P. Monsan, M. Remaud-Simeon and G. Potocki-Veronese. 2010. Functional metagenomics to mine the human gut microbiome for dietary fiber catabolic enzymes. Genome Res. 20: 1605–1612.

Tremaroli, V. and F. Bäckhed. 2012. Functional interactions between the gut microbiota and host metabolism. Nature 489: 242–249.

Turnbaugh, P.J., M. Hamady, T. Yatsunenko, B.L. Cantarel, A. Duncan, R.E. Ley, M.L. Sogin, W.J. Jones, B.A. Roe, J.P. Affourtit, M. Egholm, B. Henrissat, A.C. Heath, R. Knight and J.I. Gordon. 2009. A core gut microbiome in obese and lean twins. Nature 457: 480–484.

Vanhoutte, T., G. Huys, E. Brandt and J. Swings. 2004. Temporal stability analysis of the microbiota in human feces by denaturing gradient gel electrophoresis using universal and group-specific 16S rRNA gene primers. FEMS Microbiol. Ecol. 48: 437–446.

Wall, R., T.M. Marques, O. O'Sullivan, R.P. Ross, F. Shanahan, E.M. Quigley, T.G. Dinan, B. Kiely, G.F. Fitzgerald, P.D. Cotter, F. Fouhy and C. Stanton. 2012. Contrasting effects of *Bifidobacterium breve* NCIMB 702258 and *Bifidobacterium breve* DPC 6330 on the composition of murine brain fatty acids and gut microbiota. Am. J. Clin. Nutr. 95: 1278–1287.

Wang, Q., G.M. Garrity, J.M. Tiedje and J.R. Cole. 2007. Naive Bayesian classifier for rapid assignment of rRNA sequences into the new bacterial taxonomy. Appl. Environ. Microbiol. 73: 5261–5267.

Woese, C.R., O. Kandler and M.L. Wheelis. 1990. Towards a natural system of organisms: proposal for the domains Archaea, Bacteria, and Eucarya. Proc. Natl. Acad. Sci. U.S.A. 87: 4576–4579.

Zoetendal, E.G., A.D. Akkermans and W.M. De Vos. 1998. Temperature gradient gel electrophoresis analysis of 16S rRNA from human fecal samples reveals stable and host-specific communities of active bacteria. Appl. Environ. Microbiol. 64: 3854–3859.

Zoetendal, E.G., C.M. Plugge, A.D.L. Akkermans and W.M. de Vos. 2003. Victivallis vadensis gen. nov., sp. nov., a sugar-fermenting anaerobe from human faeces. Int. J. Syst. Evol. Microbiol. 53: 211–215.

Zoetendal, E.G., M. Rajilic-Stojanovic and W.M. de Vos. 2008. High-throughput diversity and functionality analysis of the gastrointestinal tract microbiota. Gut 57: 1605–1615.

# CHAPTER 9

# Engineering of Probiotics for Technological and Physiological Functionalities

*Roy D. Sleator*

## 1. Introduction

Patho-biotechnology focuses on equipping non-pathogenic or probiotic bacteria with the genetic constructs necessary to overcome stresses encountered in the food production and storage environment (e.g., spray and freeze drying; experienced during product formulation (Augustin and Hemar 2009, Maa and Prestrelski 2000)) as well as the antimicrobial defences faced during host transit and/or colonisation (e.g., gastric acidity, bile, low iron and elevated osmolarity (Hill et al. 2001)).

The patho-biotechnology strategy can be further divided into three distinct approaches. The first tackles the issue of probiotic storage and delivery by cloning and expression of pathogen specific stress survival mechanisms (facilitating improved survival at extremes of temperature and water availability), thus countering reductions in probiotic numbers which can occur during manufacture and storage of delivery matrices (such as foods and tablet formulations). The second approach aims to improve host colonisation by expression of host specific survival strategies (or virulence associated factors) thereby positively affecting the therapeutic efficacy of the probiotic. The final approach involves the development of so called 'designer probiotics'; strains which specifically target invading pathogens by blocking crucial ligand-receptor interactions between the pathogen and host cell (Paton et al. 2006, Sleator and Hill 2008c).

Department of Biological Sciences, Cork Institute of Technology, Rossa Avenue, Bishopstown, Cork.
E-mail: roy.sleator@cit.ie

## 2. Improving Probiotic Stress Tolerance

Perhaps the most important stresses encountered during food production and/or tablet formulation are temperature and water availability ($a_w$) (Hill 2001, Sleator and Hill 2007a). The ability to overcome such stresses is thus an extremely desirable trait in the selection of commercially important probiotic strains. A common strategy employed by bacteria to overcome both of these stress is the accumulation of protective compounds, for example the trimethyl ammonium compound glycine betaine. These compatible solutes, so called because of their compatibility with vital cellular processes at high cytoplasmic concentrations, stabilise protein structure and function at low temperatures and prevent water loss from the cell and plasmolysis under low $a_w$ conditions (Sleator and Hill 2002, Sleator et al. 2003).

Improving a strain's ability to accumulate compatible solutes is thus an obvious first step in the development of more robust probiotic strains. The foodborne pathogen *Listeria monocytogenes* (probably the best studied pathogen in terms of compatible solute accumulation; Sleator et al. 2003) possesses three distinct betaine uptake systems (BetL, Gbu and OpuC), the simplest of which in genetic terms is the secondary betaine transporter BetL (Sleator et al. 1999, 2000, 2003, Wemekamp-Kamphuis et al. 2004). By cloning the *betL* gene under the transcriptional control of the nisin inducible promoter P*nisA* we were able to assess the ability of BetL to contribute to probiotic survival under a variety of stresses encountered during food and/or tablet manufacture (Sheehan et al. 2006). We proposed that expressing the listerial *betL* gene against a *Lactobacillus salivarius* background might increase betaine accumulation by the probiotic strain, thereby improving its stress tolerance profile. As expected, following nisin induction, the *betL* complemented *L. salivarius* strain exhibited a significant increase in betaine accumulation compared to the wild type (Fig. 1a). Indeed, sufficient BetL was produced to confer increased salt tolerance, with growth of the transformed construct being observed at salt concentrations (~7% NaCl) far in excess of that observed for lactobacilli (Fig. 1b). In addition to increased osmotolerance the BetL$^+$ strain showed significantly improved resistance to both chill and cryotolerance (2 logs greater survival than the control at –20°C and 0.5 logs greater at –70°C) as well as freeze-drying (36% survival as compared to 18% for the control strain) and spray-drying (1.4% compared to 0.3%); common stresses encountered during food and/or tablet formulation. Furthermore, the presence of BetL resulted in a significant improvement in barotolerance (Fig. 1c). This is particularly significant given that high pressure processing is gaining increasing popularity as a novel non-thermal mechanism of food processing and preservation (Considine et al. 2008, Smiddy 2005, Smiddy et al. 2004).

Thus, improving the stress tolerance profile of probiotic cultures significantly improves tolerance to processing stress and prolongs survival during subsequent storage. This in turn contributes to a significantly larger proportion of the administered probiotic reaching the desired location (e.g., the gastrointestinal tract) in a bioactive form.

In addition to the *ex vivo* stresses encountered during food/tablet manufacture and storage, probiotic bacteria must also overcome the physiological defences of the host in order to survive at sufficient numbers within the gastrointestinal tract to exert

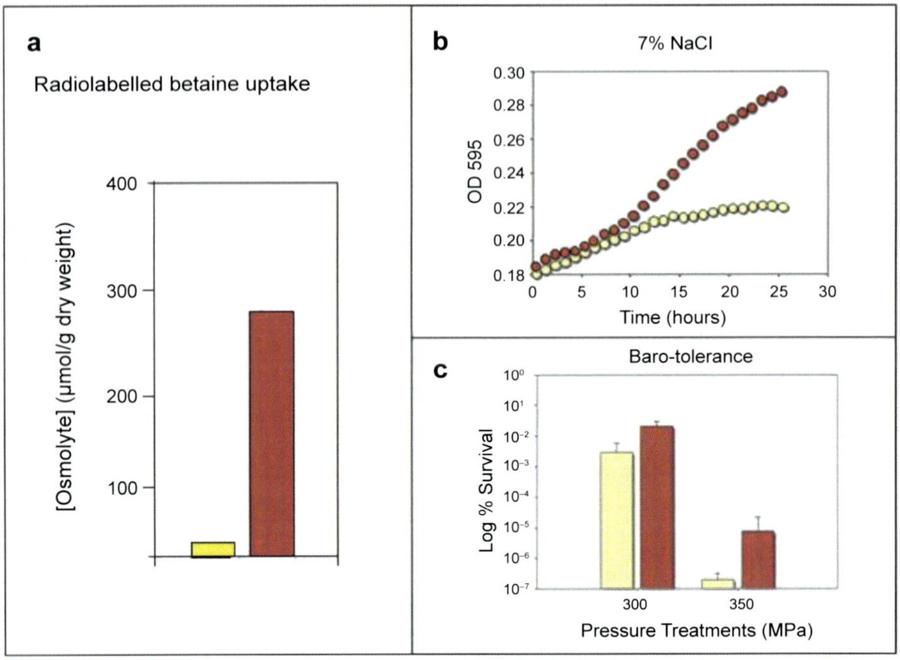


**Fig. 1.** (a) [14C] Glycine betaine uptake in the *Lactobacillus salivarius* wild type (yellow bar) and the BetL complemented strain UCC118-BetL$^+$ (red bar). (b) Growth of *L. salivarius* wild type (yellow circles) and UCC118-BetL$^+$ (red circles) in MRS broth with 7% added NaCl. (c) High-pressure-induced inactivation of *L. Salivarius* wild type (yellow bars) and UCC118-BetL$^+$ (red bars). Adapted from Sheehan et al. (2006).

a therapeutic effect. Recently, we demonstrated that cloning *betL* into *Bifidobacterium breve* UCC2003, significantly improved the tolerance of the probiotic to gastric juice (Sheehan et al. 2007). Interestingly, in support of this observation Termont et al. (2006) also reported improved tolerance to gastric juice in a *L. lactis* strain expressing the *E. coli* trehalose synthesis genes, thus suggesting a novel protective role for compatible solutes in the gastric environment. Genes encoding carnitine uptake, which we have shown to contribute to the gastrointestinal survival of *L. Monocytogenes* (Sleator et al. 2001) are another obvious candidate for this approach. Once free of the stomach, bacteria enter the upper small intestine where they are exposed to low $a_w$ conditions (equivalent to 0.3 M NaCl). Consistent with our previous observations with *L. salivarius* UCC118 (Sheehan et al. 2006), a significant osmoprotective effect was observed following the introduction of *betL* into *B. breve*, facilitating growth of the probiotic in conditions similar to those encountered *in vivo* (1.5% NaCl and 6% sucrose, both of which approximate the osmolarity of the gut). Furthermore, whilst stable colonisation of the murine intestine was achieved by oral administration of *B. breve* UCC2003, strains harbouring BetL were recovered at significantly higher levels in the faeces, intestines and caecum of inoculated animals. Finally, in addition to improved gastric transit and intestinal persistence, the addition of BetL improved the clinical efficacy of the probiotic culture; mice fed *B. breve* UCC2003 ($BetL^+$) exhibited significantly lower levels of systemic infection compared to the control strain following oral inoculation with *L. monocytogenes* (Fig. 2). This is, to the best of our knowledge, the first clear evidence of an enhanced therapeutic effect following precise bio-engineering of a probiotic strain.

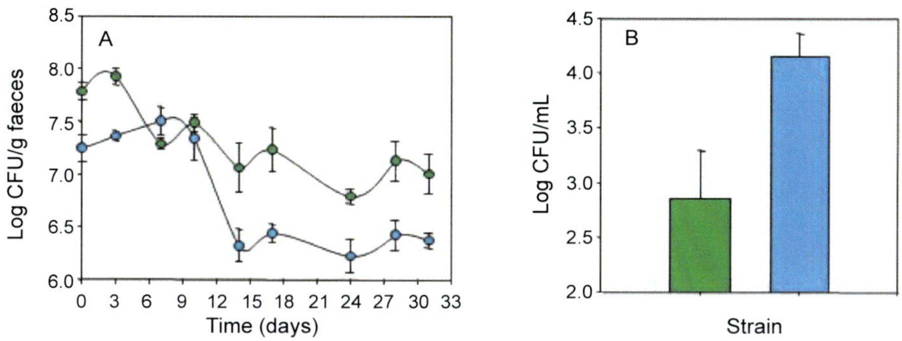


**Fig. 2.** (A) Recovery of *Bifidobacterium breve* UCC2003-$BetL^+$ (green circles) and UCC2003 wild type (blue circles) from female BALB/c mice over 32 days of analysis. Faeces for bacteriological analysis were obtained from five mice in each treatment group and viable counts of *B. breve* UCC2003 derivatives were determined. (B) Listerial infection in the spleens of BALB/c mice. Animals were fed ~$10^9$ CFU $ml^{-1}$ of either UCC2003-$BetL^+$ or UCC2003 wild type for three consecutive days. On the fourth day, all animals were infected with ~$10^{11}$ CFU $ml^{-1}$ *Listeria monocytogenes* EGD-e. Three days post infection the animals were sacrificed and the numbers of *Listeria* were determined. Adapted from Sheehan et al. (2007).

## 3. 'Designer Probiotics'

In addition to improving their physiological stress tolerance, recent studies have led to the development of 'designer probiotics' which specifically target enteric infections by blocking crucial ligand-receptor interactions between the pathogen and host cell (Paton et al. 2006, Sleator 2010a,b, Sleator 2009, Sleator 2008a, Sleator and Hill 2009, Sleator and Hill 2008a,b,d,e, Sleator and Hill 2007b, Sleator and Hill 2006). Many of the pathogens responsible for the major enteric infections exploit oligosaccharides displayed on the surface of host cells as receptors for toxins and/or adhesions, enabling colonization of the mucosa and entry of the pathogen or secreted toxins into the host cell. Blocking this adherence prevents infection, while toxin neutralization ameliorates symptoms until the pathogen is eventually overcome by the immune system. 'Designer probiotics' have been engineered to express receptor-mimic structures on their surface (Paton et al. 2000). When administered orally these probiotics bind to and neutralize toxins in the gut lumen and interfere with pathogen adherence to the intestinal epithelium (Fig. 3). One such construct consists of an *E. coli* strain expressing a chimeric lipopolysaccharide (LPS) terminating in a shiga toxin (Stx) receptor. 1 mg dry weight of this recombinant strain has been shown to neutralize >100 μg of Stx1 and Stx2 (Paton et al. 2000). Paton et al. (2001, 2005) have

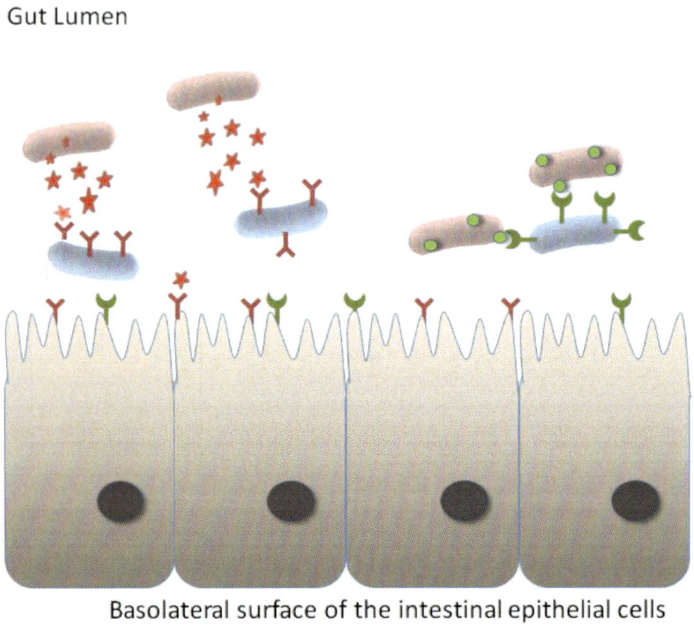


**Fig. 3.** Receptor mimic strategy employed by 'designer probiotics'. Probiotic bacteria (in blue) engineered to express surface host-receptor mimics that can bind to and neutralise toxins (red stars) in the gut lumen or interfere with the adherence of pathogens (in red)/antigens (green circles) to the intestinal epithelium. Adapted from Sleator and Hill (2008e).

also constructed probiotics with receptor blocking potential against Enterotoxigenic *E. coli* (ETEC) toxin LT and cholera toxin (Ctx).

As well as treating enteric infections, 'designer probiotics' have also been recruited to combat HIV. Rao et al. (2005) recently described the construction of a probiotic strain of *E. coli*, engineered to secrete HIV-gp41-haemolysin A hybrid peptides, which block HIV fusion and entry into target cells. When administered orally or as a rectal suppository, this 'live microbicide' colonizes the gut mucosa and secretes the peptide *in situ*, thereby providing protection in advance of HIV exposure for up to a month (Lagenaur and Berger 2005). Other anti-HIV probiotics currently in development include a genetically engineered *Streptococcus gordonii* which produces cyanovirin-N, a potent HIV-inactivating protein originally isolated from cyanobacterium, and a natural human vaginal isolate of *Lactobacillus jensenii* modified to secrete two-domain CD4 which inhibits HIV entry into target cells (Chang et al. 2003).

In addition to in infection control, probiotics (and other non-pathogenic bacteria) are also being engineered to function as novel vaccine delivery vehicles which can stimulate both innate and acquired immunity but lack the possibility of reversion to virulence which exists with more conventional pathogenic platforms. Guimarães et al. (2005) described the construction of a *L. lactis* strain expressing *inlA*, encoding internalin A, a surface protein related to invasion in *L. monocytogenes*. In this instance the otherwise non-invasive *L. lactis* strain is now capable of invading the small intestine and delivering DNA and/or protein into mammalian epithelial cells, making it a safer and more attractive alternative to attenuated *L. monocytogenes* as an antigen delivery platform. Furthermore, the authors suggest that the addition of *hlyA* (encoding listeriolysin) to *L. lactis inlA*$^+$ may promote phagosomal escape within the macrophage and induction of an immune response comparable to that of the intracellular pathogen.

Probiotic vaccine carriers administered by the mucosal route mimic the immune response elicited by natural infection and can lead to long lasting protective mucosal and systemic responses (Holmgren and Czerkinsky 2005). Mucosal vaccine delivery (those administered orally, anally or by nasal spray) also offers significant technological and commercial advantages over traditional formulations including: reduced pain and the possibility of cross contamination associated with intramuscular injection as well as the lack of a requirement for medically trained personnel to administer the vaccine.

## 4. Biological Containment

The use of genetically modified organisms in medicine raises legitimate concerns about their survival and propagation in the environment and the host and about the dissemination of antibiotic markers or other genetic modifications to other microorganisms. At least some of these concerns might be alleviated by the implementation of stringent bio-containment measures. Biological containment systems can be subdivided into active and passive forms. Active containment provides control through the conditional production of a compound which is toxic to the cells. Examples of this type of control include the phage T7 lysozome and colicin E3 in *E. coli* (Torres et al. 2003). Passive containment on the other hand is dependent on complementation of an auxotrophy by supplementation with either an intact gene or the essential metabolite.

Perhaps the most elegant biological containment strategy devised to date targets the essential thymidylate synthase (*thyA*) gene, replacing it with a transgene (encoding the desired stress survival trait) (Steidler et al. 2003). Since the *thyA* gene is essential for growth; mutant strains only grow in the presence of added thymidine or thymine. Thymine auxotrophy involves activation of the SOS repair system and DNA fragmentation, thereby constituting an indigenous suicide system. Thymine and thymidine growth dependence differs from most other auxotrophies in that absence of the essential component is bactericidal in the former and bacteriostatic in the latter. The choice of *thyA* as a target gene thus combines the advantages of passive and active containment systems with the end result being that *thyA*-deficient bacteria cannot accumulate in the environment.

This approach addresses biosafety concerns on several levels. Firstly, no resistance marker is required to guarantee stable inheritance of the transgene, thus overcoming any potential problems associated with dissemination of antibiotic resistance to the commensal populations or opportunistic pathogens. Second, accumulation of the genetically modified organism in the environment is highly unlikely given that rapid death occurs upon thymidine starvation. Finally, should an intact *thyA* be acquired by lateral gene transfer then the transgene would be lost as a result of the double cross over.

## 5. Conclusions

Although conventional molecular medical research continues to provide effective therapeutic and prophylactic compounds, their application is often complicated by both *ex vivo* and *in vivo* sensitivity and rising production costs. Engineered probiotics thus provide an effective means of circumventing the short half-life and fragility of conventional therapeutics, providing a cost effective alternative which will ultimately contribute to health and social gain, particularly in developing countries (Sleator 2007b, Culligan et al. 2009). However, while consumer acceptance of genetically engineered 'designer probiotics' still remains a significant challenge, this obstacle should eventually be overcome by the application of rigorous scientific controls, such as adequate biological containment, and proper risk-benefit analysis of the potential advantages of such a strategy.

## 6. Acknowledgements

Dr. Roy D. Sleator is an ESCMID Research Fellow and Coordinator of the EU FP7 IAPP Project ClouDx-i. This chapter is an update on a previously published review in Gene Ther. Mol. Biol. 11: 269–274, 2007.

## References

Augustin, M.A. and Y. Hemar. 2009 Nano- and micro-structured assemblies for encapsulation of food ingredients. Chem. Soc. Rev. 38: 902–12.

Chang, T.L., C.H. Chang, D.A. Simpson, Q. Xu, P.K. Martin, L.A. Lagenaur, G.K. Schoolnik, D.D. Ho, S.L. Hillier, M. Holodniy, J.A. Lewicki and P.P. Lee. 2003. Inhibition of HIV infectivity by a natural human isolate of *Lactobacillus jensenii* engineered to express functional two-domain CD4. Proc. Natl. Acad. Sci. USA 100: 11672–7.

Considine, K.M., A.L. Kelly, G.F. Fitzgerald, C. Hill and R.D. Sleator. 2008. High-pressure processing—effects on microbial food safety and food quality. FEMS Microbiol. Lett. 281: 1–9.

Culligan, E.P., C. Hill and R.D. Sleator. 2009. Probiotics and gastrointestinal disease: successes, problems and future prospects. Gut Pathog. 1: 19.

Guimaraes, V.D., J.E. Gabriel, F. Lefevre, D. Cabanes, A. Gruss, P. Cossart, V. Azevedo and P. Langella. 2005. Internalin-expressing *Lactococcus lactis* is able to invade small intestine of guinea pigs and deliver DNA into mammalian epithelial cells. Microbes and Infection 7: 836–844.

Hill, C., P.D. Cotter, R.D. Sleator and C.G.M. Gahan. 2001. Bacterial stress response in *Listeria monocytogenes*: jumping the hurdles imposed by minimal processing. International Dairy Journal 12: 273–283.

Holmgren, J. and C. Czerkinsky. 2005. Mucosal immunity and vaccines. Nat. Med. 11: S45–53.

Lagenaur, L.A. and E.A. Berger. 2005. An anti-HIV microbicide comes alive. Proc. Natl. Acad. Sci. USA 102: 12294–5.

Maa, Y.F. and S.J. Prestrelski. 2000. Biopharmaceutical powders: particle formation and formulation considerations. Curr. Pharm. Biotechnol. 1: 283–302.

Paton, A.W., M.P. Jennings, R. Morona, H. Wang, A. Focareta, L.F. Roddam and J.C. Paton. 2005. Recombinant probiotics for treatment and prevention of enterotoxigenic *Escherichia coli* diarrhea. Gastroenterology 128: 1219–28.

Paton, A.W., R. Morona and J.C. Paton. 2000. A new biological agent for treatment of Shiga toxigenic *Escherichia coli* infections and dysentery in humans. Nat. Med. 6: 265–70.

Paton, A.W., R. Morona and J.C. Paton. 2001. Neutralization of Shiga toxins Stx1, Stx2c, and Stx2e by recombinant bacteria expressing mimics of globotriose and globotetraose. Infect. Immun. 69: 1967–70.

Paton, A.W., R. Morona, R. and J.C. Paton. 2006. Designer probiotics for prevention of enteric infections. Nat. Rev. Microbiol. 4: 193–200.

Rao, S., S. Hu, L. Mchugh, K. Lueders, K. Henry, Q. Zhao, R.A. Fekete, S. Kar, S. Adhya and D.H. Hamer. 2005. Toward a live microbial microbicide for HIV: commensal bacteria secreting an HIV fusion inhibitor peptide. Proc. Natl. Acad. Sci. USA 102: 11993–8.

Sheehan, V.M., R.D. Sleator, G.F. Fitzgerald and C. Hill. 2006. Heterologous expression of BetL, a betaine uptake system, enhances the stress tolerance of *Lactobacillus salivarius* UCC118. Appl. Environ. Microbiol. 72: 2170–7.

Sheehan, V.M., R.D. Sleator, C. Hill and G.F. Fitzgerald. 2007. Improving gastric transit, gastrointestinal persistence and therapeutic efficacy of the probiotic strain *Bifidobacterium breve* UCC2003. Microbiology 153: 3563–71.

Sleator, R.D. 2007a. Improving probiotic function using a patho-biotechnology based approach. Gene Therapy and Molecular Biology 11: 269–274.

Sleator, R.D. 2007b. Probiotics as therapeutics for the developing world. Journal of Infection in Developing Countries 1: 7–12.

Sleator, R.D. 2008a. Patho-biotechnology—making new friends of old foes. Open Infectious Disease Journal 2: 39.

Sleator, R.D. 2009. Designing probiotics to control antibiotic resistant "superbugs". International Journal of Medical and Biological Frontiers 15: 1–4.

Sleator, R.D. 2010a. Probiotic therapy—recruiting old friends to fight new foes. Gut Pathog. 2: 5.

Sleator, R.D. 2010b. Probiotics—a viable therapeutic alternative for enteric infections especially in the developing world. Discov. Med. 10: 119–24.

Sleator, R.D., C.G. Gahan, T. Abee and C. Hill. 1999. Identification and disruption of BetL, a secondary glycine betaine transport system linked to the salt tolerance of *Listeria monocytogenes* LO28. Appl. Environ. Microbiol. 65: 2078–83.

Sleator, R.D., C.G. Gahan and C. Hill. 2003. A postgenomic appraisal of osmotolerance in *Listeria monocytogenes*. Appl. Environ. Microbiol. 69: 1–9.

Sleator, R.D., C.G.M. Gahan, B. O'driscoll and Hill. 2000. Analysis of the role of betL in contributing to the growth and survival of *Listeria monocytogenes* LO28. Int. J. Food Microbiol. 60: 261–8.

Sleator, R.D. and C. Hill. 2002. Bacterial osmoadaptation: the role of osmolytes in bacterial stress and virulence. FEMS Microbiol. Rev. 26: 49–71.

Sleator, R.D. and C. Hill. 2006. Patho-biotechnology: using bad bugs to do good things. Curr. Opin. Biotechnol. 17: 211–6.

Sleator, R.D. and C. Hill. 2007a. Food reformulations for improved health: A potential risk for microbial food safety? Med. Hypotheses 69: 1323–4.

Sleator, R.D. and C. Hill. 2007b. Patho-biotechnology; using bad bugs to make good bugs better. Sci. Prog. 90: 1–14.

Sleator, R.D. and C. Hill. 2008a. Battle of the bugs. Science 321: 1294–5.

Sleator, R.D. and C. Hill. 2008b. 'Bioengineered Bugs'—a patho-biotechnology approach to probiotic research and applications. Med. Hypotheses 70: 167–9.

Sleator, R.D. and C. Hill. 2008c. Designer probiotics: a potential therapeutic for Clostridium difficile? J. Med. Microbiol. 57: 793–4.

Sleator, R.D. and C. Hill. 2008d. Engineered pharmabiotics with improved therapeutic potential. Hum. Vaccin. 4: 271–4.

Sleator, R.D. and C. Hill. 2008e. New frontiers in probiotic research. Lett. Appl. Microbiol. 46: 143–7.

Sleator, R.D. and C. Hill. 2009. Rational design of improved pharmabiotics. J. Biomed. Biotechnol. 2009: 275287.

Sleator, R.D. and C. Hill. 2008b. Genetic manupulation to improve probiotic strains. Nutrafoods 7: 37–42.

Sleator, R.D., J. Wouters, C.G. Gahan, T. Abee and C. Hill. 2001. Analysis of the role of OpuC, an osmolyte transport system, in salt tolerance and virulence potential of *Listeria monocytogenes*. Appl. Environ. Microbiol. 67: 2692–8.

Smiddy, M., L. O'gorman, R.D. Sleator, C. Hill and A. Kelly. 2005. Greater high-pressure resistance of bacteria in oysters than in broth. Innovative Food Science and Emerging Technologies 6: 83–90.

Smiddy, M., R.D. Sleator, M.F. Patterson, C. Hill and A.L. Kelly. 2004. Role for compatible solutes glycine betaine and L-carnitine in listerial barotolerance. Appl. Environ. Microbiol. 70: 7555–7.

Steidler, L., S. Neirynck, N. Huyghebaert, V. Snoeck, A. Vermeire, B. Goddeeris, E. Cox, J.P. Remon and E. Remaut. 2003. Biological containment of genetically modified *Lactococcus lactis* for intestinal delivery of human interleukin 10. Nat. Biotechnol. 21: 785–9.

Termont, S., K. Vandenbroucke, D. Iserentant, S. Neirynck, L. Steidler, E. Remaut and P. Rottiers. 2006. Intracellular accumulation of trehalose protects *Lactococcus lactis* from freeze-drying damage and bile toxicity and increases gastric acid resistance. Appl. Environ. Microbiol. 72: 7694–700.

Torres, B., S. Jaenecke, K.N. Timmis, J.L. Garcia and E. Diaz. 2003. A dual lethal system to enhance containment of recombinant micro-organisms. Microbiology 149: 3595–601.

Wemekamp-Kamphuis, H.H., R.D. Sleator, J.A. Wouters, C. Hill and T. Abee. 2004. Molecular and physiological analysis of the role of osmolyte transporters BetL, Gbu, and OpuC in growth of *Listeria monocytogenes* at low temperatures. Appl. Environ. Microbiol. 70: 2912–8.

# PART III

# Preservation of Probiotics

CHAPTER 10

# Freezing of Probiotic Bacteria

*Catherine Béal** and *Fernanda Fonseca*

## 1. Introduction

Probiotic bacteria are submitted to freezing, either in resource banks and culture collections, or to allow long term preservation in a context of industrial production. In this last case, freeze-drying is more commonly employed, as these bacteria are often commercialized in dried form, after incorporation into food products (soymilk, juices, smoothies, cereal based products), as food supplements or in medicines. When they are involved in a product's fermentation, i.e., in fermented milks or in ice-cream they may be frozen. In that case, they are first cultivated in an appropriate medium and under convenient environmental conditions, and concentrated either by centrifugation or by microfiltration. They are then added with cryoprotective agents before freezing and storage at negative temperature. This procedure aims at limiting the negative effects of freezing on the cellular components and on the functional properties of the cells. However, when they are subjected to freezing, bacteria suffer from different stresses at different levels that affect their physical and biological properties. Depending on the stress magnitude, they can display either freezing tolerance or sensitivity.

A stress is defined as a change of environmental conditions that induce different responses of the bacteria. In the case of freezing, the stress is mainly caused by a temperature decrease and subsequent ice crystallization. These changes, sudden or gradual, allow the bacteria to reach negative temperatures, in order to be preserved for a long time. The kinetic at which the stress is applied to the cells (freezing rate), the intensity (freezing and storage temperature) and the length of the stress (storage duration) generate different cellular responses, which lead either to a strong sensitivity or to a resistance of the bacteria.

AgroParisTech – INRA, UMR 782 Génie et Microbiologie des Procédés Alimentaires, 78850 Thiverval-Grignon, France.
E-mails: catherine.beal@agroparistech.fr; fernanda.fonseca@grignon.inra.fr
* Corresponding author

These responses are classified into three categories:

- The passive responses are mainly observed at the physico-chemical level. They lead to a deterioration of biological characteristics and functionalities of the bacteria: cellular plasmolysis, increase in the intracellular viscosity, change of the saturation degree of fatty acids, phase change of lipids, membrane rigidification, physical deterioration of the membranes, proteins denaturation, pH variations, oxidation reactions, and increase in ionic concentrations.
- The active responses that are observed at the biological level may help the bacteria to withstand the stress. They are recognized as modifications of fatty acids and lipids composition in bacterial membranes, changes in intracellular proteins contents (stress proteins and proteins involved in general metabolic pathways), and adjustment of expression or regulation of some genes.
- The global responses are the consequence of the balance between the passive and active responses, within a bacterial population. When passive responses are predominant, a loss in cultivability or viability, a decline of metabolic properties, and a deterioration of some vital functions are observed, thus leading to cell death. On the contrary, when active responses prevail, freezing tolerance may be observed, as a result of bacterial adaptation.

Freezing tolerance is defined as a biological phenomenon, through which the bacteria or a fraction of the bacterial population, may withstand the stress. It characterizes the ability of the microorganisms to survive after the stress (e.g., to maintain a high level of cultivability or viability), to maintain a high level of metabolic activity (enzymatic activities, probiotic properties) and to insure all vital functions. In the case of freezing, this ability is recovered after thawing, which can also generate a stress. The freezing tolerance depends on the considered strain (Fonseca et al. 2000, Rault et al. 2007) and on the environmental conditions during the stress.

As an outline, Fig. 1 summarizes these different behaviors of the bacteria, when they are submitted to freezing stress.

In this context, this chapter aims at reviewing the key features related to freezing of probiotic bacteria, by identifying the stress suffered by the cells and their biological

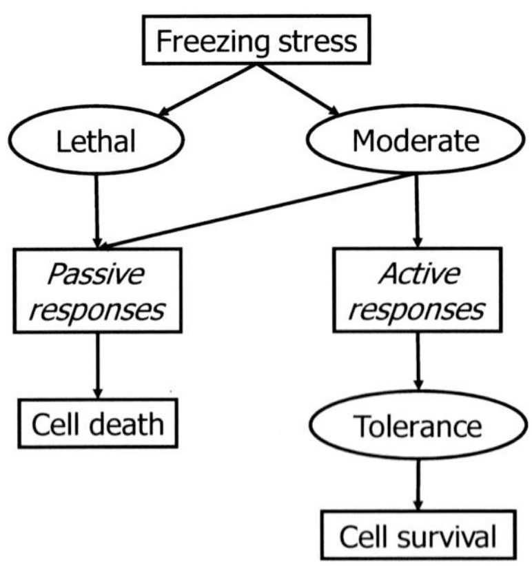


**Fig. 1.** Classification of stress responses of probiotic bacteria.

responses. It also intends to present the means used for improving their resistance to freezing, either by controlling freezing and storage conditions or by managing fermentation and post-fermentation conditions. Examples are taken from the two main genus and species of probiotic bacteria, including *Lactobacillus acidophilus* (Baati et al. 2000, Fernandez-Murga et al. 2001, Lorca and Font de Valdez 2001, Wang 2005, Wright and Klaenhammer 1981), *Lactobacillus agilis* (Basyigit et al. 2006), *Lactobacillus bulgaricus* (Rault et al. 2007, Stamatova 2010, Vasiljevic and Shah 2008), *Lactobacillus casei* (Hussain et al. 2009, Kankaanpää et al. 2004, Wu et al. 2012), *Lactobacillus fermentum* (Suutari and Laakso 1992), *Lactocacillus leichmanii* (Johannsen, 1972), *Lactobacillus plantarum* (Cohen et al. 2006), *Lactobacillus rhamnosus* (Alakomi et al. 2005, Succi et al. 2007, Volkert et al. 2008), *Lactobacillus reuteri* (Taranto et al. 2003), *Bifidobacterium animalis* subsp. *lactis* (Alakomi et al. 2005, Modesto et al. 2004, Saarela et al. 2005, Zacarías et al. 2011), *Bifidobacterium adolescentis* (Novik et al. 1998), *Bifidobacterium bifidum* (Ebel et al. 2011, Modesto et al. 2004), *Bifidobacterium breve* (Modesto et al. 2004), *Bifidobacterium catenulatum* (Modesto et al. 2004), *Bifidobacterium infantis* (Modesto et al. 2004), *Bifidobacterium lactis* (Maus and Ingham 2003, Modesto et al. 2004) and *Bifidobacterium longum* (Doleyres et al. 2004, Maus and Ingham 2003, Modesto et al. 2004), *Bifidobacterium pseudocatelunatum* (Modesto et al. 1998).

## 2. Principles of Freezing and Biological Consequences on Probiotic Bacteria

Concentrates of probiotic and lactic acid bacteria generally consist of complex media containing: micro-organisms, residual culture medium and cryo-protective additives, the whole in an aqueous environment. The physical state of all these constituents changes following freezing, from a liquid to a solid frozen state. In addition, decreasing temperature induces different complex phenomena: supercooling of the aqueous solution, ice formation, and finally glass transition of the remaining unfrozen solution containing cells and solutes at high concentration. The modifications of the cell's environment have important and direct consequences on the cellular structures and indirectly on the cell's physiological state.

### *2.1 Physical Phenomena Taking Place during Freezing (and Thawing)*

When heat is removed from aqueous media, several physical events take place, as illustrated by the two following figures. Figure 2 displays the temperature variation as a function of time inside the product.

Once the temperature is below 0°C, freezing may occur. However, supercooling or undercooling generally takes place due to the tendency of aqueous solutions to cool below their melting point (Fig. 2). The extent of supercooling is the difference between the temperature at nucleation (Tn) and the melting point temperature (Tm), and like nucleation, is a stochastic phenomenon. Ice nucleation in undercooled bacterial samples is generally facilitated by solid or liquid substrate in contact with the water, which allows groups of sorbed water molecules to take up configurations that are able to promote ice formation (heterogeneous nucleation). The ice crystal morphology and size distribution formed at Tn are determined by the nucleation temperature but not by the rate of cooling (Searles et al. 2001). Importantly, a high supercooling (low Tn) correlates well with small ice sizes and lower extent of ice crystallization.

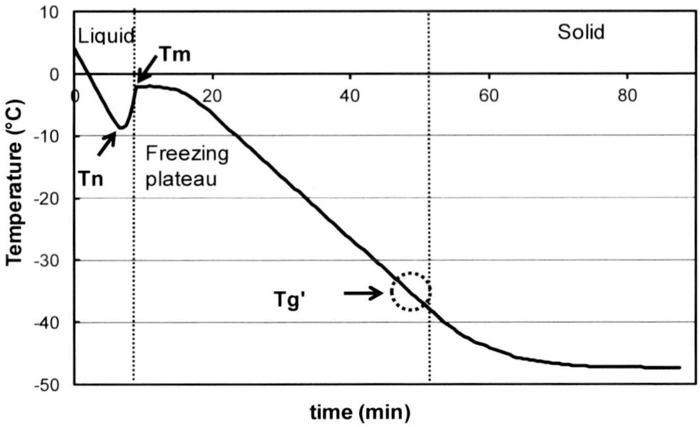


**Fig. 2.** Product temperature profile and principal phenomena taking place during freezing.
Tn: nucleation temperature (ºC); Tm: Melting temperature (ºC); Tg': Glass transition temperature of the maximally concentrated medium (°C).

Figure 3 represents the state diagram of sucrose, taken as example.

After a variable degree of supercooling, ice formation starts in the extracellular medium at the melting (solid/liquid equilibrium) temperature (Tm). As ice forms, the concentration of the unfrozen phase in contact with ice increases. Following temperature decrease, the melting temperature progressively drops off according to Raoult's law. The concomitant decrease in temperature and increase in solute concentration induces a rapid increase in viscosity. Once all freezable water has frozen,

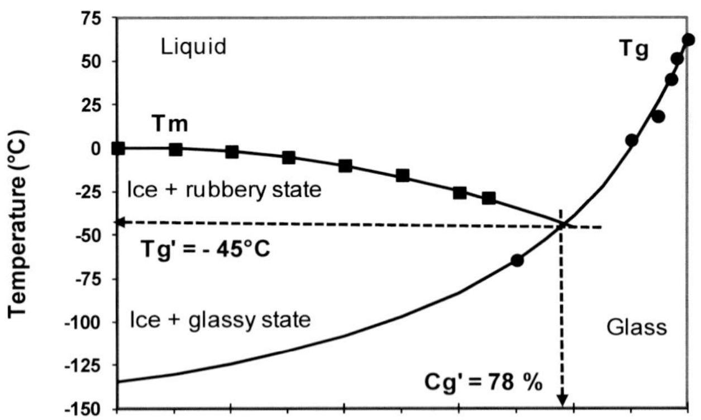


**Fig. 3.** State diagram of sucrose.
■ Tm: Melting temperature (ºC); • Tg: Glass transition temperature (ºC); Tg': Glass transition temperature of the maximally concentrated medium (°C); Cg': concentration of the maximally concentrated medium (dry matter/100 g sample).

no more ice will separate. The remaining cryoconcentrated phase containing low water content (10–30% from initial content) follows a glass transition from an extremely viscous liquid to a solid glass at the glass transition temperature (called Tg'). In the glassy state, the viscosity climbs rapidly and the molecular motions become very constrained, thus limiting all chemical reactions. Tg' is characteristic of the medium: –45°C for sucrose solution (Fig. 3) and –44°C for a bacterial concentrate in sucrose solution. If temperature remains higher than Tg', the maximally freeze-concentrated phase remains in a viscous liquid instable state.

The characteristics of the cells have only low influence on the physical behavior of bacterial suspensions following freezing. On the contrary, the nature of the compounds used for the bacterial "conditioning" medium determines the glass transition of the final mixture and the state diagram (Fonseca et al. 2001b). The physical mechanisms are similar to those observed for multi-component aqueous solutions. Then, the cooling kinetics may induce supercooling, thus maintaining the cell suspension in liquid state at a temperature lower than the equilibrium melting point. The supercooling of cell suspensions may easily reach –10°C to –15°C (Mazur 1965).

### 2.2 Degradation of Biological Structures Caused by Freezing

During freezing, cells suffer different kinds of stresses, leading to both morphological and biochemical changes and to cell deterioration or death (Fonseca et al. 2006, Klaenhammer and Kleeman 1981, Volkert et al. 2008). Four main kinds of stresses take place during freezing and trigger passive responses of cells. Thermal stress (cold or hot temperatures) induces protein denaturation and phase transitions in membrane lipids, with the concomitant modification of membrane fluidity. Osmotic stress caused by the formation of ice crystals leads to changes in intracellular concentration and pH thus inducing protein deterioration. Ice crystal formation and reduction of cellular volume generate, in turn, mechanical stress that may cause destruction of biological membrane and cellular structures. Finally, oxidative stress can happen at many steps of the process, principally during fermentation and storage steps (Fonseca et al. 2003), thus affecting membrane lipids and cell proteins (Sanders et al. 1999, Zhang et al. 2010).

Cell damages following freezing and frozen storage are highly dependent on operating conditions and difficult to dissociate from thawing induced injuries. However, all cell systems do share common cryobiological responses, which may be exploited to better understand and remediate the specific problems of freezing bacteria. In particular, when ice forms in a cellular suspension, electrolytes and proteins from the original extra-cellular solution are left within an increasingly concentrated unfrozen liquid fraction containing cells. The cells respond to the increased concentration of the unfrozen fraction by either dehydrating (transporting water out into the unfrozen liquid fraction to re-equilibrate the cytoplasmic concentration), or by the formation of intracellular ice in the cytoplasm (Mazur 1984). These two biophysical processes compete during any cooling procedure. For defined cooling conditions, the rate of water flux from the cell depends on the membrane permeability to water and on the ratio cell surface area/cell volume (Dumont et al. 2004, Fonseca et al. 2000, Mazur 1970). Intracellular ice formation occurs when the supercooling within the cytoplasm is sufficiently large to drive the nucleation and make possible the growth of an ice crystal within the cytoplasmic compartment. Intracellular ice formation is generally considered lethal if more than 10–15% of the cellular water is involved (Mazur 1990). In the case of lactic acid bacteria, losses in activity during freezing were observed to be due to the formation of intracellular ice at high cooling rates only when distilled water was used as extra-cellular medium (Fonseca et al. 2006). In the presence of cryoprotective agents, freezing injury was related to cell dehydration since plasmolysis occurred during thawing due to an osmotic imbalance between extracellular and intracellular media.

Recently, Clarke et al. (2013) have presented empirical evidence that micro-organisms undergo freeze-induced desiccation during cooling together with an intracellular glass transition (vitrification) at a temperature ranging from –10°C for thermophilic lactic acid bacteria and the ice nucleating *Pseudomonas syringae,* to –26°C for *Corynebacterium variabile* and *Arthrobacter ariletensis*. The high internal viscosity following vitrification slows down the diffusion of molecules extremely and the cellular metabolism ceases, even though extracellular medium follows a bulk glass transition at lower temperatures.

Survival of lactic acid bacteria during freezing is thus highly dependent on the cellular biophysical event of cell dehydration due to osmotic stress and cell volume contraction. Importantly, cell membrane appears as one of the most critical sites for cell injury, since it has been established that cooling and dehydration stresses following freezing, alter the physical state of lipids of different types of cells including probiotics, and thus lipid organization and membrane fluidity (Balasubramanian et al. 2009, Beney and Gervais 2001, Gautier et al. 2013). Freezing affects the physical state of membrane lipids due to changes in hydration level (lyotropism) and temperature (thermotropism) (Chapman 1975). The reorganization of lipids and changes in the membrane conformational disorder due to phase transitions, e.g., from liquid crystalline to gel phase under different thermal and hydration conditions, affect the functions of the membrane and are implicated in cell injury (Wolfe and Bryant 1999). The cellular consequences of these phase changes are thought to be related to lateral phase separation of membrane components inducing, in turn, increased membrane permeability and modifications of the protein/lipids interactions.

Finally, the oxidation of lipid and proteins can take place during high subzero temperature storage (–20°C). It has been proved that the addition of an antioxidant to the protective medium is essential for limiting oxidation reactions of *Lb. delbrueckii* subsp. *bulgaricus* CFL1 during frozen storage but also during the cryoprotection step before freezing (Fonseca et al. 2003).

## 3. Biological Responses of Probiotic Bacteria to Freezing Stress

In order to determine the freezing tolerance of a population of probiotic bacteria, different methods are commonly used at the level of global stress responses. They concern the measurements of cultivability, viability, acidification activity and resistance to gastrointestinal stress, with strong differences among considered strains.

### *3.1 Decrease in Cultivability and Viability*

#### *3.1.1 Loss in Cultivability*

Cultivability measurement is often considered as a reference method. It consists in spreading the bacterial suspension on specific agar medium, incubating the plates under convenient conditions (temperature, anaerobiosis, duration), and enumerating the number of colony forming units. By comparing the plate counts after the stress (e.g., after freezing and during storage) to that measured before the stress, this method allows determining the bacterial survival. As an illustration, Fig. 4 shows that the cultivability of a *Bifidobacterium* strain is strongly reduced after freezing at –20°C, whereas it remained quite stable at –70°C for two months.

This behavior is confirmed by many studies that exhibit a cultivability decrease during freezing and frozen storage. From Baati et al. (2000), survival of *Lb. acidophilus* after freezing was comprised between 24 and 89%, depending on the culture temperature and suspension medium. By studying the behavior of six strains of lactobacilli, Juárez Tomás et al. (2004) demonstrated that their residual cultivability after freezing was comprised between $10^4$ and $10^9$ CFU/mL, with differences among strains and cryoprotective media. The survival of six strains of *Lb. rhamnosus* was shown to vary between 84 and 97% after 90 days storage, depending on the temperature, the medium and the strain used (Succi et al. 2007). Finally, the freezing resistance of 23 strains of bifidobacteria, cryoprotected in skim milk or sucrose based media, was comprised between 20% and 100% (Modesto et al. 2004).

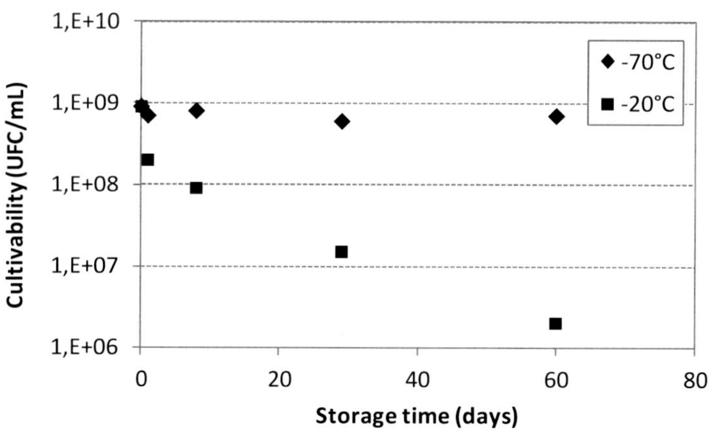


**Fig. 4.** Loss of cultivability of *Bifidobacterium* sp. grown in MRS medium added with 0.5 g/L cysteine, after freezing and storage at –20°C or –70°C.

#### *3.1.2 Loss in Viability*

Viability is commonly quantified by combining fluorescent staining with detection by fluorescent microscopy or by flow cytometry (Rault et al. 2007). The fluorescent probe carboxyfluorescein diacetate (cFDA) is a small molecule that can easily enter into bacterial cells, where it is transformed into carboxyfluorescein (cF, green fluorescence at 525 nm upon excitation at 488 nm) by the action of cellular esterases. By displaying this enzymatic activity, cFDA stained cells are thus considered as active cells. On the contrary, the fluorescent probe propidium iodide (PI, red fluorescence at 620 nm upon excitation at 488 nm) enters the cells and associates with nucleic acids, when membrane integrity is disturbed. Consequently, PI stained cells are characterized as cells without membrane integrity. The combination of these two fluorescent probes allows identifying three subpopulations in a bacterial suspension: viable cells that show simultaneously esterase activity and membrane integrity, dead cells that display no esterase activity and no membrane integrity, and altered cells, which demonstrate residual esterase activity but no membrane integrity.

As an illustration, Fig. 5 shows typical flow cytometry density charts obtained from cFDA/PI staining and flow cytometry, for a *Lb. bulgaricus* strain, before and after freezing. It shows that the percentage of viable cells shifts from 89% before freezing to 32% after freezing, whereas the percentages of dead cells and altered cells increase to 52% and 14%, respectively.

These approaches were successfully used by Rault et al. (2007) to quantify the loss in viability of four strains of lactobacilli and by Volkert et al. (2008) to compare the survival rate of *Lb. rhamnosus* GG submitted to different freezing conditions. The live/dead *Bac* light kit™, composed by Syto9, a membrane-permeant nucleic acid stain (green fluorescence at 530 nm upon excitation at 488 nm) and propidium iodide, is also used to quantify the loss of viability of probiotic starters (Alakomi et al. 2005). These authors demonstrated that viability of *Lb. rhamnosus* and *B. animalis* subsp. *lactis* was well correlated to cultivability, when applying an acid stress.

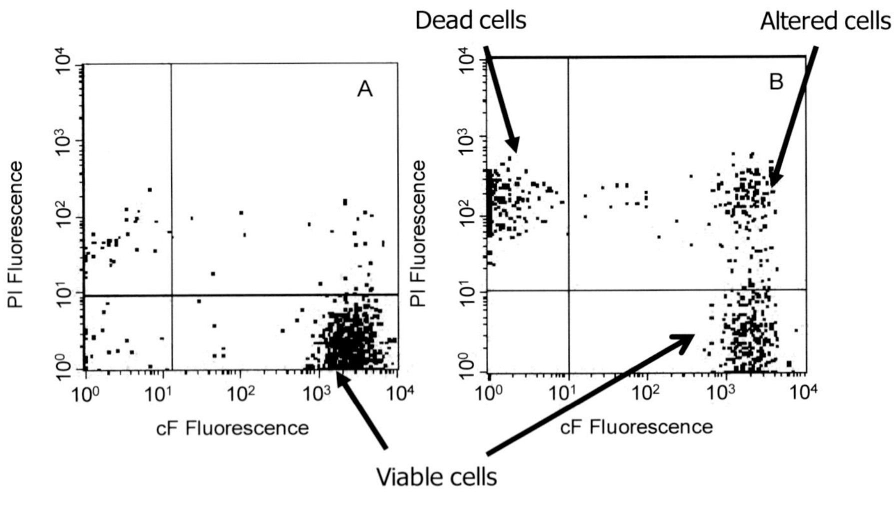


**Fig. 5.** Quantification of viability and mortality of *Lb. bulgaricus* CFL1 before (A) and after (B) freezing at –20°C, by double staining with cFDA and propidium iodide and flow cytometry detection (Rault et al. 2007). PI: Propidium iodide; cF: carboxyfluoresceine.

### 3.2 Decrease in Functional Properties

Among the main functional properties of probiotic bacteria, the acidification activity and the resistance to gastrointestinal stress have been studied, as a consequence of freezing stress.

#### 3.2.1 Loss in Acidification Activity

Central metabolism of lactic acid bacteria and bifidobacteria corresponds to the intracellular use of carbohydrates, in order to synthesize lactic acid or other organic acids, together with ATP. Excretion of organic acids leads to a pH decrease of the medium. Acidification activity corresponds to the ability of these microorganisms to decrease the pH of the medium: the more rapid the pH decrease, the higher the acidification activity. Different methods allow the quantification of this phenomenon. Among them, the Cinac system (Corrieu et al. 1988) is commonly employed in dairy companies to characterize the acidification activity of their starters. By measuring continuously and on line the pH decrease, this system allows identifying some specific descriptors, such as the time necessary to reach a given pH (tpHi, in min).

As an illustration, Fig. 6 shows the acidification kinetics of a *Bifidobacterium* strain before and after freezing at –20°C. The time to reach pH 5 was delayed by 3 hr (from 9 hr before freezing to 12 hr after freezing).

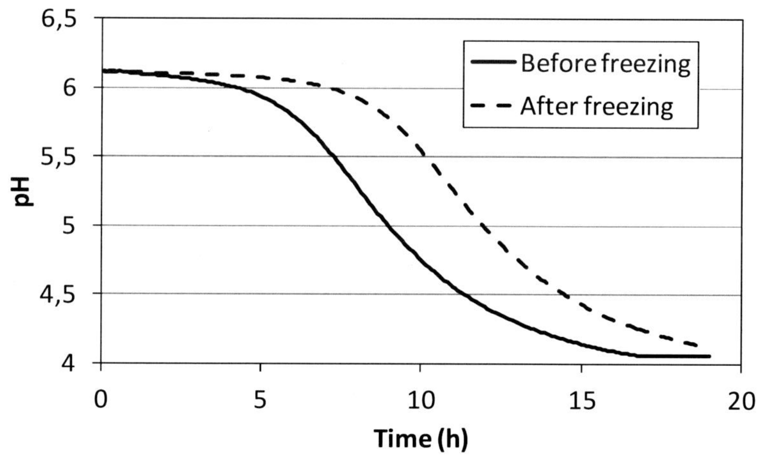


**Fig. 6.** pH decrease of fresh and frozen bacterial suspensions of bifidobacteria in milk at 37°C.

By comparing the tpHi values obtained under normalized conditions, Fonseca et al. (2000) were able to quantify the loss of acidification activity during freezing and frozen storage of *Lb. bulgaricus* and *Streptococcus thermophilus* (Fig. 7). These authors modeled the increase in tpHi by the following linear equation:

$$tpHi = t0 + dtf + k \cdot ts$$

Where t0 represents the initial acidification activity before freezing (in min), dtf is the loss in acidification activity during freezing (in min), k is the rate of loss in acidification activity during frozen storage (in min/d), and ts is the storage time (in d).

This loss of acidification activity was correlated to the loss in cultivability. A linear relationship exists between the acidification activity (tpH 5.5, in min) and the microbial counts of *Lb. acidophilus* (Wang et al. 2005a).

The loss of acidification activity can also be indirectly determined by a measurement of β-galactosidase activity (Gilliland and Lara 1988). However, these authors demonstrated that some cells of *Lb. acidophilus*, which failed to form colonies, still possessed β-galactosidase activity. This indicates that the loss of enzymatic activity was insufficient to explain the loss in viability.

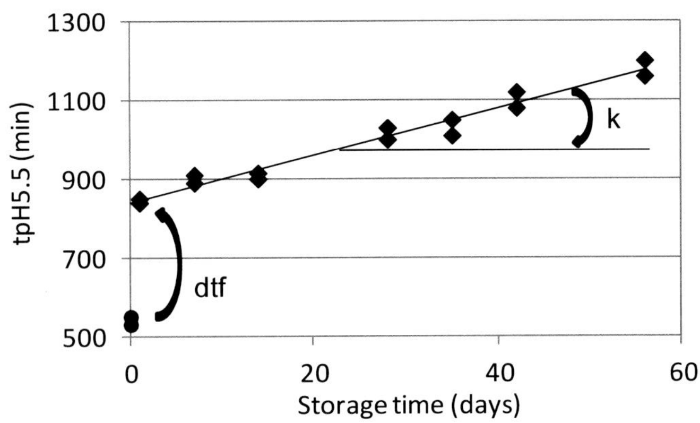


**Fig. 7.** Loss of acidification activity during freezing and frozen storage of *Lb. acidophilus* RD758 (Wang et al. 2005a). dtf: loss of acidification activity during freezing (min); k: rate of loss of acidification activity during frozen storage (min/d).

### *3.2.2 Decrease in the Resistance to Gastrointestinal Stress*

Resistance to the gastrointestinal stress is a compulsory condition for using probiotic bacteria, as they shall display a beneficial effect on the health of the gastrointestinal tract of the host. Within the environmental stress that bifidobacteria must overpass until delivery in small intestine, low pH of gastric juice in stomach (from pH 5 to pH 2), presence of bile salts in duodenum (up to 15 mmol/L) and of digestive enzymes (pepsin and pancreatin) in these two compartments are considered as the most critical factors affecting their survival (Mainville et al. 2005). Resistance to gastrointestinal stress is generally determined by comparing cultivability measurements before and after the stress. It is measured either in static conditions, e.g., by setting the cells in the presence of acidic conditions and bile salts at 37°C for about 1 hr or in dynamic conditions, by using bioreactors that simulate digestion (Madureira et al. 2011, Mainville et al. 2005, Marteau et al. 1997, Sumeri et al. 2010).

A relationship between freezing resistance and gastrointestinal stress has been demonstrated (Taranto et al. 2003). When cells of *Lb. reuteri* were grown in the presence of bile salts, such as a mix of taurocholic acid, glycocholic acid, taurochenodeoxycholic acid, cholic acid and glycochenodeoxycholic acid, their resistance to freezing was strongly reduced.

However, the effect of freezing the cells before applying the gastrointestinal stress has never been studied. By considering that freezing modifies the viability and the acidification activity of the bacteria, it may probably act on these properties.

### 3.3 Strain Effect

Depending on the bacterial strain, the resistance to freezing and frozen storage strongly differ, even for the same genus or the same species (Godward et al. 2000, Juárez Tomás et al. 2004, Succi et al. 2007). This strain dependency has been quantified by considering four *Lb. delbrueckii* strains (Rault et al. 2007), which displayed different levels of resistance to freezing and frozen storage at –80°C, as shown in Fig. 8.

These observations were confirmed by Modesto et al. (2004) with various bifidobacteria (*B. bifidum, B. longum, B. animalis, B. breve, B. infantis, B. catenulatum* and *B. pseudocatenulatum*) that exhibit different resistance percentages (from 20 to 100%) after freezing at –135°C (Modesto et al. 2004). In contrast, the freezing tolerance did not differ among six isolates of *B. animalis* subsp. *lactis* (Zacarías et al. 2011).

The preservation of acidification activity during freezing and frozen storage is also strain dependant, as observed by measuring the pH lowering after 24 hrs of storage at –20°C and –80°C with six probiotic strains of *Lb. rhamnosus* (Succi et al. 2007).

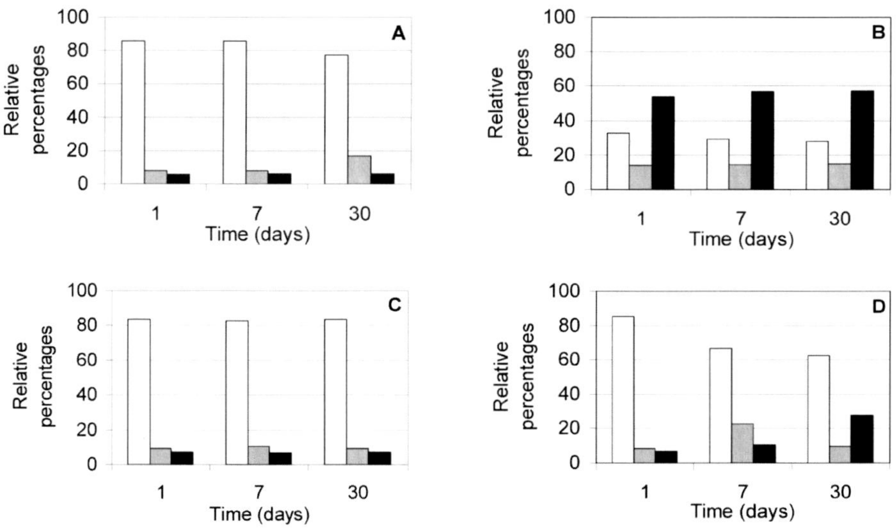


**Fig. 8.** Evolution over 30 days of frozen storage at –80°C of relative percentages of viable (□), injured (■) and dead (■) cells of *Lb. bulgaricus* CNRZ 208T (A), *Lb. bulgaricus* CFL1 (B), *Lb. bulgaricus* CIP 101027T (C) and *Lb. lactis* ITG LL57 (D), evaluated by multiparameter flow cytometry combined with PI/cFDA double staining (Rault et al. 2007).

## 4. Improvement of Freezing Resistance of Probiotic Bacteria by Controlling Freezing and Storage Conditions

After their production by fermentation and their concentration, probiotic bacteria can be stabilized by freezing, even if freeze-drying is more often used. Freezing process involves different steps, including cryoprotection, freezing, frozen storage and thawing. Cryoprotection aims at helping cells to better resist the freezing stress, while frozen storage, and thawing are unavoidable for commercialization and end-using of bacterial concentrates.

### 4.1 Cryoprotection

Complex multi-component solutions are usually added to bacterial cells before freezing in order to attenuate the damaging effects of freezing and to extend the range of optimal freezing operating conditions. Carvalho et al. (2004) distinguished three categories of protective molecules, depending on their capability of penetrating the cell: Molecules that remain in the extracellular medium (no penetration) include high molecular weight molecules like proteins, polysaccharides; Molecules that penetrate both the cell wall and the cytoplasmic membrane like dimethyl sulfoxide and glycerol; Molecules that are able to penetrate the cell wall, but not the membrane are characterized by their low molecular weight and include oligosaccharides and amino acids. The penetration ability of these cryoprotectants and the presence of specific chemical groups in the molecules define the protection feature (Font de Valdez et al. 1983, Hubalek 2003).

A complete review of different molecules used for protecting different kinds of cells, from microorganisms to animal and vegetable cells and tissues, has been published by Hubalek (1996). Briefly, penetrating agents, generally of low molecular weight (<100 Da), are used in concentrations of about 1 M. They limit cell dehydration and intracellular ice formation risk by lowering the equilibrium freezing temperature inside the cell. The non penetrating molecules of high molecular weight (generally >1000 Da) are currently used in mM concentrations. They facilitate water flux from the cell during freezing, thus limiting intracellular ice-crystal formation during freezing and also act by reducing ice crystal growth by increasing the viscosity of the extracellular medium (Leloux 1999).

#### *4.1.1 Physical Properties and Preservation Mechanisms of Protective Molecules*

The protective mechanisms of cryoprotectants are not completely elucidated, even if numerous studies have reported on this subject. The available explanations of preservation mainly come up from physical mechanisms and cellular physiological consequences. The physical mechanisms relevant for cryoprotectants effect are those that allow limiting the degradation induced by ice crystals formation and water efflux from cell. Seven principal ways of action have been identified.

- Prevention of intracellular ice crystal formation (IIF). Penetrating agents bind water through hydrogen bonding and lower the freezing point inside the cell.

Therefore, they lessen the salt concentration, in turn limiting osmotic shock (MacFarlane 1987, Mazur 1977). No penetrating agents induce water efflux from cell, thus favouring cell dehydration and consequently reducing the risks of IIF (Farrant and Morris 1973, McGann 1978);

- Limitation of ice-crystal growth. Cryoprotective molecules affect ice crystal size, by inhibiting ice crystal growth through different mechanisms: molecules adsorption at the interface water-ice, mass transfer coefficient reduction, sterical hindrance, increase in intracellular (penetrating agents) or extracellular (non penetrating agent) medium viscosity (Karlsson and Toner 1996, MacFarlane 1987);
- Limitation of solute concentration increase in the extracellular medium, thus preventing excess cell dehydration. Several authors attribute the cryoprotective role of some molecules to a colligative action, meaning reducing electrolytes concentration for a given equilibrium freezing temperature (Anchordoguy et al. 1987, Leloux 1999). Following Raoult's law, this colligative effect is proportional to the molar concentration of the cryoprotective molecule (McGann 1978);
- Prevention of injurious eutectic freezing (Chen et al. 2000), thus avoiding important pH changes. The protective action results from the reduction of ice fraction (Francks 1998) and by trapping salts (NaCl) in a highly viscous or glass-like phase;
- Vitrification of the extracellular medium containing cells. The higher the vitrification temperature (Tg), the lower is the ice fraction, due to water in the glassy amorphous state. The increasing viscosity and decreasing temperature restrain translational motion of molecules, in particular of water, and limit cell dehydration to critical volume (Branca et al. 1999, Levine and Slade 1988, McFarlane 1987). Vitrification, to be effective, is generally associated to a very fast cooling rate (Dumont et al. 2004);
- Protection of membranes by direct interaction with membrane phospholipids. Some authors suggest the formation of hydrogen bonding with polar phospholipid head groups or hydrophobic interactions with acyl chains, thus reducing membrane changes during freezing (Anchordoguy et al. 1987, Rudolph and Crowe 1985);
- Protection of membranes by modification of lipid structure and transition temperature. This temperature conditions membrane fluidity since it indicates the passage from a fluid (disordered, liquid-crystalline state) to a gel phase (ordered, rigid state), thus influencing cells capability to resist cold temperature (Beney and Gervais 2001). Some cryoprotectants (dimethyl sulfoxide, glycerol) have been shown to modify the lipid phase transition of cells following freezing (Ragoonanan et al. 2010a);
- Preventing oxidation reactions by scavenging reactive oxygen species (ROS), thus limiting oxidation of lipids and proteins of the cell wall and cytoplasmic components. Oxidation damage has been reported on *Lb. bulgaricus* before freezing and during frozen storage but not during the freezing step (Fonseca et al. 2003). The addition of sodium ascorbate as a protective medium before freezing was effective on preserving acidifying activity and viability before freezing and during frozen storage at –20°C. However, the oxidation observed during storage

could have been caused by reactive oxygen species formed during freezing. Zhang et al. (2010) showed that the intracellular accumulation of glutathione enhanced the survival and the biotechnological performance of *Lb. sanfranciscensis* when exposed to freeze-thawing, by preventing peroxidation of membrane fatty acids. Recently, Amaretti et al. (2013) pointed out the antioxidative effect of different probiotic strains (*Bifidobacterium* and *Lactobacillus*), by limiting excessive amounts of reactive radicals. Consequently, mixtures of probiotics containing this kind of strains have a potential interest for preventing oxidation damage of probiotic concentrates.

The nature and concentration of molecules composing the protective mixtures strongly condition the thermal and physical properties of the protective formula. The effect of cells remains low, principally affecting the dry matter content, as previously explained in section 2.1.

Among the principal properties of cryoprotectants, the thermal properties involve latent heat changes (water crystallization and ice melting) and heat transfer properties (like specific heat and thermal conductivity). It is the water fraction which essentially determines the magnitude of these properties. Other physical properties depend on the nature and concentration of additives, which in turn determine the product quality and the process productivity. Concerning the freeze-thawing process, the freezing/melting equilibrium temperature and the glass transition temperature are the most important. However, in the case of complex mixtures of probiotic and lactic acid bacteria, the data are very scarce (Fonseca et al. 2001b, Fonseca et al. 2004). The equilibrium freezing temperature is positively related to the molar fraction of solutes present in the mixture. The freezing temperature is linearly lowered from –3°C to –9°C, when the molar concentration of cryoprotectants increases from 0.01 to 1.6 M (Fonseca et al. 2003). Low freezing temperatures are associated with high resistance to freezing of the cells, as a consequence of reduction in ice crystal size (determined by the nucleation temperature) and lower ice fraction.

Glass transition temperature values are today well known for the most currently used additives for lactic acid and probiotic bacteria (sugars, polyols, polymers, salts) (Levine and Slade 1988). Disaccharides and high molecular weight sugars, like maltodextrin, are good stabilizers since they induce high glass transition temperatures.

In multi-component mixtures, the glass transition temperatures may be estimated with a good approximation from simple linear relationship involving the glass transition temperature of binary aqueous solutions of each constituent (Fonseca et al. 2001b). Consequently, by adding high molecular weight molecules, the glass transition temperature of the mixture increases with an evident economical advantage for storage preservation.

### *4.1.2 Protective Molecules used for Preserving Frozen cells of Probiotic and Lactic Acid Bacteria*

Different molecules are usually employed for improving resistance to freezing and frozen storage of probiotic and lactic acid bacteria. Table 1 summarizes the main molecules used, together with their concentration range of use and freezing conditions.

**Table 1.** Main cryoprotective agents successfully used for protecting probiotic and lactic acid bacteria.

| Cryoprotectants | Concentration (%) | Freezing conditions | References |
|---|---|---|---|
| Adonitol | 2 to 7 | –20°C | (Morice et al. 1992) |
| Glycerol | 10 to 20 | –20°C | (Thunell et al. 1984) |
| | 10 | –20°C | (Fonseca et al. 2000) |
| Mannitol | 10 | –80°C | (Berner and Viernstein 2006) |
| Sorbitol | 10 | –80°C | (Berner and Viernstein 2006) |
| Fructose | 10 | –80°C | (Berner and Viernstein 2006) |
| Glucose | 0.5 | –80°C | (Coppola et al. 1996) |
| | 32 | –80°C | (De Giulio et al. 2005) |
| | 10 | –80°C | (Berner and Viernstein 2006) |
| Lactose | 10 | –20°C, –40°C, –70°C | (Chavarri et al. 1988) |
| | 32 | –80°C | (De Giulio et al. 2005) |
| | 10 | –80°C | (Berner and Viernstein 2006) |
| Maltose | 32 | –80°C | (De Giulio et al. 2005) |
| | 10 | –80°C | (Berner and Viernstein 2006) |
| Sucrose | 5 to 15 | –80°C | (Berner and Viernstein 2006) |
| Sucrose, vitamins B1, B6, B12 | 10 | –70°C, –196°C (liquid $N_2$) | (Novik et al. 1998) |
| Trehalose | 6 to 20 | –60°C | (De Antoni et al. 1989) |
| | 32 | –80°C | (De Giulio et al. 2005) |
| Maltodextrines | 0.6 to 30 | –20°C | (To and Etzel 1997) |
| | 10 | –80°C | (Berner and Viernstein 2006) |
| Sodium ascorbate | 0.5 to 5 | –20°C | (Fonseca et al. 2003) |
| Sodium glutamate | 5 to 15 | –20°C | (Fonseca et al. 2003) |
| Betaine | 2.3 | –20°C | (Fonseca et al. 2003) |
| Yeast extract | 5 | –196°C (liquid $N_2$) | (Baumann and Reinbold 1966) |
| Whey | 10 | –70°C, –196°C (liquid $N_2$) | (Novik et al. 1998) |
| Skim milk | 1 to 13 | –30°C | (Jakubowska et al. 1980) |
| | 10 | –196°C (liquid $N_2$) | (Foschino et al. 1996) |
| | 10 | –70°C, –196°C (liquid $N_2$) | (Novik et al. 1998) |
| | 11 | –20°C, –80°C | (Carcoba and Rodriguez 2000) |
| | 20 | –30°C (spray freezing) | (Volkert et al. 2008) |
| Skim milk, sucrose | 10 to 20 | –20°C, –80°C | (Carcoba and Rodriguez 2000) |
| | 6 and 10 | –40°C, –196°C (liquid $N_2$) | (Tsvetkov and Shishkova 1982) |

In summary, the protective agents most currently used are polyols, mono-, di- and polysaccharides, vitamins, amino acids, peptides and proteins, frequently furnished by complex biological media (skim milk, whey).

Protective molecules are currently employed in mixtures or added to skim milk (Carcoba and Rodriguez 2000, Tsvetkov and Shishkova 1982). Each molecule of a mixture normally affords protection against a specific degradation mechanism. For example, glycerol, a penetrating polyol, helps limiting dehydration, while disaccharides like trehalose, and polysaccharides like maltodextrin help in increasing the glass transition temperature, thus facilitating the maintenance of the frozen bacterial suspension in an amorphous stable state. Antioxidants like sodium ascorbate are particularly useful for avoiding oxidative deterioration of cells specially when intermediate storage before freezing is needed or when storing frozen cells at high subzero temperatures (–20°C) (Fonseca et al. 2003).

The food matrix itself can be protective. An example is ice cream, where the milk fat and sugar content helps to protect probiotic bacteria. The number of viable probiotic cells (*Lb. acidophilus, Lb. agilis,* and *Lb. rhamnosus*) was unaltered when the cultures where directly mixed with ice-cream (Basyigit et al. 2006).

Microencapsulation has also been proved to efficiently protect probiotic cells from freezing. By coating cells with adequate hydrocolloids, cells are entrapped and isolated from the surrounding environment, thus protecting them against freezing and other stresses. When encapsulated in calcium alginate beads and incorporated into fermented frozen dairy desserts, probiotic microorganisms (*Lb. acidophilus* MJLA1 and *Bifidobacterium* spp. BDBB2) showed an improved viability in the product ($10^5$ CFU.$g^{-1}$) as compared to non-encapsulated organisms ($10^3$ CFU.$g^{-1}$) (Shah 2000). The immobilization in Ca-alginate beads of other lactic acid bacteria strains (*Lb. delbrueckii* subsp. *bulgaricus* DSM20081, *Lb. acidophilus* DSM20079 and *S. salivarius* subsp. *thermophilus*) induced a good survival rate after freezing but it was lower as compared to the trehalose effect (De Giulio et al. 2005).

The initial cell density, characterized by the relative proportion of cells to protective medium, is another factor that has to be considered. High initial cell concentration ($10^{10}$ CFU $ml^{-1}$) generally yields high survival after freezing (Berner and Viernstein 2006). However, these authors indicate that the final recovery after freezing is dependent on the protectant used.

Finally, from this information it is still not possible to give general rules for cryoprotection of probiotic bacteria. The numerous works on the subject highlight the variability of the results that differ according to bacterial species and strains, fermented medium composition and operating conditions (centrifugation, freeze-thawing kinetics and storage temperature). Mechanical and physicochemical events taking place during freezing determine the survival and activity recovery of cells (Fonseca et al. 2001a). It is then necessary to identify the freezing conditions allowing long term frozen storage with maximal functionality recovery, while considering the transport and storage conditions. Apart from the cryoprotectants formulation, undoubtedly of major importance, the freezing and thawing kinetics and the storage temperature, are essential factors influencing the freezing-resistance of probiotic bacteria. Cryoprotection, freezing, storage and thawing need to be considered in an integrated approach for the successful preservation of bacterial cells.

### 4.2 Freezing Kinetics

From observation of the freezing rate effect on different kind of cells, Mazur (1970) postulated the existence of an optimal cooling rate, as a result of two damage mechanisms. At low freezing rates, cellular damage is principally caused by osmotic stress to cells (so-called "solution effects"). The formation of extracellular ice induces high solute concentrations in the extracellular medium, exposing the cells to an increase in ionic concentration, changes in pH, etc., for extended periods of time. In contrast, at very high cooling rates, damage is due to the formation of intracellular ice (Karlsson et al. 1993, Mazur 1977), but only when freezing is done in distilled water (Fonseca et al. 2006). As an alternative mechanism, these authors demonstrated that cell damage occurring after rapid cooling rates is caused by an osmotic imbalance encountered during thawing or high subzero temperature storage, but not by the formation of intracellular ice.

Frozen probiotic bacteria, such as *Lb. rhamnosus* GG, are usually prepared at very high cooling rates (>200°C/min) by using liquid $N_2$ (Volkert et al. 2008). However, this method can conduct to low survival for *Lb. bulgaricus* (6–27%) depending on the strain (Fonseca et al. 2006, Smittle et al. 1972). Furthermore, interactions between freezing kinetics and subsequent storage temperatures have been demonstrated by Fonseca et al. (2006), together with their effects on the biological activity of the bacteria. High cooling rates (2500°C/min) obtained by direct immersion in liquid nitrogen resulted in the minimum loss of acidification activity and viability only for samples protected with glycerol and stored at –80°C. When samples were stored at –20°C, a maximal loss of viability and acidification activity was observed with rapidly cooled cells. By scanning electron microscopy, these cells did not contain intracellular ice but they were plasmolyzed. By considering *Lb. plantarum*, Peter and Reichart (2001) advised a low cooling rate (2.1°C/min) associated with inositol and horse serum as cryoprotectants, while Dumont et al. (2004) achieved high survival rates whatever the cooling rate applied. The presence of cryoprotectants (sucrose with vitamin B, whey, skim milk) allowed lowering about 100 times the optimal cooling rate (100–400°C/min to 4°C/min) for *B. adolescentis* MS-42 (Novik et al. 1998). However, even if some interesting results were obtained with probiotic bacteria, generalization is still not possible as it clearly appears that optimal freezing kinetics are extremely dependent on cryoprotection and storage conditions, as well as on considered strains.

### 4.3 Storage of Frozen Starters

Temperature and duration are strictly controlled during storage of frozen starters. The final holding temperature at which the frozen microorganisms are stored is the first essential factor for the stability of the starters. The lower this temperature, the higher the shelf life is. Frozen storage in the range of –20°C to –40°C causes high loss of viability and metabolic activity (Fonseca et al. 2000, Novik et al. 1998). On the contrary, storage temperatures lower than –40°C improve frozen storage resistance of cells protected with glycerol, skim milk or lactose (Chavarri et al. 1988, Fonseca et al. 2000, Succi et al. 2007). The positive effect of these low temperatures is explained by the Tg' values of most currently used media that are lower than –30°C. Low storage temperatures are then needed for obtaining a glassy extracellular medium ensuring minimal cell degradation kinetics. If storage at –20°C is required, the extracellular medium will be in a viscous instable state and, as recently shown by Clarke et al. (2013), the intracellular medium can also be in a viscous state depending on the strain. Consequently, the protective medium should be optimized, including for example glycerol or a sugar for increasing viscosity, together with an antioxidant to reduce oxidative damage (Fonseca et al. 2003).

The duration of the frozen storage determines the survival and residual activity of the cells in final use. The viability of frozen starters continuously declined during storage of *Streptococcus thermophilus* and *Lb. bulgaricus* at –75°C (Fonseca et al. 2000) and of *B. adolescentis* at –20°C (Novik et al. 1998). The acidification activity of probiotic lactobacilli also diminishes linearly according to the storage time (Wang et al. 2005a). The rate of loss of acidifying activity can be quantified by the Cinac system (cf. section 6.4). It varies from about 1 to 5 min/day for different strains of lactic acid bacteria protected with 5% glycerol, and can reach 11.4 min/day in the absence of cryoprotectant (Fonseca et al. 2000).

### 4.4 Thawing Conditions

Less information is available on thawing processes of probiotic bacteria, even though this process is involved in cell degradation. A devitrification phenomenon can take place during warming depending on the cooling rate. At high cooling rates, ice has no opportunity to completely crystallize and grow and, upon warming, migratory recrystallization occurs when molecular mobility increases (Mazur 1977, McFarlane 1986). Such crystallization of residual water that was kinetically inhibited during cooling has been detected by differential scanning calorimetry (DSC) on *Lb. bulgaricus* suspension (Fonseca et al. 2006). Consequently, it is recommended to apply warming rates as high as possible during thawing of bacterial suspensions (Morice et al. 1992, Piatkiewicz and Mokrosinska 1995). Such high warming rates limit cellular degradation caused by the dilution of the medium during thawing (Farrant and Morris 1973, McGann and Farrant 1976). Rapid warming in a water bath at 20–45°C was better than slow thawing at 4°C, for preserving viability of mesophilic streptococci (Shurda et al. 1980), while heating to 30–48°C gave best survival levels for *Lb. leichmanii* (Johannsen 1972). In contrast to these data, a rapid cooling of *Lb. acidophilus* (from 37°C to –80°C) induces considerable loss of cell viability, while slow cooling leads to a survival rate of 75% (Baati et al. 2000). Consequently, different thawing temperatures and thawing rates seem advisable according to the strain and the associated protective media.

## 5. Improvement of Freezing Resistance of Probiotic Bacteria by Controlling Fermentation and Post-fermentation Conditions

### 5.1 Definition of Biological Adaptation

Adaptation of probiotic bacteria to adverse environment is another interesting approach to increase their cryotolerance. Biological adaptation is obtained by an intentional action on the environment or on the process, in order to generate a moderate stress. This moderate stress may allow the bacteria to develop active responses, thus helping them to better resist the further stress. Homologous adaptation is achieved when the moderate stress is of the same nature than the real stress, whereas heterologous adaptation combines a different moderate stress (e.g., osmotic stress, acid stress…) before the freezing stress, thus leading to development of cross-resistance (Lacroix and Yildirim 2007). Such adaptation is generally associated with the induction of a large number of genes, leading to the alteration of some specific metabolisms, the synthesis of stress proteins, and the modification of membrane lipids and fatty acid composition. Adaptation can be carried out either during the bacterial growth, or at the end of the growth, or during any unit operation before applying the freezing stress. A combination of these different actions may also be accomplished.

## 5.2 Biological Mechanisms Involved during Biological Adaptation to Cold Stress

When bacterial cells are submitted to a moderate stress before freezing, they are able to develop different biological responses. At the membrane level, these responses concern the membrane lipid content and fatty acid composition, as well as lipid transition and fluidity. Changes also occur at the molecular level, when the bacteria are submitted to a cold stress. These changes happen at the transcriptomic and the proteomic levels, but only when a moderate stress is applied, as they require an active metabolism of the bacteria.

### 5.2.1 Modification of Lipid Content

As the membrane is a selective and permeable barrier between extracellular medium and cytoplasm, many changes occur at the membrane level, when the bacteria are submitted to a cold stress. By considering Gram positive bacteria, the membrane is composed by lipids (30–40%) and proteins (60–70%). Some carbohydrates are present that are associated with proteins (glycoproteins) or lipids (glycolipids). Among the membrane lipids that comprise phospholipids and glycolipids, the phospholipids are the most important as they are responsible for the lipid bilayer according to the fluid mosaic model. This lipid bilayer allows the physical separation between intracellular and extracellular media, facilitates the transport of substrates and metabolites among the inside and the outside of the cells, and maintains the cell potential. However, when probiotic bacteria are submitted to a cold shock, some properties of the biological membranes are affected, in terms of lipids and fatty acid composition thus modifying their relationship with their environment.

The lipid composition of the membranes is influenced by temperature. After extraction in methanol chloroform, the balance between phospholipids and glycolipids in the membranes of *Lb. acidophilus* CRL 640 was shown to be affected at 25°C, in comparison to 37°C (Fernandez-Murga et al. 1999). More particularly, they contained more glycolipids, namely diglycosylic and triglycosylic fractions, than phospholipids at low temperature. The authors suggested that these modifications have an effect on the ability of the bacteria to counteract the cold stress.

### 5.2.2 Modification of Membrane Fatty Acid Composition

Membrane phospholipids are amphilic molecules that are divided into glycerophospholipids and sphingophospholipids, which are composed of fatty acids linked to phosphoric acid, glycerol and other alcohols, amino acids or sphingosine. The membrane fatty acids of probiotic bacteria generally comprised chains of

14 to 22 carbons, saturated, unsaturated (with double bounds either in *cis* or *trans* position), with a cyclopropane, or with ramifications. The length and the degree of saturation are two essential factors that affect membrane fluidity, by altering the composition, structure and organization of lipids within the membrane. The characterization of the fatty acid composition of bacterial membranes involved five main steps: extraction and methylation of the fatty acids that may be coupled, separation by gas chromatography analysis associated with detection by flame ionization, identification with standards or mass spectrometry, and quantification with a suitable software.

The fatty acid composition of *Lb. rhamnosus* GG and *Lb. casei* Shirota, two recognized probiotic strains, was determined by Kankaanpää et al. (2004). These authors identified the most abundant bacterial fatty acids as oleic (C18:1 cis-9), vaccenic (C18:1 trans-11) and dihydrosterculic (cycC19:0 cis-9, 10) acids. The conjugated linoleic (C18:2, cis-9, trans-11), α-linolenic (C18:2, cis-9, 12), eicosapentaenoic (C20:5) and docosahexaenoic acids (C22:6) were also identified. The fatty acid composition of *Lb. casei* Zhang was characterized by Wu et al. (2012). The saturated fatty acids and unsaturated and cyclic fatty acids were well balanced (52% and 48%, respectively), the most abundant being pentadecylic acid (C15:0), palmitic acid (C16:0), oleic, vaccenic and dihydrosterculic acids. More recently, we identified the fatty acid composition of *B. animalis* subsp. *lactis* BB12 (Florence 2013). Cell membranes mainly contained saturated fatty acids (almost 70%), as a result of myristic (C14:0), palmitic (C16:0) and stearic (C18:0) acids. Monounsaturated fatty acids (27–28%) corresponded to palmitoleic (C16:1 cis-9), oleic (C18:1 cis-9), trans-vaccenic (C18:1 trans-11) and elaidic (C18:1 trans-9) acids. Polyunsaturated fatty acids (2.5–3.5%) comprised α-linoleic (C18:2, cis-9, 12), conjugated linoleic (C18:2, cis-9, trans-11) and α-linolenic acids (C18:3, cis-9,12,15).

The effect of freezing on the membrane fatty acid composition has never been studied by considering probiotic bacteria. However, some previous works exist by considering other lactic acid bacteria submitted to a moderate cold stress. By cultivating *Lb. acidophilus* at sub-optimal temperature (30°C instead of 37°C), Wang et al. (2005a) observed that the relative concentrations of C16:0 and cycC19:0 increased, to the detriment of C18:0 relative content in the membranes. This behavior led to a higher ratio between unsaturated and saturated fatty acids. These results partially confirmed those obtained by Fernandez-Murga et al. (1999) with *Lb. acidophilus* CRL 640 strains grown at 25°C by considering C16:0. However, these authors observed an increase in C18:2 but a decrease in cycC19:0 and they showed that the balance between phospholipids and glycolipids became more favorable to the first ones. These last findings differed from those obtained by Suutari and Laakso (1992) with *Lb. fermentum*. By decreasing the culture temperature from 35°C to 20°C, these authors indicated that the relative contents in C16:0, C16:1 and cycC19:0 decreased, whereas those of C18:1 increased. These divergent observations indicate that the modifications of membrane fatty acid composition as a function of temperature are strongly strain dependent.

More recently, the fatty acid composition of *Lb. acidophilus* was determined after cooling the cells at temperatures comprised between 33.8°C and 15°C for 1 to 8 hr (Wang et al. 2005b). These authors demonstrated that adaptation to low temperatures was related to an increase in the unsaturated to saturated fatty acid ratio and in the relative cycC19:0 fatty acid concentration.

Finally, this information points out that the fatty acid composition of bacterial membranes is affected by the temperature applied during growing and during cooling. Nevertheless, there is still no publication concerning the effect of freezing on these characteristics, by considering probiotic bacteria.

### *5.2.3 Membrane Lipid Transition and Fluidity*

By modifying culture conditions (medium composition, temperature, etc.), membrane lipid composition changes, which in turn, modulates membrane lipid phase behavior (lipid transition temperature and hysteresis) and fluidity, and consequently affects the bacterial resistance to freezing.

The membrane fluidity is usually determined by fluorescence anisotropy that is inversely proportional to membrane fluidity (Beney and Gervais 2001, Cao-Hoang et al. 2008, Tymczyszyn et al. 2007). This measurement needs the insertion of a fluorescent probe in the cell membrane, like diphenylhexatriene (DPH) or its trimethylammonium analog (TMA-DPH). In addition, FTIR spectroscopy makes it possible to study, *in situ* and during the freeze-thawing process, the lipid transitions of cell membranes (Crowe et al. 1989, Oldenhof et al. 2005, Ragoonanan et al. 2010b). The characterization of the state and organization of membrane lipids can be related to membrane fluidity: a gel state to a rigid membrane and a liquid crystalline state to a fluid membrane.

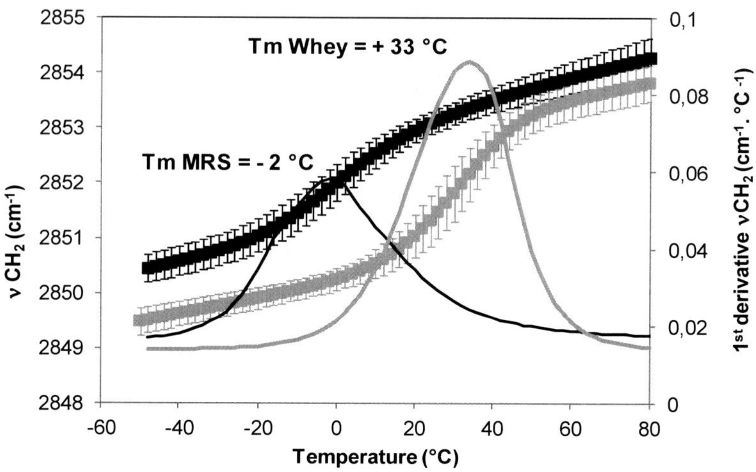


**Fig. 9.** Membrane lipid phase behavior ($CH_2$ symmetric) versus temperature plots of *Lb. delbrueckii* ssp. *bulgaricus* CFL1 cells after culture in de Man, Rogosa and Sharpe (MRS) broth (■) or in whey based medium (■). Cells were frozen from 50°C to −50°C and they heated from −50°C to 80°C at 2°C/min. The maximum of the first derivative of the wave number of the symmetric $CH_2$ stretching peak ($\nu CH_2$) versus temperature plot (solid line) allows the determination of the lipid phase transition temperature (Adapted from Gautier et al. 2013).

Gautier et al. (2013) recently characterized by FTIR spectroscopy the biophysical behavior of *Lb. delbrueckii* subsp. *bulgaricus* CFL1 cell membranes during freezing and thawing, after growing in different media. Following culture in a rich medium (MRS), cell membranes exhibited a higher content of cyclic and unsaturated fatty acids and a lower lipid transition temperature (Tm = –2°C) than cell membranes cultivated in whey based medium (Fig. 9). The maintenance of high membrane fluidity all along the freeze-thawing process was related to high resistance to freezing.

### *5.2.4 Modification of Gene Expression*

Transcriptomic approach has never been used to characterize the changes in gene expression following a cold stress by freezing in probiotic bacteria. However, previous works demonstrated that a temperature downshift from 30°C to 15°C modified the expression of 546 genes (217 were induced and 329 were repressed) in *Shewanella oneidensis* (Gao et al. 2006). Among them, genes encoding cold shock proteins, membrane proteins, DNA metabolism and translation apparatus components, metabolic proteins and regulatory proteins were up-regulated. These authors also observed that most of the affected metabolic proteins were implicated in catalytic processes that generate NADH or NADPH.

From this information, gene expression is affected at low temperature, thus modifying the ability of the cells to resist to a cold stress. Even if this phenomenon has never been revealed in probiotic bacteria, it may be involved in their ability to withstand the cold stress.

### *5.2.5 Modification of Intracellular Protein Synthesis*

Reducing the temperature triggers changes at the proteomic level in lactic acid bacteria, on the condition that temperature remains positive. The proteomic response to low temperature is determined by using a comparative proteomic analysis, implicating two-dimensional gel electrophoresis and identification of up- or down-regulated proteins by Matrix-assisted laser desorption/ionization-Time of flight (Maldi-TOF) or by liquid chromatography-tandem mass spectrometry (LC-MS/MS). By using Maldi-TOF, an increased synthesis of four specific proteins was demonstrated in *Lb. acidophilus* RD758 at low temperature (28°C instead of 37°C) by Wang et al. (2005b). It included the stress protein ATP-dependent ClpP and two cold induced proteins: pyruvate kinase and a putative glycoprotein endopeptidase. More recently, the response of *Lactococcus piscium* to cold acclimation at 5°C and cold shock at 0°C was analyzed (Garnier et al. 2010). These authors demonstrated that both cold shock and cold acclimation induced the up-regulation of proteins involved in general and oxidative stress responses (5 proteins), in gene expression (10 proteins) and in fatty acid (3 proteins) and energetic metabolisms (5 proteins).

Consequently, low temperatures act on protein synthesis in lactic acid bacteria. Even if no information concerns probiotic bacteria, such approaches are of great interest as they help understanding some mechanisms involved in cold stress.

## 5.3 Adaptation by Changing Environmental Conditions during Fermentation

By considering that probiotic bacteria respond to cold stress at the membrane, proteomic and transcriptomic levels, carrying out a moderate stress before freezing was shown to help the cells to better survive and maintain their metabolic properties after freezing and/or during frozen storage (Streit et al. 2008). The effect of different moderate environmental or stress conditions during fermentation was established by many authors. These conditions include the culture medium composition as well as the fermentation temperature and pH.

Two other factors may be involved in adaptation of probiotic bacteria. The atmosphere during the culture influences the growth of *B. bifidum* in milk (Ebel et al. 2011), but no work has still been published on the effect of this factor of the cryotolerance of this bacterium. In addition, the culture mode, combining continuous mixed cultures and cell immobilization, was shown to improve the resistance to freeze-drying of *B. longum* ATCC 15707 (Doleyres et al. 2004), but not yet to freezing.

### *5.3.1 Composition of Culture Medium*

The culture medium is generally defined with the aim of optimizing bacterial growth. However, some works demonstrated that bacterial cryotolerance may be improved by incorporating some specific compounds in the medium. These compounds may lead the cells to modify their structural and biochemical properties, thus helping them to better tolerate the freezing stresses.

The addition of calcium 0.1%, in the form of calcium carbonate, calcium chloride or calcium phosphate, enhanced the cryotolerance of *Lb. acidophilus* during storage at –20°C (Wright and Klaenhammer 1981), as a result of morphological transition of the lactobacilli, from filamentous to bacilloid form. This modification leads to an increase of the surface/volume ratio of the cells, which was related to improved cryotolerance in other lactic acid bacteria (Fonseca et al. 2000).

By considering that membrane fatty acid composition is the first target for biological response to stress, modification of membrane fatty acid composition was searched by adding Tween 80 (oleic acid, 1 g/L) in the culture medium of *S. thermophilus* (Béal et al. 2001). These authors demonstrated that this compound allowed increasing the relative cycC19:0 content in the membrane, which was related to improved cryotolerance after freezing at –20°C. However, such an approach was never confirmed with probiotic bacteria.

### *5.3.2 Fermentation Temperature*

Different authors improved the cryotolerance of *Lb. acidophilus* by controlling the fermentation temperature at a sub-optimal value (Fernandez-Murga et al. 2000, 2001, Maus and Ingham 2003, Wang et al. 2005a).

*Lb. acidophilus* CRL 640 grown at 25°C, was more resistant to the freeze-thaw process than after growth at 30, 37 and 40°C (Fernandez-Murga et al. 2001). The mortality was higher (87% of the cells died) at 37°C than at 25°C (33%). These authors related this behavior to the increase of palmitoleic (C16:0) and oleic (C18:0) fatty acids in cardiolipids, phosphatidyl-glycerol and diglycosyl-diglyceride fractions and to the decrease of dihydrosterculic (cycC19:0) and 10-hydroxyoctadecanoic (C18:0, 10-OH) fatty acids in neutral lipid and cardiolipid fractions of cells grown at 25ºC. They also demonstrated a significant increase in phosphatidyl-glycerol to the detriment of cardiolipid content, with respect to total phospholipids, and in diglycosyl-diglyceride level to the detriment of triglycosyl-diglyceride, with respect to the total glycolipids. In addition, a higher glycolipids to phospholipids ratio was observed at 25ºC as compared to 37ºC (Fernandez-Murga et al. 1999), which enhance the lipid membrane stability.

These results were confirmed and completed (Wang 2005, Wang et al. 2005a), as the cryotolerance of *Lb. acidophilus* RD758 was improved by controlling the fermentation temperature at a sub-optimal value (30ºC instead of 37ºC). By quantifying the ability of the cells to recover their acidification activity after freezing and frozen storage at –20ºC, these authors revealed that no loss in acidification activity occurred during freezing, and that the rate of loss in acidification activity was reduced during frozen storage, as compared to cells grown at their optimal temperature (37ºC). This improvement was firstly ascribed to a modification in the membrane fatty acid composition during the freezing step. The improved cryotolerance was associated to a high ratio of unsaturated to saturated fatty acids, a low C18:0 content together with high C16:0 and cycC19:0 relative concentrations. During frozen storage, the better resistance was related to a high cycC19:0 relative concentration in the membranes. Secondly, some changes occurred in the cytoplasm of *Lb. acidophilus* RD758, as demonstrated by the changes in protein synthesis. By comparing the density level of the cytoplasmic proteins detected at 37ºC and 30ºC, 11 proteins were over-synthesized, five proteins were under-synthesized and 10 proteins were not synthesized, when the cells were cultivated at 30ºC. Some of these proteins were identified. Among them, the synthesis of the ribosomal protein S2 and the GTPase TrmE was activated, thus facilitating translation at low temperature. In contrast, the enzymes glutamate N-acetyltransferase (amino acid biosynthesis), pyruvate dehydrogenase (glycolysis) and elongation factor (protein translation) were less synthesized, thus indicating that the general metabolism was reduced at 30ºC.

From this information, the fermentation temperature is a key factor to improve the cryotolerance of probiotic bacteria, as a consequence of modifications of their proteome and their membrane composition.

### *5.3.3 Fermentation pH*

Controlling the pH during the culture, at the optimal value for the growth of the bacteria, is commonly used in order to improve bacterial concentrations at industrial scale (Savoie et al. 2007). However, when *Lb. acidophilus* RD758 was cultivated at sub-optimal pH (pH 5), the loss of acidification activity during freezing and frozen storage was reduced, by comparison with the optimal pH (pH 6) (Wang et al. 2005a). This improvement was explained by modifications at the membrane level, as the relative concentrations of five fatty acids were influenced by the fermentation pH. Notably, higher C16:0, C18:1 and cycC19:0 concentrations were observed at acidic pH compared with pH 6, to the detriment of C14:0 and C18:0 relative contents.

These modifications allowed enhancing the ratio between unsaturated and saturated fatty acids, thus helping the cells to better resist the freezing step and frozen storage. Changes at the membrane level have been recently confirmed by Wu et al. (2012), who developed an acid-resistant mutant of *Lb. casei* Zhang that exhibited higher membrane fluidity, higher proportions of unsaturated fatty acids, and higher mean chain length of fatty acids, together with better cell integrity and lower membrane permeability.

Even if this approach is interesting, conducting the fermentation at a lower pH significantly enhances the duration of the culture, which is most of the time not compatible with industrial constraints.

## 5.4 Adaptation by Changing Environmental Conditions at the end of Fermentation

The effect of different environmental or stress conditions at the end of the fermentation, on the resistance of probiotic bacteria to freezing, was determined by some authors. They mainly studied the effects of harvesting time, associated or not to nutritional deprivation, as well as cooling and concentration conditions.

### 5.4.1 Harvesting Time

The time at which the cells are recovered at the end of the fermentation strongly affects the resistance to freezing of probiotic and lactic acid bacteria (Johannsen 1972, Rault et al. 2010), the better resistance being obtained when the cells were harvested in stationary phase, by considering both cultivability and acidification activity. These last authors demonstrated that the loss in cultivability of *Lb. delbrueckii* subsp. *bulgaricus* CFL1 was reduced by 2 log (CFU/mL) and that the loss in acidification activity was limited to 150 min instead of 350 min, after harvesting the cells in late stationary phase (Rault 2009). They explained this behavior by physiological changes that occurred in the stationary phase. By considering the membrane level, stable frozen starters were associated with high relative contents in C16:0 and cycC19:0 but low C18:1 percentages in membrane lipids. At the cytosolic level, a reduction of nucleotide metabolism together with transcription and translation processes, and a stimulation of energetic metabolism as well as amino acids, lipids and envelopes metabolisms, together with over-synthesis of stress proteins were observed. These results confirmed and completed those obtained with *Lb. acidophilus* (Lorca and Font de Valdez 2001) and *Lb. plantarum* WCFS1 (Cohen et al. 2006), as these works demonstrated that seven and six proteins, respectively, with a potential role for survival during starvation or involved in stress response, were significantly up-regulated in the late stationary phase.

### 5.4.2 Nutritional Deprivation

Besides the growth phase, nutritional starvation also acts on the ability of the bacteria to withstand the freezing stress. By considering *B. longum* and *B. lactis*, Maus and Ingham (2003) demonstrated that starved *B. lactis* cells displayed enhanced acid-tolerance that persisted through 8 to 11 weeks of storage at –80°C. In addition, Hussain et al.

(2009) demonstrated that numerous glycolytic enzymes were differentially regulated under lactose starvation of *Lb. casei*, thus suggesting a potential survival strategy under harsh growth conditions. Finally, a recent study correlated the physiological responses that followed starvation to the ability of *Lb. acidophilus* RD758 to survive after freezing and during storage (Wang et al. 2011). These authors explained this improved cryotolerance by a reduced synthesis of proteins involved in transcription and translation, cell wall biogenesis, amino acid and protein biosynthesis, but an over-synthesis of enzymes implicated in carbohydrate and energetic metabolisms and in pH homeostasis. These modifications were associated to a higher recovery of acidification activity and cultivability after freezing. Moreover, as observed under other stress conditions, the relative contents of branched (x5.57), unsaturated (x1.15) and cyclic (x1.42) fatty acids in membrane lipids were enhanced under starvation conditions, thus probably decreasing the solid-to-fluid transition temperature of membrane lipids and then maintaining the membrane fluidity under stress conditions (Wang et al. 2011).

In addition to these biological modifications, some changes were pointed out at the transcriptomic level, when a glucose starvation was applied to *Lactococcus lactis* IL1403 (Redon et al. 2005). Even if this work did not relate these changes to the cryotolerance of the bacteria, one could suggest that such modifications may also be implicated in cryotolerance, even in probiotic bacteria.

### *5.4.3 Cooling Conditions*

At the end of the fermentation step and before stabilization by freezing, probiotic starters are cooled from their growth temperature to about 10–15°C, in order to stop biological reactions. Some authors studied the effect of the cooling conditions on the resistance to freezing of *Lb. acidophilus* RD758 (Wang et al. 2005b). They demonstrated that a cold adaptation occurs when the cells were maintained at moderate temperature (28°C) for 8 hrs before being cooled to the final temperature of 15°C, as the rate of loss of acidification activity during frozen storage was ten times reduced. This adaptation to cold temperatures was related to an increase in the unsaturated to saturated fatty acid ratio and in the relative cycC19:0 fatty acid concentration in bacterial membranes. In addition, four specific proteins were over-synthesized at this intermediate temperature, including the stress protein ATP-dependent ClpP and two cold induced proteins: pyruvate kinase and a putative glycoprotein endopeptidase.

Besides the effect of cooling temperature, another study demonstrated that an acidification of the starters before the cooling step allowed improving the cryotolerance of *Lb. bulgaricus* (Streit et al. 2007). These authors indicated that decreasing the pH to 5.15 for 30 min just before cooling the cells at 15°C improves their resistance to freezing and frozen storage by 30%. This cryotolerance improvement was associated to the synthesis of 21 proteins, involved in energy metabolism, nucleotide and protein synthesis as well as cold stress response (Streit et al. 2008).

From this information, a physiological adaptation of probiotic bacteria to freezing happens during the cooling step, but the responses depend on the bacterial strain or species.

### *5.4.4 Concentration Conditions*

Before freezing, a partial removal of water, metabolites and residual substrates is performed, in order to reduce the volume of the starters to be stabilized and handled, to increase the bacterial density and to improve the cryotolerance of the bacteria. The concentration factor is generally comprised between 10 and 40, in order to reach bacterial concentrations comprised between $10^9$ to $10^{12}$ CFU/g (Béal and Corrieu 1994). This concentration step is generally done by centrifugation or tangential filtration.

The centrifugation conditions affect the resistance to freezing and frozen storage of *Lb. bulgaricus* (Streit et al. 2010). More precisely, these authors indicated that centrifugation speed and duration slightly alter the cell resistance to freezing and frozen storage, whereas no effect of temperature (4°C or 15°C) was demonstrated.

Microfiltration has been used for concentrating probiotic and lactic acid bacteria during starter production (Persson et al. 2001, Streit et al. 2011). These authors pointed out that the bacterial cryotolerance was higher when *Lb. bulgaricus* cells were microfiltered instead of centrifuged. They also demonstrated that cells microfiltered at a cross-flow velocity of 2 m/s and a transmembrane pressure of 0.15 Mpa displayed an improved cryotolerance. This improvement was explained by increased unsaturated to saturated and cyclic to saturated fatty acid ratios in bacterial membranes, which enhanced membrane fluidity, thus helping the cells to better resist freezing and frozen storage (Streit et al. 2011). In addition, the synthesis of proteins involved in energetic metabolism was enhanced, whereas that of nitrogen metabolism was reduced (Streit 2008).

Finally, these data indicated that it is possible to reduce the negative effect of concentration conditions, by selecting appropriate downstream processes together with adequate conditions before the freezing step.

## 6. Freezing Technology and Control of Quality during Probiotic Starter Production

By considering the physico-chemical and biological responses of probiotic bacteria to freezing stress, together with the routes used to improve their cryotolerance, procedures have been defined for their cryoprotection, freezing and storage at large scale.

### *6.1 Cryoprotection of Probiotic Starters*

At an industrial scale, cryoprotection is done by mixing the concentrated bacteria with a well defined cryoprotective mix. As previously described (section 4.1.2), this cryoprotective medium is generally composed of polysaccharides, disaccharides, skim milk, polyols and antioxidants. Carbohydrates make it possible to raise the glass transition temperature (Tg'), thus lowering the cellular damages. The polyols contribute to enhance the medium viscosity, and then to reduce biological reactions and water mobility. Antioxidants, such as sodium ascorbate, are frequently added to limit oxidation phenomena. The mixing of the concentrated bacteria and the cryoprotective medium is carried out gently, in order to avoid incorporation of air, and under aseptic conditions to avoid any contamination of the product.

### 6.2 Freezing and Packaging of Frozen Probiotic Starters

Frozen probiotic starters have to display long term stability to allow commercial use. In this respect, they are packed up in specific packages and frozen at ultra low temperature, by freezing in liquid nitrogen (–196ºC) or in vapor nitrogen (–130ºC), after concentration and cryoprotection. These operations are processed in clean rooms and under hygienic standards, in order to avoid any contamination and to ensure the quality of the products. They shall meet all Good Manufacturing Practices (GMP) guidelines in the country concerned.

Two main routes are employed. According to the first one, the concentrated and cryoprotected bacteria are frozen in bulk form, after packaging in aluminum tin cans, which are immersed in liquid nitrogen. Even if this method is simple to operate, it doesn't allow mixing different strains in the same can after freezing and is less and less carried out. In the second way, the cryoprotected bacteria are frozen as small pellets (2 to 5 mm diameter), either by pulverization of the bacterial concentrates in vapor nitrogen, or by distribution on an alveolated surface cooled with liquid nitrogen. The frozen pellets are then conditioned in aseptic Tetra Brik® containers. As this method leads to the manufacture of frozen beads, mixing different strains in the form of different beads in the final product is facilitated, as compared to the cans. In addition, it permits the use of fractional doses. However, it results in a high volume to weight ratio, thus engendering higher surfaces of storage and then, higher storage costs.

In the case of fermented milk manufacturing, the capacity of the packaging (aluminum tin cans or Tetra Brik® containers) is define to inoculate either 500 L or 1,000 L of milk, thus corresponding to inoculation amounts comprised between 0.01 and 0.05% (w/v). These inoculation rates generally correspond to bacterial concentrations ranging between $10^6$ and $10^7$ CFU/mL.

### 6.3 Storage and use of Frozen Probiotic Starters

Storage of frozen bacteria requires a very rigorous respect of the cold chain. The storage temperature of the starters has to be maintained under –60ºC at the supplier stage, by using freezers equipped with temperature monitoring and $CO_2$/$LN_2$ back-up systems. Accurate shipping conditions include the use of insulated packaging and dry ice in the secondary packaging. Once delivery is completed at the end-user, the product shall be stored at –45ºC or less, thus allowing 3 to 12 months preservation without loss of viability or acidification activity.

Commercial starters have to be thawed and handled according to the supplier's recommendations. Starters packed in cans require partial thawing and decontamination with chloride products (20ºC for 10 min) before being used. After the lid is open, the bulk starter is directly introduced in the manufacturing tank, previously placed at the right temperature. In contrast, pelleted frozen starter cultures don't need thawing and can be directly added to the tank.

### *6.4 Control of Quality of Frozen Probiotic Starters*

The quality of probiotic starters is defined by their ability to exhibit their technological and probiotic properties during their use, which includes different aspects. In all cases, a good stability of the bacterial concentration in the commercial starter is required at expiration date. This characteristic is commonly established from cultivability measurements, by using plate counts (Shah 2000). In order to anticipate the results, it is possible to store the frozen starters at a higher temperature to accelerate the degradation of the products. A relationship with the results obtained at the right storage temperature should be first established and validated.

In a context of fermented milk manufacture, acidification activity is also controlled, although this characteristic is often ascribed to the other lactic acid bacteria used in combination with the probiotic starters. This trait is generally determined by using an automatic system such as the Cinac system (Corrieu et al. 1988). By measuring on-line and continuously the pH decrease in a convenient medium and in well defined conditions, this system allows quantifying different descriptors to characterize the starter, for example, the time necessary to reach a given pH or the pH variation during a given time. Currently, this system is utilized by most suppliers and end-users, and it generally forms part of the specifications of the products.

Finally, the temperature of the product is controlled all along the supply chain, and during the whole storage of the product.

## 7. Conclusion

Freezing probiotic bacteria, in addition to freeze-drying and drying, is one of the main ways for keeping these starters preserved for a long time. By decreasing the temperature of the product below 0ºC, this process allows slowing down the physico-chemical and biological reactions. This practice engenders some physical and chemical stress that lead to partial cell death. However, the bacteria are able to develop some biological responses, at the membrane and cytoplasmic levels. They help them to counteract the cold stress, thus leading to partial cell survival, with strong differences among the various strains.

In order to help the probiotic bacteria to better survive and maintain their biological properties, two main routes are employed: the use of adequate cryoprotection, freezing rate and freezing temperature conditions; the carrying out of adaptation procedures, by applying a moderate stress during fermentation, cooling and/or concentration steps, in order to prepare the cells before the real freezing stress.

Finally, at the industrial level, freezing is mainly done in liquid or vapor nitrogen, in order to reduce the physico-chemical and biological stress. However, some works are still conducted in order to better understand the mechanisms involved in fast freezing stress resistance and to improve the cryotolerance of probiotic bacteria.

## References


Alakomi, H.L., J. Matto, I. Virkajarvi and M. Saarela. 2005. Application of a microplate scale fluorochrome staining assay for the assessment of viability of probiotic preparations. Journal of Microbiological Methods 62(1): 25–35.

Amaretti, A., M. Nunzio, A. Pompei, S. Raimondi, M. Rossi and A. Bordoni. 2013. Antioxidant properties of potentially probiotic bacteria: *in vitro* and *in vivo* activities. Applied Microbiology and Biotechnology 97(2): 809–817.

Anchordoguy, T.J., A.S. Rudolph, J.F. Carpenter and J.H. Crowe. 1987. Modes of interaction of cryoprotectants with membrane phospholipids during freezing. Cryobiology 24(4): 324–331.

Baati, L., C. Fabre-Gea, D. Auriol and P.J. Blanc. 2000. Study of the cryotolerance of *Lactobacillus acidophilus*: effect of culture and freezing conditions on the viability and cellular protein levels. International Journal of Food Microbiology 59(3): 241–247.

Balasubramanian, S.K., W.F. Wolkers and J.C. Bischof. 2009. Membrane hydration correlates to cellular biophysics during freezing in mammalian cells. Biochimica et Biophysica Acta (BBA)-Biomembranes 1788(5): 945–953.

Basyigit, G., H. Kuleasan and A. Karahan. 2006. Viability of human-derived probiotic lactobacilli in ice cream produced with sucrose and aspartame. Journal of Industrial Microbiology and Biotechnology 33(9): 796–800.

Baumann, D.P. and G.W. Reinbold. 1966. Freezing of lactic cultures. Journal of Dairy Science 49(6): 259–263.

Béal, C. and G. Corrieu. 1994. Viability and acidification activity of pure and mixed starters of *Streptococcus salivarius* ssp. *thermophilus* 404 and *Lactobacillus delbrueckii* ssp. *bulgaricus* 398 at the different steps of their production. Lebensmittel Wissenschaft und Technologie 27(1): 86–92.

Béal, C., F. Fonseca and G. Corrieu. 2001. Resistance to freezing and frozen storage of *Streptococcus thermophilus* is related to membrane fatty acid composition. Journal of Dairy Science 84(11): 2347–2356.

Beney, L. and P. Gervais. 2001. Influence of the fluidity of the membrane on the response of microorganisms to environmental stresses. Applied Microbiology and Biotechnology 57(1-2): 34–42.

Berner, D. and H. Viernstein. 2006. Effect of protective agents on the viability of *Lactococcus lactis* subjected to freeze-thawing and freeze-drying. Scientia Pharmaceutica 74(3): 137–149.

Branca, C., S. Magazu, G. Maisano and P. Migliardo. 1999. Anomalous cryoprotective effectiveness of trehalose: Raman scattering evidences. Journal of Chemical Physics 111(1): 281–287.

Cao-Hoang, L., F. Dumont, P.A. Marechal, M. Le-Thanh and P. Gervais. 2008. Rates of chilling to 0°C: implications for the survival of microorganisms and relationship with membrane fluidity modifications. Applied Microbiology and Biotechnology 77(6): 1379–1387.

Carcoba, R. and A. Rodriguez. 2000. Influence of cryoprotectants on the viability and acidifying activity of frozen and freeze-dried cells of the novel starter strain *Lactococcus lactis* ssp. *lactis* CECT 5180. European Food Research and Technology 211(6): 433–437.

Carvalho, A.S., J. Silva, P. Ho, P. Teixeira, F.X. Malcata and P. Gibbs. 2004. Relevant factors for the preparation of freeze-dried lactic acid bacteria. International Dairy Journal 14(10): 835–847.

Chapman, D. 1975. Phase transitions and fluidity characteristics of lipids and cell membranes. Quaterly Reviews of Biophysics 8(2): 185–235.

Chavarri, F.J., M. De Paz and M. Nunez. 1988. Cryoprotective agents for frozen concentrated starters from non-bitter *Streptococcus lactis* strains. Biotechnology Letters 10(1): 11–16.

Chen, T., A. Fowler and M. Toner. 2000. Literature review: Supplemented phase diagram of the trehalose-water binary system. Cryobiology 40(3): 277–282.

Clarke, A., G.J. Morris, F. Fonseca, B.J. Murray, E. Acton and H.C. Price. 2013. A low temperature limit for life on earth. Plos One 8(6): e66207.

Cohen, D.P.A., J. Renes, F.G. Bouwman, E.G. Zoetendal, E. Mariman, W.M. de Vos and E.E.Vaughan. 2006. Proteomic analysis of log to stationary growth phase *Lactobacillus plantarum* cells and a 2-DE database. Proteomics 6(24): 6485–6493.

Coppola, R., M. Iorizzo, A. Sorrentino, E. Sorrentino and L. Grazia. 1996. Resistenza al congelamento di lattobacilli mesofili isolati da insaccati e paste acide. Industrie Alimentari 35-347: 349–351.

Corrieu, G., H.E. Spinnler, Y. Jomier, D. Picque and INRA, assignee. 1988. Automated system to follow up and control the acidification activity of lactic acid starters. French patent FR 2 629–612.

Crowe, J.H., F.A. Hoekstra, L.M. Crowe, T.J. Anchordoguy and E. Drobnis. 1989. Lipid phase transitions measured in intact cells with Fourier Transform Infrared Spectroscopy. Cryobiology 26(1): 76–84.

De Antoni, G.L., P. Perez, A. Abraham and M.C. Añon. 1989. Trehalose, a cryoprotectant for *Lactobacillus bulgaricus*. Cryobiology 26(2): 149–153.

De Giulio, B., P. Orlando, G. Barba, R. Coppola, M. De Rosa, A. Sada, P.P. De Prisco and F. Nazzaro. 2005. Use of alginate and cryo-protective sugars to improve the viability of lactic acid bacteria after freezing and freeze-drying. World Journal of Microbiology and Biotechnology 21(5): 739–746.

Doleyres, Y., I. Fliss and C. Lacroix. 2004. Increased stress tolerance of *Bifidobacterium longum* and *Lactococcus lactis* produced during continuous mixed-strain immobilized-cell fermentation. Journal of Applied Microbiology 97(3): 527–539.

Dumont, F., P.A. Marechal and P. Gervais. 2004. Cell size and water permeability as determining factors for cell viability after freezing at different cooling rates. Applied and Environmental Microbiology 70(1): 268–272.

Ebel, B., F. Martin, L.D.T. Le, P. Gervais and R. Cachon. 2011. Use of gases to improve survival of *Bifidobacterium bifidum* by modifying redox potential in fermented milk. Journal of Dairy Science 94(5): 2185–2191.

Farrant, J. and G.J. Morris. 1973. Thermal shock and dilution shock as the causes of freezing injury. Cryobiology 10(2): 134–140.

Fernandez-Murga, M.L., D. Bernik, G. Font de Valdez and A.E. Disalvo. 1999. Permeability and stability properties of membranes formed by lipids extracted from *Lactobacillus acidophilus* grown at different temperatures. Archives of Biochemistry and Biophysics 364(1): 115–121.

Fernandez-Murga, M.L., G.M. Cabrera, G. Font de Valdez, A. Disalvo and A.M. Seldes. 2000. Influence of growth temperature on cryotolerance and lipid composition of *Lactobacillus acidophilus*. Journal of Applied Microbiology 88(2): 342–348.

Fernandez-Murga, M.L., G. Font de Valdez and E.A. Disalvo. 2001. Effect of lipid composition on the stability of cellular membranes during freeze-thawing of *Lactobacillus acidophilus* grown at different temperatures. Archives of Biochemistry and Biophysics 388(2): 179–184.

Florence, A.C.R. 2013. Physiological responses of bifidobacteria subjected to acid, cold and gastro-intestinal stress in organic and conventional milks [Ph.D. thesis]. Paris (France) and São Paulo (Brazil): AgroParisTech and University of São Paulo. 203 p.

Fonseca, F., C. Béal and G. Corrieu. 2000. Method of quantifying the loss of acidification activity of lactic acid starters during freezing and frozen storage. Journal of Dairy Research 67(1): 83–90.

Fonseca, F., C. Béal and G. Corrieu. 2001a. Operating conditions that affect the resistance of lactic acid bacteria to freezing and frozen storage. Cryobiology 43(3): 189–198.

Fonseca, F., C. Béal, F. Mihoub, M. Marin and G. Corrieu. 2003. Improvement of cryopreservation of *Lactobacillus delbrueckii* subsp. *bulgaricus* CFL1 with additives displaying different protective effects. International Dairy Journal 13(11): 917–926.

Fonseca, F., M. Marin and G.J. Morris. 2006. Stabilization of frozen *Lactobacillus delbrueckii* subsp. *bulgaricus* in glycerol suspensions: freezing kinetics and storage temperature effects. Applied and Environmental Microbiology 72(10): 6474–6482.

Fonseca, F., J.P. Obert, C. Béal and M. Marin. 2001b. State diagrams and sorption isotherms of bacterial suspensions and fermented medium. Thermochimica Acta 366(2): 167–182.

Fonseca, F., S. Passot, P. Lieben and M. Marin. 2004. Collapse temperature of bacterial suspensions: the effect of cell type and concentration. Cryo-Letters 25(6): 425–434.

Font de Valdez, G., G. Savoy de Giori, A. Pesce de Ruiz Holgado and G. Oliver. 1983. Comparative study of the efficiency of some additives in protecting lactic acid bacteria against freeze-drying. Cryobiology 20(5): 560–566.

Foschino, R., E. Fiori and A. Galli. 1996. Survival and residual activity of *Lactobacillus acidophilus* frozen cultures under different conditions. Journal of Dairy Research 63(2): 295–303.

Franks, F. 1998. Freeze-drying of bioproducts: putting principles into practice. European Journal of Pharmaceutics and Biopharmaceutics 45(3): 221–229.

Gao, H.Y., Z.K. Yang, L.T. Wu, D.K. Thompson and J. Zhou. 2006. Global transcriptome analysis of the cold shock response of Shewanella oneidensis MR-1 and mutational analysis of its classical cold shock proteins. Journal of Bacteriology 188(12): 4560–4569.

Garnier, M., S. Matamoros, D. Chevret, M.F. Pilet, F. Leroi and O. Tresse. 2010. Adaptation to cold and proteomic responses of the psychrotrophic biopreservative *Lactococcus piscium* strain CNCM I-4031. Applied Environmental Microbiology 76(24): 8011–8018.

Gautier, J., S. Passot, C. Penicaud, H. Guillemin, S. Cenard, P. Lieben and F. Fonseca. 2013. A low membrane lipid phase transition temperature is associated with a high cryotolerance of *Lactobacillus delbrueckii* subspecies *bulgaricus* CFL1. Journal of Dairy Science 96(9): 5591–5602.

Gilliland, S.E. and R.C. Lara. 1988. Influence of storage at freezing and subsequent refrigeration temperatures on b-galactosidase activity of *Lactobacillus acidophilus*. Applied and Environmental Microbiology 54(4): 898–902.

Godward, G., K. Sultana, K. Kailasapathy, P. Peiris, R. Arumugaswamy and N. Reynolds. 2000. The importance of strain selection on the viability and survival of probiotic bacteria in dairy foods. Milchwissenschaft 55(8): 441–445.

Hubalek, Z. 1996. Cryopreservation of microorganisms at ultra-low temperatures. Pragues: Academia. 286 p.

Hubalek, Z. 2003. Protectants used in the cryopreservation of microorganisms. Cryobiology 46(3): 205–229.

Hussain, M.A., M.I. Knight and M.L. Britz. 2009. Proteomic analysis of lactose-starved *Lactobacillus casei* during stationary growth phase. Journal of Applied Microbiology 106(3): 764–773.

Jakubowska, J., Z. Libudzisz and A. Piatkiewicz. 1980. Evaluation of lactic acid Streptococci for the preparation of frozen concentrated starter cultures. Acta Microbiologica Polonica 29(2): 135–144.

Johannsen, E. 1972. Influence of various factors on the survival of *Lactobacillus leichmannii* during freezing and thawing. Journal of Applied Bacteriology 35(4): 415–421.

Juárez Tomás, M.S., V.S. Ocaña and M.E. Nader-Macías. 2004. Viability of vaginal probiotic lactobacilli during refrigerated and frozen storage. Anaerobe 10(1): 1–5.

Kankaanpää, P., B. Yang, H. Kallio, E. Isolauri and S. Salminen. 2004. Effects of polyunsaturated fatty acids in growth medium on lipid composition and on physicochemical surface properties of lactobacilli. Applied and Environmental Microbiology 70(1): 129–136.

Karlsson, J.O.M., E.G. Cravalho and M. Toner. 1993. Intracellular ice formation: causes and consequences. Cryo-Letters 14: 323–334.

Karlsson, J.O.M. and M. Toner. 1996. Long-term preservation of tissues by cryopreservation: critical issues. Biomaterials 17(3): 243–256.

Klaenhammer, T.R. and E.G. Kleeman. 1981. Growth characteristics, bile sensitivity, and freeze damage in colonial variants of *Lactobacillus acidophilus*. Applied and Environnmental Microbiology 41(6): 1461–1467.

Lacroix, C. and S. Yildirim. 2007. Fermentation technologies for the production of probiotics with high viability and functionality. Current Opinion in Biotechnology 18(2): 176–183.

Leloux, M.S. 1999. The influence of macromolecules on the freezing of water. Journal of Macromolecular Science. Reviews in Macromolecular Chemistry and Physics C39(1): 1–16.

Levine, H. and L. Slade. 1988. Principles of "cryostabilization" technology from structure/property relationships of carbohydrate/water systems—A review. Cryo-Letters 9: 21–63.

Lorca, G.L. and G. Font de Valdez. 2001. A low-pH-inducible, stationary-phase acid tolerance response in *Lactobacillus acidophilus* CRL 639. Current Microbiology 42(1): 21–25.

Madureira, A.M., M. Amorim, A.M. Gomes, M. Pintado and F.X. Malcata. 2011. Protective effect of whey cheese matrix on probiotic strains exposed to simulated gastrointestinal conditions. Food Research International 44(1): 465–470.

Mainville, I., Y. Arcand and E.R. Farnworth. 2005. A dynamic model that simulates the human upper gastrointestinal tract for the study of probiotics. International Journal of Food Microbiology 99(3): 287–296.

Marteau, P., M. Minekus, R. Havenaar and J.H.J. Huis In't Veld. 1997. Survival of lactic acid bacteria in a dynamic model of the stomach and small intestine: validation and the effects of bile. Journal of Dairy Science 80(6): 1031–1037.

Maus, J.E. and S.C. Ingham. 2003. Employment of stressful conditions during culture production to enhance subsequent cold- and acid-tolerance of bifidobacteria. Journal of Applied Microbiology 95(1): 146–153.

Mazur, P. 1965. Causes of injury in frozen and thawed cells. Federation Proceedings 24: S175–S182.

Mazur, P. 1970. Cryobiology: The freezing of biological systems. Science 168: 939–949.

Mazur, P. 1977. The role of intracellular freezing in the death of cells cooled at supraoptimal rates. Cryobiology 14(3): 251–272.

Mazur, P. 1984. Freezing of living cells: Mechanisms and implications. American Journal of Physiology C16: 125–142.

Mazur, P. 1990. Equilibrium, quasi-equilibrium, and nonequilibrium freezing of mammalian embryos. Cell Biochemistry and Biophysics 17(1): 53–92.

McFarlane, D.R. 1986. Devitrification in glass-forming aqueous solutions. Cryobiology 23(3): 230–244.

McFarlane, D.R. 1987. Physical aspects of vitrification in aqueous solutions. Cryobiology 24(3): 181–195.

McGann, L.E. 1978. Differing actions of penetrating and non penetrating cryoprotective agents. Cryobiology 15(4): 382–390.

McGann, L.E. and J. Farrant. 1976. Survival of tissue culture cells frozen by a two-step procedure to −196°C. II. Warming rate and concentration of dimethyl sulphoxide. Cryobiology 13(3): 261–268.

Modesto, M., P. Mattarelli and B. Biavati. 2004. Resistance to freezing and freeze-drying storage processes of potential probiotic bifidobacteria. Annals of Microbiology 54(1): 43–48.

Morice, M., P. Bracquart and G. Linden. 1992. Colonial variation and freeze-thaw resistance of *Streptococcus thermophilus*. Journal of Dairy Science 75(5): 1197–1203.

Novik, G.I., N.I. Astapovitch, N.G. Kadnikova and N.E. Ryabaya. 1998. Retention of the viability and physiological properties of bifidobacteria during storage by cryopreservation and lyophilisation. Microbiology 67(5): 525–529.

Oldenhof, H., W.F. Wolkers, F. Fonseca, S. Passot and M. Marin. 2005. Effect of sucrose and maltodextrin on the physical properties and survival of air-dried *Lactobacillus bulgaricus*: An *in situ* Fourier Transform Infrared Spectroscopy study. Biotechnology Progress 21(3): 885–892.

Persson, A., A.S. Jonsson and G. Zacchi. 2001. Separation of lactic acid-producing bacteria from fermentation broth using a ceramic microfiltration membrane with constant permeate flow. Biotechnology and Bioengineering 72(3): 269–277.

Peter, G. and O. Reichart. 2001. The effect of growth phase, cryoprotectants and freezing rates on the survival of selected micro-organisms during freezing and thawing. Acta Alimentaria 30(1): 89–97.

Piatkiewicz, A. and K. Mokrosinska. 1995. Effect of thawing rate on survival and activity of lactic acid bacteria. Polish Journal of Food and Nutrition Sciences 4(2): 33–46.

Ragoonanan, V., A. Hubel and A. Aksan. 2010a. Response of the cell membrane cytoskeleton complex to osmotic and freeze/thaw stresses. Cryobiology 61(3): 335–344.

Ragoonanan, V., T. Wiedmann and A. Aksan. 2010b. Characterization of the effect of NaCl and trehalose on the thermotropic hysteresis of DOPC lipids during freeze/thaw. The Journal of Physical Chemistry B 114(50): 16752–16758.

Rault, A. 2009. Incidence de l'état physiologique des lactobacilles d'intérêt laitier sur leur cryotolérance [Ph.D. thesis]. Paris (France): AgroParisTech. 274 p.

Rault, A., C. Béal, S. Ghorbal, J.C. Ogier and M. Bouix. 2007. Multiparametric flow cytometry allows rapid assessment and comparison of lactic acid bacteria viability after freezing and during frozen storage. Cryobiology 55(1): 35–43.

Rault, A., M. Bouix and C. Béal. 2010. Cryotolerance of *Lactobacillus delbrueckii* subsp. *bulgaricus* CFL1 is influenced by the physiological state during fermentation. International Dairy Journal 20(11): 792–799.

Redon, E., P. Loubiere and M. Cocaign-Bousquet. 2005. Transcriptome analysis of the progressive adaptation of *Lactococcus lactis* to carbon starvation. Journal of Bacteriology 187(10): 3589–3592.

Rudolph, A.S. and J.H. Crowe. 1985. Membrane stabilisation during freezing: the role of two natural cryoprotectants, trehalose and proline. Cryobiology 22(4): 367–377.

Saarela, M., I. Virkajärki, H.L. Alakomi, T. Mattila-Sandholm, A. Vaari, T. Suomalainen and J. Mättö. 2005. Influence of fermentation time, cryoprotectant and neutralization of cell concentrate on freeze-drying survival, storage stability, and acid and bile exposure of *Bifidobacterium animalis* ssp. *lactis* cells produced without milk-based ingredients. Journal of Applied Microbiology 99(6): 1330–1339.

Sanders, J.W., G. Venema and J. Kok. 1999. Environmental stress responses in *Lactococcus lactis*. FEMS Microbiology Reviews 23(4): 483–501.

Savoie, S., C.P. Champagne, S. Chiasson and O. Audet. 2007. Media and process parameters affecting the growth, strain ratios and specific acidifying activities of a mixed lactic starter containing aroma-producing and probiotic strains Journal of Applied Microbiology 103(1): 163–174.

Shurda, G.G. 1980. Influence of profound oxidation [freezing] on certain properties of lactic acid bacteria. Applied Biochemistry and Microbiology 16(1): 11–16.

Searles, J.A., J.F. Carpenter and T.W. Randolph. 2001. The ice nucleation temperature determines the primary drying rate of lyophilization for samples frozen on a temperature-controlled shelf. Journal of Pharmaceutical Sciences 90(7): 860–871.

Shah, N.P. 2000. Probiotic bacteria: Selective enumeration and survival in dairy foods. Journal of Dairy Science 83(4): 894–907.

Smittle, R.B., S.E. Gilliland and M.L. Speck. 1972. Death of *Lactobacillus bulgaricus* resulting from liquid nitrogen freezing. Applied Microbiology 24(4): 551–554.

Stamatova, I.V. 2010. Probiotic activity of *Lactobacillus delbrueckii* subsp. *bulgaricus* in the oral cavity. An *in vitro* study [Ph.D. thesis]. Helsinki (Finland): University of Helsinky. 82 p.

Streit, F. 2008. Influence des conditions de récolte et de concentration sur l'état physiologique et la cryotolérance de *Lactobacillus delbrueckii* subsp. *bulgaricus* CFL1 [Ph.D. thesis]. Paris (France): AgroParisTech. 222 p.

Streit, F., V. Athes, A. Bchir, G. Corrieu and C. Béal. 2011. Microfiltration conditions modify *Lactobacillus bulgaricus* cryotolerance in response to physiological changes. Bioprocess and Biosystems Engineering 34(2): 197–204.

Streit, F., G. Corrieu and C. Béal. 2007. Acidification improves cryotolerance of *Lactobacillus delbrueckii* subsp. *bulgaricus* CFL1. Journal of Biotechnology 128(3): 659–667.

Streit, F., G. Corrieu and C. Béal. 2010. Effect of centrifugation conditions on the cryotolerance of *Lactobacillus bulgaricus* CFL1. Food and Bioprocess Technology 3(1): 36–42.

Streit, F., J. Delettre, G. Corrieu and C. Béal. 2008. Acid adaptation of *Lactobacillus delbrueckii* subsp. *bulgaricus* induces physiological responses at membrane and cytosolic levels that improves cryotolerance. Journal of Applied Microbiology 105(4): 1771–1780.

Succi, M., P. Tremonte, A. Reale, E. Sorrentino and R. Coppola. 2007. Preservation by freezing of potentially probiotic strains of *Lactobacillus rhamnosus*. Annals of Microbiology 57(4): 537–544.

Sumeri, I., L. Arike, J. Stekolstsikova, R. Uusna, S. Adamberg, K. Adamberg and T. Paalme. 2010. Effect of stress pretreatment on survival of probiotic bacteria in gastrointestinal tract simulator. Applied Microbiology and Biotechnology 86(6): 1925–1931.

Suutari, M. and S. Laakso. 1992. Temperature adaptation in *Lactobacillus fermentum*: interconversions of oleic, vaccenic and dihydrosterculic acids. Journal of General Microbiology 138(3): 445–450.

Taranto, M.P., M.L. Fernandez-Murga, G. Lorca and G. Font de Valdez. 2003. Bile salts and cholesterol induce changes in the lipid cell membrane of *Lactobacillus reuteri*. Journal of Applied Microbiology 95(1): 86–91.

Thunell, R.K., W.E. Sandine and F.W. Bodyfelt. 1984. Frozen starters from internal-pH-control-grown cultures. Journal of Dairy Science 67(1): 24–36.

To, B.C.S. and M.R. Etzel. 1997. Spray-drying, freeze-drying, or freezing of three different lactic acid bacteria species. Journal of Food Science 62(3): 576–578 & 585.

Tsvetkov, T. and I. Shishkova. 1982. Studies on the effects of low temperatures on lactic acid bacteria. Cryobiology 19(2): 211–214.

Tymczyszyn, E.E., M. Del Rosario Díaz, A. Gómez-Zavaglia and E.A. Disalvo. 2007. Volume recovery, surface properties and membrane integrity of *Lactobacillus delbrueckii* subsp. *bulgaricus* dehydrated in the presence of trehalose or sucrose. Journal of Applied Microbiology 103(6): 2410–2419.

Vasiljevic, T. and N.P. Shah. 2008. Probiotics—From Metchnikoff to bioactives. International Dairy Journal 18(7): 714–728.

Volkert, M., E. Ananta, C. Luscher and D. Knorr. 2008. Effect of air freezing, spray freezing, and pressure shift freezing on membrane integrity and viability of *Lactobacillus rhamnosus* GG. Journal of Food Engineering 87(4): 532–540.

Wang, Y. 2005. Preadaptation and cryotolerance in *Lactobacillus acidophilus*: Effect of operating conditions [Ph.D. thesis]. Paris (France): Institut National Agronomique Paris-Grignon. 214 p.

Wang, Y., G. Corrieu and C. Béal. 2005a. Fermentation pH and temperature influence the cryotolerance of *Lactobacillus acidophilus* RD758. Journal of Dairy Science 88(1): 21–29.

Wang, Y., J. Delettre, G. Corrieu and C. Béal. 2011. Starvation induces physiological changes that act on the cryotolerance of *Lactobacillus acidophilus* RD758. Biotechnology Progress 27(2): 342–350.

Wang, Y., J. Delettre, A. Guillot, G. Corrieu and C. Béal. 2005b. Influence of cooling temperature and duration on cold adaptation of *Lactobacillus acidophilus* RD758. Cryobiology 50(3): 294–307.

Wolfe, J. and G. Bryant. 1999. Freezing, drying, and/or vitrification of membrane-solute-water systems. Cryobiology 39(2): 103–129.

Wright, C.T. and T.R. Klaenhammer. 1981. Calcium-induced alteration of cellular morphology affecting the resistance of *Lactobacillus acidophilus* to freezing. Applied and Environmental Microbiology 41(3): 807–815.

Wu, C., J. Zhang, M. Wang, C. Du and J. Chen. 2012. *Lactobacillus casei* combats acid stress by maintaining cell membrane functionality. Journal of Industrial Microbiology and Biotechnology 39(7): 1031–1039.

Zacarías, M.F., A. Binetti, M. Laco, J. Reinheimer and G. Vinderola. 2011. Preliminary technological and potential probiotic characterisation of bifidobacteria isolated from breast milk for use in dairy products. International Dairy Journal 21(8): 548–555.

Zhang, J., G.C. Du, Y.P. Zhang, X.Y. Liao, M. Wang, Y. Li and J. Chen. 2010. Glutathione protects *Lactobacillus sanfranciscensis* against freeze-thawing, freeze-drying, and cold treatment. Applied and Environmental Microbiology 76(9): 2989–2996.

# CHAPTER 11

# Freeze-drying of Probiotics

*Mathias Aschenbrenner,*[1,2,*] *Petra Foerst*[2] and *Ulrich Kulozik*[2]

## 1. Introduction

Today, the preservation of sensitive biomaterial like proteins, liposomes and microorganism with minimum losses in quality is one of the major challenges of drying research. The state of the art in dewatering for this kind of susceptible material is freeze-drying. Due to the low product temperatures and the solid state during drying, a highly native product quality can be maintained. However, as in case of other drying processes, drying and the subsequent storage go along with significant losses in the biomaterial fraction. The addition of so-called lyo-protectants (e.g., disaccharides) leads to significantly reduced drying damage and improved storage stability. The protective mechanisms are complex and are not fully understood. One of the most discussed protection mechanism is the so-called vitrification hypothesis. According to this hypothesis, biomaterial, embedded in a glassy sugar matrix, shows significantly improved stability due to the intrinsic low mobility that retards detrimental chemical and physical reactions. The validity of this hypothesis could be confirmed for a number of products from the area of food material science, pharmaceutical science and biotechnology. A special case is the preservation of living bacteria like probiotics and bacterial starter cultures for the food industry. The demand for preserved living bacteria continuously increased over the last years. For food applications, probiotics are mainly employed alone or together with starter cultures in fermented dairy products (e.g., yogurts and cultured drinks). In addition, probiotics are also potentially used as nutraceuticals and dry adjunct in non-fermented and non-dairy products, such as fruit juices, cereals, dried powder foods (Rivera-Espinoza and Gallardo-Navarro 2010). In

[1] Weihenstephaner Berg 1, 85354 Freising, Germany.
[2] TU München, Chair for Food Process Engineering and Dairy Technology, Research Center for Nutrition and Food Sciences (ZIEL), Department Technology.
E-mails: mathias.aschenbrenner@mytum.de; petra.foerst@tum.de; ulrich.kulozik@tum.de
* Corresponding author

either case, high viability of bacteria is of great importance in order to ensure immediate recovery of fermentation activity (in case of starter cultures) and to meet the minimal requirements for health-promoting effects (in case of probiotics).

Remarkable in this context is the discrepancy between high economical relevance on the one side, and the specific process relevant knowledge concerning inactivation and stabilization of the biomaterial on the other side. In particular, the knowledge concerning protection of bacteria by glassy protectants is very scarce and fragmentary. This is due to the fact that the relevance of the glassy state has not been evaluated explicitly for bacterial systems, but has simply been transferred from more or less similar model systems, such as bimolecular reaction systems in food or pharmaceutical matrices that are significantly different from bacterial systems. The observed simple deteriorative reactions do not adequately represent the complex inactivation mechanism that apply for bacterial systems. In other model systems the composition and the observed degradation pathways may be more complex (e.g., pharmaceutically relevant proteins). However, the nature of degradation and inactivation criteria for proteins and bacteria are again different. The applicability of findings from known model systems has to be questioned, simply due to the much more complex set-up of cell walls and the entire metabolite system of living cells as compared to liposomes or even more simple systems. Thus, in the context with real living bacteria the available data is scarce, rather of phenomenological nature or not comprehensive and detailed. The actual knowledge concerning the protective mechanisms (particularly the glassy state) for freeze-dried bacteria to a great extent bases on assumptions and simplifications. It is the aim of this chapter to provide deeper insights into bacteria specific protective and detrimental effects that apply during the drying process. The following findings shall underline the unique and individual interdependence and requirements between bacteria and their potential protectants.

## 2. The Freeze-Drying Process

In the field of food- and pharmaceutical industry and research, freeze-drying or lyophilization is the method of choice if sensitive and valuable products have to be stabilized in a dry solid state. Oetjen (2004) generally differentiates between three main areas of application that are pharmaceutical/biological/medical products, luxury food products and specialty products in the chemical industry. According to Jennings and Duan (1995) the classical freeze-drying process is defined as a process, where the solution that has to be dried is frozen in an initial freezing step and subsequently dried under a reduced pressure. Compared to other drying processes, the resulting products can generally be characterized as highly native in structure, biological activity and chemical and physical composition. In the pharmaceutical area, products are typically dried within their final container system (vial or syringe). This is necessary to assure absolute sterility and safety of the pharmaceutical product. In order to avoid contamination from outside the containers become closed within the freeze-dryer under vacuum. In the food industry, where sterility is not a requisite, the product can be dried in shelves (batch process) or on belts (continuous process) allowing significantly higher production rates. Typical product examples are bacterial starter cultures (shelve drying) and instant coffee (belt drying). After having passed the freeze-drying process, these types of products are further processed and finally packed (Oetjen 2004).

The freeze-drying process generally consists of three process steps, namely freezing, primary drying and secondary drying (see Fig. 1). During freezing the solvent (usually water) crystallizes under atmospheric conditions and separates from the residual sample. As a consequence the unfrozen fraction concentrates. In the subsequent primary drying step the frozen solvent is removed by transferring it directly from the solid to the gaseous state (sublimation) and concomitant avoidance of the liquid aggregate state. In the final secondary drying step additional non-frozen solvent is removed from the sample via desorption and the sample reaches its final water content. An exact description of these processes, the sample transitions and the underlying mechanisms are explained in detail in the following.

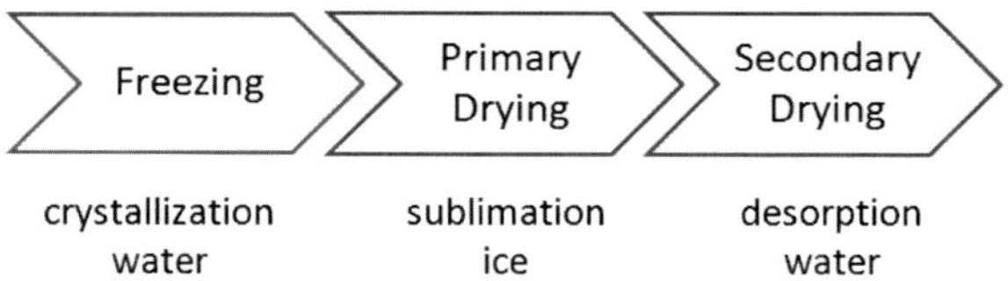


**Fig. 1.** Overview on the three process steps of freeze-drying.

A generic scheme of a freeze-drier system more or less is depicted in Fig. 2. Independent of the scale, typical freeze-drying equipment consists of a drying chamber that can be evacuated to a certain vacuum level. The initial vacuum can be achieved by means of the connected vacuum pump system. Within the drying chamber one or more heating shelves are located. These shelves provide the samples with the required heat during drying. The condenser acts as water vapor trap that maintains the vacuum level during the drying process by condensing and freezing the water vapor being formed during drying. The condenser can be located in the drying chamber (single chamber system) or in a separate condenser chamber (two chamber system) as shown in Fig. 2. Moreover, the drying system is equipped with an electronic control unit and a set of different sensors that allow monitoring of the drying progress.

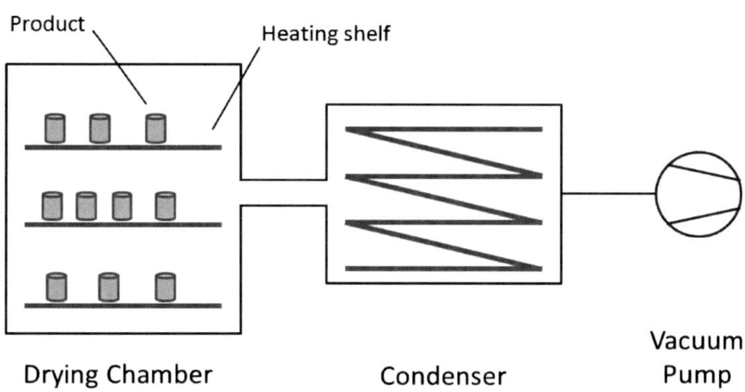


**Fig. 2.** Design example of a typical two chamber freeze-dryer.

## 2.1 Freezing

The type of the "raw material" for freeze-drying can vary from simple solutions to highly complex and semi-solid material like plants in the agro/food industry. Typically, products that are to be freeze-dried are aqueous multiphase systems. Besides the solvent water, typical standard samples further contain buffer salts, sugars (mono-, di-, or oligomers) and the biological product that has to be stabilized. Prior to freezing, the product is transferred into adequate containers, varying in size from large shelves/plates to small glass vials of few mL. Cooling and freezing of the product can be realized outside or inside the dryer, depending on the respective requirements in terms of final temperature or cooling velocity. As already mentioned above, during freezing the water crystallizes and thus is not available as solvent anymore. As a consequence the unfrozen fraction of the sample concentrates to a degree corresponding to the applied freezing temperature. The freezing-induced transitions in the sample can be retraced in a specific state diagram. Figure 3 shows an example for an aqueous sugar solution. The state diagram includes three curves, the freezing line, the solubility line and the glass transition line, which show the sample state in dependence of the solid concentration and temperature. The freezing line depicts the equilibrium between the product temperature and the concentration of the non-frozen sample fraction. The trend of the freezing line as shown in Fig. 3 is typical and a result of the so-called freezing point depression. The higher the solute concentration in the non-frozen phase, the steeper is the corresponding section of the freezing line. The solubility curve, in contrast, shows the equilibrium solubility of the respective solute in dependence of temperature. On the left hand of this solubility line the solution is in an undersaturated, on the right hand in the supersaturated state. This supersaturated state is unstable and the solute

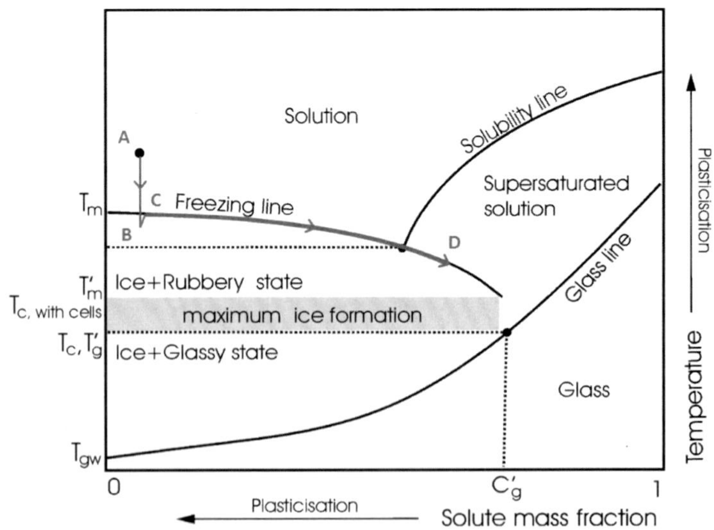


**Fig. 3.** Example of complete state diagram for an aqueous sugar solution showing the freezing process (adapted and modified from Santivarangkna et al. 2011).

tends to crystallize. Crystallization, however, only occurs at sufficiently long residence times in this unstable state. In contrast to the freezing and solubility lines, the glass transition line does not represent an equilibrium state, but a so-called iso-viscosity line (Kramer 1999) that shows combinations of temperature and concentration at which the sample experiences a transition from the non-glassy to the glassy state. Generally, by means of these three curves all sample states that are relevant in freeze-drying can be identified. The respective sample states, which depend on solute concentration and temperature, are also presented in Fig. 3.

Typically, a freezing process for a solution starts above the freezing line. In the initial step the liquid sample has to be cooled from room temperature to below its freezing temperature (A → B). Usually, ice formation does not start immediately when crossing the freezing line, but several degrees below. This phenomenon is called super-cooling. The size of the temperature interval below, until the crystallization starts, is called degree of super-cooling. This super-cooling process is a consequence of a delayed initial ice crystal formation, called nucleation. The degree of this super-cooling process depends on several factors like cooling velocity, sample formulation, sample purity, sample size and the freezing container (Blond and Simatos 1998) and has a significant influence on the generated ice crystal size and shape. For a given sample, the degree of super-cooling increases with cooling velocity. The exact prediction and control of this super-cooling phenomenon, however, is still an unsolved problem in pharmaceutical research (Kramer 1999). When crystallization starts, the sample temperature returns back to the freezing line (B → C) due to the exothermic heat of crystallization and further follows its course until it reaches the final destination temperature (C → D). On the way along the freezing line the sample passes the eutectic point or temperature $T_e$. This is where the solubility line meets the freezing line. The unfrozen fraction turns from an undersaturated to a thermodynamically unstable and supersaturated state. Due to the fast temperature decrease and the resulting short residence time in this unstable state, crystallization of sugars does not occur. In this case the system is kinetically stabilized (Roos 1995). In contrast to sugars, buffer salts easily crystallize during freezing and may cause undesired pH-shifts. Much care has to be taken in choosing an adequate buffer system that is minimally affected by the freezing process (Shalaev 2005). Usually, by means of the freezing process around 70 to 90% of the water in the sample can be crystallized (Roos 1997, Slade et al. 1991). The maximum amount of water which can be frozen out is reached at the temperature $T_m$', at the end of the freezing line. Due to the high viscosity in the highly concentrated solution, further decrease in temperature does not cause additional ice formation. The water fraction that remains within the maximally freeze-concentrated solution is called unfreezable water (UFW). Below $T_m$' but above the $T_g$-line the freeze-concentrated solution resides in the so-called rubbery state, a highly viscous, visco-elastic but unstable state (Roos 1997, Sablani et al. 2010, Franks 1998). Further cooling of the solution below the $T_g$-line causes a transition from the rubbery to the glassy state. The temperature at which this transition occurs is called $T_g$', the glass transition temperature of the maximally freeze-concentrated sample. The solute concentration at which this transition occurs is named $c_g$', the concentration of the maximally freeze-concentrated solution (see also Fig. 3). For the initial freezing process these two parameters are of high importance. A long-term stability of frozen products is only given in a glassy freeze-concentrated solution below $T_g$'. As a result, it is also common for freeze-drying to freeze the sample below its $T_g$' in order to obtain a stable glassy sample matrix. Simultaneously, these conditions guarantee maximum ice formation in the sample. In order to achieve an ideal drying behaviour in the following process steps, this maximum ice formation is necessary. Both parameters, $T_g$' and $c_g$', are product properties that are characteristic for the initial sample solution. Knowledge of these product properties is essential in order to adequately adapt the process parameters. Values for the most relevant binary aqueous solutions in the field of freeze-drying are available in literature (see Roos 1993). In case of more complex products the individual estimation of $T_g$' and $c_g$' is inevitable.

Generally, the freezing process plays an important role in freeze-drying. This is due to the fact, that the structure of the product, that has to be freeze-dried, is formed during freezing. One important parameter besides the final freezing temperature is the cooling rate. High cooling rates go along with a high degree of super-cooling and cause the formation of small ice crystals. With decreasing cooling rates, the degree in super-cooling diminishes and the size of the resulting ice crystals increases. The size of the ice crystals plays an essential role. First of all, the ice crystal size determines the drying velocity in the subsequent drying step. The larger the ice crystals in the product, the higher are the resulting drying velocities (Adams 2007, Oetjen 2004, Goldblith et al. 1975). Also, final product properties like porosity, specific surface area and sample dissolution behaviour are determined by the ice crystals formed during freezing (Searles 2004). However, the ideal ice crystal size is not the maximum size. Especially for the incorporated biomaterial as large ice crystals can cause severe damage. In case of bacteria, large ice crystals may cause mechanical damage of the cell membrane. Similar effects are reported for the freezing of vegetables (Voda et al. 2012). Besides a possible mechanical damage of the sensitive biomaterial, low cooling rates induce the danger of significant pH-shifts. These pH-shifts are caused by a partial crystallization of one of the buffer components. Compared to the added saccharides, the buffer salts are generally very susceptible to crystallization. Typical buffers are multi-component systems. Their buffering capability depends on the interaction of these components. Cooling and freezing, however, can easily cause crystallization of one of the components. As a result the sample experiences a significant shift in pH (Shalaev 2005, Franks 2008). This risk can be minimized by applying high cooling rates that avoid long-term exposure to unstable and oversaturated states.

Generally, the best freezing conditions have to be determined individually for each sample type. Due to the fact that in most cases the biomaterial is a valuable component, rather preservation of the sensitive material in its native state, than process optimization (reduced drying time due to large ice crystals) lies in the focus of process design. Within this book a comprehensive review on the freezing process is given in chapter "3.1 Freezing of probiotic bacteria".

### 2.2 Primary Drying

During the primary drying step frozen water is removed from the sample via sublimation. In contrast to the freezing, which can be accomplished outside the freeze-dryer, the two subsequent drying steps require a vacuum within the drying chamber. By lowering the chamber pressure below the so-called triple point $T_r$ (p = 6.104 mbar; T = 0.0099°C), ice crystals sublimate under concomitant avoidance of the liquid aggregate state. As shown in Fig. 4, the aggregate state of water or a solution depends on the combination of pressure and temperature. At atmospheric conditions, for pure water a freezing point $T_f$ around 0°C and a boiling point $T_b$ around 100°C can be observed. As a result, the aggregate state depends on the existing temperature. Immediately at the triple point all three aggregate states (solid, liquid and gaseous) coexist in equilibrium. However, at pressures below the triple point, water can be directly transferred from the solid to the liquid state. As a result, undesired melting, foaming or softening phenomena can be avoided. A typical freeze-drying process is shown in the aggregate state diagram (Fig. 4). First of all, liquid water is frozen by cooling it at atmospheric pressure below its freezing point $T_f$. In the subsequent step the chamber pressure is lowered below the triple point $T_r$. Compared to pure water, solutions show a decreased $T_f$ and $T_r$ as well as an increased $T_b$-value. Finally, crossing the border between the solid and gaseous

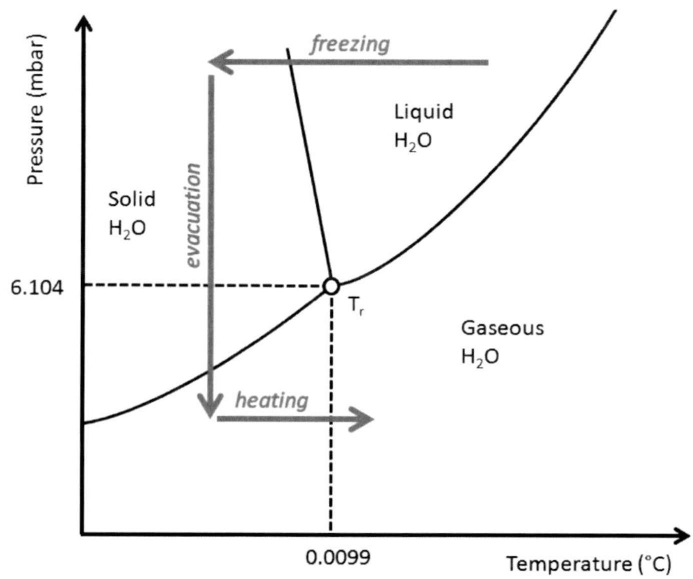


**Fig. 4.** State diagram for water showing the aggregate state in dependence of temperature and pressure. The arrows indicate the course of a typical freeze-drying process.

aggregate state initiates sublimation. By carefully increasing the product temperature, the sublimation process can be significantly accelerated. The temperature at which sublimation occurs is dependent on the respective chamber pressure and given by the sublimation curve (border between solid and gaseous aggregate state). It can be seen, that the sublimation temperature decreases with chamber pressure. Table 1 exemplarily shows sublimation temperature ($T_{sub}$) values for some selected chamber pressures.

**Table 1.** Selected chamber pressures $p_{chamber}$ and the corresponding sublimation temperatures $T_{sub}$ for pure ice.

| $T_{sub}$ (°C) | $p_{chamber}$ (mbar) |
|---|---|
| –50 | 0.04 |
| –40 | 0.12 |
| –30 | 0.37 |
| –20 | 1.03 |

These theoretical sublimation temperatures in Table 1, however, do only apply for pure ice or water and cannot be directly transferred to a real drying process. Due to the coupled heat and mass transfer in freeze-drying, products may exhibit significantly different temperatures. Nevertheless, these values give a first orientation for the identification of the appropriate process conditions.

Besides a vacuum pump that enables chamber pressures below the triple point, the dryer has to be equipped with a condenser and a heat source. The driving force for every sublimation process is the difference in water vapour pressure ΔP between the sublimation front and the chamber pressure (Tang and Pikal 2004). In this context the sublimation rate dm/dt can be described as a function of the water vapour pressure at the sublimation front $P_{ice}$, the chamber pressure $P_c$, and the resistance of the dry product layer $R_p$ and the stopper $R_s$ against the water vapour transport (see equation (1)).

$$\frac{dm}{dt} = \frac{P_{ice} - P_c}{R_p + R_s} \qquad (1)$$

Due to the low product temperatures during primary drying the resulting $P_{ice}$ value is relatively low. Nevertheless, it is possible to maintain a sufficient sublimation rate by continuous removal of the formed water vapour from the surrounding atmosphere (Presser 2003). This removal of vapour is accomplished by the condenser, which is located directly in the drying chamber or in an adjacent condenser chamber. Due to the low condenser temperatures, around –60°C, water vapour condenses and immediately freezes at the condenser surface. As a result, the chamber pressure $P_c$ can be kept at its low level and thus guarantees a sufficiently high ΔP (see numerator of equation (1)).

As shown by equation (1), the sublimation rate is further influenced by the resistance of the product layer $R_p$ and the resistance of the vial stopper $R_s$. The higher these resistance values, the lower is the resulting sublimation rate. As already mentioned above, in classical pharmaceutical freeze-drying, at the end of the drying process, the vial has to be closed under vacuum. Due to this fact, freeze-drying vials are typically equipped with a rubber stopper that allows two possible positions. During drying the stoppers on top of the vials are in a half-opened position. Due to their special design water vapour can still escape through lateral channels. At the end of the drying process vials can be completely closed by pushing the stopper completely into the vial neck. With this procedure, escape of water vapour during drying is possible. However, compared to vials without stopper the water vapour experiences a slightly increased resistance $R_s$. This $R_s$ is a constant value which is dependent on the stopper design.

However, compared to the product resistance $R_p$, the $R_s$-values more or less can be neglected. The main resistance is governed by the properties of the dry product layer.

Independent of the sample size or shape, sublimation initially starts on the sample surface. In the course of the sublimation, the sublimation front continuously moves into the product. In case of vials the sublimation front moves top down, whereas in case of spherical products from outside to the inside. Examples for both product types are depicted in Fig. 5. It can be seen that in the course of drying, the sublimation front moves inwards and leaves behind a solid, porous matrix. Water vapour coming from the inner sublimation front has to escape through this top layer. In the course of sublimation, the thickness of the dry product layer, and thus its resistance to water vapour migration, increases. Besides the thickness, its resistance to water vapour is mainly determined by its porosity. The larger the pores within this layer, the lower is the corresponding $R_p$-value. As mentioned above (see section 2.1), porosity in turn can be influenced by means of the freezing conditions.

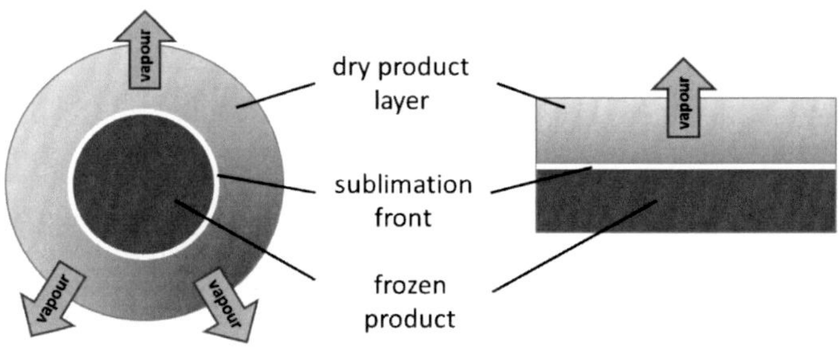


**Fig. 5.** Schematic drawing of water vapour transport in case of a sphere (frozen droplet) or a cake (vial). In both cases the water vapour has to diffuse from the sublimation front to the product surface. The dry product layer acts as barrier with certain resistance.

As already mentioned above, an additional heat source in freeze-drying is inevitable. Due to the coupled heat and mass transfer, the sample experiences a decrease in temperature immediately with start of sublimation. A decrease in product temperature, however, leads to a decline of $P_{ice}$ and thus decreases the sublimation rate (see equation (1)). In order to overcome this problem, additional heat has to be transferred to the sample. Usually, heating shelves serve as source of heat. In freeze-drying, the samples that have to be dried are placed onto these heating shelves. The amount of heat that can be transported from the shelf to the product in a certain time interval dQ/dt depends on the difference between shelf temperature $T_s$ and the product temperature $T_p$, the total area of the vial bottom $A_v$ and the vial heat transfer coefficient $K_v$ (see equation (2)). For a given freeze-drying process the parameters $A_v$ and $K_v$ are set. The amount of heat transferred to the product depends on the temperature difference between shelf and product. The product temperature $T_p$ cannot be controlled directly. Due to the coupled heat and mass transport, the product temperature $T_p$ changes throughout the whole sublimation process. Several influencing parameters like chamber pressure and the continuously changing product resistance complicate the prediction of a resulting product temperature. In classical freeze-drying processes the shelf temperature $T_s$ is controlled in order to influence product temperature and sublimation rate.

$$\frac{dQ}{dt} = A_v\, K_v\, (T_s - T_p) \tag{2}$$

The total vial heat transfer coefficient $K_v$ is determined by three components (see equation (3)): the heat transfer via direct contact between shelf and vial $K_c$, the heat transfer via radiation from the surroundings $K_r$ and the heat transfer from gas conduction (convection) $K_g$ (Rambhatla and Pikal 2005). Here, Fig. 6 gives an overview of the possible applying heat transfer mechanism. $K_c$ and $K_r$ are not dependent on chamber pressure. $K_g$, however, decreases with decreasing chamber pressure. Generally, the relevance of the respective heat transfer mechanisms strongly depends on the chosen drying parameters, the loading of the dryer and constructional details of the dryer equipment (Presser 2003).

$$K_v = K_c + K_r + K_g \tag{3}$$

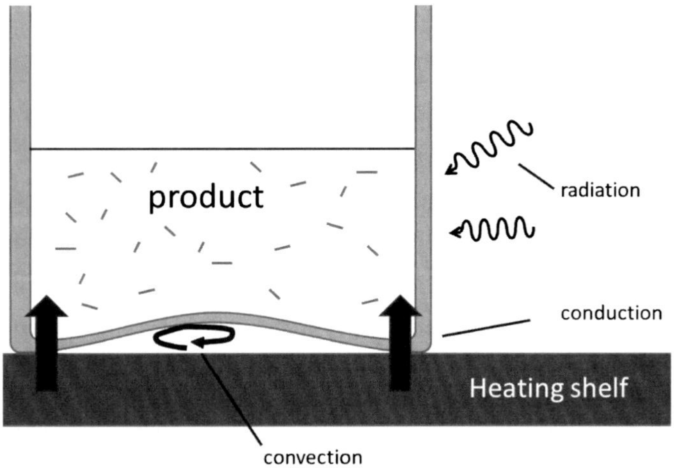


**Fig. 6.** Possible mechanisms of heat transfer for a freeze-drying process with vial.

The heat required for sublimation of the ice fraction is almost completely transferred via radiation and conduction (see Fig. 6). Within the frozen sample the absorbed heat is transferred further to the sublimation front. Due to the fact, that product temperature cannot be directly controlled it is suggested by Tang and Pikal (2004) to continuously adapt the shelf temperature $T_s$ in order to obtain the desired product temperature $T_p$. Assuming that the transferred heat is completely dissipated by the sublimation process (see equation (4)), the theoretically required shelf temperature can be further expressed by means of equation (2) and (4). This interrelation of shelf and product temperature is given by equation (5),

$$\frac{dQ}{dt} = \Delta H_s \frac{dm}{dt} \tag{4}$$

$$T_s = T_p + \frac{1}{A_v}\frac{dQ}{dt}\left(\frac{1}{K_v} + \frac{l_{ice}}{k_i}\right) \quad (5)$$

where $\Delta H_s$ is the heat of ice sublimation, $l_{ice}$ is the ice thickness and $k_i$ is the thermal conductivity of ice. However, in order to calculate the shelf temperature according to equation (5) the heat transfer coefficient $K_v$ has to be determined for the respective dryer setting.

The target product temperature for the sublimation process depends on several factors. On the one hand, $T_p$ should be high enough to enable effective sublimation. On the other hand, the sample should not be heated above a certain critical temperature. In this context, a universal critical temperature does not exist, but has to be determined individually for each sample type. Typically, in pharmaceutical freeze-drying, this critical temperature that should not be exceeded is the sample collapse temperature $T_c$. This is the temperature, where the sample experiences structural changes due to the drop in sample viscosity. Usually, $T_c$ is located within the temperature interval between $T_g$' and $T_m$' (see section 2.1), but in most cases lies only few degrees above the sample $T_g$' (Meister 2009, Schersch 2009). During sublimation, the sample consists of two fractions: a "dry" porous product layer outside and a frozen product layer inside. The product parameters from the freezing process ($T_g$' and $T_m$') are still valid during sublimation. Both, the dry and the frozen layers are relatively stable. The direct environment of the sublimation front, however, is very instable. In case the critical temperature is exceeded, possible structural changes occur in its direct vicinity. The question, if structural changes during sublimation are acceptable or not cannot be generally answered. In terms of product quality and drying performance, it has been reported that collapse may have a beneficial (Ekdawi-Sever et al. 2003, Tang and Pikal 2004) as well as a detrimental (Passot et al. 2007, Fonseca et al. 2004a) effect. Comprehensive investigations concerning this topic are given by Meister (2009) and Schersch (2009).

Reaching the end of the sublimation process, no ice crystals remain in the sample. The sample, however, still contains around 5–30% of non-freezeable water, bound in the maximally freeze-concentrated sample fraction (Sheehan and Liapis 1998). This residual water has to be removed in the secondary drying step. Due to the fact that the sample during sublimation reacts very sensitive to changes regarding the process conditions, it is reasonable to identify the so-called endpoint before adjusting the process parameters. The endpoint is the moment where freeze-drying turns from sublimation to desorption. Several methods in the field of process analytical technology (PAT) have been developed in order to identify the transition from sublimation to desorption. Within this book, section "3.5 Process Analytical Technology (PAT) in Freeze Drying" provides a comprehensive overview regarding devices and methods.

### 2.3 Secondary Drying

At the end of the primary drying step the sample still contains between 5 and 30% of non-freezeable water, bound in the maximally freeze-concentrated sample fraction (Sheehan and Liapis 1998). Residual water can be adsorbed on surfaces, bound as hydrate water in crystalline form or bound in the amorphous matrix (Liapis et al. 1996,

Nail and Gatlin 1993). In order to achieve a sufficiently high storage stability of the sample, its residual water content has to be further lowered. Usually, already at the end of primary drying the saccharide sample matrix exists in an amorphous glassy state. Therefore, water removal during the secondary drying stage involves diffusion of the water in the amorphous glass, evaporation of water at the solid/vapour boundary, and water vapour flow through the pore structure of the dry product (Rambhatla and Pikal 2005). As a consequence, this process, also known as desorption, is mainly governed by diffusion processes. Compared to sublimation, desorption is a relatively slow process. Depending on the desired residual water content and the respective practicable product temperatures, desorption phase may even last several days (Nail and Gatlin 1993).

The main influencing parameter for the desorption process is the product temperature. The higher the product temperature, the faster the rate limiting diffusion process precedes. The chamber pressure, in contrast, plays a minor role (Sadikoglu et al. 1998). Consequently, with the beginning of secondary drying the shelf temperature has to be raised. It is again the freezing process that indirectly determines the sample drying behaviour. The higher the specific surface area of the sample the shorter is the required secondary drying step. In case of collapsed samples (due to increased sublimation temperatures) desorption behaviour may be significantly decreased.

In contrast to sublimation, where the reference temperature $T_g$' stays constant, the sample $T_g$ during secondary drying continuously increases due to on-going desorption. Due to this continuously increasing stability of the amorphous sugar matrix it is reasonable to adapt the shelf temperature throughout the whole secondary drying process. In order to optimize secondary drying time the temperature difference between the actual and the maximally allowable product temperature should be minimized. It is suggested by Rambhatla and Pikal (2005) that the shelf temperature should be increased stepwise at least every 6 h in order to ideally adapt to the continuously increasing $T_g$ (see Fig. 7). Generally, the higher the number of these temperature adaptions the closer the product can be brought to its maximum allowable temperature. The optimum

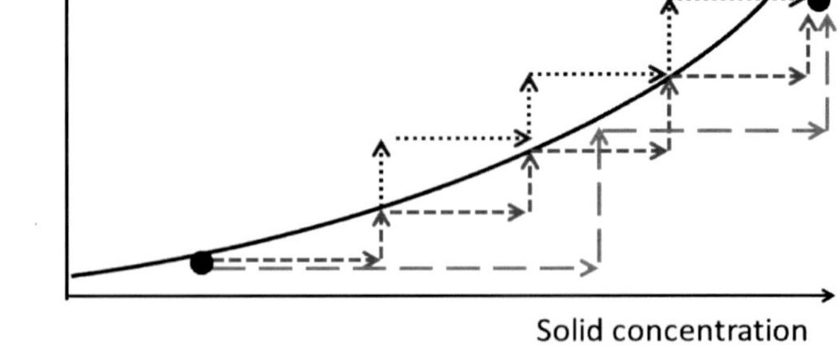


**Fig. 7.** State diagram showing three integrated drying courses. The drying velocity increases from process A to process C due to the higher applied product temperature.

distance between the actual and the maximally allowable product temperature is to be found by weighting between the target criteria process safety and process efficiency and thus to be assessed case by case. Also the maximally allowable product temperature strongly depends on the respective product and its corresponding attributes. It also may be a temperature different than $T_g$ that allows product temperatures above the glass transition line (see process C in Fig. 7).

The optimum residual water content for ideal sample stability strongly depends on the respective sample type and storage conditions. In case of most pharmaceutical products (e.g., proteins) residual water contents below 1% are recommended (Tang and Pikal 2004). In case of bacterial lyophilizates, however, residual water contents up to 5% (strain dependent) may be sufficient (Potts 1994, Santivarangkna et al. 2008b).

## 3. Protection Mechanism during Freeze-Drying[1]

Within the single steps of the lyophilization process and the subsequent storage, the biomaterial is exposed to harsh conditions and a significantly changing environment. Potts (1994) provides a comprehensive review concerning the desiccation tolerance of prokaryotes. In case of bacteria the most sensitive and thus the most critical component regarding desiccation is the bacterial cell membrane. In order to avoid undesired damage and to achieve sufficiently high yields, so-called lyo-protectants have to be added to the biomaterial prior to drying. Common lyo-protectants are disaccharides (saccharose, lactose, trehalose) polyols (mannitol, sorbitol) and polymers (maltodextrin, dextran, inulin) (Santivarangkna et al. 2008a). Characteristic for a lyo-protectant is that it can be easily vitrified and provides protection to the embedded biomaterial throughout the whole production process (freezing, primary drying and secondary drying) and during subsequent storage. Among these protection mechanisms are preferential hydration, the water replacement and the formation of a glassy matrix. The available information is mainly based on investigations with simple model systems. Their individual contribution to an overall protective effect in case of complex systems like bacteria is almost impossible to determine.

### *3.1 Preferential Hydration*

As mentioned above, the structure of a native membrane or protein is dependent on the availability of sufficient water. Here, the formation of hydrogen bonds between the biomaterial and the solvent water is essential. Removal of water goes in parallel with destabilisation of these structures and an increased probability in undesired protein interactions (Timasheff 2002). It could be shown that in presence of the above mentioned solutes (lyo-protectants), water molecules and proteins arrange in a specific order. Preferential hydration describes the phenomenon that the solute is excluded from the macromolecule surface, which in turn causes an increased accumulation of

[1] Acknowledgements: This information has been previously published in: Santivarangkna, C., M. Aschenbrenner, U. Kulozik and P. Foerst. 2011. Role of Glassy State on Stabilities of Freeze-Dried Probiotics. J. Food Sc. 8: R152–R156.

water molecules at the macromolecule surface (see Fig. 8). The exclusion mechanism itself can be induced by different specific and non-specific effects like steric hindrance, solvophobic effects and increases in water surface tension (Timasheff et al. 1989, Cioci and Lavecchia 1997). Due to the preferred interaction between the solute and the solvent the chemical potential of the protein increases. This increase in chemical potential is proportional to the area of the interface between protein and water (Thimasheff 2002). Denaturation through unfolding of the structure would further increase the chemical potential. Due to its smaller interface and lower chemical potential the native state is thermodynamically favoured. The consequence is a shift of the equilibrium between native and denatured towards the native protein state (Timasheff 2002). Due to the required participation of water, this protective mechanism, however, only applies to systems with water contents above 0.3 $g_{H20}/g_{dry\ matter}$ (Hoekstra et al. 2001).

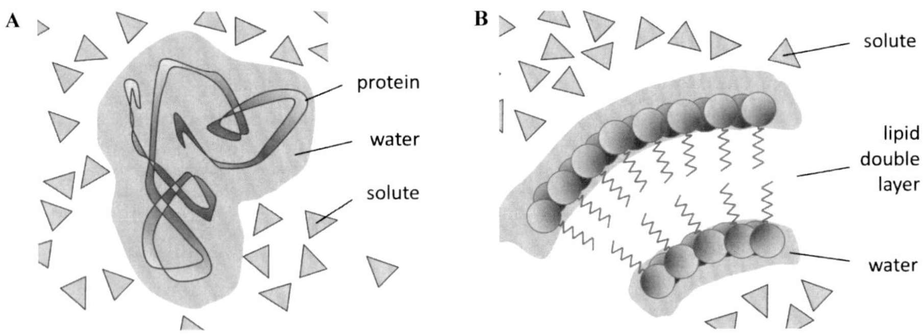


**Fig. 8.** Schematic drawing of the preferential hydration effect for single proteins (A) and lipid double layers (B).

### *3.2 Water Replacement*

The native liquid-crystalline state of the bacterial cell membrane requires sufficient hydration. The cell membrane to a great extent consists of phospholipids. It is these phospholipids that mainly determine the behaviour of the cell membrane during water removal. Phospholipids consist of a polar hydrophilic phosphate head group and a hydrophobic acyl chain. In the native hydrated liquid-crystalline state water molecules surround the phospholipid heads and act as spacers between the individual phospholipids. These spacers are essential for the fluidity of the membrane because they avoid an approach of the acyl chains between the different phospholipids. Removal of these spacers, as it is the case during drying, causes an approach of the acyl chains, which induces a significant increase in van-der-Waals interactions. Consequence of this increase in van-der-Waals forces is a transition of the cell membrane from the liquid-crystalline to the highly viscous gel phase (see Fig. 9). The appearance of the gel phase itself is not detrimental, in contrast to the transition from the gel phase back to the liquid-crystalline state. This is exactly what happens during rehydration of dried cells. Due to the inhomogeneity in phospholipid composition, the cell during rehydration

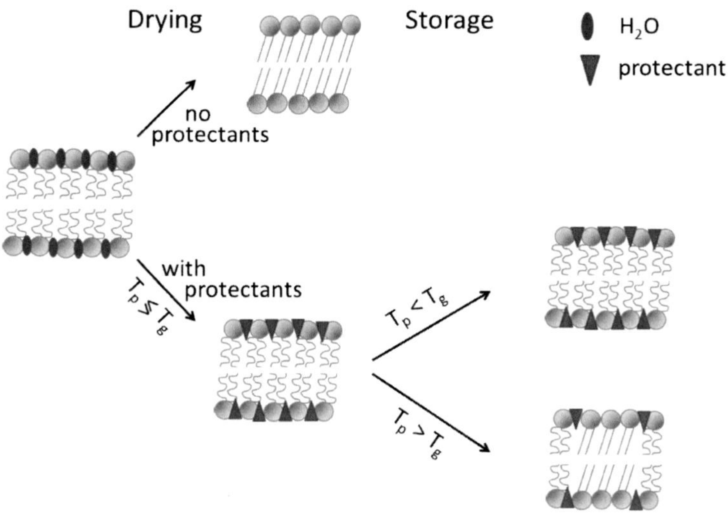


**Fig. 9.** Schematic drawing of the extended Water Replacement Theory (xWR) for freeze-drying of bacteria. Drying without protectant causes an immediate transition of the cell membrane from the liquid-crystalline to the gel phase. Freeze-drying at temperatures above and below $T_g$ allows water replacement. Storage above $T_g$ causes detrimental transitions of the cell membrane. Only at storage temperatures below $T_g$ phase detrimental transitions can be inhibited.

does not experience an overall homogeneous transition but several local transitions (Crowe et al. 1989). This inhomogeneous transition, however, causes packing defects due lateral phase separation between gel and liquid-like domains (Santivarangkna et al. 2008a). The consequence is a loss in cell membrane integrity, which is known to cause cell death. A characteristic property of a cell membrane is its membrane phase transition temperature ($T_m$). At temperatures above $T_m$ the membrane is fluid, whereas below $T_m$ it turns into a gel-like state. The more water is removed during drying, the higher is the packing density of the acyl chains and the resulting $T_m$ (Crowe et al. 1989). Consequently, at room temperature a hydrated membrane may appear fluid, whereas a dry membrane is already in the gel-like state.

The addition of lyo-protectants to the cells prior to drying can avoid a drying-induced increase in $T_m$ and thus prevent the initial transition of the membrane into the gel phase (see Fig. 9). Due to their hydroxyl groups the protectants can interact via H-bonds with the phospholipid heads and replace the water molecules as spacers during drying (Champion et al. 2000, Lee et al. 1986). In this context it could be shown that this water replacement mechanism could be best achieved with small protectant molecules like mono- and disaccharides that easily fit between the head groups (Diaz et al. 1999, Luzardo et al. 2000, Villarreal et al. 2004). In case of larger polymers the ability to interact strongly increases with the flexibility of the polymer chain (Vereyken et al. 2001, Vereyken et al. 2003, Cacela and Hincha 2006). It was further found that the water replacement mechanism requires an amorphous non-crystalline protectant (Crowe et al. 1996).

### 3.3 Vitrification

The high viscosity of an amorphous glassy sugar matrix can retard undesired chemical and physical reactions for incorporated biomaterial. This fact can be explained based on a simple bimolecular reaction model. Bimolecular reactions typically involve a diffusive step, in which the potential reactants diffuse together in prior to the reaction (Parker et al. 2002). This fact very likely also applies to other more complex multi-molecular reactions (see Fig. 10). Independently of the type of reaction the diffusion process becomes the rate limiting step within the reaction chain. This effect typically applies in case of freeze-dried products. A diffusional limitation, however, only applies in case of amorphous matrix material. Crystalline protectants are not able to act in a protective manner. According to Crowe et al. (1998), disaccharides are the ideal lyo-protectants due to the fact that they can easily be vitrified, show relatively high $T_g$-values but still being able to intensively interact with biomolecule surfaces due to their compact molecule size. Due to the fact that sugar glasses are water soluble, the water content of the sample as well as the storage relative humidity is of highest importance. Water acts as plasticiser and hence lowers the $T_g$-value. In case the sample $T_g$ drops below the product temperature, the sample matrix experiences a transition from the glassy to the non-glassy rubbery state. As a consequence the effect of diffusional limitation is reduced.

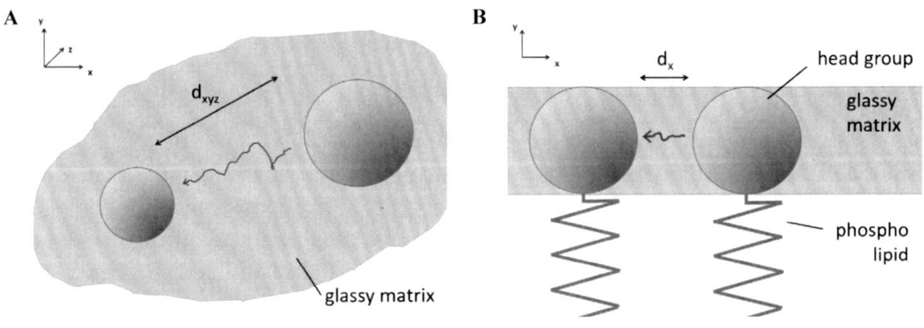


**Fig. 10.** The protective effect of a surrounding glassy matrix through hindered diffusion for simple bimolecular reactions in a three-dimensional space (A) and for lateral diffusing phospholipids within a cell membrane (B).

## 4. Protection and Inactivation of Probiotics during Freeze-Drying

### 4.1 Probiotics and the Freezing Process

The typical freeze-drying or lyophilization process consists of 3 steps: freezing, primary drying or sublimation, and secondary drying or desorption. As already mentioned in section 2.1, it is important to note that already with the initial freezing step the solvent water is separated from the residual sample. For incorporated sensitive biomaterial the desiccation stress, typical for drying processes, already applies during freezing.

Besides its influence on the biomaterial, the freezing step is the key process step in freeze-drying that determines sample structure formation.

The ice crystals formed during freezing determine the morphology and distribution of the pores (macroscopic cake structure) that are formed during removal of ice crystals. Thus, the freezing process significantly determines the drying behaviour of the sample throughout the following process steps. In order to ensure rapid drying and to facilitate vapour migration during drying, ice crystals in the suspension should be large and contiguous. On the one hand, large ice crystals can only be achieved by relatively slow freezing rates (not with liquid $N_2$). On the other hand, these large ice crystals can induce cell damage due to mechanical stress. Additionally, slow freezing may induce the eutectic crystallization of buffer salt components, which is also suspected to cause cell membrane damage (Morgan et al. 2006).

In this context it could be shown by Gehrke (1991) that each cell type species has its own optimum cooling rate. This individual optimum cooling rate is a result of its ability to release water. This property, again, depends on several factors like cell size and shape, water permeability of the cell membrane and the presence of a cell wall (Dumont et al. 2004). The most widely accepted theory states that applying cooling rates below the optimum causes extreme cell dehydration (solution injury), whereas above the optimum cooling rates lead to intracellular ice formation (IIF) (Mazur 1970, Mazur et al. 1972). Both of these phenomena are detrimental and should be avoided. In other words, optimum freezing conditions often require a compromise between the requirements of the bacteria and the drying performance.

For a frozen solution, at $T_g$', a maximally freeze-concentrated solution is formed (Roos 1997). The high viscosity in this maximum freeze-concentrated phase inhibits further detrimental physical and chemical reactions like ice crystal formation and eutectic crystallization of buffer salts and thus guarantees long-term stability. It has been suggested that the formation of a maximally freeze-concentrated matrix with entrapped microbial cells is essential for the survival during freezing (Pehkonen et al. 2008). As a result, for the industrial production of probiotics frozen granules are commonly produced by distributing the cell suspension through a droplet disc with pores or through a nozzle into liquid $N_2$. In addition to direct freezing in liquid $N_2$, innovative rapid freezing methods were explored by Volkert et al. (2008). The frozen granules are produced by the spray freezing, where cell suspension is sprayed in an air blast freezer to get droplets (for example, size of ca. 5 to 30 μm). Given the $T_g$' of skim milk (–50°C), sucrose (–46°C), and trehalose (–40°C), which are the most common suspending medium used in many studies, and $T_g$' of pure water (–135°C), the maximally freeze-concentrated matrix should be obtained by common industrial freezing protocols, for example, immersion in liquid $N_2$ (–196°C). Otherwise, the freezing temperature should be selected with care to ensure a maximally freeze-concentrated state and formation of a glassy structure of the unfrozen continuous phase.

In freeze-drying formulations, cryo-protectants are used to prevent cell injury. Some viscous cryo-protectants (for example, glycerol, sugars, and polymers) increase the viscosity of the freeze-concentrated solution or cytoplasm depending on their respective membrane permeability. Therefore, the glassy state can be reached at a lower cooling rate and a higher freezing temperature (Morris et al. 2006). An annealing step is commonly used for the freeze-drying of proteins and pharmaceuticals. However, it is not applied for the production of freeze-dried probiotics. Annealing is a process, where frozen samples are kept isothermally at a temperature between $T_g$' and the onset of ice melting temperature. During annealing, freezeable water entrapped in amorphous regions (due to rapid freezing) is crystallized into larger and more uniform ice crystals. This is especially true for the frozen probiotics, which are commonly produced by rapid freezing using liquid nitrogen ($LN_2$). A pellet of 2 mm diameter is cooled from 0 to –50°C in approximately 10 s or at a freezing rate of approximately 300°C/min (Oetjen 2004). The increased content and uniform ice formation improve the efficacy of the subsequent drying process. The knowledge of the influence of annealing on probiotics' survival and stability is very limited. In a study by Ekdawi-Sever et al. (2003), it was reported that the annealing does not cause cell death, but improves storage stability and reduces browning reaction of *L. acidophilus*.

Concerning bacterial inactivation, it was stated by some authors (To and Etzel 1997, Tsvetkov and Brankova 1983) that the bacterial fraction, which survives the freezing process, does not show additional inactivation in the subsequent drying step. The authors argued that the detrimental dewatering process and the accompanying inactivation would already take place during the initial freezing step. However, Gehrke (1991) showed that this assumption does only apply for the sublimation step in which the frozen water is removed. In the subsequent desorption phase further non-frozen water is removed causing additional losses. Observation of these losses only can be made in case a critical residual water content is finally reached. According to Gehrke (1991) this critical residual water content lies around 5%.

Generally, it can be concluded that the freezing protocol has the potential to significantly influence the structure of the respective sample as well as the survival of the incorporated biomaterial. A universal freezing protocol, however, does not exist. In most cases the chosen freezing protocol will be a compromise between drying behaviour on the one, and preservation of sensitive biomaterial on the other hand. Depending on the individual situation the focus may lie on structure formation or high yields in survival. Besides the chosen ideal cooling rate, it should be the aim of each process to reach a stable maximally freeze-concentrated sample state as fast as possible. A long-term stability in the frozen state or structure stability during subsequent sublimation is only guaranteed for product temperatures below $T_g$'.

### 4.2 Probiotics and the Drying Process

It is generally acknowledged that each drying process poses a stress to bacteria and to some extent causes inactivation. The availability of sufficient water is a prerequisite not only for the stability of proteins, DNA and lipids, but also for the preservation of the cell structure (Potts 1994). Removal of water generally goes hand-in-hand with a so-called dehydration inactivation. It could be shown by implementing several techniques (Brennan et al. 1986, Santivarangkna et al. 2007, Selmer-Olsen et al. 1999) that the principal site of inactivation is the cell membrane (see Fig. 11). The cell membrane, which is mainly composed of phospholipids, suffers a detrimental phase transition from the liquid crystalline to the gel phase (Wolfe 1987). Due to the fact that each bio-membrane consists of lipids of different type and chain lengths, these phase transitions do not occur for all lipids simultaneously. Inhomogeneous phase

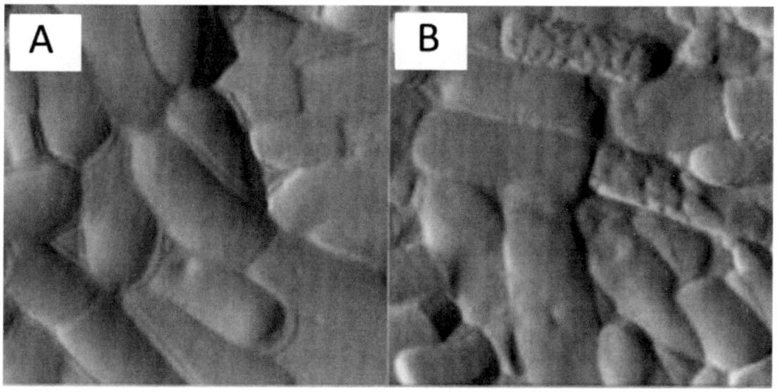


**Fig. 11.** AFM image of *Lactobacillus helveticus* before drying (A) and after vacuum drying of 12 h at 43°C and 100 mbar (B) showing characteristic cracks on cell surface (Santivarangkna et al. 2011) (with courtesy of John Wiley and Sons).

transition causes packing defects from lateral phase separation between liquid-like and solid-like domains, a process known as leakage. Detrimental leakage may also occur during rehydration of the dried cells due to the phase transition from the gel to the liquid crystalline phase (Crowe et al. 1989). Besides a general dehydration damage, drying specific detrimental effects play a role. In the case of freeze-drying the formation of ice crystals during the freezing step causes so-called cryo-injuries. Whilst during spray-drying the high temperatures may cause thermal injuries.

In addition to free water, which is removed during primary drying, cells contain around 0.25 g water/g dry weight of strongly associated water that is unfreezeable. This strongly associated water is removed by desorption. In order to promote desorption of water, the shelf temperature has to be raised stepwise during the second stage of freeze-drying. At the end of the secondary drying, the sample should have a water content that is optimal for storage. It has been suggested that the moisture content of 1% or less is required for a long-term shelf life (Nakamura 1996). However, Gardiner et al. (2000) showed that moisture removal below 4% may be considered practically low enough. It was also reported in a study by Zayed and Roos (2004) that the optimal moisture content for storage of freeze-dried *Lactobacillus salivarius* ssp. *salivarius* is in the range of 2.8 to 5.6%. Therefore, it is still debatable whether cells should be dried to moisture contents as low as possible. An argument against is the fact that lipid oxidation is enhanced at very low water contents, and water acts as a protectant against oxidation (Schmidt 2004).

As mentioned above, a drying-induced phase transition of the bacterial cell membrane can be avoided through addition of lyo-protectants to the cell suspension prior to drying. Due to their ability for water replacement bacterial membranes can be maintained in their native state. In this context it is known that only an amorphous protectant is able to interact with the membrane components and to avoid membrane leakage. Crystalline sugars cannot act in a protective manner (Pikal and Rigsbee 1997, Izutsu and Kojima 2002). The fact, that long-term stability of sensitive biomaterial in the frozen state only is only ensured for storage below the sample $T_g$' has been adapted to drying and finally led to the conviction, that during freeze-drying the sample has to stay in the glassy state. Additional results from liposome research (Crowe et al. 1998, Wolkers et al. 2004, Hincha et al. 2003) confirmed this assumption by showing that drying above $T_g$' results in liposome leakage and significant product losses. Considering the liposome as a kind of simple bacteria model, the relevance of $T_g$' as maximally allowable temperature is further underlined. Finally, drying above $T_g$' has been reported to induce losses in macroscopic lyophilizate structure. In the field of pharmaceutical production, however, changes in lyophilizate structure are unacceptable due to the high safety standards and very sensible consumer perception (Schersch 2009). As a consequence, in the field of classical freeze-drying it is more or less a given rule to dry at product temperatures below $T_g$'.

In this respect, the process conditions (pressure and shelf temperature during sublimation), must be controlled to avoid product temperatures above $T_g$' during drying (see section 2). For this reason, solutes with high $T_g$ are often added to increase $T_g$' so that drying can be carried out at elevated temperatures, and a final product with high $T_g$ and consequently high storage stability can be achieved. Disaccharide sugars and oligomeric sugars are preferred as additives for freeze-drying not only because they exhibit a high $T_g$ (Adams and Ramsay 1996), but also because they can be easily vitrified (Franks 1998, Ward et al. 1999). For sugar alcohols such as mannitol, caution must be exercised. Mannitol can easily separate from a frozen solution in the form of a crystalline phase (Adams and Ramsay 1996, Kim et al. 1998), resulting in a loss of the product stability after freeze-drying (Izutsu et al. 1994, Izutsu and Kojima 2002). This effect has been made responsible for the little or no protection conferred by mannitol on freeze-dried malolactic cultures (Zhao and Zhang 2005). In this context, it was reported that the protective potential of protectants is not only dependent on the chosen process conditions but also on the respective microorganism species (Valdez et al. 1983), which since has been confirmed by the results of many studies, that show variant results, depending on these different factors.

Generally, freeze-drying is known to be a relatively slow and hence expensive process (Sadikoglu et al. 1998). Due to the fact that conservative freeze-drying cycles usually take several days, there is still high potential in optimization of lyophilization. Within the last years, several modifications have led to promising and innovative technologies like evaporative freezing (Crespi et al. 2008), microwave freeze-drying (Duan et al. 2010, 2008a, 2008b, Huang et al. 2009, Cui et al. 2008), atmospheric freeze-drying (Rahman and Mujumdar 2008, Claussen et al. 2007a, 2007b, 2007c) and foam-mat freeze-drying (Muthukumaran et al. 2008a, 2008b).

Even with standard freeze-dryer equipment drying velocity can be simply raised by drying at elevated product temperatures. Any increase in product temperature leads to a significant decrease of primary drying time (Pikal and Shah 1990, Barresi et al. 2009). By either formulation or process development (Pikal and Shah 1990, Passot et al. 2005, Colandene et al. 2007, Schersch et al. 2010) several attempts have been made in the past to speed up the process. This tendency of applying higher product temperatures during primary drying finally led to the so-called collapse drying technique. In this case a product temperature significantly above the glass transition temperature of the maximum freeze-concentrated sample $T_g$' is chosen. The consequence is collapse, i.e., loss of the macroscopic sample structure. The critical (maximum) temperature which is associated with this phenomenon is called collapse temperature ($T_c$). It was proposed by Fonseca et al. (2004a) to use $T_c$ as the critical temperature for freeze-drying of formulations with cells. In comparison, $T_c$ and $T_g$' are measured with techniques based on different principles. Differential Scanning Calorimetry (DSC) has been commonly used over decades to determine the $T_g$ as a mid- or onset point of the temperature range where the endothermic shift in heat capacity appears. The DSC measurement is carried out with a representative frozen sample *ex situ* at atmospheric pressure. In contrast to the calorimetric principle, $T_c$ is determined visually by means of a freeze-drying. In other words, $T_c$ reflects the sample state under realistic drying conditions, whereas $T_g$' rather reflects general material specific sample properties. A state diagram showing glass transition temperature and possible collapse temperatures is depicted in Fig. 12. In the absence of cells, the difference between $T_g$' and $T_c$ is very small. Although the presence of cells does not clearly influence $T_g$ (Fonseca et al. 2001, Schoug et al. 2006), it significantly increases $T_c$ (Fonseca et al. 2004a). Bacteria can give some kind of structure and thus reduce or avoid viscous flow when $T_g$ of pure sugar solution is reached. The increase in stability depends on the cell types, that is, size, shapes, and cell chain formation and concentration (Fonseca et al. 2004b). As a result, a freeze-drying matrix with cells is more robust, and when $T_c$ is taken as the critical temperature, it allows a drying stage with a higher product temperature. This is of economic importance because it is estimated that the increase in 1°C of product temperature will decrease primary drying time by about 13% (Tang and Pikal 2004).

The benefits and disadvantages of drying above $T_g$' are still a matter of debate. In the case of protein formulations, Passot et al. (2005, 2007) observed decreased storage stability for samples freeze-dried above $T_g$'. In case of bacteria like *Lactobacillus* ssp. Fonseca et al. (2004) reported a decreased activity. Colandene et al. (2007) and

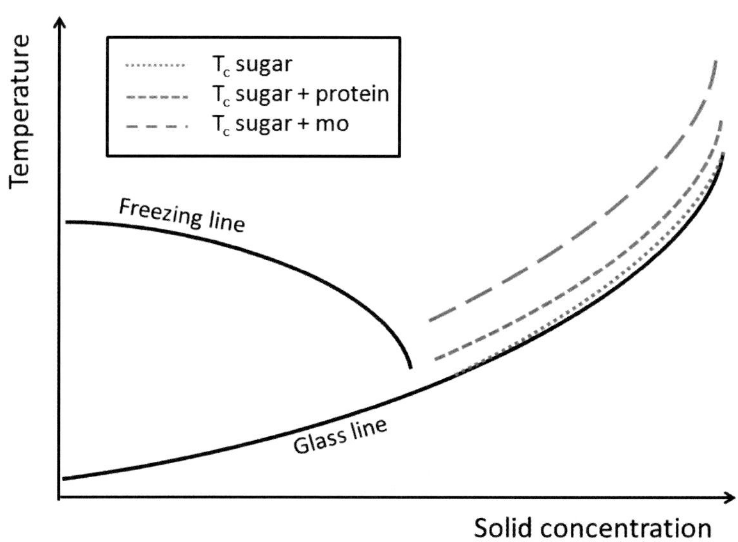


**Fig. 12.** The influence of sample composition on its collapse temperature ($T_c$). The glass transition line for all three samples is identical. Microorganism (mo) and proteins do not significantly influence $T_g$. The collapse temperature, however, is significantly increased.

Schersch et al. (2010) however, did not see a significant negative effect of sample collapse on storage stability or rehydration behaviour of protein formulations.

The opinion that cells should be retained in the glassy state during drying lacks clear empirical evidence. The physical state has been measured mostly by analyzing the frozen and dry sample before and after drying, while the physical state during drying changes with drying time and conditions. Higl (2008) compared bacterial survival for different drying conditions above and below sample $T_g$'. The study showed that the glassy state during drying may not play the essential role for the viability of cells. The results are in agreement with unexpected results from a recent study by Schersch et al. (2010) with pharmaceutically relevant proteins. Loss of biological activity of the proteins was neither observed in collapsed nor in non-collapsed cakes. Generally, a collapsed lyophilizate, the appearance of which is unacceptable for many pharmaceutical products, is not critically important for probiotic products. Very often the dried lyophilizate even has to be milled to a powder in order to mix it with other food components, with other fillers for probiotic tablets or capsules, or to use it directly as starter culture. When the collapsed structure as a result of drying does not negatively affect the viability of cells, the proposed concept "collapse drying", which is expected to substantially reduce drying time (Schersch et al. 2010) can be possibly applied for the production of freeze-dried probiotics.

Generally, the relevance of the glassy state for storage stability of probiotics is widely accepted whereas its relevance for the freeze-drying process is still not clear. Common standard methods more or less are based on assumptions and experiences with similar systems like liposomes or have been transferred from other disciplines like pharmaceutical research. Comprehensive and systematic investigations concerning the inactivation and protection mechanism in relation to the glassy state, however, are rare.

## 5. Process Development in Freeze-Drying

As already mentioned above, the inactivation and protection mechanisms strongly depend on the individual properties of the respective probiotic culture. In order to achieve optimum results, product and process have to be aligned. In this context two main strategies can be differentiated: development of a freeze-drying process for a given sample composition and adaption of sample formulation for a given freeze-drying process.

### *5.1 Process Development for given Formulations*

If the formulation of a probiotic mixture due to certain reasons is given, the optimum process parameters have to be identified in order to achieve best possible results. The two most prominent sample properties in this context are its $T_g$' and $T_c$.

Determination of the glass transition temperature of the maximally freeze-concentrated solution ($T_g$') is essential in order to identify the minimum required freezing temperature and the maximally allowable product temperature during sublimation that is required to obtain a glassy sample matrix throughout the whole freeze-drying process. Typically, $T_g$' is detected via Differential Scanning Calorimetry (DSC). In DSC measurements a small amount of the probiotic suspension is cooled down to a temperature between –60 and –80°C and heated back to ambient temperature. Within the resulting thermogram the $T_g$' can be identified as a step change. In order to obtain a maximally freeze-concentrated sample, the chosen freezing temperature should lie at least 5°C below the detected $T_g$'. However, to characterize and fully understand the sample behaviour it is strongly recommended to determine a complete state diagram (see Fig. 3).

If a glassy matrix during freeze-drying is not necessarily required but structural changes of the product have to be prevented, the collapse temperature $T_c$ needs to be identified. Especially bacterial lyophilizates are known to show collapse temperature several degrees above $T_g$'. This is due to the fact that the bacteria in the μ-scale stabilize the structure of the sample by a non-specific steric effect (see Figs. 13 and 14).

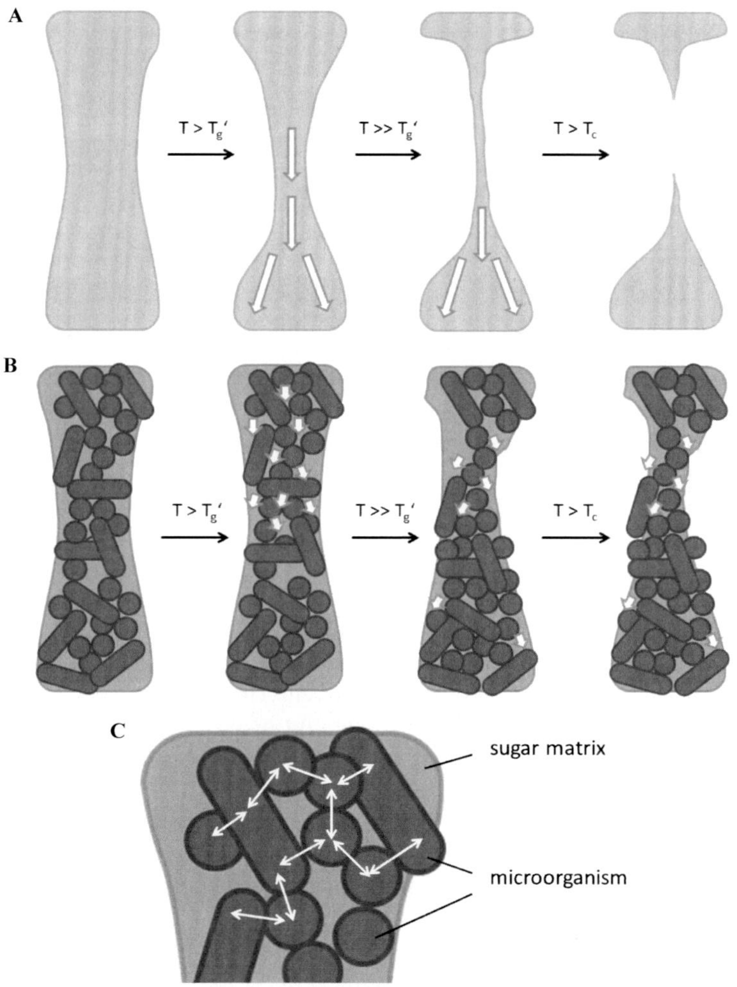


**Fig. 13.** Schematic drawing of structural changes of pure sugar lyophilizates (A) and sugar-bacteria lyophilizates (B) in case of exceeding the individual sample $T_g$' and $T_c$ during drying. Due to the steric effects of the incorporated bacteria (C) bacteria-sugar-lyophilizates show increased stability.

Even at temperatures above $T_g$' where the sugar matrix experiences a drop in viscosity, the sample retains its structures. Identification of the collapse temperature can be accomplished by means of a freeze-drying microscope. In a study by the authors the freeze-drying process was simulated below a special microscope device and the temperature where structural changes appear can be identified. The decision about which of these parameters is the more relevant has to be answered from case to case.

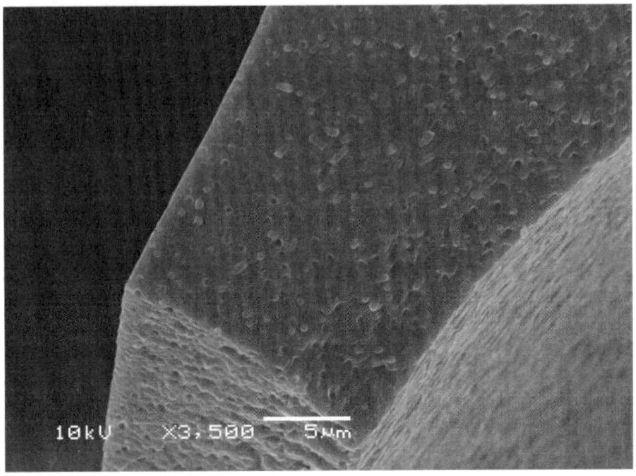


**Fig. 14.** SEM photograph of a freeze-dried sample consisting of F19, buffer and trehalose. A cross section of a thin wall within the lyophilizate shows that bacteria are incorporated throughout the whole sugar matrix.

### 5.2 Adaption of a Formulation to a given Process

There also may be cases where the process parameters are given or at least cannot be adapted arbitrarily to the respective sample needs. In this case the formulation has to be changed. It may be that the $T_g$ of a sample is too low and that drying leads to sometimes undesired results like sample collapse. One possibility is to replace a certain fraction or the whole amount of protectant with another protective substance showing higher $T_g$ values. However, one has to make sure that in case of full replacement the protective function of the new protectant is given. This may be the case for large protectant molecules that show reduced capability of water replacement along the cell membrane. If only a certain fraction of the given protectant is replaced (for example in order to increase the sample $T_g$), the miscibility of the two protectants has to be considered. Depending on the respective case and the protectant miscibility a homogeneous sample matrix with one increased $T_g$ or a heterogeneous sample matrix with at least two separate fractions having two different $T_g$ could be the result. In this case development of ternary state diagrams may be useful. Besides increasing the sample $T_g$ through addition of further protectants, washing and purification of the biomass directly after the fermentation process may be reasonable. Several fermentation broth constituents (buffers, salts, low-$T_g$ carbon sources,...) are known to have an undesirable $T_g$-lowering effect for the subsequent freeze-drying process.

From the freeze-drying perspective, view washing and purification of cells is strongly recommended. If washing and purification is no option, especially in case of complex fermentation media, uncontrolled and thus undesired fluctuations in media composition have to be avoided in order to assure constant $T_g$'-values.

A further interesting question is the choice of the buffer-system. Some common buffer-systems (e.g., phosphate buffers) are simply not well suited for freezing processes. These buffers may experience a dramatic shift of up to 2 pH-units due to partially salting out. This effect has a significant impact on freezing of protein solutions and is of highest importance in the field of pharmaceutical freeze-drying. Within our own investigations with bacterial systems we could not observe a significant contributing effect of the buffer system. This finding however may vary with bacterial species and individual sample composition. Generally, one should keep in mind that during freezing, pH and ionic milieu of a sample may significantly change from optimum to severe conditions.

## 6. Bacteria Specific aspects of Freeze-Drying

Finally, we would like to further extend the chapter with some interesting bacteria-specific aspects that may be more or less helpful for freeze-drying of bacteria without going into detail:

- Differences between liposomes and bacteria
- Residual water content
- Extra- vs. intracellular space
- Rehydration of dry bacteria

A lot of the available information on protective interactions between sugars and liposomes (e.g., the water replacement mechanism) serves as a basis for many general conclusions on microorganism drying. However, the relevance of $T_g$ during drying is a different one. For liposomes, drying temperatures below sample $T_g$ are essential due to the fact that contact between liposomes causes detrimental fusion processes. Further, for liposomes a glassy matrix avoids fusion through immobilization. In case of bacteria, however, this detrimental reaction does not apply and contact between cells does not cause undesired fusion processes. From our experience the protective impact (like water replacement) also applies to the drying process where the sample enters the glassy state after a certain time. Consequently, for bacteria drying above $T_g$ can be an option. Nevertheless, for the subsequent storage process $T_g$ is of high relevance.

In contrast to many pharmaceutically relevant proteins a close to zero water content for bacteria is not reasonable. Here, water contents around 2 to 5% seem to cause the highest stability. This may be due to the fact that oxidation reactions show highest reaction rates at close to zero water contents. A certain amount of residual water seems to protect against oxidation.

Bacterial lyophilizates are highly heterogeneous and complex systems showing a different extra- and intracellular space. Almost nothing is known about the intracellular space and its transitions during freeze-drying. Till date it is not known how much of the added components from a formulation reach the intracellular space.

Finally, one aspect which should not be forgotten is the rehydration process. Here the temperature of the rehydration media is of highest importance. If possible, bacteria should be rehydrated at temperatures above their membrane phase transition temperature $T_m$. $T_m$ depending on the respective species as well as on the drying process and can be determined by FTIR spectroscopy. For freeze-dried bacteria the optimum rehydration temperature may even lie above 40°C.

## References

Adams, G. 2007. The principles of freeze-drying. pp. 15–38. *In*: J.G. Day and G.N. Stacey (eds.). Cryopreservation and Freeze-Drying Protocols. Humana Press, Totowa, USA.

Adams, G.D. and J.R. Ramsay. 1996. Optimizing the lyophilization cycle and the consequences of collapse on the pharmaceutical acceptability of *Erwinia L-asparaginase*. J. Pharm. Sci. 12: 1301–1305.

Barresi, A.A., S. Ghio, D. Fissore and R. Pisano. 2009. Freeze drying of pharmaceutical excipients close to collapse temperature: Influence of the process conditions on process time and product quality. Dry. Technol. 6: 805–816.

Blond, G. and D. Simatos. 1998. Optimized thermal treatments to obtain reproducible DSC thermograms with sucrose + dextran frozen solutions. Food hydrocolloid 2: 133–139.

Brennan, M., B. Wanismail, M.C. Johnson and B. Ray. 1986. Cellular-damage in dried *Lactobacillus acidophilus*. J. Food Prot. 1: 47–53.

Cacela, C. and D.K. Hincha. 2006. Monosaccharide composition, chain length and linkage type influence the interactions of oligosaccharides with dry phosphatidylcholine membranes. BBA-Biomembranes 5: 680–691.

Champion, D., M. Le Meste and D. Simatos. 2000. Towards an improved understanding of glass transition and relaxations in foods: molecular mobility in the glass transition range. Trends Food Sci. Tech. 2: 41–55.

Cioci, F. and R. Lavecchia. 1997. Molecular thermodynamics of heat-induced protein unfolding in aqueous media. AIChE journal 2: 525–534.

Claussen, I.C., T. Andresen, T.M. Eikevik and I. Strømmen. 2007a. Atmospheric freeze drying—Modelling and simulation of a tunnel dryer. Dry. Technol. 12: 1959–1965.

Claussen, I.C., I. Strømmen, A.K.T. Hemmingsen and T. Rustad. 2007b. Relationship of product structure, sorption characteristics, and freezing point of atmospheric freeze-dried foods. Dry. Technol. 5: 853–865.

Claussen, I.C., T.S. Ustad, I. Strømmen and P.M. Walde. 2007c. Atmospheric freeze drying—A review. Dry. Technol. 6: 947–957.

Colandene, J.D., L.M. Maldonado, A.T. Creagh, J.S. Vrettos, K.G. Goad and T.M. Spitznagel. 2007. Lyophilization cycle development for a high-concentration monoclonal antibody formulation lacking a crystalline bulking agent. J. Pharm. Sci. 6: 1598–1608.

Crespi, E., A. Capolongo, D. Fissore and A.A. Barresi. 2008. Experimental investigation of the recovery of soaked paper using evaporative freeze drying. Dry. Technol. 3: 349–356.

Crowe, J.H., J.F. Carpenter and L.M. Crowe. 1998. The role of vitrification in anhydrobiosis. Annu. Rev. Physiol. 1: 73–103.

Crowe, J.H., L.M. Crowe and F.A. Hoekstra. 1989. Phase transitions and permeability changes in dry membranes during rehydration. J. Bioenerg. Biomembr. 1: 77–91.

Crowe, J.H., F.A. Hoekstra, K.H.N. Nguyen and L.M. Crowe. 1996. Is vitrification involved in depression of the phase transition temperature in dry phospholipids? BBA-Biomembranes 2: 187–196.

Cui, Z.-W., C.-Y. Li, C.-F. Song and Y. Song. 2008. Combined microwave-vacuum and freeze drying of carrot and apple chips. Dry. Technol. 12: 1517–1523.

Diaz, S., F. Amalfa, A. Biondi de Lopez and E.A. Disalvo. 1999. Effect of water polarized at the carbonyl groups of phosphatidylcholines on the dipole potential of lipid bilayers. Langmuir 15: 5179–5182.

Duan, X., M. Zhang, X. Li and A.S. Mujumdar. 2008a. Microwave freeze drying of sea cucumber coated with nanoscale silver. Dry. Technol. 4: 413–419.

Duan, X., M. Zhang, X. Li and A.S. Mujumdar. 2008b. Ultrasonically enhanced osmotic pretreatment of sea cucumber prior to microwave freeze drying. Dry. Technol. 4: 420–426.

Duan, X., M. Zhang, A.S. Mujumdar and R. Wang. 2010. Trends in microwave-assisted freeze drying of foods. Dry. Technol. 4: 444–453.

Dumont, F., P.A. Marechal and P. Gervais. 2004. Cell size and water permeability as determining factors for cell viability after freezing at different cooling rates. Appl. Environ. Microb. 1: 268–272.

Ekdawi-Sever, N., La Goentoro and J.D. Pablo. 2003. Effects of Annealing on Freeze-Dried *Lactobacillus acidophilus*. J. Food Sci. 8: 2504–2511.

Fonseca, F., J.P. Obert, C. Beal and M. Marin. 2001. State diagrams and sorption isotherms of bacterial suspensions and fermented medium. Thermochim. Acta 2: 167–182.

Fonseca, F., S. Passot, O. Cunin and M. Marin. 2004a. Collapse Temperature of Freeze-Dried *Lactobacillus bulgaricus* Suspensions and Protective Media. Biotechnol. Progr. 1: 229–238.

Fonseca, F., S. Passot, P. Lieben and M. Marin. 2004b. Collapse temperature of bacterial suspensions: the effect of cell type and concentration. Cryo-Lett. 6: 425–434.

Franks, F. 1998. Freeze-drying of bioproducts: putting principles into practice. Eur. J. of Pharm. Biopharm. 3: 221–229.

Franks, F. 2008. Freeze-drying of Pharmaceuticals and Biopharmaceuticals. Royal Society of Chemistry, Cambridge. UK.

Gardiner, G.E., E. O'Sullivav, J. Kelly, M.A.E. Auty, G.F. Fitzgerald, J.K. Collins, R.P. Ross and C. Stanton. 2000. Comparative Survival Rates of Human-Derived Probiotic *Lactobacillus paracasei* and *L. salivarius* Strains during Heat Treatment and Spray Drying. Appl. Environ. Microbiol. Appl. Environ. Microbiol. pp. 2605–2612.

Gehrke, H.H. 1991. Untersuchungen zur Gefriertrocknung von Mikroorganismen. Ph.D. Thesis. Universität Braunschweig, VDI-Verlag GmbH, Düsseldorf.

Goldblith, S.A., L. Rey and W.W. Rothmayr. 1975. Freeze drying and advanced food technology. Academic Press, London, New York.

Higl, B. 2008. Bedeutung der Verfahrenstechnik und des Glaszustands für die Stabilität von Mikroorganismen während der Lyophilisation und der Lagerung. Ph.D. Thesis. TU München, Freising.

Hincha, D.K., E. Zuther and A.G. Heyer. 2003. The preservation of liposomes by raffinose family oligosaccharides during drying is mediated by effects on fusion and lipid phase transitions. Biochim. Biophys. Acta 2: 172–177.

Hoekstra, F.A., E.A. Golovina and J. Buitink. 2001. Mechanisms of plant desiccation tolerance. Trends Plant Sci. 9: 431–438.

Huang, L.-l., M. Zhang, A.S. Mujumdar, D.-f. Sun, G.-w. Tan and S. Tang. 2009. Studies on decreasing energy consumption for a freeze-drying process of apple slices. Dry. Technol. 9: 938–946.

Izutsu, K. and S. Kojima. 2002. Excipient crystallinity and its protein-structure-stabilizing effect during freeze-drying. J. Pharm. Pharmacol. 8: 1033–1039.

Izutsu, K., S. Yoshioka and T. Terao. 1994. Effect of mannitol crystallinity on the stabilization of enzymes during freeze-drying. Chem. Pharm. Bull. 1: 5–8.

Jennings, T.A. and H. Duan. 1995. Calorimetric monitoring of lyophilization. PDA J. Pharm. Sci. Tech. 6: 272–282.

Kim, A.I., M.J. Akers and S.L. Nail. 1998. The physical state of mannitol after freeze-drying: Effects of mannitol concentration, freezing rate, and a noncrystallizing co-solute. J. Pharm. Sci. 8: 931–935.

Kramer, M. 1999. Innovatives Einfrierverfahren zur Minimierung der Prozesszeit von Gefriertrocknungszyklen. Ph.D. Thesis. Friedrich-Alexander-Universität, Erlangen-Nürnberg.

Lee, C.W.B., J.S. Waugh and R.G. Griffin. 1986. Solid-state NMR study of trehalose 1, 2-dipalmitoyl-sn-phosphatidylcholine interactions. Biochemistry 13: 3737–3742.

Liapis, A.I., M.L. Pim and R. Bruttini. 1996. Research and development needs and opportunities in freeze drying. Dry. Technol. 6: 1265–1300.

Luzardo, M.D.C., F. Amalfa, Am Nunez, S. Diaz, A.C. Biondi de Lopez and E.A. Disalvo. 2000. Effect of trehalose and sucrose on the hydration and dipole potential of lipid bilayers. Biophys. J. 5: 2452–2458.

Mazur, P. 1970. Cryobiology: the freezing of biological systems. Science 3934: 939–949.

Mazur, P., S.P. Leibo and E.H.Y. Chu. 1972. 2-Factor hypothesis of freezing injury—evidence from chinese-hamster tissue culture cells. Exp. Cell Res. 2: 345–355.

Meister, E. 2009. Methodology, Data Interpretation and Practical Transfer of Freeze-Dry Microscopy Methodik, Dateninterpretation und praktische Übertragbarkeit der Gefriertrocknungsmikroskopie. Ph.D. Thesis. Friedrich-Alexander Universität Erlangen-Nürnberg.

Morgan, C.A., N. Herman, P.A. White and G. Vesey. 2006. Preservation of micro-organisms by drying; a review. J. Microbiol. Meth. 2: 183–193.

Morris, G.J., M. Goodrich, E. Acton and F. Fonseca. 2006. The high viscosity encountered during freezing in glycerol solutions: Effects on cryopreservation. Cryobiology 3: 323–334.

Muthukumaran, A., C. Ratti and V.G.S. Raghavan. 2008a. Foam-mat freeze drying of egg white and mathematical modelling Part I optimization of egg white foam stability. Dry. Technol. 4: 508–512.

Muthukumaran, A., C. Ratti and V.G.S. Raghavan. 2008b. Foam-mat freeze drying of egg white—Mathematical modelling part II: Freeze drying and modelling. Dry. Technol. 4: 513–518.

Nail, S.L. and L.A. Gatlin. 1993. Freeze drying: principles and practice. pp. 163–233. *In*: K.E. Avis, H.A. Lieberman and L. Lachman (eds.). Pharmaceutical Dosage Forms: Parenteral Medications, vol. 2: Informa Healthcare.

Nakamura, L.K. 1996. Preservation and Maintenance of Eubacteria. pp. 65–84. *In*: J. Hunter-Cevera and A. Belt (eds.). Maintaining Cultures for Biotechnology and Industry. Academic Press, Inc., London.

Oetjen, G. 2004. Freeze-Drying. Wiley Online Library, Weinheim. Germany.

Parker, R., Y.M. Gunning, B. Lalloue, T.R. Noel and S.G. Ring. 2002. Glassy state dynamics, its significance for biostabilisation and the role of carbohydrates. pp. 73–87. *In*: H. Levine (ed.). Amorphous food and pharmaceutical systems. Royal Society of Chemistry, Cambridge. UK.

Passot, S., F. Fonseca, M. Alarcon-Lorca, D. Rolland and M. Marin. 2005. Physical characterisation of formulations for the development of two stable freeze-dried proteins during both dried and liquid storage. Eur. J. of Pharm. Biopharm. 3: 335–348.

Passot, S., F. Fonseca, N. Barbouche, M. Marin, M. Alarcon-Lorca, D. Rolland and M. Rapaud. 2007. Effect of product temperature during primary drying on the long-term stability of lyophilized proteins. Pharm. Dev. Technol. 6: 543–553.

Pehkonen, K.S., Y.H. Roos, S. Miao, R.P. Ross and C. Stanton. 2008. State transitions and physicochemical aspects of cryoprotection and stabilization in freeze-drying of *Lactobacillus rhamnosus GG* (LGG). J. Appl. Microbiol. 6: 1732–1743.

Pikal, M.J. and D.R. Rigsbee. 1997. The stability of insulin in crystalline and amorphous solids: Observation of greater stability for the amorphous form. Pharmaceut. Res. 10: 1379–1387.

Pikal, M.J. and S. Shah. 1990. The collapse temperature in freeze drying: Dependence on measurement methodology and rate of water removal from the glassy phase. Int. J. Pharm. 2: 165–186.

Potts, M. 1994. Desiccation tolerance of prokaryotes. Microbiol. Rev. 4: 755–805.

Presser, I. 2003. Innovative Online Messverfahren zur Optimierung von Gefriertrocknungsprozessen. Ph.D. Thesis. Ludwigs-Maximilian Universität München.

Rahman, S.M. and A.S. Mujumdar. 2008. A novel atmospheric freeze-drying system using a vibro-fluidized bed with adsorbent. Dry. Technol. 4: 393–403.

Rambhatla, S. and M.J. Pikal. 2005. Heat and mass transfer issues in freeze-drying process development. pp. 75–109. *In:* H.R. Costantino and M.J. Pikal (eds.). Lyophilization of biopharmaceuticals. American Assoc. of Pharm. Scientists, Arlington. USA.

Rivera-Espinoza, Y. and Y. Gallardo-Navarro. 2010. Non-dairy probiotic products. Food Microbial. 1: 1–11.

Roos, Y. 1993. Melting and glass transitions of low molecular weight carbohydrates. Carbohyd. Res. 39–48.

Roos, Y.H. 1995. Phase transitions in foods. Academic Press, London.

Roos, Y.H. 1997. Frozen state transitions in relation to freeze drying. Journal of thermal analysis 3: 535–544.

Sablani, S.S., R.M. Syamaladevi and B.G. Swanson. 2010. A review of methods, data and applications of state diagrams of food systems. Food Eng. Rev. 3: 168–203.

Sadikoglu, H., A.I. Liapis and O.K. Crosser. 1998. Optimal control of the primary and secondary drying stages of bulk solution freeze drying in trays. Dry. Technol. 5: 399–431.

Santivarangkna, C., M. Aschenbrenner, U. Kulozik and P. Först. 2011. Role of Glassy State on Stabilities of Freeze-Dried Probiotics. J. Food Sci. 8: R152–156.

Santivarangkna, C., B. Higl and P. Foerst. 2008a. Protection mechanisms of sugars during different stages of preparation process of dried lactic acid starter cultures. Food Microbiol. 3: 429–441.

Santivarangkna, C., U. Kulozik and P. Foerst. 2008b. Inactivation mechanisms of lactic acid starter cultures preserved by drying processes. J. Appl. Microbiol. 1: 1–13.

Santivarangkna, C., M. Wenning, P. Foerst and U. Kulozik. 2007. Damage of cell envelope of *Lactobacillus helveticus* during vacuum drying. J. Appl. Microbiol. 3: 748–756.

Schersch, K.B. 2009. Effect of collapse on pharmaceutical protein lyophilizates. Ph.D. Thesis. Ludwigs-Maximilian Universität München.

Schersch, K., O. Betz, P. Garidel, S. Muehlau, S. Bassarab and G. Winter. 2010. Systematic investigation of the effect of lyophilizate collapse on pharmaceutically relevant proteins I: Stability after freeze-drying. J. Pharm. Sci. 5: 2256–2278.

Schmidt, S.J. 2004. Water and solids mobility in foods. Adv. Food Nutr. Res. 48: 1–101.

Schoug, Å., J. Olsson, J. Carlfors, J. Schnürer and S. Håkansson. 2006. Freeze-drying of *Lactobacillus coryniformis Si3*—effects of sucrose concentration, cell density, and freezing rate on cell survival and thermophysical properties. Cryobiology 1: 119–127.

Searles, J.A. 2004. Freezing and annealing phenomena in lyophilization. pp. 109–146. *In*: L. Rey and J.C. May (eds.). Freeze-drying/lyophilization of Pharmaceutical and Biological Products. Marcel Dekker, New York.

Selmer-Olsen, E., S.E. Birkeland and T. Sorhaug. 1999. Effect of protective solutes on leakage from and survival of immobilized *Lactobacillus* subjected to drying, storage and rehydration. J. Appl. Microbiol. 3: 429–437.

Shalaev, E.Y. 2005. The impact of buffer on processing and stability of freeze-dried dosage forms, part 1: solution freezing behavior. American Pharmaceutical Review 5: 80–87.

Sheehan, P. and A.I. Liapis. 1998. Modeling of the primary and secondary drying stages of the freeze drying of pharmaceutical products in vials: Numerical results obtained from the solution of a dynamic and spatially multi-dimensional lyophilization model for different operational policies. Biotechnol. Bioeng. 6: 712–728.

Slade, L.H. Levine and D.S. Reid. 1991. Beyond water activity: recent advances based on an alternative approach to the assessment of food quality and safety. Crit. Rev. Food Sci. 2-3: 115–360.

Tang, X.C. and M.J. Pikal. 2004. Design of freeze-drying processes for pharmaceuticals: practical advice. Pharmaceut. Res. 2: 191–200.

Timasheff, S.N. and T. Arakawa. 1989. Stabilization of protein structure by solvents. pp. 349–364. *In*: S.N. Timasheff, T. Arakawa and T.E. Creighton (eds.). Protein Structure: a Practical Approach vol. 2. Oxford University Press, Oxford. USA.

Timasheff, S.N. 2002. Protein-solvent preferential interactions, protein hydration, and the modulation of biochemical reactions by solvent components. Proceedings of the National Academy of Sciences 15: 9721–9726.

To, B. and M.R. Etzel. 1997. Spray drying, freeze drying, or freezing of three different lactic acid bacteria species. J. Food Sci. 3: 576–578.

Tsvetkov, T. and R. Brankova. 1983. Viability of micrococci and lactobacilli upon freezing and freeze drying in the presence of different cryoprotectants. Cryobiology 3: 318–323.

Valdez, G.F. de, G.S. Degiori, A.P.D. Holgado and G. Oliver. 1983. Comparative-Study of the Efficiency of Some Additives in Protecting Lactic-Acid Bacteria against Freeze-Drying. Cryobiology 5: 560–566.

Vereyken, I.J., V. Chupin, R.A. Demel, S. Smeekens and B. de Kruijff. 2001. Fructans insert between the headgroups of phospholipids. BBA-Biomembranes 1: 307–320.

Vereyken, I.J., V. Chupin, F.A. Hoekstra, S. Smeekens and B. de Kruijff. 2003. The effect of fructan on membrane lipid organization and dynamics in the dry state. Biophys. J. 6: 3759–3766.

Villarreal, M.A., S.B. Díaz, E.A. Disalvo and G.G. Montich. 2004. Molecular dynamics simulation study of the interaction of trehalose with lipid membranes. Langmuir 18: 7844–7851.

Voda, A., N. Homan, M. Witek, A. Duijster, G. van Dalen, R. van der Sman, J. Nijssea, L. Vlietc, H. Asb, and J. Duynhovena. 2012. The impact of freeze-drying on microstructure and rehydration properties of carrot. Food Res. Int. 2: 687–693.

Volkert, M., E. Ananta, C. Luscher and D. Knorr. 2008. Effect of air freezing, spray freezing, and pressure shift freezing on membrane integrity and viability of *Lactobacillus rhamnosus GG*. J. Food Eng. 4: 532–540.

Ward, K.R., G.D. Adams, H.O. Alpar and W.J. Irwin. 1999. Protection of the enzyme L-asparaginase during lyophilisation—a molecular modelling approach to predict required level of lyoprotectant. Int. J. Pharm. 2: 153–162.

Wolfe, J. 1987. Lateral stresses in membranes at low water potential. J. Plant Physiol. 3: 311–318.

Wolkers, W.F., H. Oldenhof, F. Tablin and J.H. Crowe. 2004. Preservation of dried liposomes in the presence of sugar and phosphate. Biochim. Biophys. Acta-Biomembr. 2: 125–134.

Zayed, G. and Y.H. Roos. 2004. Influence of trehalose and moisture content on survival of *Lactobacillus salivarius* subjected to freeze-drying and storage. Process Biochem. 9: 1081–1086.

Zhao, G. and G. Zhang. 2005. Effect of protective agents, freezing temperature, rehydration media on viability of malolactic bacteria subjected to freeze-drying. J. Appl. Microbiol. 2: 333–338.

# CHAPTER 12

# Alternative Drying Processes for Probiotics and Starter Cultures

*Foerst P.*

## 1. Introduction

In the production of fermented food, starter cultures are used to prevent fermentation failure, and to ensure high product quality. Starter cultures are cultures with well-defined properties that ensure a fast, safe and defined fermentation and lead to fermented food products with high and constant product quality. In many food sectors defined cultures are already state-of-the art; especially in the dairy and beverage industry. Defined cultures replace more and more wild and uncontrolled fermentations as well as fermentations with "inhouse" cultures in the production of many food products such as meat, bakery and vegetable products. In contrast to starter cultures, which are prone to affect product properties, the primary goal for probiotic culture preparations from a technological point of view is survival in the product. This is mainly due to the fact that the research on the effect of probiotic organisms is still at the beginning and not a lot of structure-function relationships are known by now. Bacterial structures have to be elucidated that exert probiotic effects (see chapter 7 (Charalampopoulos et al. 2009)). Therefore it is not yet possible to design the process in such a way that the relevant structure (e.g., surface protein) is preserved. Hence maximum survival is still a reasonable goal for preserving probiotic microorganisms by means of drying. In ideal case there may be even growth of probiotic bacteria in the colon. Therefore, viability is an essential prerequisite for probiotic function.

Both starter cultures and probiotics are mainly produced by highly specialized suppliers and distributed world-wide. As in case of powder the distribution is much easier and the transportation costs are much cheaper (see chapter 11), so the trend goes towards dried cultures. Especially probiotics are preferably dispensed in dry form especially for pharmaceutical applications.

Chair of Process Engineering of Disperse Systems, TU München, Maximus von Imhof Forum 2, 85356 Freising.
Email: petra.foerst@tum.de

In order to protect the probiotic functionality during drying, gentle and energy efficient drying processes have to be established. The drying process has to be designed such that the losses in viable cell number are minimized during drying and the sorption behaviour and residual water content after drying enables a long-term stability of the culture, ideally at ambient environmental conditions with regard to humidity and temperature. Ideally, the storage stability is maintained for one year. It is important to note that due to extended storage times in comparison to drying time the loss during storage is in general the main contribution to overall absolute viability loss along the production process chain (Ananta 2005, Foerst et al. 2012, Higl et al. 2006, 2007, Lievense et al. 1994). Therefore investigations on storage stability are of highest relevance. As the drying step also influences the storage stability, this impact must also be included in the studies.

Especially for starter cultures, freeze drying is the standard procedure for preservation as it is regarded as a gentle process for sensitive products due to the low product temperatures during drying. However this process is very energy-intensive and requires large investment (Peighambardoust et al. 2011, Regier et al. 2004). Therefore it is mainly applied for speciality products with low production volume where the preservation of biological activity or aroma is the most decisive parameter for quality (e.g., herbs, fruit, coffee and probiotic and starter cultures).

For probiotic products the dosage per gram product is higher than for starter cultures as the probiotics are added in their final concentration and are not expected to grow in the food product. At time of consumption the viable cell number should be higher than $10^6$ cells/gram product to ensure a number of $10^8$ viable cells in one portion of 100 gram (Kosin and Rakshit 2006). Therefore the total annual production volume of probiotics is likely to be much higher than for starter cultures.

The freeze drying process is usually carried out in batch mode and the capacity is limited. Therefore many probiotics are stabilised by other drying technologies that allow higher throughput and even continuous processing. Another disadvantage of the freeze drying process is that the product has to be frozen prior to drying. The freezing process itself can lead to cell damages (e.g., disruption of cell membrane) especially for freeze-sensitive strains (see chapter 10 Freezing). Especially for freeze sensitive strains the alternative thermal drying technologies are an potential alternative that may lead to even higher yield than with freeze drying as a preservation method (Bauer et al. 2012).

The alternative drying processes that are mainly used for the drying of bacteria or yeasts are convective drying processes such as spray drying and fluidized bed drying, contact drying processes such as vacuum drying and radiative drying processes such as microwave drying. The working areas of the different drying processes are depicted in the phase diagram of water in Fig. 1.

The alternative drying processes are described in more detail in the sections 3 and 4 in this chapter. In order to better understand the impact of the processing parameters during drying on the inactivation of cells, the inactivation kinetics of micro-organisms under different environmental conditions has to be understood separately from the drying process. The mathematical description of microbial inactivation during drying is explained in the following section.

## 2. Inactivation Kinetics of Micro-organisms

In order to better understand the inactivation of viable micro-organisms during drying, the inactivation kinetics must be described dependent on the strain, the matrix composition and the environmental conditions such as humidity and temperature. The decrease in viable cell number over time for constant environmental conditions is frequently described as a first-order reaction:

$$\frac{dN(t)/N_0}{dt} = -k \cdot \frac{N(t)}{N_0} \qquad \text{eq. 1}$$

where $N(t)$ is the viable cell number at time $t$, $N_0$ is the viable cell number at the beginning of the treatment and $k$ is the inactivation rate. For a given strain with given matrix composition, the inactivation rate $k$ is in the most simple case a function of the water content $X$ or the water activity $a_w$, respectively. The relationship between water content and water activity at constant temperature is given by the sorption isotherm.

$$k = f(T,X) \text{ with } X = f(a_w) \qquad \text{eq. 2}$$

During drying, the water content is decreases and the temperature increases with time (see, for an example, chapter 14 and schematically Fig. 2). Therefore the inactivation rate $k$ is different at each time step of drying. Several attempts were made to establish the function described in eq.1 for different bacterial strains. The inactivation kinetics depends on the specific inactivation mechanism in each temperature regime (Santivarangkna et al. 2008). It was shown that in low temperature regime inactivation

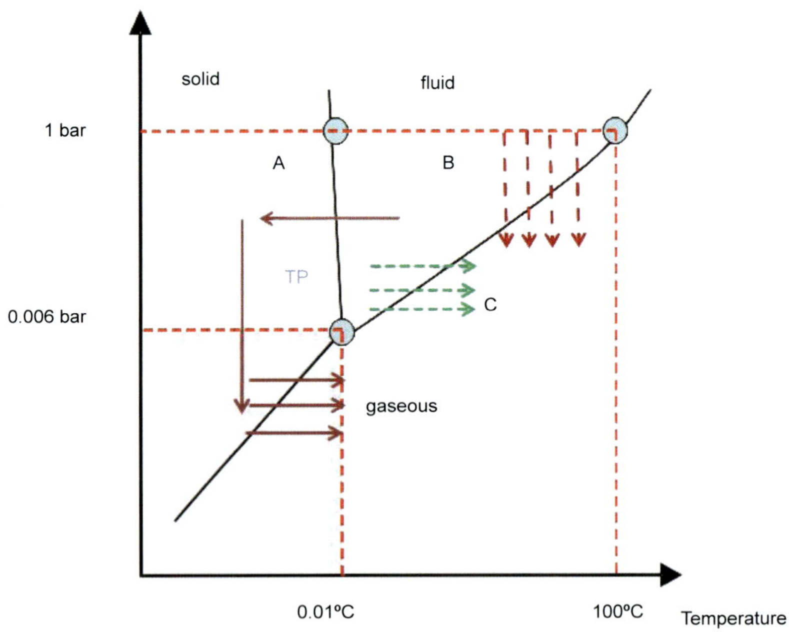


**Fig. 1.** Phase diagram of water and working ranges of different thermal drying processes (Solid lines A: Freeze Drying; Dashed lines B: spray/fluidized bed drying; Dotted lines C: Vacuum drying.

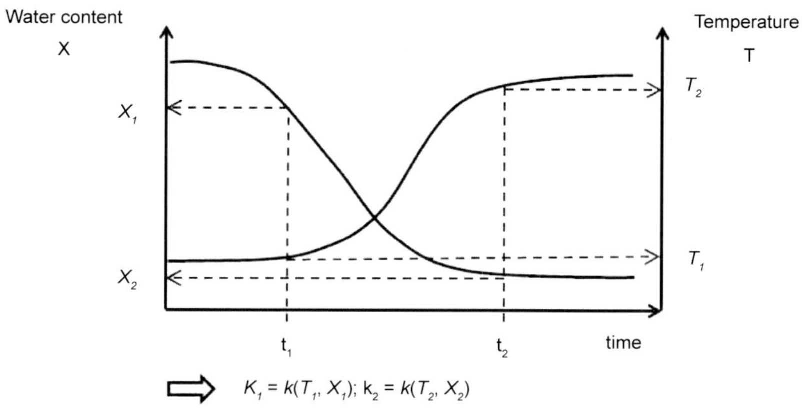


**Fig. 2.** Coupling of the inactivation model $k = f(T,X)$ with the drying kinetics X(t) and product temperature T(t) to derive $k_i = f(X_i, T_i)$.

mechanisms are different to the high temperature regime (Aljarallah and Adams 2007, Tymczyszyn et al. 2007a). Therefore the right inactivation model has to be chosen with regard to the temperature range during drying. Foerst and Kulozik (2012) established the inactivation kinetics for *L. paracasei* (F19) in the moderate temperature range between 4 and 50°C and described the inactivation rate *k* as

$$k(a_W, T) = k_1\, e^{(m \cdot T)} + k_2\, e^{\left(\frac{E_a}{R(T+273)} - 0.5\left(\frac{a_W - a_{W0}}{b}\right)^2\right)} \qquad \text{eq. 3}$$

Here, $k_1$ and $k_2$ are rate constants describing the impact of $a_w$ and temperature, $E_a$ is the activation energy that describes the temperature dependence of the inactivation and $a_{w0}$ and $b$ are coefficients that describe the $a_w$-dependence. The coefficients for *L. paracasei* in a matrix of PBS buffer are given in Table 1.

**Table 1.** Model coefficients of the model described by eq. 3 for *L. paracasei* ssp. *paracasei* (F19) in buffer solution.

| $k_1/h^{-1}$ | $k_2/h^{-1}$ | Ea/(kJ/mol) | $a_{w0}/–$ | m/– | b/– | $r^2$ |
|---|---|---|---|---|---|---|
| 4.1E-3 | 1.91E5 | 33.88 | 0.52 | –4.08E3 | 5.25E-2 | 0.98 |

In this model, the inactivation rate exhibits a maximum value in a water activity range between 0.3 and 0.7. This means that inactivation is highest in that water activity range and the impact of the drying process is most detrimental in this drying stage. Therefore the respective drying stage must be avoided at this temperature or accelerated as much as possible in this water activity range. This leads to the suggestion of a drying process with dynamically adapted process parameters (Foerst and Kulozik 2012).

Next to our own model for vacuum drying at moderate temperatures, other models exist in literature for the description of the influence of water content *X* and temperature *T* on the inactivation rate *k*. Li et al. (2006) have studied the inactivation of two probiotic bacteria in a single milk droplet and established a model for the inactivation rate $k$ that not only takes into account the temperature and moisture but also the rate of change of temperature and moisture during drying.

$$k(X,T) = k_0 \left(1 + a \cdot \left|\frac{dX}{dt}\right| + b \cdot \left|\frac{dX}{dt}\right|^2\right) \cdot e^{\left(-\frac{E_a}{R \cdot T}\right)} \qquad \text{eq. 4}$$

With this model they were able to predict the inactivation of *Bifidobacterium infantis* and *Streptococcus thermophilus* during convective drying of a single droplet of the bacteria embedded in the milk matrix, which provides the basis for the prediction of residual survival rate after spray drying.

Lievense et al. (1992) also gave a model for the rate constant of the metabolic activity of *L. plantarum* with regard to moisture $X$ and temperature $T$. The model is based on the assumption that thermal and dehydration inactivations are the two independent factors that affect the metabolic activity during drying. The comparison to experimental results showed that for low temperatures the dehydration inactivation is the dominating effect and for higher temperature (higher than 55°C) it is the thermal inactivation. The impact of the rate of change in temperature and moisture, however, was found to be negligible. The choice of a valid inactivation model is essential for drying process optimization. Depending on the inactivation rate, the drying process parameters can be adapted such that the inactivation is minimized whereas at the same time the drying process is accelerated as much as possible. The acceleration of the drying process can be either achieved by higher energy input (higher temperature), by a larger surface to volume ratio or a lower chamber pressure. If the inactivation model is known, it can be coupled with the drying kinetics and product temperature. This is schematically depicted in Fig. 2.

## 3. Spray Drying

Spray drying is an interesting alternative to freeze drying for the preservation of probiotics as it is a prevalent drying technology in food industry and enables continuous processing with high throughput. Furthermore, spray drying is considered to be a sensitive drying technique as the maximum product temperature always stays below the air outlet temperature. In addition, the residence time is short for optimal design of the dryer and hence thermal degradation is low (Fu and Chen 2011, Peighambardoust et al. 2011, Perdana et al. 2012, Schutyser Maarten et al. 2012).

For spray drying, the cell suspension is pumped as liquid concentrate to the dryer and separated in small droplets (≈10–100 µm) by means of a pressure or pneumatic nozzle (one or two liquid nozzle) or a rotary disc (Peighambardoust et al. 2011). The type of the nozzle and the fluid properties determine the droplet size distribution. The droplets are then mixed with a large amount of hot and dry air for rapid drying. There are different ways of guiding the air through the drier. If the air is guided parallel to the product feed, the dryer is called co-current spray drier. If the air is guided antiparallel, the drier is called counter current spray drier. As in counter current mode the dry particles at the outlet of the drier come in contact with hot and dry air, the counter current spray drier is not suitable for the drying of temperature sensitive products such as probiotics. The majority of the industrial spray driers are therefore co-current spray driers (Kessler 1996). A schematic picture of a co-current spray drier is shown in Fig. 3.

After spraying, the droplets are mixed with hot air of high temperature (inlet air temperatures generally between 150 and 200°C) and are immediately heated up until wet bulb temperature is reached. The air is heated with a damper register to inlet temperature. The air directly in contact with the droplets starts cooling down to wet bulb temperature by the evaporative cooling effect. The wet bulb temperature can be derived from the Mollier-h-x-diagram for wet air (Mollier 1923) and is dependent on the initial moisture of the drying air and the inlet temperature. The absolute moisture of the drying air is given by

$$x = \frac{m_W}{m_{dA}} \qquad \text{eq. 5}$$

Where $m_w$ is the mass of water and $m_{dA}$ is the mass of dry air. For instance, the wet bulb temperature is below 50°C for an inlet air of 200°C and x = 0.006 (that refers to relative moisture of 40% at a temperature of 20°C). Therefore the temperature at the

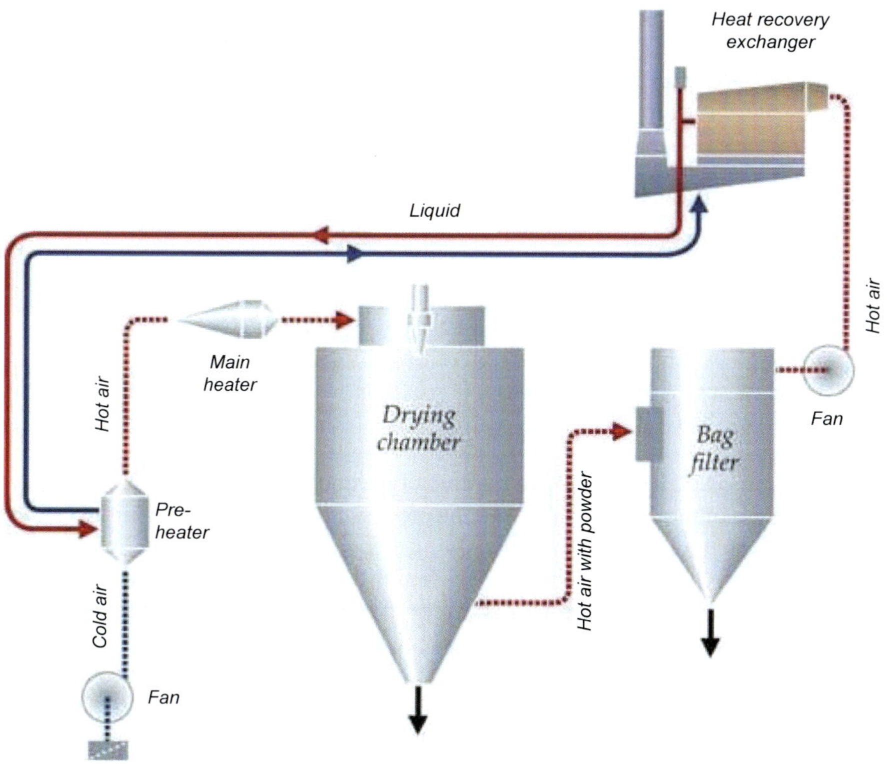


**Fig. 3.** Schematic drawing of a co-current spray drier ©GEA Process Engineering A/S.

beginning of the drying is not in the lethal range for many micro-organisms and thus the spray drying process is considered as sensitive drying process (Fu and Chen 2011, Perdana et al. 2013). The evaporation cooling effect holds as long as free water is at the surface of the droplet. This stage is called the first drying stage and is characterized by constant drying rate. The Mollier-h-x-diagram is depicted in Fig. 4. The heating process of the air is depicted by AB, and the evaporative cooling by BC. The heating of the droplet from wet bulb temperature in the first drying stage to end temperature at the outlet is given by DE.

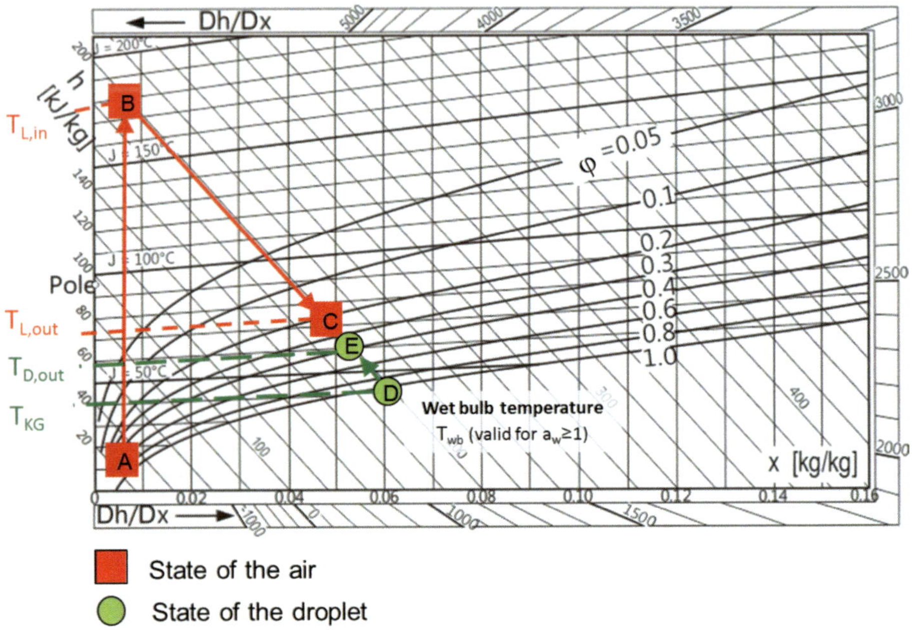


**Fig. 4.** Mollier-h-x-Diagram for the state of humid air with exemplary droplet/particle and air temperatures during spray drying ($T_{L,in}$: Air inlet temperature, $T_{L,out}$: Air outlet temperature, $T_{KG}$: Wet bulb temperature, $T_{D,out}$: Particle outlet temperature).

When the water at the surface is completely evaporated, additional water has to be transported out of the droplet, therefore the process becomes diffusion limited and the drying rate is no longer constant. This drying stage is called the second drying stage. In this stage, the droplet temperature rises and the evaporation rate decreases. In this drying stage the temperature may rise above the lethal temperature of the bacteria. The most decisive process and system parameters for cell survival in this drying stage are the residence time distribution and the droplet size distribution. It has been shown that the residence time distribution is strongly dependent on the design of the drier and the air to feed ratio (Schmitz-Schug et al. 2013). For lysine blockage in infant formula it has been shown that a short residence time with a narrow distribution leads to negligible reactions. This behavior is also applicable to the spray drying of micro-organisms (Fu and Chen 2011). As explained in section 2 above, the main decisive factors of microbial survival during drying are temperature and time. Therefore the exposition time to lethal temperatures must be minimized by short residence times and narrow residence time distribution.

The dried particles are discharged from the bottom of the spray drier. The small particles still remaining in the air are separated in a cyclone and a dust filter. The fine particles are sometimes returned to the drier to agglomerate with larger particles, which must be avoided for probiotics in order to minimize thermal stresses for the bacteria (Kessler 1993, Schmitz-Schug et al. 2013). If the particles that are discharged from the drier are not completely dry, therefore a post-drying step is often used for final drying. This drying step is often carried out in a fluidized bed drier (Kessler 1996). This setup of two subsequent drying processes is especially interesting for probiotics as the first drying stage in the spray drier is fast and gentle and produces small droplets with large surface. The second drying stage where the conditions are harsher and may be detrimental for the bacteria is preferably carried out in a drier where the conditions can be better controlled and are milder (Chávez and Am Ledeboer 2007, Santivarangkna et al. 2007). The possible combinations of drying processes are addressed later in this chapter.

As explained in the previous section, the product temperature, time and water content are the most decisive factors affecting survival. Especially critical is the maximum product temperature. Usually product temperatures at the outlet of the drier lie about 15ºC below the air outlet temperature (Písecký 1997). However, it is important to know that the product temperature during drying can rise above the outlet temperature (Fu and Chen 2011). Nonetheless the outlet temperature is a good indicator for the thermal load during drying (Ananta et al. 2005, O'Riordan et al. 2001, Sunny-Roberts and Knorr 2009). Therefore next to the residence time, the outlet air temperature is the most important process parameter during spray drying. However the outlet air temperature is not an independent process parameter but depends on many other process parameters such as air flow rate (Wang et al. 2004), inlet air conditions (moisture and temperature) (O'Riordan et al. 2001, Sunny-Roberts and Knorr 2009), liquid feed rate (Gardiner et al. 2000, Lian et al. 2002) and solids concentration in the feed as well as droplet size (Boza et al. 2004, Santivarangkna et al. 2007, 2008). Therefore the outlet temperature is difficult to control. It is important that the initial moisture (absolute moisture) of the air is controlled and kept low. However, the final water content of the particles also increases with increasing outlet air temperature. A low outlet temperature leads to high residual water content when the residence time is not increased (Ananta et al. 2005, Desmond et al. 2002, Kim and Bhowmik 1990, To and Etzel 1997a). Therefore post drying steps such as fluidized bed or vacuum drying are necessary to produce a shelf stable product.

A narrow residence time distribution can be achieved when the air is guided in plug flow through the drier and vortexes are minimized. A short residence time is achieved when the air to liquid ratio (ALR) is high. However this means a high energy consumption to produce a large amount of hot air which is not fully loaded with water at the outlet (relative humidity $\varphi \ll 1$). Therefore a recirculating air operation would make sense in this case. In ideal design of drier and process very short residence times could be achieved with the particle moisture being in equilibrium with the air when leaving the drier (around 3 to 4% residual moisture). This leads to a long term stable product.

Typical survival rates for different strains after spray drying are listed in Santivarangkna et al. (2007) and Fu and Chen (2011). For optimum conditions they can be close to 100%. However, storage stability of spray dried probiotics is mostly lower than for probiotics preserved with other drying technologies (Ananta et al. 2005, Desmond et al. 2002, Lievense and van Riet 1994, Lievense et al. 1994, To and Etzel 1997a, 1997b).

Several attempts have been made to overcome the disadvantages of the spray drying process, i.e., the high inactivation in the falling rate period due to high temperatures and the long residence time when lower air temperatures are applied. Semyonov et al. (2011) developed an ultrasonic vacuum spray drier combined with fluidized bed drier to produce highly viable probiotics under reduced oxidative and thermal stresses. Chávez and Am Ledeboer (2007) combined spray drying with vacuum drying. The spray drying was carried out mainly in the constant rate period where the temperature is low and the vacuum drying was carried out in the second drying stage to prevent thermal and oxidative stresses. Stadhouders et al. (1969) also used vacuum drying as a post drying step after spray drying.

## 4. Vacuum Drying

As described in the previous section above, vacuum drying is sometimes used as a post-drying step because of the gentle drying conditions due to reduced pressure. Vacuum drying can also be used as the sole drying method and is explored by several research groups as a potential alternative to freeze drying. The vacuum drying process is mostly carried out as a conductive drying process where the heat is transferred via conduction, but can also be combined with microwaves to the microwave vacuum drying process. According to the phase diagram of water, the boiling point is reduced at low pressure leading to the possibility to carry out the drying process at low temperatures. Furthermore the oxygen partial pressure is reduced and therefore oxidative reactions are minimized. The working area for vacuum drying is schematically depicted in Fig. 1.

A vacuum dryer consists of a vacuum pump connected to a chamber with heated shelves where the samples to be dried are placed. Next to batch vacuum driers, continuous driers also exist. For continuous driers the liquid concentrate is spread in a thin layer on a belt or a drum which is under vacuum. The belt and the drum are rotating and therefore the residence time of the sample in the dryer is given by the belt or the drum speed and the geometric dimensions. Continuous driers have to be equipped with a vacuum lock. The advantage of continuous drying is that due to the thin sample layers, drying is usually very fast.

If the vacuum drying process is driven close to the triple point of water ($p_{tr}$ = 6.1 mbar, $T_{tr}$ = 273,16 K) then the process is referred to as controlled low temperature vacuum drying. Depending on the process parameters the damages during vacuum drying may be reduced to only dehydration damages leading to a high viability of dried microorganisms with high storage stability. King and Su (1993) have shown that the survival rate of *L. acidophilus* after low temperature vacuum drying is comparable to that of freeze drying. Similar results were found by (Higl 2008). It is shown for *L. paracasei* ssp. *paracasei* that the survival rate at optimum conditions is comparable for freeze and vacuum drying and that the optimum working range is in the range of the triple point of water (for freeze drying just below the triple point and vacuum drying just above the triple point) (Higl et al. 2008). Bauer et al. (2012) demonstrated further that the optimum process conditions for vacuum drying depend on the previous fermentation step. This characteristic makes the optimization procedure to produce highly viable starter cultures a very complex task. An interrelation between fermentation conditions and the effect of protectants during vacuum drying was also found by Tymczyszyn et al. (2007b).

The survival rate after vacuum drying can be increased by the use of an appropriate drying medium. As the most suitable protectants are dependent on the drying process, and the protective mechanisms are not yet fully understood, the most appropriate protectants must be found in elaborative experiments. For vacuum drying, the following protectants were shown to be effective: sorbitol (Foerst et al. 2010, Santivarangkna et al. 2006), trehalose and sucrose (Tymczyszyn et al. 2007b) and skim milk powder (King et al. 1998). Foerst et al. (2010) have shown through the low resolution NMR experiments that the protective effect of sorbitol only occurs at the end of drying where the water content is below $X = 0.2$. They also showed that below this water content sorbitol is incorporated in the biomass suggesting that the water replacement mechanism plays a role in stabilizing cells during vacuum drying (Foerst and Kulozik 2009). Higl et al. (2008) also showed for *L. paracasei* ssp. *Paracasei* that the protective effect is dependent on the drying process conditions. If the drying conditions are detrimental (i.e., high shelf temperature) then the protectants are an essential prerequisite for survival. For optimum drying conditions, the protectant does not lead to a further improvement of survival during drying. It is important to note, however, that the protectant may, nevertheless, be necessary for the stabilization of the cells during storage. The aspects of storage of probiotics are addressed in chapter15.

Due to the fact that vacuum drying is mainly carried out as a conductive drying method, one limiting factor is the heat transfer in the product. Another limiting factor that makes the vacuum drying process a lengthy process is the mass transfer of vapour out of the sample. Due to the missing freezing step the sample is not frozen before drying and therefore the liquid sample shrinks during drying (Sansiribhan et al. 2012). This characteristic leads to a very compact structure of the vacuum dried sample. The diffusion limitation leads to a long falling rate period (cf. section 3) making the vacuum drying process lengthy. As the inactivation during drying is a time dependent process (cf. section 2), very long drying times should be avoided. Even after long drying times the residual water contents after vacuum drying are higher than for freeze drying (Bauer et al. 2013, Higl et al. 2008). As described in Chapter 2, the optimum process is controlled according to the change of inactivation rate with moisture and temperature. In order to optimize the process, the process parameters have to be adapted dynamically. Therefore on the one hand the impact of the process parameters shelf temperature and chamber pressure on the drying kinetics have to be known. On the other hand online control possibilities must be available to influence the drying kinetics. Due to heat and mass transfer limitations, the process control possibilities are limited.

In order to speed up the vacuum drying process (especially in the detrimental region) and to get a less compact structure and a larger inner surface similar to freeze drying, the vacuum drying process can be combined with radiative heat transfer to the microwave vacuum drying process. With microwaves the energy is brought volumetrically into the sample and high energy densities may be possible, leading to a volume expansion (Kraus et al. 2013). Another advantage is the selective heating of water in the sample given that the water content of the product is high and the solid fraction has a lower dielectric loss factor than water. The energy input $P$ by microwaves is dependent on:

$$P = 2 \cdot \pi \cdot f \cdot E^2 \cdot \varepsilon_0 \cdot \varepsilon'' \qquad \text{eq. 6}$$

Here, $f$ is the frequency of the microwaves, $E$ the electric field strength, $\varepsilon_0$ the dielectric constant and $\varepsilon$ the dielectric loss factor.

No temperature gradient is therefore needed for heating up the sample and thus it is possible to control the process such that no overheating occurs. The drying rate during microwave drying can be well controlled via the chamber pressure and the absorbed energy in the sample. The absorbed energy depends on the dielectric properties of the sample components (see eq. 5). In order to avoid the formation of hot-spots, it is important to bring in the radiation energy evenly. This is mainly done by rotating the sample by a turntable. With microwaves, the vacuum drying process can be accelerated up to 5 times (Ahmad et al. 2012). Furthermore the control strategies are more efficient than for conductive vacuum drying. It is possible to control the process by the amount of absorbed energy in the sample (Mujumdar 2014).

The survival rates after microwave vacuum drying are comparable to that after vacuum drying if optimal process parameters are chosen (Ahmad et al. 2012, Kim et al. 1997). It has been shown that the power input is the most decisive factor for survival as it is directly correlated to product temperature (Ahmad et al. 2012). As the microwave drying process can be controlled by the absorbed energy in the sample and the energy input is more uniform than for contact drying (lower temperature gradients within the sample) the dynamic adaptation of process parameters seems to be possible. This process is subject of actual research.

## 5. Conclusions

The survival of probiotics after drying depends on the strain, the fermentation conditions, the protectants and the drying process. Freeze drying is the most common method for the preservation of living bacteria as it is considered as a gentle drying method (cf. chapter 11). However it is also a lengthy and energy intensive process and some strains that are freeze sensitive may be damaged and therefore not suitable for freeze drying. This chapter has shown that alternative drying processes such as spray drying, vacuum drying and microwave vacuum drying exist that may lead to comparable survival rates than for freeze drying when applied in the appropriate manner. Even more, in some cases they may be even more suitable than the traditional conservation process. This may lead to a larger spectrum of commercial probiotics, because so far only the robust strains but not the strains with the best probiotic effect are selected by screening. The alternative drying technologies lead to comparable survival rates than with freeze drying assuming that the inactivation kinetics is understood and the process parameters are dynamically adapted to the inactivation kinetics. The alternative drying technologies are characterized by lower investment costs, lower energy costs, shorter processing times and the possibility for continuous processing. A comparison between freeze drying and alternative drying technologies is made in Table 2. A very important aspect is the impact of the drying technology on the storage stability. This aspect is of utmost importance as the major loss occurs during storage due to extended storage times. Therefore the drying process is the starting point for the storage process. The aspect of storage is addressed in chapter 15.

**Table 2.** Comparison between freeze drying and alternative drying technologies with regard to investment costs and energy requirement (according to (Regier et al. 2004)).

| Drying Process | Relative Investment costs at comparable throughput/% | Spec. Energy requirement kWh/kg Water |
|---|---|---|
| Freeze Drying | 100 | 2.0 |
| Vacuum Drying | 65.2 | 1.3 |
| Spray Drying | 52.2 | 1.6 |
| Air Drying | 43.5 | 1.9 |

## References


Ahmad, S., P. Yaghmaee and T. Durance. 2012. Optimization of Dehydration of *Lactobacillus salivarius* Using Radiant Energy Vacuum. Food and Bioprocess Technology 5(3): 1019–27.

Aljarallah, K.M. and M.R. Adams. 2007. Mechanisms of heat inactivation in Salmonella serotype Typhimurium as affected by low water activity at different temperatures. J. Appl. Microbiol. 102(1): 153–60.

Ananta, E. 2005. Impact of environmental factors on viability and high pressure pretreatment on stress tolerance of Lactobacillus rhamnosus GG (ATCC 53103) during spray drying.

Ananta, E., M. Volkert and D. Knorr. 2005. Cellular injuries and storage stability of spray-dried *Lactobacillus rhamnosus* GG. Int. Dairy J. 15(4): 399–409.

Bauer, S.A.W., U. Kulozik and P. Foerst. 2013. Drying Kinetics and Survival of Bacteria Suspensions of *L. paracasei* F19 in Low-Temperature Vacuum Drying. Drying Technology 31(13-14, SI): 1497–503.

Bauer, S.A.W., S. Schneider, J. Behr, U. Kulozik and P. Foerst. 2012. Combined influence of fermentation and drying conditions on survival and metabolic activity of starter and probiotic cultures after low-temperature vacuum drying. Journal of Biotechnology 159(4): 351–357.

Boza, Y., D. Barbin and A.R. Scamparini. 2004. Effect of spray-drying on the quality of encapsulated cells of *Beijerinckia* sp. Process Biochemistry 39(10): 1275–84.

Charalampopoulos, D. and R. Rastall (eds.). 2009. Prebiotics and Probiotics Science and Technology. New York: Springer.

Chávez, B.E. and A.M. Ledeboer. 2007. Drying of probiotics: Optimization of formulation and process to enhance storage survival. Drying Technology 25(7-8): 1193–201.

Desmond, C., R.P. Ross, E. O'Callaghan, G. Fitzgerald and C. Stanton. 2002. Improved survival of *Lactobacillus paracasei* NFBC 338 in spray-dried powders containing gum acacia. Journal of Applied Microbiology 93(6): 1003–11.

Foerst, P. and U. Kulozik. 2009. A low resolution 1H NMR study to investigate the protective mechanism of sorbitol during vacuum drying of a probiotic microorganism. *In*: Maria Guojonsdottir(Editor) PSB(AW(editor. Magnetic resonance in food science: challenges in a changing world. P. 73–80.
Foerst, P. and U. Kulozik. 2012. Modelling the Dynamic Inactivation of the Probiotic Bacterium *L. Paracasei* ssp. *paracasei* During a Low-Temperature Drying Process Based on Stationary Data in Concentrated Systems. Food and Bioprocess Technology 5(6): 2419–27.
Foers, P., U. Kulozik, M. Schmitt, S. Bauer and C. Santivarangkna. 2012. Storage stability of vacuum-dried probiotic bacterium *Lactobacillus paracasei* F19. Food and Bioproducts Processing 90(2): 295–300.
Foerst, P., J. Reitmaier and U. Kulozik. 2010. The role of sorbitol for the molecular mobility and survival of vacuum dried preparations of *Lactobacillus paracasei* ssp. *paracasei*. Journal of Applied Microbiology submitted (108): 841–50.
Fu, N. and X.D. Chen. 2011. Towards a maximal cell survival in convective thermal drying process. Food Research International (44).
Gardiner, G.E., E. O'Sullivan, J. Kelly, M.A. Auty, G.F. Fitzgerald, J.K. Collins, R.P. Ross and C. Stanton. 2000. Comparative survival rates of human-derived probiotic *Lactobacillus paracasei* and L. salivarius strains during heat treatment and spray drying. Appl. Environ. Microbiol. 66(6): 2605–12.
Higl, B. 2008. Bedeutung der Verfahrenstechnik und des Glaszustandes für die Stabilität von Mikroorganismen während der Lyophilisation und der Lagerung: Technische Universität München.
Higl, B., P. Foerst and U. Kulozik. 2006. Determination of the physical state of cell-sugar suspensions throughout a freeze drying process and its influence on the survival of probiotic microorganisms. Drying 2006 15th International Drying Symposium pp. 1065–72.
Higl, B., L. Kurtmann, C.U. Carlsen, J. Ratjen, P. Forst, L.H. Skibsted, U. Kulozik and J. Risbo. 2007. Impact of water activity, temperature, and physical state on the storage stability of *Lactobacillus paracasei* ssp. *paracasei* freeze-dried in a lactose matrix. Biotechnol. Prog. 23(4): 794–800.
Higl, B., C. Santivarangkna and P. Foerst. 2008. Bewertung und Optimierung von Gefrier-und Vakuumtrocknungsverfahren in der Herstellung von mikrobiellen Starterkulturen 80(8): 1157–64.
Kessler, H.G. 1996. Lebensmittel- und Bioverfahrenstechnik. 4th ed. München: A. Kessler.
Kessler, U. 1993. Experimentelle Untersuchung und Modellierung der Überlebensrate von Milchsäurebakterien bei der Thermischen Trocknung: Technische Universität München.
Kim, S.S. and S.R. Bhowmik. 1990. Survival of Lactic-Acid Bacteria During Spray Drying of Plain Yogurt. Journal of Food Science 55(4): 1008–11.
Kim, S.S., S.G. Shin, K.S. Chang, S.Y. Kim, B.S. Noh and S.R. Bhowmik. 1997. Survival of lactic acid bacteria during microwave vacuum-drying of plain yoghurt. Food Science and Technology-Lebensmittel-Wissenschaft & Technologie 30(6): 573–7.
King, V.A.E., H.J. Lin and C.F. Liu. 1998. Accelerated storage testing of freeze-dried and controlled low-temperature vacuum dehydrated *Lactobacillus acidophilus*. Journal of General and Applied Microbiology 44(2): 161–5.
King, V.A.E. and J.T. Su. 1993. Dehydration of *Lactobacillus acidophilus*. Process Biochemistry 28(1): 47–52.
Kosin, B. and S.K. Rakshit. 2006. Microbial and processing criteria for production of probiotics: A review. Food Technology and Biotechnology 44(3): 371–9.
Kraus, S., K. Solyom, H.P. Schuchmann and V. Gaukel. 2013. Drying Kinetics and Expansion of Non-Predried Extruded Starch-based Pellets during Microwave Vacuum Processing. Journal of Food Process Engineering 36(6): 763–73.
Li, X., S.X.Q. Lin, X.D. Chen, L. Chen and D. Pearce. 2006. Inactivation Kinetics of Probiotic Bacteria during the Drying of Single Milk Droplets. Drying Technology 24: 695–701.
Lian, W.C., H.C. Hsiao and C.C. Chou. 2002. Survival of bifidobacteria after spray-drying. International Journal of Food Microbiology 74(1-2): 79–86.
Lievense, L.C. and K. van 't Riet. 1994. Convective drying of bacteria. II. factors influencing survival. Adv.Biochem. Eng. Biotechnol. 51(0724-6145 (Print)): 71–89.
Lievense, L.C., M.A.M. Verbeek, T. Taekema, G. Meerdink and K. Vantriet. 1992. Modeling the inactivation of *Lactobacillus plantarum* during a drying process. Chem. Eng. Sci. 47(1): 87–97.
Lievense, L.C., M.A.M. Verbeek, K. Vantriet and A. Noomen. 1994. Mechanism of Dehydration Inactivation of *Lactobacillus plantarum*. Applied Microbiology and Biotechnology 41(1): 90–4.
Mollier, R. 1923. A new diagram for air-water-vapor mixtures. VDI Zeitschrift 67: 869–72.
Mujumdar, A. (ed.). 2014. Handbook of Industrial Drying. 3rd ed. Boca Raton: CRC Press.

O'Riordan, K., D. Andrews, K. Buckle and P. Conway. 2001. Evaluation of microencapsulation of a Bifidobacterium strain with starch as an approach to prolonging viability during storage. Journal of Applied Microbiology 91(6): 1059–66.

Peighambardoust, S., A. GoshanTafti and J. Hesari. 2011. Application of spray drying for preservation of lactic acid starter cultures: a review. Trends in Food Science & Technology (22): 215–24.

Perdana, J., M.B. Fox, M.A.I. Schutyser and R.M. Boom. 2012. Enzyme inactivation kinetics: Coupled effects of temperature and moisture content. Food Chemistry 133(1): 116–23.

Perdana, J., M.B. Fox, M.A.I. Schutyser, A.I. Maarten and R.M. Boom. 2013. Mimicking Spray Drying by Drying of Single Droplets Deposited on a Flat Surface. Food and Bioprocess Technology 6(4): 964–77.

Písecký, J. 1997. Handbook of milk powder manufacture. Copenhagen, Denmark: Niro A/S.

Regier, M., K. Knörzer and U. Erle. 2004. Mikrowellen- und Mikrowellen-Vakuumtrocknung von Lebensmitteln. ChemieIngenieurTechnik 76(4): 424–32.

Sansiribhan, S., S. Devahastin and S. Soponronnarit. 2012. Generalized microstructural change and structure-quality indicators of a food product undergoing different drying methods and conditions 109(1): 148–54.

Santivarangkna, C., U. Kulozik and P. Foerst. 2006. Effect of carbohydrates on the survival of *Lactobacillus helveticus* during vacuum drying. Lett. Appl. Microbiol. 42(3): 271–6.

Santivarangkna, C., U. Kulozik and P. Foerst. 2007. Alternative drying processes for the industrial preservation of lactic acid starter cultures. Biotechnol. Prog. 23(2): 302–15.

Santivarangkna, C., U. Kulozik and P. Foerst. 2008. Inactivation mechanisms of lactic acid starter cultures preserved by drying processes. Journal of Applied Microbiology 105(105): 1–13.

Schmitz-Schug, I., P. Foerst and U. Kulozik. 2013. Impact of the spray drying conditions and residence time distribution on lysine loss in spray dried infant formula. Dairy Science & Technology 93(4-5, SI): 443–62.

Schutyser, M.A.I., A.I. Maarten, J. Perdana and R.M. Boom. 2012. Single droplet drying for optimal spray drying of enzymes and probiotics. Trends in Food Science & Technology 27(2): 73–82.

Semyonov, D., O. Ramon and E. Shimoni. 2011. Using ultrasonic vacuum spray dryer to produce highly viable probiotics. LWT Food Science and Technology 2011(44): 1844–52.

Stadhouders, J., L.A. Janson and G. Hup. 1969. Preservation of Starters and Mass Production of Starter Bacteria. Neth. Milk Dairy Journal 23: 182–99.

Sunny-Roberts, E.O. and D. Knorr. 2009. The protective effect of monosodium glutamate on survival of <i> *Lactobacillus rhamnosus* GG and<i> *Lactobacillus rhamnosus* E-97800 (E800) strains during spray-drying and storage in trehalose-containing powders. International Dairy Journal 19(4): 209–14.

To, B.C.S. and M.R. Etzel. 1997a. Spray drying, freeze drying, or freezing of three different lactic acid bacteria species. Journal of Food Science 62(3): 576–85.

To, B.C.S. and M.R. Etzel. 1997b. Survival of Brevibacterium linens (ATCC 9174) after spray drying, freeze drying, or freezing. Journal of Food Science 62(1): 167–89.

Tymczyszyn, E.E., D.M. Del Rosario, A. Gomez-Zavaglia and E.A. Disalvo. 2007a. Volume recovery, surface properties and membrane integrity of *Lactobacillus delbrueckii* subsp. *bulgaricus* dehydrated in the presence of trehalose or sucrose. J. Appl. Microbiol. 103(6): 2410–9.

Tymczyszyn, E.E., A. Gomez-Zavaglia and E.A. Disalvo. 2007b. Effect of sugars and growth media on the dehydration of *Lactobacillus delbrueckii* ssp. *bulgaricus*. J. Appl. Microbiol. 102(3): 845–51.

Wang, Y.C., R.C. Yu and C.C. Chou. 2004. Viability of lactic acid bacteria and bifidobacteria in fermented soymilk after drying, subsequent rehydration and storage. Int. J. Food Microbiol. 93(2): 209–17.

# CHAPTER 13

# Industrial Aspects of Probiotic Production

*Anders Clausen** and *Susanne Grøn*

## 1. Introduction

This chapter reviews the different aspects of probiotic cultures production. The complete value chain from bacterial isolate to bulk product will be described and analyzed. Moreover, production of the finished item will also be depicted in some details.

A number of different probiotic products are available on the market, each of them require a specific probiotic culture solution. Probiotic products offered to the consumer are diverse and include dairy products (yogurts, drinking yogurts, cheese, etc.), non-dairy food and beverages (fruit juices, fermented soy products, energy or chocolate bars among many others), dietary supplement—typically in a format containing high concentrations of probiotics cultures, like powder, tablets or capsules including probiotic and prebiotic supplements.

In addition to the regulatory status (food, dietary supplement, medical device or pharmaceutical) each of the particular products will impose different requirements regarding product quality, documentation level and production facilities.

Furthermore, different cultural/geographical, political and religious areas will have special requirements or restrictions for the production process. Typically the same production facility should be able to support all the different culture solutions in an effective, cost efficient and flexible manner. Additionally, as many suppliers of probiotic cultures also provide non probiotic dairy starter cultures using the same production facilities, a very high degree of flexibility is required.

Department Manager Process Technology, Innovation Process, Culture & Enzyme Division, Chr Hansen Inc., 10-12 Boegeallé, DK-2970 Hoersholm, Denmark.
E-mail: dkacl@chr-hansen.com
* Corresponding author

Throughout this chapter many of the aspects discussed will be valid for both probiotic culture production and production of dairy starter cultures, however there are at least five areas where major differences exist:

1. In probiotic culture production, the culture producer most often produce an intermediate product which is then further processed by another company producing product for the final application. This could for example be probiotic cultures for supporting the intestinal microbial flora (for example for preventing or reducing the symptoms of traveler's diarrhea), where the final product could be a tablet, typically sold by a pharmacy or health store. Here the starter company produces the dried bulk, whereas the tablets are produced by a different company; typically a company producing dietary supplements (like vitamin pills/tablets/capsules).
2. Very limited information on the mode of action of probiotic cultures is available and no reliable biological markers are available for assessing product quality on a routine basis in a production environment. In probiotic culture production, therefore performance is most often measured as Colony Forming Units per gram (CFU/g) although this is not a measure for the desired biological function. In contrast, for example *Streptococcus thermophilus* strains (whose role during yogurt production is to acidify the milk) product performance is evaluated by measuring acidification activity.
3. In probiotic production the actual probiotic strain **is** the final product. Dairy starter cultures, in contrast, are **utilized** to produce the final products which are typically yogurt or cheese.
4. The storage stability of the probiotic culture itself is extremely important, both in a wet (e.g., yogurt) or dry (powder, tablets, and capsules) application.
5. Depending on product type considered, the regulatory requirements will be different (food supplement, food additive, dietary supplement, pharmaceutical ingredient, etc.), and different levels of production process documentation are required.

This chapter will start with a description of the production process for probiotic culture production from culture banks to the final step of cleaning and sterilization of the production equipment. Development of the production processes will often be a compromise between the optimally developed process and what is actually practically possible in the factory. These aspects will be discussed in two separate sections entitled Product quality and Process development. Regulatory requirements for probiotic production will be described and finally a discussion of future challenges of probiotic production (where production of probiotic products is expected to be heading in the coming years) will be presented.

## 2. Production of Probiotic Cultures

### *2.1 Microbial Culture Bank*

Maintaining the microbial culture bank is an integrated part of production and production planning. The inoculation material is the basis and starting point of all production batches and significant attention should be given to the task of maintaining the cell culture bank. The bank is typically kept at –80°C in temperature controlled rooms and backups are maintained at different physical locations Quality control of cell banks should include both identity and performance tests. It is essential to ensure that, not only is it the correct microbial strain, but also that the genetic material is maintained unchanged no "genetic drift" occurs in the inoculation material.

## 2.2 Manufacturing

A typical production process is illustrated in Fig. 1. The production process consists of the following steps: (a) media preparation, (b) handling of inoculation material, (c) cultivation in fermentation vessel, (d) concentration of the propagated biomass, (e) freezing, (f) freeze drying, (g) grinding, (h) blending, (i) packaging and (j) storage.

Probiotic cultures are often referred to as one "class" of cultures. In reality they might be phylogenetically only very distantly related organisms and therefore from a production perspective should also be treated in that respect. Each culture will represent different challenges and will require specific production parameters. In the following section the description and discussion is therefore kept in general terms.

Growth media for the culture production are sterilized either by UHT treatment (Ultra High Temperature) or batch sterilized *in situ* (in the fermentation tank) to ensure sterility and cooled to the desired fermentation temperature before inoculation (typically 30–40°C). Fermentation parameters (pH, temperature, agitation rate, headspace gases, etc.) depending on the optimal growth conditions of the probiotic species considered are set according to optimized product performance (typically biomass yield and storage stability). During fermentation the lactic acid bacteria are producing acids (primarily lactic acid) and pH will decrease. To increase biomass yield, pH is maintained at a constant level by addition of a base. The most commonly used base is either sodium hydroxide (NaOH) or ammonium hydroxide ($NH_4OH$). After fermentation, which normally takes place in 10–50 $m^3$ stainless steel fermentation tanks, the fermentate is cooled and the biomass is harvested either by centrifugation or by membrane filtration. Biomass, which represents the active product, is concentrated to minimize the volume to be frozen, freeze dried and subsequently stored. The biomass in the fermentate will typically be concentrated 10–20 times; resulting in 500–5000 liters of concentrated product before formulation and freeze drying. The concentrated biomass is then formulated with cryo additives to protect cells from the freezing process and eventually also formulated with lyo additives to protect the cells during the freeze drying (dehydration) process and to ensure storage stability. Subsequently the formulated material is frozen, for example, by pelletizing into small frozen pellets using liquid nitrogen or by drum freezing. The frozen material is then transferred to freeze drying trays and freeze dried. In another production set-up the formulated concentrate might be transferred as a liquid to the freeze drying trays but in this case the flexibility, of being able to store the frozen material as an intermediate before freeze drying, is lost. On the other hand considerable freezing room space (and energy) can be saved. The freeze dried material is finally ground/milled to obtain smaller particle

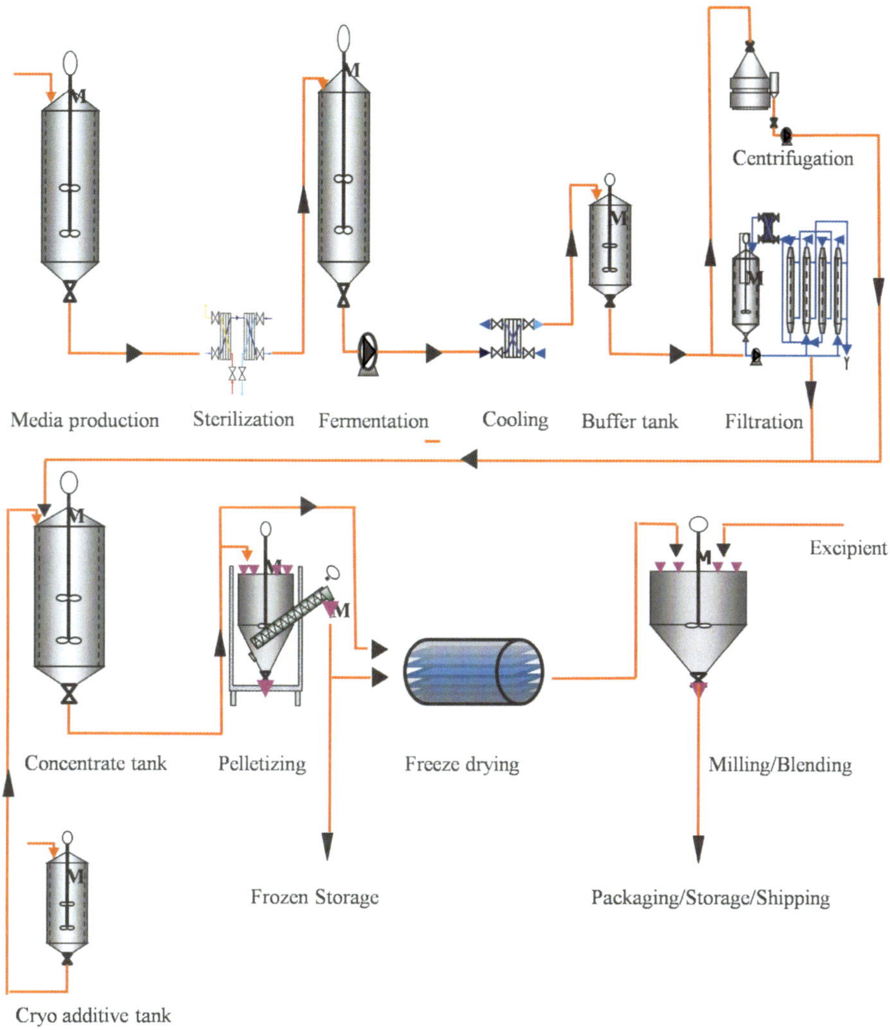


**Fig. 1.** Typical production process for probiotic cultures.

size suitable for blending with excipients (for example maltodextrin) in an inert atmosphere. These final blends will be kept in temperature controlled storage until shipping, in order to minimize degradation.

### *2.2.1 Cleaning and Sterilization*

A full production cycle usually consists of: production, cleaning, and sterilization and then the equipment is ready for a new production cycle. Both cleaning and sterilization are done, whenever possible, "in place" as Cleaning In Place/CIP and Sterilisation In Place/SIP. Cleaning and sterilization are essential to produce consecutively different probiotic species and can be challenging and lengthy processes, depending on complexity and size of the equipment. It is therefore important to have these processes in mind when developing new production processes or when optimizing existing processes.

### *2.2.2 Product Quality and Process Control*

Product quality is described in terms of specifications and strain performance and it is defined in collaboration with product developers prior to process development. For a probiotic bulk product specification will generally be the number of cells per amount product (often Colony Forming Unit/g) and shelf life (defined by a CFU/g value after a number of months at a given temperature, atmosphere and water activity). In addition control of product quality typically includes analysis, and demand for, the absence of contaminating organisms, water activity ($a_w$) and an evaluation of the physical appearance of the material. However defining the right process control parameters to ensure that the product will always perform according to specifications is an important task in process development, and often the process development will be a very integrated part of product development. Parameters like temperature and pH can readily be assessed. A number of other factors can however contribute to variations in product quality. These are, for example, batch to batch variations in raw materials (e.g., complex raw material like yeast extract), inoculation material, and raw material handling. These parameters can also be addressed, but time and economic constraints often mean that sourcing several alternative raw materials and producing more than one inoculation material batch (and testing), before launch of a new product, are not possible. The revenue from probiotics are surely higher than from traditional dairy starter cultures (due to customer added value from the health benefits) but still much lower than from pharmaceuticals; thus process development is still shifted towards a more pragmatic approach.

Quality control typically takes place after the production process, it is off-line measurements, time consuming and consequently there are no possibilities of using this information for adjusting the production process. In addition only the apparent cell concentration can be measured, e.g., it is not possible to perform a real time assessment of the shelf life of the product before release. This parameter will have to be taken into account during process development. Many production plants are operated 24/7 typically by having three operator shifts (day, evening and night). This means that a number of different employees with different skills will alternatively be operating the equipment and running the processes. Therefore from a practical operational point of view the process control must be as simple as possible but also very robust.

## 3. Process Development

Process development for industrial scale production originates several issues that have to be accounted for and are listed below.

### 3.1 Scalability and Robustness of the Production Process

First of all the final product has to perform according to specifications. Then the process must be scalable from laboratory equipment to industrial scale fermentation in a 10 to 50 $m^3$ fermenter, large scale separation equipment and large industrial freeze driers. Lastly, the process must be robust. This means that relevant and likely deviations from the optimal process should be accounted for and incorporated into the process design. This could for example be batch to batch variations, variations in inoculation material, variations in growth medium (e.g., due to alternative suppliers of medium components), pH variations, holding times between process steps, temperature changes of frozen material (due to handling of, for example, 2000 kg frozen pellets, etc.).

### 3.2 Process Economy and Capacity Utilization

Cost of production (COP). The developed process must support the desired revenue of the particular product. This parameter is extremely relevant especially for dairy cultures but also for the probiotic products. Production capacity and production planning have also to be addressed as most culture production plants are operated 24/7 and optimized utilization of the production capacity is essential to the overall business. Most culture production plants are multipurpose plants. This means that no or only very little dedicated equipment is available and all processes have to run on the same equipment; eventually adding some extra constraints on the process development.

### 3.3 Process Interfaces

Process development work is done in small scale, which often does not resemble the technical scale process. Being in control of the up-scaling process is therefore a key to ensure efficient, fast and robust process development and hereafter implementation. Surely the scaling of each individual unit operation is important (media preparation, fermentation, separation, freeze drying, grinding and blending), but also the interfaces between process steps need to be taken into consideration. For example, how long does it take to cool 50 $m^3$ of fermentate at 37 to 5–10°C, and which effect will this have on storage stability on the freeze dried product? How can 2000 kg of frozen pellets be transferred to a freeze dryer and still be frozen, especially if the melting point is very low? Or in order to obtain the desired storage stability, a low final water activity in the freeze dried product is important. However if the area around the freeze dryer does not support a low $a_w$, then emptying the freeze dryer might compromise $a_w$ of the product and the desired storage stability cannot be obtained. These are issues that are equally as important as for example identifying the optimal amount of a particular cryo additive in the freeze dried formulation. In practice, often tools such as "scale-down models" or risk analysis are applied in order to identify critical parameters. Depending on the amount of historic data within the same field, a process simulation can be performed either as an experimental set-up or as a computer simulation.

### 3.4 Storage Stability

Due to demands for short "time to market" and resource/budget considerations, most often assessing real time stability is not possible during process development if specification is, for example, 12 months stability (or longer). Therefore fast analysis that can predict long term storage stability (accelerated stability tests) tools will have to be used. Temperature and $a_w$ are among the most important factors determining storage stability and both factors will have to be accounted for. However the exact correlation between real time and accelerated stability can be difficult to determine and often the analysis will have to be adjusted for the specific application. This is a very time consuming task and often a pragmatic approach based on experience will have to be applied.

### 3.5 Novel Technologies and Equipment

Commonly, process development involves utilization of existing equipment, however a number of promising new technologies exist for the various production steps and are worth being tested. Developing processes involving novel equipment not previously used originate specific issues and is often time and resource consuming.

New type of equipment can be rented or tested at the supplier's facilities and consequently the equipment will often not be perfectly suited for the specific application and a number of important issues cannot be addressed. Therefore, specially designed equipment has to be acquired and implemented in a pilot plant in order to be properly tested and evaluated but such an investment will only be approved if the new process is very promising and essential risks can be eliminated

### 3.6 Regulatory Requirement for the Production of Probiotic Cultures

The manufacturing facilities will have to be approved for producing food (e.g., in Europe according to *Regulation (EC) No. 178/2002*) and Active Pharmaceutical Ingredients (API's) if pharmaceutical products are in scope. The facilities must be inspected at regular intervals by the local food and health authority as applicable, to verify compliance with the specified requirements.

A quality system based on HACCP principles will have to be in place; alternatively if pharmaceuticals are in scope, the quality system will have to comply with pharma GMP (Good Manufacturing Practice) requirements according to *ICH Q7a.*

All raw materials must be suitable for use in foods and specific restrictions on type of raw materials apply if the final product is intended, for example, for infants (EU Regulation 609/013).

Food allergens is also a very important topic to consider in connection with raw materials as a specific list of the most serious allergens will have to be avoided or specifically labelled on the product (*Directive 2000/13/EC*).

GMO free raw materials must be used if GMP labelling is undesirable. This is not a regulatory requirement but typically a customer requirement.

If the product is intended to bear health claims, these must be assessed by the European Food and Safety Authority (EFSA) and approved by the EU Commission based on comprehensive safety and efficacy documentation (Regulation (EC) No 1924/2006).

A number of other restrictions may apply both on raw materials, equipment and facility setup as well as manufacturing procedures if the production is to be Kosher and/or Halal certified.

## 4. Future Perspectives

The burden of well documented probiotic products in consort with profitability and pricing will drive the future development within production processes. In practice this implies higher requirements for documentation of robust production processes and more economical solutions. Quality by Design (QbD) and process analytical technology (PAT), will be very important approaches to successfully achieve these goals.

At present the production process is mostly optimized for maximum cell yield, typically measured as colony forming units per gram (CFU/g). Bio-markers representing the desired probiotic effect would be a more relevant parameter to be monitored during process development. However, there is currently very little information regarding the mode of action of probiotics strains and the correlation to measurable bio-markers is to a large extent lacking. Subsequently the correlation to process parameters is also lacking. In addition it is possible that these links will have to be established for each product or at least for each target area (GI health, weight management, women's health, etc.). However, in line with the principles of PAT; linking Critical Process Parameters (CPP) and Critical Quality Attributes (QCA) to actual probiotic function will be an important step forward in probiotic production.

Within fermentation technology, immobilized cell cultures and/or continuous fermentation can potentially provide some advantages with respect to cell survival, their stability and the functionality of the product. But immobilization and continuous process is relatively more expensive compared to batch cultivation. It takes several days to produce enough material, and it most likely requires investment in completely new production equipment. Can this investment be justified by the selling price of probiotics? More solid evidence for more important health benefits (moving towards pharmaceuticals) are necessary to increase prices whereby additional production costs can be justified. Alternatively less expensive new production methods must be developed.

Similar to fermentation technologies, drying technologies, investment cost and cost of production are key issues. A new drying technology will have to fit several products in order to be economically attractive, as turnover from single products within probiotics seldom justify investment in new key technology like drying technology. This adds extra complexity to implementation of new equipment and renders the industry somewhat conservative.

CHAPTER 14

# Process Analytical Technology (PAT) in Freeze Drying

*Davide Fissore*

## 1. Introduction

The pharmaceutical industry has been intensely regulated due to the quality standards required for the drugs. The introduction of Good Manufacturing Practice (GMP) in the 90's resulted in a very rigid organizational structure, both in the stage of product development and when manufacturing (Barker 1975, Levchuk 1991): the production process of each drug has to be well defined, all important parameters of the process have to be specified, and each manufacturing run has to be carried out similarly to the original (and approved) process (Donawa 1995).

In the field of freeze-drying this means that once the operating conditions (i.e., the values of the temperature of the heating shelf and of the pressure in the drying chamber vs. time) that result in an acceptable product have been identified in the stage of process development, they cannot be modified during manufacturing. Any deviation from design parameters results in lot rejection, and permanent changes to a process requires prior acceptance by the regulatory agencies (and, in some cases, extensive additional testing).

Unfortunately, in any production process there are inherent sources of variability. This holds true for the freeze-drying process, where different quality characteristics of the active pharmaceutical ingredients and/or of the excipients, as well as of the packaging (vials, syringes, etc.), can affect product quality in case the cycle is not modified. As a consequence, very conservative operating conditions are generally selected, in such a way that they can be suitable for the identified sources of variability, although this leads to time consuming and expensive processes (Tang and Pikal

Dipartimento di Scienza Applicata e Tecnologia, Politecnico di Torino, corso Duca degli Abruzzi 24, 10129 - Torino, Italy.
E-mail: davide.fissore@polito.it

2004). Besides, according to the GMP approach rigorous testing of the final product is required. This, in turn, supports the lack of process understanding, i.e., a proper control of the critical parameters during manufacturing.

As this framework became limiting for the introduction of new drugs, the US Food and Drug Administration (FDA) was motivated to introduce a new legislation aimed at modifying the pharmaceuticals development and production processes, and in September 2004 the "Guidance for Industry: PAT—A Framework for Innovative Pharmaceutical Development, Manufacturing, and Quality Assurance" was issued. This document describes a regulatory framework encouraging the voluntary design and implementation of innovative pharmaceutical development, manufacturing, and quality assurance, in such a way that product quality is built-into the process, or is soby design, and no longer tested into products at the end of the manufacturing process. This result can be obtained using Process Analytical Technology (PAT) tools for designing, analyzing, and controlling manufacturing through timely measurements (i.e., during processing) of critical quality and performance attributes of raw and in-process materials and processes. As it is highlighted in the Guidance, the term "analytical" in PAT is viewed broadly to include chemical, physical, microbiological, mathematical, and risk analysis conducted in an integrated manner. Critical parameters have to be identified for both process and product. They have to be measured during the manufacturing process, and their influence on product quality has to be assessed. Process control strategies have to be introduced to prevent (or mitigate) the risk of producing a poor quality product. Besides, production processes have to be optimized, thus reducing costs and the time between production and release.

Although the "Guidance for Industry: PAT" was, in most cases, promptly accepted by the pharmaceutical industry (DePalma 2011), the reaction to this initiative seems to be rather weak and in very few cases PAT tools were implemented, mainly due to time and cost reasons (Shanley 2009, 2010, Kristan and Horvat 2012). In fact, although it was recognized that PAT tools allow achieving a thorough understanding of the production process and of product characteristics during manufacturing, thus getting desired quality characteristics at the end of the process, this requires using expensive devices and, in some cases, modifying existing equipment, thus impeding their implementation.

In a freeze-drying process the temperature of the heating shelf and the pressure in the drying chamber are the most important process parameters. The critical product parameter is, in most cases, product temperature, which has to remain below a limit value in order to prevent product denaturation, as well as meltback (in case of crystalline products) or the structural collapse of the dried cake (in case of amorphous products). These defects are critical for the release and acceptance of a freeze-dried product (Pikal 1994, Franks 1998). Dried cake collapse is due to viscous flow of the glassy matrix. The collapse increases cake density, eventually resulting in pore blockage, as well as cake resistance to vapor flow, and this provokes a decrease of the sublimation rate, thus delaying the end of the primary drying stage (Tsourouflis et al. 1976). Moreover, a higher amount of residual moisture in the product, a higher reconstitution time, and, in some cases, the loss of activity of the active pharmaceutical ingredient can result from dried cake collapse (Bellows and King 1972, Tsourouflis et al. 1976, Adams and Irons 1993, Wang et al. 2004).

Besides product temperature, the other critical product parameter of interest is the residual amount of ice. It is in fact important to promptly identify the ending point of the primary drying stage, with the goal to modify the operating conditions to those selected for the secondary drying stage (in most cases this means increasing the temperature of the heating shelf) when ice sublimation is completed. In case the operating conditions are modified too early, product temperature could overcome the maximum allowed value, while in case they are modified too late, the duration of the process would be unnecessarily prolonged.

An additional critical product parameter is the sublimation flux. A choking flow will occur in the duct connecting the drying chamber and the condenser when the sublimation flux is higher than a critical value, and this causes the loss of pressure control in the chamber (Searles 2004, Nail and Searles 2008, Patel et al. 2010a).

PAT tools for a freeze-drying process have thus to monitor product temperature, the residual amount of ice, and the sublimation flux, in particular in the primary drying stage where the limit temperature can be very low (due to the higher amount of residual moisture), and the sublimation flux is higher. Besides, it can be useful to estimate in-line the value of the heat transfer coefficient ($K_v$), used to express the dependence of the heat flux to the product ($J_q$) on the difference between the temperature of the heating shelf ($T_{shelf}$) and of the product at the bottom of the container ($T_B$):

$$J_q = K_v\left(T_{shelf} - T_B\right) \tag{1}$$

Similarly, it can be useful to estimate in-line the value of the dried cake resistance ($R_p$) used to express the dependence of the mass flux from the interface of sublimation ($J_w$) on the difference between the water partial pressure at the interface of sublimation ($p_{w,i}$) and in the drying chamber ($p_{w,c}$):

$$J_w = \frac{1}{R_p}\left(p_{w,i} - p_{w,c}\right) \tag{2}$$

where $p_{w,c}$ can be assumed to be coincident with total chamber pressure ($P_c$). Once $K_v$ and $R_p$ are known, they can be used with a mathematical model of the process to calculate the design space of the primary drying stage, i.e., the values of $T_{shelf}$ and $P_c$ that allow maintaining product temperature (and the sublimation flux) below the limit (or desired) value (Sundaram et al. 2010, Giordano et al. 2011, Koganti et al. 2011, Fissore et al. 2011a, Pisano et al. 2013a).

In this framework a challenging issue is represented by the fact that product dynamics in the freeze-dryer is different according to the position of the product over the shelf. This is due to the uncontrolled nucleation temperature in the freezing stage, which is responsible for the non-uniform size of the ice crystals in the frozen product and, thus, of the pores in the dried cake, and of different heat transfer mechanisms to the product, as the product at the edge of the shelf can be heated not only by the heating fluid flowing into the shelves but also by radiation from chamber walls (Gan et al. 2005, Rasetto et al. 2010, Barresi et al. 2010).

Various PAT tools have been designed and tested, and they can be grouped according to the physical variable that is measured, namely product temperature, chamber pressure, sublimation flux, composition of the atmosphere in the drying chamber, and other variables. Advantages and limitations of these devices are discussed in the following sections.

## 2. Use of Temperature Measurement as PAT Tool

Product temperature can be measured using thermocouples or resistance thermal detectors. Thermocouples are mainly used in laboratory freeze-dryers, but they are difficult to handle, less robust, and susceptible to sterility problems. On the contrary, resistance thermal detectors are mainly used in manufacturing freeze-dryers. The measurement is reliable, and they are easy to sterilize. However, their thermo-sensitive region is larger than that of thermocouples. This makes temperature measurements at one point in the vial impossible, and accurate positioning is difficult. Moreover, they require a power source that produces heat during the measurement, and this can heat also the product, thus altering product dynamics in the monitored vial with respect to the rest of the batch (Willemer 1991, Oetjen 1999, Presser 2003).

Obviously, the insertion of a thermocouple (or of a resistance thermal detector) in the vial can affect product dynamics in the freeze-drying cycle. It is generally recognized that the drying rate is higher in the vials where the thermocouple has been inserted, and this can be due to the fact that the thermocouple affects the nucleation temperature and, in particular, it lowers the under-cooling, thus resulting in the formation of larger ice crystals. Therefore, when ice sublimation occurs, the size of the pores of the dried cake is larger, and the resistance of the dried cake to vapor flux is lower. Product temperature in this case can thus be different from that of the product in the other vials of the batch. This is particularly true in manufacturing freeze-dryers, operating in a GMP environment, while the effect at lab scale is poorer, and product temperature measured with the thermocouple is almost the same as the remaining part of the batch (Bosca et al. 2013a). Another reason proposed to explain the effect of the insertion of the thermocouple on product dynamics is the increased heat transfer rate to the product. Recently, a plasma sputtered sensor has been proposed with the goal to minimize the interactions between the thermocouple and the product. It is based on a thin film of sub-micrometric wires embedded into the vials wall and coated with a glass like thin film. By this method the product is not in contact with the temperature probe (Grassini et al. 2013).

Another problem related to the use of thermocouples is that, generally, they are connected to a PC using wires. In this framework various types of wireless sensors were proposed in the last decade, using active or passive transponders. Corbellini et al. (2010) designed a wireless system for temperature measurement using an active transponder, where the main limitation is constituted by the capacity of its battery that can limit the operating time. In case of passive transponders the battery is replaced by an electromagnetic field that supplies the energy required for data transmission (Hammerer 2007), as in the system designed by Vallan et al. (2005), or by Schneid and Gieseler (2008).

Figure 1 shows typical dynamics of product temperature measured with thermocouples (graphs A and C). When primary drying starts, the temperature increases

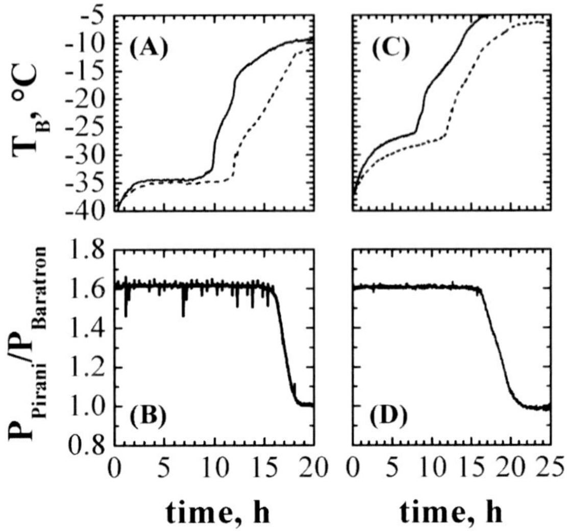


**Fig. 1**. Evolution of product temperature (graphs A and C) in a vial in the centre of the shelf (dashed line) and at the edge of the shelf (solid line) and of the ratio between Pirani and Baratron pressure measurements (graphs B and D) during freeze-drying of a 5% by weight sucrose solution (graphs A and B, vial type: ISO 8362-1 8R, $T_{shelf} = -20$°C, $P_c = 10$ Pa) and of a 5% by weight polyvinylpyrrolidone solution (graphs C and D, vial type: ISO 8362-1 8R, $T_{shelf} = -14$°C, $P_c = 10$ Pa).

from the value reached in the freezing stage. At first the rate of temperature increase is high, as sublimation flux is lower, due to the lower temperature, and the heat supplied to the product is mainly used to increase its temperature. Then, in the central part of the primary drying stage the temperature increases, but at a lower rate. Finally, at a certain moment, the measured temperature suddenly increases, and it rapidly reaches the temperature of the heating shelf (or, in particular in lab-scale units, a higher temperature due to radiation heat flux from chamber walls). This could be interpreted as the ending point of primary drying in the monitored vial: as the heat supplied to the product is no longer used for ice sublimation, then the product temperature increases. Recent studies carried out in a non-GMP environment evidenced that there is still a certain amount of ice in the product when the sudden temperature rise occurs (Bosca et al. 2013a). This is confirmed by the analysis of the curve of pressure ratio. The ratio between the pressure measurement obtained with a thermoconductivity gauge (Pirani type) and with a capacitive gauge (Baratron type) can be used to identify the ending point of the primary drying stage (more details will be discussed in the following section). This occurs in the time interval when the pressure ratio moves from about 1.6 to 1.0. Thus, when comparing graphs B and A (or D and C) it appears that when the sudden temperature rise occurs the ice sublimation is not completed in the batch. This means that from that point the measurement of product temperature is no longer reliable. The reasons proposed to explain this behavior are numerous, e.g., it could be due to the loss of thermal contact between the ice and the thermocouple, or the fact that the sublimation front advances past the thermocouple tip.

As highlighted in the Introduction section, the dynamics of the product placed in different vials in the freeze-dryer can be different due to different heat transfer mechanisms to the product. When using thermocouples (or resistance thermal detectors) to monitor product dynamics, it is possible to place the probes in different positions and, thus, to get information about product temperature in different positions in the batch. As shown in Fig. 1 (graphs A and C) the temperature of the product in the vials at the edges of the shelf is higher than that of the product in the central part of the batch, and this is due to radiation effect from chamber walls.

When using a temperature probe it is not possible to measure the residual amount of ice in the container, as well as the heat and mass transfer coefficients ($K_v$ and $R_p$). Anyway, in case, $K_v$ is known from previous experiments, then using the measurement of product temperature it is possible to calculate the heat flux to the product using eq. (1) once the temperature of the heating shelf is measured. Then, assuming that no heat accumulation occurs in the product, the sublimation flux can be calculated using the following equation (i.e., the energy balance of the frozen product at steady-state):

$$J_q = \Delta H_s J_w \tag{3}$$

where $\Delta H_s$ is the heat of sublimation of the frozen solvent. Once the sublimation flux $J_w$ is calculated, then it can be integrated over time, thus calculating the total amount of ice sublimated from the onset of the primary drying stage (and, thus, the residual ice content in the batch). Finally, assuming that the temperature gradient in the vial is negligible, the temperature of the product at the interface of sublimation can be assumed to be equal to $T_B$: this allows calculating $p_{w,i}$ (as vapor pressure of the solvent is a well known and tabulated function of the temperature) and, finally, $R_p$ using eq. (2). The simple temperature measurement, coupled with a preliminary estimation of $K_v$ and with some hypothesis about product dynamics, can thus be used as a full PAT tool, as both $J_w$ and $R_p$ can be estimated.

A different approach is that proposed by Velardi et al. (2009, 2010): it is based on the use of a soft-sensor, an algorithm that couples a mathematical model of the process and the temperature measure obtained with a thermocouple to estimate in line the residual amount of ice in the vial and the heat and mass transfer coefficients ($K_v$ and $R_p$). The idea at the basis of the soft-sensor is very simple: when using a model of a process it can be possible to "predict" the evolution of the system given the values of the operating conditions and solving the system of differential equations of the model:

$$\dot{\mathbf{x}} = \mathbf{f}(\mathbf{x}, u) \tag{4}$$

$$y = h(\mathbf{x}, u) \tag{5}$$

where **x** represents the state variables array (in this case product temperature and the residual amount of ice), $u$ is the manipulated variable (in this case the temperature of the heating shelf), $y$ is the measured variable (the temperature of the product at the bottom of the vial), and **f** and $h$ are non-linear functions of input and state variables. Obviously, a mathematical model of a process is not a "perfect" description of the heat and mass transfer processes occurring in the system, due to various simplifications that are generally required when writing heat and mass balance equations; moreover, model parameters are not perfectly known. Thus, when comparing the calculated value of product temperature with the measured one the two values are not coincident. The idea at the basis of the soft-sensor is to "use" the difference between the calculated and the measured value of a certain process variable (in this case $T_B$), i.e., the error of the model, to "correct" model equations. The algorithm of a soft-sensor is composed of the following equations:

$$\dot{\mathbf{x}} = \mathbf{f}\left(\hat{\mathbf{x}}, u\right) + \mathbf{K}\left(\hat{y} - y\right) \tag{6}$$

$$\hat{y} = h\left(\hat{\mathbf{x}}, u\right) \tag{7}$$

where $\hat{\mathbf{x}}$ is the estimate of the state variable array, $\hat{y}$ is the calculated value of the process variable ($T_B$), whose measured value is $y$, $(\hat{y} - y)$ is the estimation error, and $\mathbf{K}$ is the parameter whose value has to be selected in order to get the convergence of the algorithm, i.e., to drive the estimation error to zero.

When using a soft-sensor a rough estimation of model parameters ($K_v$ and $R_p$) is required (the initial value of product temperature is that measured by the thermocouple, while the initial amount of ice depends on the composition of the formulation being processed): then, using the difference between calculated and measured values of $T_B$ the algorithm is able to refine the initial estimations and, after a short transient, fairly accurate values of desired variables (ice content, $K_v$ and $R_p$) are obtained.

Bosca and Fissore (2011) proposed an algorithm that can be used to monitor product dynamics in products characterized by different types of dependence of dried cake resistance on product thickness, and Bosca et al. (2014) modified this algorithm in order to get a more robust monitoring system and to cope with the problem of the temperature measurement, which is not reliable from a certain moment during the primary drying stage because the convergence of the algorithm is quite rapid, i.e., accurate estimations of model parameters $K_v$ and $R_p$ are obtained after few hours from the onset of the primary drying stage. Then, when the value of product temperature measured with a thermocouple is no longer reliable, the model of the process is used to calculate product dynamics until the ending point of the primary drying stage.

Figure 2 shows an example of the results that can be obtained when using the soft-sensor to monitor a freeze-drying process: dashed lines identify the variables estimated by the soft-sensor, while symbols correspond to the variables calculated by means of the mathematical model once reliable temperature measurements are no longer available. With respect to product temperature, the soft-sensor allows estimating the temperature of the interface of sublimation, which appears to be slightly lower than the temperature of the product at the bottom of the container, at least until accurate measurements are obtained (graph A). Besides, it is possible to estimate the thickness of the dried layer (graph B), as well as the sublimation flux (graph D), and, thus, the ending point of the primary drying stage, which appears to be in fairly good agreement with the value that can be obtained considering the pressure ratio (graph C). Besides, the soft-sensor allows estimating the values of $R_p$ as a function of dried layer thickness (graph E), and of $K_v$ (graph D), which appears to be in good agreement

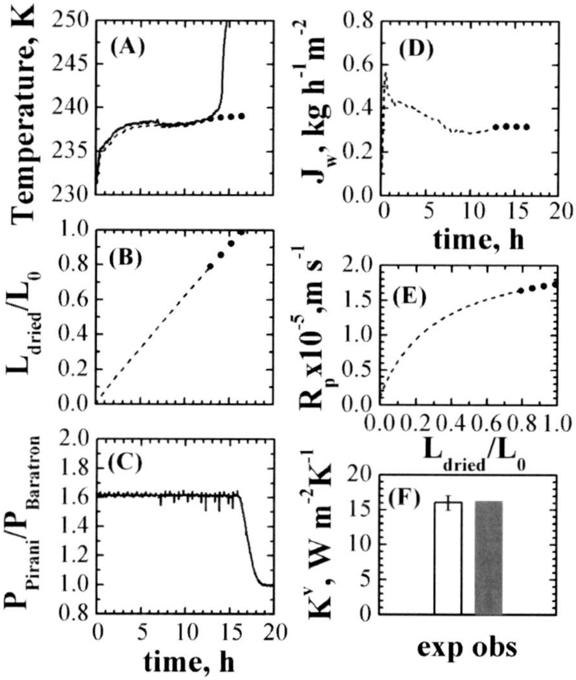


**Fig. 2.** Results obtained when monitoring the freeze-drying of a 5% by weight sucrose solution (vial type: ISO 8362-1 8R, $T_{shelf}$ = −20°C, $P_c$ = 10 Pa) with the soft-sensor (observer: dashed line, mathematical model: symbols). Graph A: comparison between product temperature measured with a thermocouple at the bottom of the vial (solid line), and estimated with the soft-sensor. Graph B: estimated evolution of dried layer thickness. Graph C: ratio between Pirani and Baratron pressure measurements. Graph D: estimated evolution of the sublimation flux. Graph E: estimated values of dried cake resistance. Graph F: comparison between the value of $K_v$ measured experimentally with the gravimetric test and the value estimated by the observer.

with the value that can be obtained using different techniques (e.g., the gravimetric test). It should be now clear that the soft-sensor allows getting the same information about product dynamics and model parameters that can be obtained using the simple temperature measurement, but it is no longer required to carry out any preliminary (and time consuming) investigation to get the value of $K_v$.

## 3. Use of Pressure Measurement as PAT Tool

The measurement of chamber pressure can be used to get information about product dynamics during a freeze-drying process and, in particular, about the ending point of the primary drying stage. This requires using two different pressure gauges, namely a capacitance gauge (Baratron type) and a thermoconductive gauge (Pirani type). The capacitance gauge provides a pressure measure that is independent on the atmosphere composition in the drying chamber, while the measure provided by the thermoconductive gauge depends on the gas composition. As the latter sensor is generally calibrated in air (or nitrogen), and the gas composition in the drying chamber is mostly composed of water vapor during the primary drying stage, the values measured by the two sensors are different (the ratio of the measurements remains almost constant and close to 1.6). When ice sublimation is completed there is obviously no more water vapor in the chamber and, thus, the ratio of the two measurements becomes close to 1. When drying is going to be completed, the pressure ratio decreases from 1.6 to 1 as the gas composition in the drying chamber changes from mostly water to nitrogen. Patel et al. (2010b) identified three points in the pressure ratio curve:

- the onset, given by the intersection between the higher asymptote of the curve and the tangent to the curve in the inflection point;
- the midpoint, coincident with the inflection point;
- the offset, given by the intersection between the lower asymptote of the curve and the tangent to the curve in the inflection point.

Patel et al. (2010b) showed that ice sublimation can be considered complete at the offset in the case of sucrose solutions, and at the onset in case of mannitol solutions. Nevertheless, water concentration in the drying chamber is affected and by the operating conditions (e.g., sublimation rate and nitrogen flux) and by freeze-dryer characteristics (e.g., chamber volume), in particular at the end of the primary drying stage. Thus, it is not obvious to identify the ending point of ice sublimation, and the onset and the offset of the curve identify the uncertainty range of the ending point.

The pressure measurements obtained during the pressure rise test, with a mathematical model of the process, can be used to get information such as product temperature, residual amount of ice, sublimation flux and heat and mass transfer coefficients ($K_v$ and $R_p$). The pressure rise test consists of isolating the drying chamber from the condenser for a short time interval (generally 30 s, or less): a valve in the duct connecting the chamber to the condenser can be closed to get this result. By this technique, as heating is not stopped, chamber pressure increases due to water vapor accumulation. The mass balance in the drying chamber during the pressure rise test is given by the following equation:

$$\frac{dP_c}{dt} = \frac{N_v A_p R T_c}{V_c M_w} J_w + F_{leak} \qquad (8)$$

where $F_{leak}$, the leakage rate, is a known characteristic of the freeze-dryer, $N_v$ is the number of vials of the batch, $A_p$ is the cross surface of each vial, $V_c$ is the chamber volume, $M_w$ is water molecular mass, $R$ is the ideal gas constant and $T_c$ is chamber temperature (that can be assumed to be coincident with $T_{shelf}$). Using a mathematical model of the process it is possible to calculate $J_w$ as a function of model parameters ($K_v$ and $R_p$), as well as of product state (temperature and residual amount of ice). Therefore, the values of model parameters and of product state are obtained looking for best fit between calculated and measured values of chamber pressure during the test. It has to be highlighted that all methods based on the pressure rise test assume that product dynamics are the same in all the vials of the batch, although it can be different due to non-uniform nucleation temperatures and heat transfer processes. Therefore, the estimated values of product temperature, residual ice content, $K_v$ and $R_p$ have to be considered as "average" values for the batch.

Various models were proposed in the past to perform these calculations, e.g., the Manometric Temperature Measurement (MTM) proposed by Milton et al. (1997), the Pressure Rise Analysis (PRA) proposed by Chouvenc et al. (2004), the Dynamic Parameters Estimation (DPE) proposed by Velardi et al. (2008). What differentiates one method from the others is the model used to calculate the pressure rise in the chamber, and the estimated parameters. In all cases $p_{w,i}$, the partial pressure of water at the interface of sublimation, and $R_p$ (or some equivalent parameter), are the variables obtained by solving the following least-square problem:

$$\min_{p_{w,i}, R_p} \sum_k \left(P_{c,k} - P_{c,meas,k}\right)^2 \tag{9}$$

Product temperature at the interface of sublimation ($T_i$) can then be calculated as the equilibrium water pressure ($p_{w,i}$), which is a well known function of $T_i$. The Goff-Gratch equation (Goff and Gratch 1946) can be used to this purpose, as the values obtained with this equation are in perfect agreement with those given by the International Association for the Properties of Steam. The sublimation flux is calculated using eq. (2) and the values of $p_{w,i}$ and $R_p$, and by time integration of $J_w$ it is possible to calculate the residual amount of ice, as well as the thickness of the frozen product ($L_f$):

$$\frac{dL_f}{dt} = -\frac{1}{\rho_f - \rho_d} J_w \tag{10}$$

where $\rho_f$ and $\rho_d$ are, respectively, the density of the frozen and of the dried product. Finally, the heat transfer coefficient $K_v$ is calculated using eq. (3). This requires knowing the value of $T_B$: to this purpose PRA and DPE algorithms solve the heat balance for the frozen layer.

A different approach is used by DPE+ algorithm (Fissore et al. 2011b). In this case only $p_{w,i}$ is calculated looking for best fit between calculated and measured values of chamber pressure during the pressure rise test, and the dried cake resistance is calculated from the slope of the pressure rise curve at the beginning of the test. In fact the mass balance for the drying chamber at the beginning of the test can be written as:

$$J_w = \frac{V_c M_w}{N_v A_p R T_c} \frac{dp_{w,c}}{dt} \tag{11}$$

and, writing eq. (2) for the beginning of the test:

$$J_w = \frac{1}{R_p}\left(p_{w,i,0} - p_{w,c,0}\right) \tag{12}$$

it is possible to get the following equation:

$$R_p = \frac{N_v A_p R T_c}{V_c M_w} \left(\frac{dp_{w,c}}{dt}\right)_{t=t_0}^{-1} \left(p_{w,i,0} - p_{w,c,0}\right) \tag{13}$$

The steps required by DPE+ algorithm are thus the following:

- Initial guess of $T_{i,0}$, product temperature at the sublimation interface at the beginning of the test (i.e., when $t = t_0$);
- Calculation of $p_{w,i,0}$ when product temperature is $T_{i,0}$;
- Calculation of the first derivative of the pressure rise curve at $t = t_0$;
- Calculation of $R_p$ using eq. (13);
- Calculation of $J_w$ using eq. (12);
- Determination of $L_f$ integrating numerically eq. (10);
- Determination of $K_v$ from eq. (3) (where $J_q$ is given by eq. (1));
- Integration of model equations describing pressure rise in the time interval $(t_0, t_f)$, where $t_f - t_0$ is the duration of the PRT, in order to calculate $P_{c,k}$.
- Determination of $T_{i,0}$ solving the following least-square problem:

$$\min_{T_{i,0}} \sum_k \left(P_{c,k} - P_{c,meas,k}\right)^2 \tag{14}$$

Problem conditioning is improved by the reduction of the number of the estimated variables, thus resulting in more accurate estimations, in particular in the second half of the primary drying stage (Fissore et al. 2011b), although in the ending part of the primary drying stage a decreasing value of product temperature is estimated also by DPE+ algorithm, mainly as a consequence of the non-uniformity of the batch. In fact, when using the pressure rise test-based methods the product in all the vials is assumed to have the same dynamics, but at the ending of primary drying the sublimation flux is lower due to the fact that ice sublimation is completed in vials of the first rows that have received heat also by radiation from chamber walls, and this results in a lower estimated product temperature.

Figure 3 shows the results that can be obtained when using the pressure rise test and DPE+ algorithm to monitor a freeze-drying process. Product temperature is estimated accurately in the first part of the primary drying stage (graph A), while close to the ending point the temperature estimated by this method appears to be decreasing. As previously explained, this is due to the fact that the method estimates a "mean" temperature for the batch, where the product is assumed to have the same temperature in all the vials, while in the second part of the primary drying stage ice sublimation is already completed in the edge vials, as they have received heat also by radiation. The estimated value of dried cake thickness (graph B), as well as of the sublimation flux (graph D), can be used to identify the ending point of the primary drying stage, and this value appears to be in good agreement with the value obtained from the pressure ratio curve (graph C). With respect to model parameters, $R_p$ is estimated accurately only in the first part of the primary drying stage, when reliable temperature estimations are obtained (graph E). Care must be taken when examining the estimated values of $K_v$ (graph F). In fact, the heat transfer coefficient $K_v$ does not change during the test (as it depends on the pressure of the chamber and on the type of vial and its position on the shelf), but as $K_v$ is not measured, but estimated, then different values can be estimated at different time instants. It is therefore difficult to identify the correct value

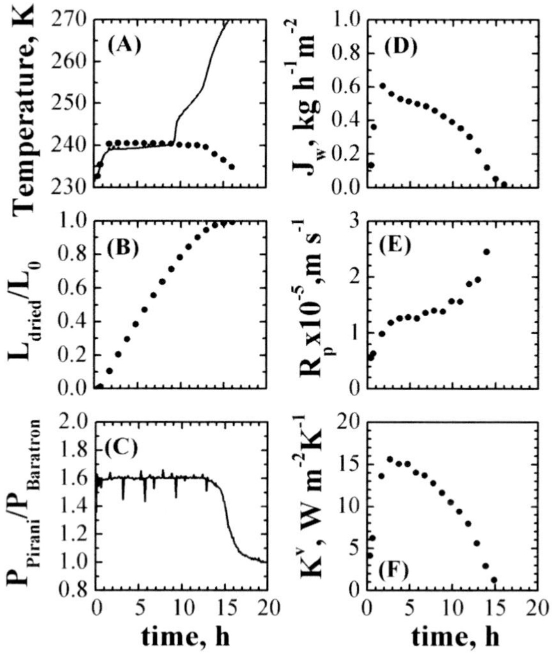


**Fig. 3.** Results obtained when monitoring the freeze-drying of a 10% by weight sucrose solution (vial type: ISO 8362-1 8R, $T_{shelf}$ = –8°C, $P_c$ = 10 Pa) with the pressure rise test and DPE+ algorithm (symbols). Graph A: product temperature measured with a thermocouple at the bottom of the vial (solid line), and estimated with the DPE+ algorithm. Graph B: estimated evolution of dried layer thickness. Graph C: ratio between Pirani and Baratron pressure measurements. Graph D: estimated evolution of the sublimation flux. Graph E: estimated values of dried cake resistance. Graph F: estimated values of heat transfer coefficient.

of $K_v$ looking at a graph like that of Fig. 3 (graph F). Generally, it is assumed that the plateau central value is the correct one, but this plateau can be narrower or larger depending on the overall duration of the primary drying stage and on the values of the operating parameters and of vial and product characteristics.

## 4. Use of Sublimation Flux Measurement as PAT Tool

Various devices were proposed in the past to measure the sublimation flux and, from this value, the other variables of interest in a freeze-drying process. The Tunable Diode Laser Absorption Spectroscopy (TDLAS) sensor is a real-time and non-invasive device based on Doppler-shifted near infrared absorption spectroscopy. It can be used to get the sublimation flux from the measurement of the water vapor concentration and gas flow velocity in the duct connecting the freeze dryer chamber and the condenser (Gieseler et al. 2007).

The measurement of the sublimation flux can be used to identify the ending point of the primary drying stage (when the water flux goes to zero), as well as the product temperature, in case the heat transfer coefficient $K_v$ has been previously determined (Schneid et al. 2009). This requires using eq. (3) to get the heat flux $J_q$, and then eq. (1) to get:

$$T_B = T_{shelf} - \frac{J_q}{K_v} = T_{shelf} - \frac{\Delta H_s J_w}{K_v} \tag{15}$$

Obviously, in this case a "mean" value of product temperature is determined, assuming that product temperature is the same in all the vials of the batch and, thus, the method is not able to account for batch heterogeneity. In case product temperature is measured, it is possible to use the measure of the sublimation flux obtained through the TDLAS sensor to estimate the heat transfer coefficient $K_v$ (Kuu et al. 2009) using the following equation:

$$K_v = \frac{J_q}{\left(T_{shelf} - T_B\right)} = \frac{\Delta H_s J_w}{\left(T_{shelf} - T_B\right)} \tag{16}$$

Similarly, it is possible to estimate in-line dried cake resistance $R_p$. In case $K_v$ is known, then it is possible to calculate at first $T_B$ using eq. (15) and, then, assuming for example that $T_i$ is equal to $T_B$ (i.e., that temperature gradient in the frozen product is negligible), $R_p$ can be calculated using the following equation (Kuu et al. 2011):

$$R_p = \frac{J_w}{\left(p_{w,i} - p_{w,c}\right)} \tag{17}$$

In case $T_B$ is measured, e.g., by using a thermocouple, then any preliminary investigation about $K_v$ is not required.

It has to be remarked that a TDLAS sensor can be installed in both lab-scale and manufacturing-scale freeze-dryers, although it can be difficult to retrofit existing equipment as it should be placed in the duct connecting chamber and condenser. The main drawbacks are the cost and the difficulty of calibration, as fluid flow modeling is required to get reliable mass flow measurements (Kessler et al. 2008).

As an alternative, it is possible to measure in-line the sublimation flux by placing a weighing device in the drying chamber. Various balances were proposed in the past to measure in-line the weight loss in one or more vials in the drying chamber (Pikal et al. 1983, Bruttini et al. 1991, Rovero et al. 2001). In this manner it could be possible to identify the ending point of the primary drying stage, as well as product temperature (in case $K_v$ is known), or model parameters $K_v$ and $R_p$ (in case product temperature is measured), as previously discussed for the case where the sublimation flux is determined using the TDLAS sensor. Unfortunately, the sublimation flux obtained using these balances is not representative of the batch as the weighed vials are exposed to different heating conditions from the rest of the batch. Moreover, in some cases the proposed balances use vials with a specific geometry, which could be different from the geometry of the other vials of the batch (Roth et al. 2001, Gieseler and Lee 2008a, 2008b). A different device is that designed by Carullo and Vallan (2012). It can weight a group of vials with the same geometrical characteristics of the other vials of the batch, and the vials are almost always in contact with the heating shelf, being lifted only during the measurement. Moreover, using a miniaturized radio-controlled

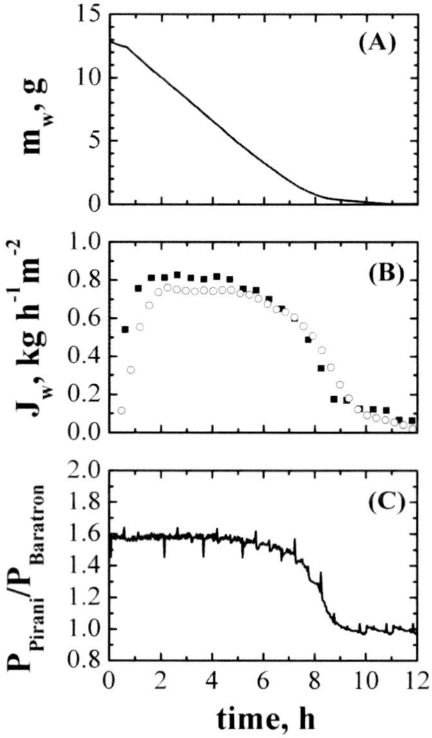


**Fig. 4.** Results obtained when monitoring the freeze-drying of a 10% by weight sucrose solution (vial type: DIN 58378-AR10, $T_{shelf}$ = −20°C, $P_c$ = 10 Pa) with the weighing device designed by Carullo and Vallan (2012). Graph A: Evolution of the water mass left in the 15 monitored vials measured by the balance. Graph B: Comparison between the ice sublimation flux calculated from the pressure rise curves, using DPE+ algorithm (•), and from the mass measurements (O). Graph C: ratio between Pirani and Baratron pressure measurements.

thermometer it is possible to measure the temperature in one or more of the weighed vials without altering the mass measurement because of the force transmitted by the thermocouple wires. If the weighed vials and the balance case are properly shielded, radiation effects can be minimized, and the heat input for the monitored vials is similar to those for the other vials of the batch (Barresi et al. 2009). Figure 4 shows an example of the results that can be obtained when using this balance to monitor a freeze-drying process (graph A). In this case weighed vials were accurately shielded, and, thus, the measured sublimation flux is close to that of the other vials of the batch, as shown in graph B where the measured sublimation flux is compared to that obtained from the pressure rise curve using DPE+ algorithm. The ending point of the primary drying stage that can be estimated from graphs A and B is in fairly good agreement with that estimated from the pressure ratio (graph C). With respect to model parameters, in case the balance is not accurately shielded the value of $K_v$ that can be obtained from the measurement of sublimation flux and product temperature is comparable with the value observed for vials placed at edge of the shelf (Pisano et al. 2008). The values of $R_p$ as a function of dried cake thickness are close to those of the other vials of the batch, obtained for example using the pressure rise test (Fissore et al. 2012). This is due to the fact that freezing conditions for the product in the monitored vials are the same as those of the other vials of the batch. Differently from TDLAS, the balance is a simple device that can be placed easily even in manufacturing freeze-dryers without requiring any modification.

## 5. Other PAT Tools

Various PAT tools are based on the measurement of the composition of the atmosphere in the drying chamber. They can be used only to detect the ending point of the primary drying stage.

The measure of the dew point of the chamber atmosphere can be used to this purpose as the dew point decreases when gas composition moves from mostly water to nitrogen, as in the ending part of the primary drying stage. Roy and Pikal (1989) proposed a moisture sensor based on the variation of the capacity of a thin film of aluminum oxide with moisture, while Bardat et al. (1993) used a gold spattered foil material whose dielectric constant changed with gas composition.

Mass spectrometry, or Residual Gas Analysis, was firstly used in freeze-drying to detect pump oil and residual solvents in the drying chamber (Jennings 1980). This tool has been then applied to detect the ending point of the primary drying stage (Connelly and Welch 1993, Barresi et al. 2009) monitoring the mass spectrum signal of water at 18 g/mol (i.e., the ionic current corresponding to the fragment of mass 18 g/mol). Using an aseptic sterile filter between the drying chamber and the mass spectrometer it becomes possible to use this device also in commercial scale freeze-dryers, where aseptic conditions are mandatory (Wiggenhorn et al. 2005). Although the mass spectrometer exhibits a high sensitivity and can be used to monitor also freeze-drying processes with organic solvents, it is a very expensive sensor and in case of aqueous solvents the same information can also be obtained from less expensive devices.

A cold plasma ionization device was also proposed to monitor water concentration in the chamber (Mayeresse et al. 2007). This device is robust and has a good sensitivity that allows detection of ice in less than 1% of the vials. The main drawback is represented by the generation of radicals by the plasma, as these radicals may cause product degradation, in particular for protein formulations. Besides, the response of the system can be affected by probe location (the device should be placed in the duct connecting the drying chamber to the condenser).

Figure 5 compares the measurements obtained by some of previously described devices when they are used together to monitor the primary drying stage of a freeze-drying process. It is possible to see that both the chamber composition detected by means of a moisture analyzer (graph A) and the signal obtained in a mass spectrometer for the ionic current of water (graph B) decrease when primary drying is going to be completed, as detected by the pressure ratio (graph C).

Another group of PAT tools for monitoring product dynamics in a freeze-drying process is based on spectroscopy. The near-infrared (NIR) spectroscopy was successfully used to detect the ending point of the primary drying stage as water

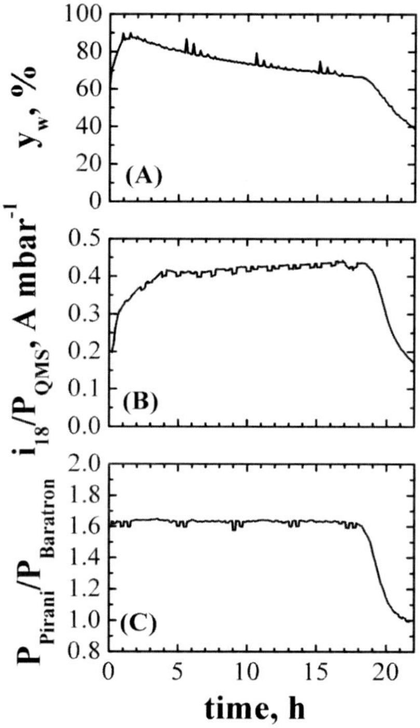


**Fig. 5.** Results obtained when monitoring the freeze-drying of a 5 % by weight lactose solution (vial type: DIN 58378-AR10, $T_{shelf} = -10$°C, $P_c = 10$ Pa) with various devices. Graph A: moisture content in the chamber measured by Panametrics Moisture Analyzer. Graph B: ratio between the ionic current of mass 18 and the pressure measured by a Quadrupole Mass Spectrometer. Graph C: ratio between Pirani and Baratron pressure measurements.

molecules can be easily monitored using NIR technology (Ciurczak 2002, 2006). NIR probes were used also for the characterization of the freeze-dried material (Lin and Hsu 2002, Brulls et al. 2003, Bai et al. 2004). In most cases the sample was measured non-invasively through the vial bottom (Stokvold et al. 2002), and the moisture content was estimated from the measured spectrum. Unfortunately, NIR spectra are highly dependent on the formulation, and, thus, the method has to be adapted to the formulation processed.

Raman spectroscopy was also proposed to monitor a freeze-drying process. It is a method similar to IR spectroscopy, but yielding complementary results, and, in particular, critical product and process information such as water to ice conversion, product crystallization, and solid state characteristics of the product (Romero-Torres et al. 2007, De Beer et al. 2007, 2009). Generally the probe is placed on top of one vial, over the cake. Unfortunately, water and ice produce very weak signals in Raman spectra and, thus, NIR spectroscopy appears to be a more sensitive tool for detecting the ending point of the primary drying stage (De Beer et al. 2009).

## 6. Process Optimization using PAT Tools

Using some of the previously presented PAT tools it is possible to design a freeze-drying process, i.e., to identify the values of the operating conditions that allow fulfilling the operation constraints, as well as to optimize the process, i.e., to minimize also the duration of the primary drying stage. Such results can be obtained either off-line or in-line (Pisano et al. 2013b).

Off-line cycle design (and optimization) requires using a mathematical model to calculate product evolution for a given set of operating conditions ($T_{shelf}$ and $P_c$). PAT tools are thus necessary to determine model parameters (e.g., $K_v$ and $R_p$) in such a way that the model can be used off-line to calculate the design space of the process (Giordano et al. 2011, Koganti et al. 2011, Fissore et al. 2011a, Pisano et al. 2013a).

With the goal to save time it is possible to design, in-line, the freeze-drying process. In this case a mathematical model of the process is used, and suitable PAT tools are required with the goal to estimate in-line model parameters. Two different groups of tools for in-line cycle design and optimization can be identified, based on the use of the pressure rise test or on temperature measure as the monitoring tool. The SMART™ Freeze-Dryer is an expert system proposed by Tang et al. (2005) that manipulates the temperature of the heating shelf and the pressure in the drying chamber using a simple model of the process and the results obtained by means of the MTM algorithm. LyoDriver uses the DPE+ algorithm (Fissore et al. 2011b) to estimate $K_v$ and $R_p$, besides the product state from the pressure rise test and uses a simplified one-dimensional model of the primary drying stage (Velardi and Barresi 2008) to optimize the value of $T_{shelf}$ for a given value of $P_c$ (Pisano et al. 2010). Once the cycle has been designed, the test can be repeated at different values of $P_c$ with the goal to also optimize this operating parameter. As an alternative it is possible to use the temperature measurement and the soft-sensor to calculate in-line the design space of the primary drying stage (Bosca et al. 2013b), and to modify accordingly the operating conditions.

Figure 6 compares the results obtained when using LyoDriver and the soft-sensor to design the freeze-drying cycle for a 5% sucrose solution at 10 Pa. Considering curve (1) in graph (A), i.e., the evolution of the temperature of the heating shelf determined by LyoDriver, it appears that at the beginning of the primary drying stage a high value is calculated (0ºC, corresponding to the maximum allowed value set by the user) and, then, the temperature is firstly decreased to about –10ºC, and then to about –20ºC and, finally to –28ºC. Drying time determined from the pressure ratio (graph B) is about 22 h, and only in the first 7 hours $T_{shelf}$ is manipulated by LyoDriver. When using the soft-sensor to calculate the design space and, thus, the cycle, it appears that drying time can be reduced (graph B, curve 2) to about 17 hours, as higher values of $T_{shelf}$ are selected. Both when using LyoDriver and when the cycle is designed using the soft sensor, product temperature remains below the limit value (–32ºC), but it appears that when using the soft-sensor product temperature is slightly higher (as a consequence of the higher temperature of the heating shelf), as shown in graphs C and D. It has to be highlighted that LyoDriver uses the pressure rise test to estimate model parameters: product temperature increases during the test, and, thus, the target temperature of LyoDriver is lower than the limit temperature used to design the cycle using the

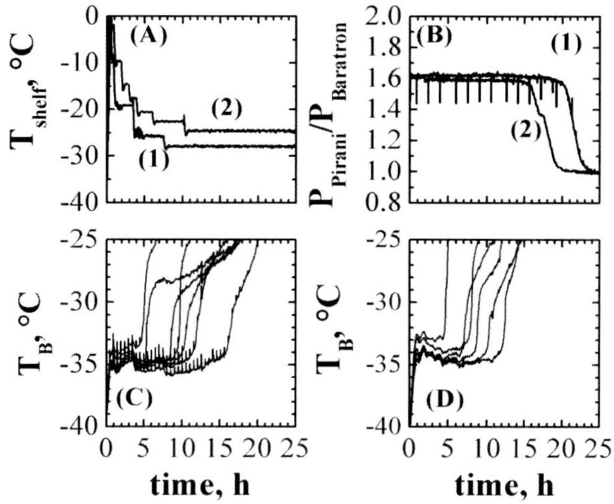


**Fig. 6.** Comparison between heating shelf temperature (graph A) and pressure ratio (graph B) when using LyoDriver (1) or soft-sensor (2) to optimize in-line the freeze-drying of a 5% sucrose solution. Product temperature measured by thermocouples when using LyoDriver or soft-sensor to design the cycle is shown in graph C and D respectively.

soft-sensor. For the case studies investigated the temperature increase during the pressure rise test ranges from 0.6 to 1.1°C, and the limit temperature is further decreased by 0.5°C for safety reasons. Thus, product limit temperature is decreased by 1.3°C on average. As a consequence, lower values of $T_{shelf}$ are calculated for a given value of $L_{dried}$ with respect to the case when the soft-sensor is used to design the cycle.

## 7. Conclusions

In this chapter various PAT tools have been described, and their advantages and drawbacks have been highlighted. Taking into account that a PAT tool for a freeze-drying process should be able to monitor product dynamics without interfering with it, both in lab-scale and in manufacturing-scale freeze-dryers, in a GMP and sterile environment, with the non-uniformity of the batch and able to provide also various information such as the product state (temperature and residual amount of ice) and model parameters (in such a way that it can be used also for process design and optimization), it appears that there is no unique tool that is able to fulfill all these requirements. The freeze-drying practitioners have thus to be aware of the characteristics of the PAT tool available on the market, and they have to use the most suitable device (or various devices) according to the desired goal.

Moreover, it has to be evidenced that no PAT tools exist for the freezing stage, although it has a crucial influence on the following drying stages. In fact, the size of the ice crystals, affected by nucleation temperature (i.e., by subcooling), corresponds to the size of the pores of the dried cake, and this influences cake resistance to vapor flow and, finally, product temperature and sublimation flux during the primary drying stage.

## 8. Acknowledgments

The author would like to acknowledge the coworkers of the LyoLab Research Team of the Politecnico di Torino, and, in particular, Antonello Barresi, Roberto Pisano, Serena Bosca, and Alberto Vallan, for the contribution to the results presented in this work.

## References

Adams, G.D.J. and L.I. Irons. 1993. Some implications of structural collapse during freeze-drying using Erwinia caratovora Lasparaginase as a model. J. Chem. Technol. Biotechnol. 58: 71–76.

Bai, S.J., M. Rani, R. Suryanarayanan, J.F. Carpenter, R. Nayar and M.C. Manning. 2004. Quantification of glycine crystallinity by near-infrared (NIR) spectroscopy. J. Pharm. Sci. 93: 2439–2447.

Bardat, A., J. Biguet, E. Chatenet and F. Courteille. 1993. Moisture measurement: a new method for monitoring freeze-drying cycles. PDA J. Parent. Sci. Technol. 47: 293–299.

Barker, L.F. 1975. Food and Drug Administration regulations and licensure. Fed. Proc. 34: 1522–1524.

Barresi, A.A., R. Pisano, D. Fissore, V. Rasetto, S.A. Velardi, A. Vallan, M. Parvis and M. Galan. 2009. Monitoring of the primary drying of a lyophilization process in vials. Chem. Eng. Proc. 48: 408–423.

Barresi, A.A., R. Pisano, V. Rasetto, D. Fissore and D.L. Marchisio. 2010. Model-based monitoring and control of industrial freeze-drying processes: Effect of batch non-uniformity. Drying Technol. 28: 577–590.

Bellows, R.J. and C.J. King. 1972. Freeze-drying of aqueous solutions: Maximum allowable operating temperature. Cryobiology 9: 559–561.

Bosca, S. and D. Fissore. 2011. Design and validation of an innovative soft-sensor for pharmaceuticals freeze-drying monitoring. Chem. Eng. Sci. 66: 5127–5136.

Bosca, S., A.A. Barresi and D. Fissore. 2013a. Use of a soft-sensor for the fast estimation of dried cake resistance during a freeze-drying cycle. Int. J. Pharm. 451: 23–33.

Bosca, S., A.A. Barresi and D. Fissore. 2013b. Fast freeze-drying cycle design and optimization using a PAT based on the measurement of product temperature. Eur. J. Pharm. Biopharm. 85: 253–262.

Bosca, S., A.A. Barresi and D. Fissore. 2014. Use of soft-sensors to monitor a pharmaceuticals freeze-drying process in vials. Pharm. Dev. Tech. 19: 148–159.

Brulls, M., S. Folestad, A. Sparen and A. Rasmuson. 2003. *In situ* near-infrared spectroscopy monitoring of the lyophilization process. Pharm. Res. 20: 494–499.

Bruttini, R., G. Rovero and G. Baldi. 1991. Experimentation and modeling of pharmaceutical lyophilization using a pilot plant. Chem. Eng. J. Bioch. Eng. 45: B67–B77.

Carullo, A. and A. Vallan. 2012. Measurement uncertainty issues in freeze-drying processes. Measurement 45: 1706–1712.

Chouvenc, P., S. Vessot, J. Andrieu and P. Vacus. 2004. Optimization of the freeze-drying cycle: a new model for pressure rise analysis. Drying Technol. 22: 1577–1601.

Ciurczak, E.W. 2002. Growth of Near-Infrared Spectroscopy in pharmaceutical and medical sciences. Eur. Pharm. Rev. 5: 68–73.

Ciurczak, E.W. 2006. Near-Infrared Spectroscopy: Why it is still the number one technique in PAT. Eur. Pharm. Rev. 9: 19–21.

Connelly, J.P. and J.V. Welch. 1993. Monitor lyophilization with mass spectrometer gas analysis. J. Parenter. Sci. Tech. 47: 70–75.

Corbellini, S., M. Parvis and A. Vallan. 2010. In-process temperature mapping system for industrial freeze-dryers. IEEE Trans. Instrum. Meas. 59: 1134–1140.

De Beer, T.R., M. Alleso, F. Goethals, A. Coppens, Y.V. Heyden, H.L. De Diego, J. Rantanen, F. Verpoort, C. Vervaet, J.P. Remon and W.R. Baeyens. 2007. Implementation of a process analytical technology system in a freeze-drying process using Raman spectroscopy for in-line process monitoring. Anal. Chem. 79: 7992–8003.

De Beer, T.R., P. Vercruysse, A. Burggraeve, T. Quinten, J. Ouyang, X. Zhang, C. Vervaet, J.P. Remon and W.R. Baeyens. 2009. In-line and real-time process monitoring of a freeze drying process using Raman and NIR spectroscopy as complementary process analytical technology (PAT) tools. J. Pharm. Sci. 98: 3430–3446.

DePalma, A. 2011. PAT slowly yielding to broader QbD. Genet. Eng. Biotechnol. News 31: 38–41.

Donawa, M.E. 1995. Taking control: the Validation Master Plan. Med. Device. Technol. 6: 12–14.

Fissore, D., R. Pisano and A.A. Barresi. 2011a. Advanced approach to build the design space for the primary drying of a pharmaceutical freeze-drying process. J. Pharm. Sci. 100: 4922–4933.

Fissore, D., R. Pisano and A.A. Barresi. 2011b. On the methods based on the Pressure Rise Test for monitoring a freeze-drying process. Drying Technol. 29: 73–90.

Fissore, D., R. Pisano and A.A. Barresi. 2012. A model-based framework for the analysis of failure consequences in a freeze-drying process. Ind. Eng. Chem. Res. 51: 12386–12397.

Franks, F. 1998. Freeze-drying of bioproducts: putting principles into practice. Eur. J. Pharm. Biopharm. 45: 221–229.

Gan, K.H., O.K. Crosser, A.I. Liapis and R. Bruttini. 2005. Lyophilisation in vials on trays: effects of tray side. Drying Technol. 23: 341–363.

Gieseler, H. and G. Lee. 2008a. Effects of vial packing density on drying rate during freeze-drying of carbohydrates or a model protein measured using a vial-weighing technique. Pharm. Res. 25: 302–312.

Gieseler, H. and G. Lee. 2008b. Effect of freeze-dryer design on drying rate of an amorphous protein formulation determined with a gravimetric technique. Pharm. Dev. Tech. 13: 463–472.

Gieseler, H., W.J. Kessler, M. Finson, S.J. Davis, P.A. Mulhall, V. Bons, D.J. Debo and M.J. Pikal. 2007. Evaluation of Tunable Diode Laser Absorption Spectroscopy for in-process water vapor mass flux measurement during freeze drying. J. Pharm. Sci. 96: 1776–1793.

Giordano, A., A.A. Barresi and D. Fissore. 2011. On the use of mathematical models to build the design space for the primary drying phase of a pharmaceutical lyophilization process. J. Pharm. Sci. 100: 311–324.

Goff, J.A. and S. Gratch. 1946. Low-pressure properties of water from –160 to 212 F. Trans. Am. Soc. Heat. Vent. Eng. 52: 95–122.

Grassini, S., M. Parvis and A.A. Barresi. 2013. Inert thermocouple with nanometric thickness for lyophilization monitoring. IEEE T. Instrum. Meas. 62: 1276–1283.

Hammerer, K. 2007. Wireless temperature measurement as an innovative PAT-method. *In*: Proceedings of 2nd Congress on Life Science Process Technology, Nuremberg, Germany.

Jennings, T.A. 1980. Residual gas analysis and vacuum freeze drying. J. Parenter. Drug. Assoc. 34: 62–69.

Kessler, W.J., G. Caledonia, M. Finson, J. Cronin, D. Paulsen, S.J. Davis, P.A. Mulhall, H. Gieseler, S. Schneid, M.J. Pikal and A. Schaepman. 2008. TDLAS-based mass flux measurements: critical analysis issues and product temperature measurements. *In*: Proceedings of Freeze-drying of Pharmaceuticals and Biologicals Conference, Breckenridge (CO), USA.

Koganti, V.R., E.Y. Shalaev, M.R. Berry, T. Osterberg, M. Youssef, D.N. Hiebert, F.A. Kanka, M. Nolan, R. Barrett, G. Scalzo, G. Fitzpatrick, N. Fitzgibbon, S. Luthra and L. Zhang. 2011. Investigation of design space for freeze-drying: Use of modeling for primary drying segment of a freeze-drying cycle. AAPS Pharm. Sci. Tech. 12: 854–861.

Kristan, K. and M. Horvat. 2012. Rapid exploration of curing process design space for production of controlled-release pellets. J. Pharm. Sci. 101: 3924–3935.

Kuu, W.Y., S.L. Nail and G. Sacha. 2009. Rapid determination of vial heat transfer parameters using tunable diode laser absorption spectroscopy (TDLAS) in response to step-changes in pressure set-point during freeze-drying. J. Pharm. Sci. 98: 1136–1154.

Kuu, W.Y., K.R. O'Bryan, L.M. Hardwick and T.W. Paul. 2011. Product mass transfer resistance directly determined during freeze-drying cycle runs using Tunable Diode Laser Absorption Spectroscopy (TDLAS) and pore diffusion model. Pharm. Dev. Technol. 16: 343–357.

Levchuk, J.W. 1991. Good manufacturing practices and clinical supplies. J. Parenter. Sci. Technol. 45: 152–155.

Lin, T.P. and C.C. Hsu. 2002. Determination of residual moisture in lyophilized protein pharmaceuticals using a rapid and non-invasive method: near infrared spectroscopy. PDA J. Pharm. Sci. Tech. 56: 196–205.

Mayeresse, Y., R. Veillon, P.H. Sibille and C. Nomine. 2007. Freeze-drying process monitoring using a cold plasma ionization device. PDA J. Pharm. Sci. Technol. 61: 160–174.

Milton, N., M.J. Pikal, M.L. Roy and S.L. Nail. 1997. Evaluation of manometric temperature measurement as a method of monitoring product temperature during lyophilization. PDA J. Pharm. Sci. Tech. 51: 7–16.

Nail, S.L. and J.A. Searles. 2008. Elements of Quality by Design in development and scale-up of freeze-dried parenterals. Biopharm. Int. 21: 44–52.

Oetjen, G.W. 1999. Freeze-Drying. Wiely-VHC, Weinheim.

Patel, S.M., C. Swetaprovo and M.J. Pikal. 2010a. Choked flow and importance of Mach I in freeze-drying process design. Chem. Eng. Sci. 65: 5716–5727.

Patel, S.M., T. Doen and M.J. Pikal. 2010b. Determination of the end point of primary drying in freeze-drying process control, AAPS Pharm. Sci. Tech. 11: 73–84.

Pikal, M.J., S. Shah, D. Senior and J.E. Lang. 1983. Physical chemistry of freeze-drying: measurement of sublimation rates for frozen aqueous solutions by a microbalance technique. J. Pharm. Sci. 72: 635–650.

Pikal, M.J. 1994. Freeze-drying of proteins: Process, formulation, and stability. ACS Symp. Ser. 567: 120–133.

Pisano, R., V. Rasetto, M. Petitti, A.A. Barresi and A. Vallan. 2008. Modelling and experimental investigation of radiation effect in a freeze-drying process. pp. 394–398. *In*: F. Scura, M. Liberti, G. Barbieri and E. Drioli (eds.). Proceedings of 5th Chemical Engineering Conference for Collaborative Research in Eastern Mediterranean Countries—EMCC5, Cetraro CS (Italy), May 24–29.

Pisano, R., D. Fissore, S.A. Velardi and A.A. Barresi. 2010. In-line optimization and control of an industrial freeze-drying process for pharmaceuticals. J. Pharm. Sci. 99: 4691–4709.

Pisano, R., D. Fissore, A.A. Barresi, P. Brayard, P. Chouvenc and B. Woinet. 2013a. Quality by Design: optimization of a freeze-drying cycle via design space in case of heterogeneous drying behavior and influence of the freezing protocol. Pharm. Dev. Tech. 18: 280–295.

Pisano, R., D. Fissore and A.A. Barresi. 2013b. In-line and off-line optimization of freeze-drying cycles for pharmaceutical products. Drying Technol. 31: 905–919.

Presser, I. 2003. Innovative online measurements procedures to optimize freeze-drying processes. PhD Thesis, University of Munich, Germany.

Rasetto, V., D.L. Marchisio, D. Fissore and A.A. Barresi. 2010. On the use of a dual-scale model to improve understanding of a pharmaceutical freeze-drying process. J. Pharm. Sci. 99: 4337–4350.

Romero-Torres, S., H. Wikstrom, E.R. Grant and L.S. Taylor. 2007. Monitoring of mannitol phase behavior during freeze-drying using non-invasive Raman spectroscopy. PDA J. Pharm. Sci. Technol. 61: 131–145.

Roth, C., G. Winter and G. Lee. 2001. Continuous measurement of drying rate of crystalline and amorphous systems during freeze-drying using an *in situ* microbalance technique. J. Pharm. Sci. 90: 1345–55.

Roy, M.L. and M.J. Pikal. 1989. Process control in freeze drying: determination of the end point of sublimation drying by an electronic moisture sensor. J. Parent. Sci. Technol. 43: 60–66.

Rovero, C., S. Ghio and A.A. Barresi. 2001. Development of a prototype capacitive balance for freeze-drying studies. Chem. Eng. Sci. 56: 3575–3584.

Schneid, S. and H. Gieseler. 2008. Evaluation of a new wireless temperature remote interrogation system (TEMPRIS) to measure product temperature during freeze-drying. AAPS Pharm. Sci. Tech. 9: 729–739.

Schneid, S., H. Gieseler, W.J. Kessler and M.J. Pikal. 2009. Non-invasive product temperature determination during primary drying using Tunable Diode Laser Absorption Spectroscopy. J. Pharm. Sci. 98: 3406–3418.

Searles, J. 2004. Observation and implications of sonic water vapour flow during freeze-drying. Am. Pharm. Rev. 7: 58–69.

Shanley, A. 2009. Pharmaceutical process control: Is the great divide growing? Pharma Manufacturing. Accessed August 14th 2013, at: http://www.pharmamanufacturing.com/articles/2009/125.html.

Shanley, A. 2010. Time and cost identified as key roadblocks to PAT and QbD; experts disagree. Pharma Manufacturing. Accessed August 14th, 2013, at: http://www.pharmamanufacturing.com/articles/2010/112.html.

Stokvold, A., K. Dyrstad and F.O. Libnau. 2002. Sensitive NIRS measurement of increased moisture in stored hygroscopic freeze dried product. J. Pharm. Biomed. Anal. 28: 867–873.

Sundaram, J., Y.H.M. Shay, C.C. Hsu and S.U. Sane. 2010. Design space development for lyophilization using DOE and process modeling. BioPharm International 23: 26–36.

Tang, X. and M.J. Pikal. 2004. Design of freeze-drying processes for pharmaceuticals: practical advice. Pharm. Res. 21: 191–200.

Tang, X.C., S.L. Nail and M.J. Pikal. 2005. Freeze-drying process design by manometric temperature measurement: design of a smart freeze-dryer. Pharmaceut. Res. 22: 685–700.

Tsourouflis, S., J.M. Flink and M. Karel. 1976. Loss of structure in freeze-dried carbohydrates solutions: Effect of temperature, moisture content and composition. J. Sci. Food. Agric. 27: 509–519.

Vallan, A., S. Corbellini and M. Parvis. 2005. Plug &Play architecture for low-power measurement systems. In: Proceedings of Instrumentation and Measurement Technology Conference—IMTC 2005, Ottawa, Canada, 565–569.

Velardi, S.A. and A.A. Barresi. 2008. Development of simplified models for the freeze-drying process and investigation of the optimal operating conditions. Chem. Eng. Res. Des. 87: 9–22.

Velardi, S.A., V. Rasetto and A.A. Barresi. 2008. Dynamic Parameters Estimation Method: advanced Manometric Temperature Measurement approach for freeze-drying monitoring of pharmaceutical. Ind. Eng. Chem. Res. 47: 8445–8457.

Velardi, S.A., H. Hammouri and A.A. Barresi. 2009. In line monitoring of the primary drying phase of the freeze-drying process in vial by means of a Kalman filter based observer. Chem. Eng. Res. Des. 87: 1409–1419.

Velardi, S.A., H. Hammouri and A.A. Barresi. 2010. Development of a high gain observer for in-line monitoring of sublimation in vial freeze-drying. Drying Technol. 28: 256–268.

Wang, D.Q., J.M. Hey and S.L. Nail. 2004. Effect of collapse on the stability of freeze-dried recombinant factor VIII and a-amylase. J. Pharm. Sci. 93: 1253–1263.

Wiggenhorn, M., G. Winter and I. Presser. 2005. The current state of PAT in freeze-drying. Am. Pharm. Rev. 8: 38–44.

Willemer, H. 1991. Measurement of temperature, ice evaporation rates and residual moisture content in freeze-drying. Dev. Biol. Stand. 74: 123–136.

CHAPTER 15

# Storage Stability of Probiotic Powder

*Chalat Santivarangkna*

## 1. Introduction

Many factors can potentially impact the *in vivo* beneficial properties of probiotics. In addition to host parameters such as microbiota, immunity, gastrointestinal transit and survival at the site of action, the viability and functional expression of probiotics are influenced by external factors, which include the preservation methods (e.g., freezing, freeze drying, and spray drying) and storage conditions. Probiotic cells can be maintained temporarily by chill storage usually for a day (Juárez Tomás et al. 2004) but may also last for a month at the most (Lee 2008). Frozen storage enables longer storage time of probiotics, and the lower the storage temperatures, the longer shelf life is. Only a small reduction in viability of cells occurs during storage at –70ºC to –80ºC for about one or two years (Foschino et al. 1996, Juárez Tomás et al. 2004), and commercial frozen concentrate of probiotic is generally recommended to be stored at a temperature lower than –45ºC in order to have a shelf life that lasts for at least 12 months (Bylund 2003). The refrigerated and frozen probiotics require a cold chain for storage and distribution, which is costly and inconvenient. Probiotic powder has advantages for industrial applications and long-distance trades. The dry powder is light, easy to handle, and has cheaper transportation costs. Moreover, in addition to using it as starter cultures for liquid and moist foods, probiotic powder can be added and mixed with dry or semi-dry foods such as cereal, chocolate, probiotic tablets and capsules. However, an important drawback of probiotic powder is the storage stability, which is often lower than frozen probiotics. Furthermore, for probiotics powder that is added as a functional ingredient in liquid and dry foods, and used in preparation forms, the stability is even more important because the number of viable cells depends largely on the initial cell concentration and storage stability since there is no fermentation or

Institute of Nutrition, Mahidol University, 999 Phutthamonthon 4 Rd., Salaya, Phutthamonthon, NakhonPathom 73170, Thailand.

ripening step to increase the number of viable cells. To ensure the high number of cells in these probiotic products, overdosing of probiotic powder is then often employed.

Taking the above into account, storage stability remains the topical issue for studying and discussion. Therefore, the content of this chapter will primarily cover issues related to the storage stability of dry probiotic, especially lactic acid bacterial probiotics and *bifidobacteria*. The different implications of stability, factors affecting stability of probiotic powder during storage will be described. In addition to the stability of probiotic powder, the stability of probiotics in food and product matrices will be discussed as well.

## 2. Stability of probiotic powder beyond Viability

Probiotics in a dry form can be used for different purposes. Major applications of probiotic powder are, for example, starter cultures or adjunct cultures for the fermented foods; functional food ingredient (non-starter) for mixing with dry foods such as infant formulae, and cereals or for liquid foods such as soy milk, and fruit juices; and oral administrative tablet or capsule (Rivera-Espinoza and Gallardo-Navarro 2010, Burgain et al. 2011). Accordingly, the stability can be evaluated by one or all of the following approaches depending on the applications.

i) Viability: the determination of cell viability is based on the ability to reproduce and form visible colonies on optimal solid growth media. Viability as determined by plate count (i.e., cfu $g^{-1}$ or $ml^{-1}$ is the criterion commonly stated in regulations or claims related to probiotics). The plate count method cannot provide a real-time result of viability because it requires long incubation time (e.g., 24–72 h). Another important drawback of the plate count is that it fails to include injured cells that are still viable but nonculturable (VNBC). These cells may still retain metabolic activity or have health beneficial effects. Importantly, the probiotic powder is likely to contain injured cells from lethal or sub-lethal stresses during the production. The stressful processes during the powder production are, for example, cell concentration (shear and oxygen stresses), freezing (cryo-injury and osmotic stress), spray or fluidized bed drying (osmotic, dehydration, thermal-and/or oxygen stresses) (Santivarangkna et al. 2008b). Therefore, the establishment of alternative viability assays that enable fast and reproducible characterization of the heterogeneity of total cell population with respect to different physiological and metabolic states is of high relevance.These assays are, e.g., flow cytometry (FC), fluorescent activated cell sorting (FACS), real time-quantitative polymerase chain reaction (RT-qPCT or qPCR), and direct imaging and visual enumeration. Major physiological states and the aforementioned culture-independent methods for probiotics and their limitations were well described by Davis (2014).

   Viability of cells can be significantly influenced by the reconstitution of the probiotic powder before plating (Muller et al. 2010, Champagne et al. 2011). Therefore, some points must be taken into account to get consistent and reliable results. Low temperature of rehydration media causes membrane leakage, and thus probiotic powder should be rehydrated in warm rehydration media (Mille et al. 2004, Wang et al. 2004). The viability can be underestimated due to the chains or clumps of cells. Hence, cell suspension should be incubated for 30 min before plating (Muller et al. 2010). Sonication of the sample (Ding and Shah 2009) and homogenization with handheld homogenizer (Lamboley et al. 2003), and stomacher (Chan and Zhang 2005) can efficiently break down chains or clumps of probiotics. Homogenization or sonication is also needed for encapsulated probiotic powder and probiotic tablet (Champagne et al. 2011). Some probiotic bacteria are sensitive to oxygen (Talwalkar and Kailasapathy 2004, Simpson et al. 2005), but little information is available on the effects of oxygen during rehydration and dilution of the samples prior to plating. Therefore, time for these steps should be as short as possible; the delay before and after plating should be avoided; and anaerobic chamber or cabinet may be necessary for strains that are highly sensitive to oxygen. The determination of viability is more complicated when a probiotic product contains multiple and closely related species with similarity in growth requirements and overlapping biochemical characteristics. Differential enumeration with multiple selective media and growth conditions is necessary. For example, selective enumeration of *L. acidophilus* in mixed culture could be made in Rogosa agar added with 5-bromo-4-chloro-3-indolyl-β-D-glucopyranoside (X-Glu) or MRS containing maltose (MRSM) and incubation in a 20% $CO_2$ atmosphere. The suitable selective media was recommended and discussed by Ashraf and Shah (2011) and Davis (2014).

ii) Fermentation activity: the determination is based on the rate of acid production or of pH decrease due to the carbohydrate fermenting activity of probiotics. The fermentation activity is very important for probiotic that are used as starter cultures. Fermentation contributes to organoleptic properties of fermented foods. Lactic acid bacterial probiotics produce organic acids, especially lactic acid as the key metabolite. Because acid can be measured easily, the fermentation activity is often assessed by the time necessary to reach the maximum acidification rate in milk or by the time required by a probiotic to reduce pH of milk or medium to a set pH value. An example is the Cinac system, which is a reference analyzer to control fermentation activity of starter cultures. In this system, milk is inoculated by a constant volume of cell suspension. For each sample, the time necessary to reach the maximum acidification rate in milk ($t_m$), is used to characterize the acidification activity. The higher the $t_m$, the longer the lag phase is, and thus the lower the acidification activity (Fonseca et al. 2006). Even though the cell viability is closely related to the fermentation activity, it cannot provide the number of VNBC cells that are metabolically active and still contribute to the fermentation process during the fermented food and beverage production (Bunthof and Abee 2002, Bensch et al. 2014). For example, in the malic fermentation, the specific malate uptake rate of the VBNC subpopulation was approximately 50% of that found for the viable population (Quirós et al. 2009).

iii) Physiological functionality: the determination is based on the related health promoting properties. The capability of probiotics to remain functionally active during long-term storage is one of the most important requirements. Preservation processes and storage of probiotic may negatively affect the functionalities of a probiotic without changing its viability accordingly (Saarela et al. 2005).

Although studies reported unchanged physiological functions of probiotic powder after storage (Silva et al. 2002, Montel Mendoza et al. 2014), some studies also reported the reduced functions of probiotics (Zarate and Nader-Macias 2006, Juarez Tomas et al. 2009, Borges et al. 2013). On the one hand, the reduction in the functionality can occur simply due to the decline of the viability during storage. On the other hand, the change in the functionality can be due to the probiotic powder production, storage, and food matrix without this fact being evidenced by plate counts (Vinderola et al. 2012). For these reasons, it was proposed by Sanders et al. (2014) that the performance of probiotics should be tested when imposing substantial changes in probiotic powder preparation, including preservation and storage of cells. The key measures of equivalency in the performance of probiotics are equivalent or improved viability of the probiotic in the final product, and equivalent results from phenotypic performance mapping. A performance map comprises an array of functional and physiological tests of biological activity but does not entail full clinical endpoint evaluations. The proposed performance map of functionality are *in vitro* or animal model assays such as lactase activity, bacteriocin production, short chain fatty acid production, $H_2O_2$ production, anti-inflammatory profile, growth curves, stomach acid resistance, bile resistance, *in vitro* immune profiling, and fecal recovery (for probiotics targeting intestinal function) (Sanders et al. 2014).

As mentioned, probiotics are used in different applications and stability should be considered in connection to their uses, with viability and physiological functionality as the key measures. For example, for stability in term of fermentation activity, viability, and physiological function is of vital importance for fermented foods, whereas the stability in term of viability and physiological functionality is of vital importance for probiotic tablet and capsule. Nevertheless, it should be noted that even though the use of the term "probiotic" is currently limited to viable cells at the time of consumption, it is becoming increasingly evident that even non-viable cells can confer health benefits (Makinen et al. 2012). New terms like "postbiotic" and "paraprobiotic/ghost probiotic" have been introduced and have become the topics of discussion. Postbiotic is considered being any factor from the metabolic activity of a probiotic (e.g., short chain fatty acids and polysaccharides) or any released molecule (e.g., lactoceptin and the p40molecules) that are capable of conferring beneficial effects. Paraprobiotic or ghost probiotic is considered to be the inactivated microbial cells or cell fractions (e.g., cell wall components and capsular polysaccharides) (Foligné et al. 2013). Advantages of products based on inactivated cells or metabolites over those based on viable cells are, for example easier to standardize, more processing flexibility, and higher storage stability.

## 3. Storage of Probiotic Powder

The storage stability of probiotic powder depends on several factors along with the dried cell preparation (Fig. 1). The stability of probiotics therefore must be considered holistically. For example, heat stress induction during the fermentation decreases cell concentration but increase viability after drying (Desmond et al. 2002). Spray drying

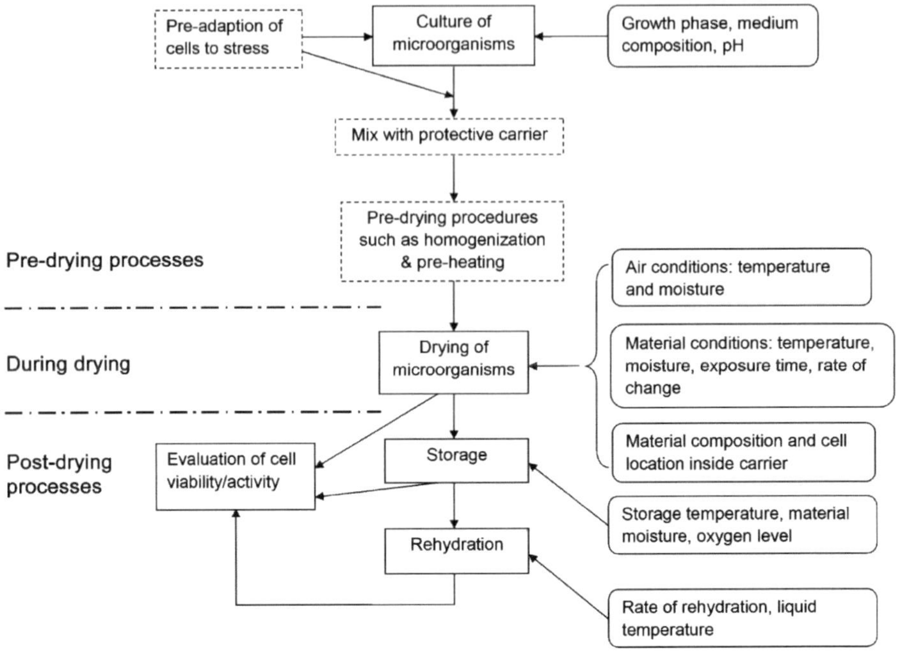


**Fig. 1.** Major processing steps during the dehydration of microorganisms and important extrinsic factors in each step. Dotted squares represent optional steps and extrinsic factors are shown in round-cornered squares Reproduced from (Fu and Chen 2011) with permission from Elsevier.

with a low outlet temperature results in a product with high moisture content and high viability after drying but low stability of cells during the storage (Ananta et al. 2005, Golowczyc et al. 2010).

## 3.1 Storage Conditions

Although studies have tried to correlate cell physiology and the storage stability of probiotic (Miyamoto-Shinohara et al. 2008), several experimental evidences have shown that the stability is specie or strain dependent (Corcoran et al. 2004, Chaluvadi et al. 2012, Montel Mendoza et al. 2014). Storage stability of dried cells is inversely related to the moisture content and storage temperature (Santivarangkna et al. 2011a, Foerst et al. 2012) and can be affected by the cell concentration (Higl et al. 2007, Foerst et al. 2012), the presence of protectants (Pehkonen et al. 2008, Strasser et al. 2009) and oxygen (To and Etzel 1997b, Coulibaly et al. 2009), as well as the physical state of the powder (Kurtmann et al. 2009b, Santivarangkna et al. 2011b). While freeze-dried probiotic powder can retain viability at 4°C for months (Wang et al. 2004, Coulibaly et al. 2009), a temperature below –18°C is required for the shelf life of at least a year for probiotics used for the preparation of bulk starter and freeze-dried DVS cultures (Bylund 2003) in order to ensure prompt fermentation activities. The stability of the dried cultures decreases greatly (e.g., 10 times higher) at close to room-temperature

(Champagne et al. 1996, Foerst et al. 2012). The stability at ambient lowers the storage and transportation costs for probiotic manufactures food and feed industries, and pharmaceuticals. Moreover, it provides convenience for consumers.

Although the stability increases with decreasing moisture content, it seems to be relevant only to a certain moisture content or water activity ($a_w$). The maximum stability of *L. salivarius* ssp. *salivarius* was at the moisture content of 2.8–5.6% (Zayed and Roos 2004). In studies on *L. bulgaricus* (Teixeira et al. 1995b, Passot et al. 2012) and *Bifidobacterium animalis* ssp. *lactis* Bb-12 (Celik and O'Sullivan 2013), the optimal $a_w$ for the storage of the dried cells was at ca. 0.1–0.2. An increase in storage $a_w$ from 0.11 to 0.22 had a relatively less negative impact on the viability than did an increase in $a_w$ from 0.22 to 0.32 (Kurtmann et al. 2009a). For probiotics, it is generally recommended that $a_w$ should be below 0.25 for long term storage (Chávez and Ledeboer 2007, Manojlović et al. 2010). However, at very low $a_w$ (<0.1), a slight increase in the degradation rate probably occur by oxidative mechanisms and slow Maillard reaction (Kurtmann et al. 2009c, Passot et al. 2012). Taken together, the storage of dried probiotics at water activity of ca. 0.1 should be practically low enough for long term storage. Overview of storage stability based on literature data collected by Makinen et al. (2012) is shown in Fig. 2. From the figure, *Bifidobacteria*, which are more sensitive to oxygen than *Lactobacilli*, are unexpectedly the most stable genus during storage. Although optimal $a_w$ varies among probiotics, common

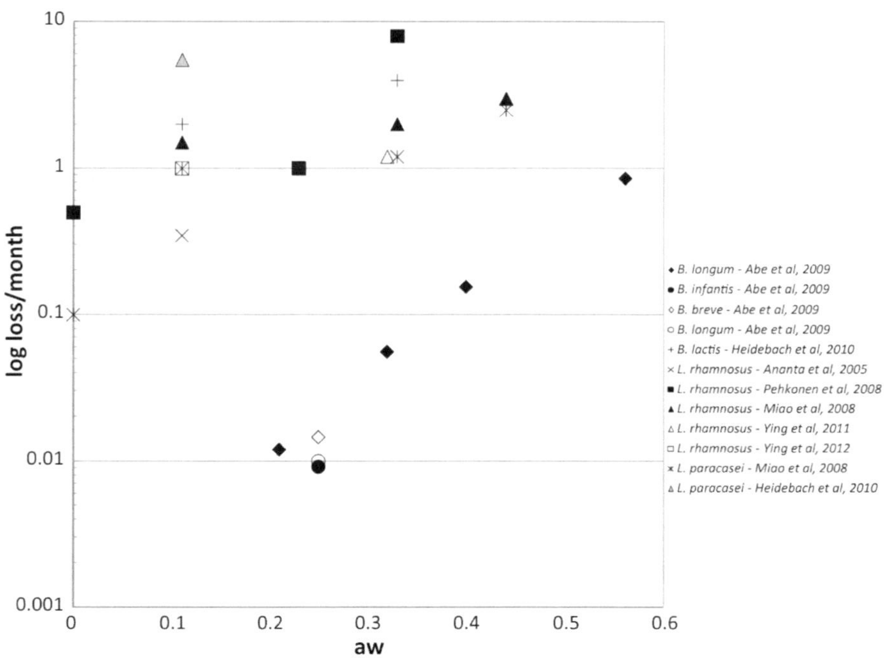


**Fig. 2.** Overview of storage inactivation rate of probiotic bacteria at 25°C based on literature data. In papers where various protective formulations were used, only the most stable preparation was included in this graph Reproduced from (Makinen et al. 2012) with permission from Elsevier.

$a_w$ ranges that provide stable storage can be noticed from the figure. The data were collected from 12 studies. It is interesting to use larger numbers of data from studies with different probiotics and conduct meta-analyses to improve the power to detect the direction, consistency of an effect, and common characteristics of the strains that have high stability.

The impact of moisture on the stability of probiotic powder is temperature dependent. At higher storage temperatures, the moisture has a higher impact on stability (Higl et al. 2007, Aschenbrenner et al. 2012). Packages for the storage of probiotic powder should be moisture and air impermissible. The removal of air by vacuum packing or the replacement of air with nitrogen can increase the storage stability (Celik and O'Sullivan 2013). A point to be considered is that the moisture content (as well as package's atmosphere) of probiotic powder is changed once the package is opened. The change in moisture content can dramatically affect the storage stability. An increase in $a_w$ of a dairy-based product from 0.1 to 0.3 resulted in only a 2% increase in moisture content but 10 times decrease in the storage stability (Ishibashi et al. 1985). The change in moisture content and $a_w$ depends on factors such as relative humidity of the environment, sorption isotherm of the powder, seal tightness of the opened package, and the storage temperature. Equilibrium of ambient conditions may take a few hours (Kurtmann et al. 2009b) or couple months (Fonseca et al. 2001). This infers that when a package containing probiotic powder is opened, the sachet or can must be closed tightly and without unnecessary delay and should be re-stored at a low temperature. Alternatively, dry probiotic products may be packed in a single use format to protect the moisture absorption until the point of consumption or use. Another strategy is the protection by micro-encapsulation of probiotics, which has been described in chapter 16.

The storage conditions are highly influenced by protectants, which can be divided roughly into two groups, i.e., complex and undefined. A good example of complex protectants is skim milk, which is one of the most investigated protectants. Skim milk can be used alone (Simpson et al. 2005) or in combination with other complex protectants such as polysaccharides (Yang et al. 2006) or with defined protectants such as mono-, di, and alcohol sugars. Sugars reported to have protective effects during storage are, for example, sucrose for *L. delbrueckii* ssp. *Bulgaricus* (Carvalho et al. 2003); sorbitol, inositol, trehalose, and fructose for *L. plantarum*, *L. rhamnosus* (Carvalho et al. 2002); and glucose, fructose, lactose, mannose, and sorbitol for *L. delbrueckii* ssp. *Bulgaricus* (Carvalho et al. 2004a). The popularity and demand for nondairy-based probiotic foods and products have increased, and therefore nondairy-based protectants are important for consumers who suffer from lactose intolerance or from milk-protein allergy (Saarela et al. 2005).

In addition to sugars, prebiotics can be used also as nondairy-based protectants. Although researches on probiotic–prebiotic combinations (synbiotic) have largely been on the health promoting aspects, their technological performance as protectants was investigated as well. The addition of fructo-oligosaccharides, inulin and pectic-oligosaccharides (10 mg $ml^{-1}$) in an alginate matrix protected and retained viability (at least $10^7$cfu $ml^{-1}$) of probiotic bacteria *B. breve* ATCC 15698, *B. longum* ATCC 15697, *L. acidophilus* Luchansky 1426, and *L. reuteri* Luchansky 1428 during storage at 4°C for 3 weeks (Chaluvadi et al. 2012). Exponential-phase cells of *L. reuteri*

TMW1.106 retained high viability during storage for 14 days at room temperature when the cells were freeze-dried with inulin (Schwab et al. 2007). Protective effects of several prebiotics (oat flour with 10 and 20% β-glucan, apple fibre, wheat dextrin, polydextrose, and inulin) were tested for their capability to protect *L. rhamnosus* cells during freeze-drying, storage in powdery form, and after formulation into apple juice and chocolate-coated oat breakfast cereals (Saarela et al. 2006). The stability of freeze-dried *L. rhamnosus* cells at 20°C was higher in chocolate-coated breakfast cereals compared with low pH apple juice. While Oat flour with 20% β-glucan was able to protect fresh cells during storage at low pH juice, wheat dextrin and polydextrose were better carriers in chocolate-coated breakfast cereals. The study suggested that it is possible to develop probiotic-fibre pairs to suitably maintain the stability of the probiotic in the food products (Saarela et al. 2006). To reduce damage from lipid oxidation, which occurs with dried cells during storage (Yao et al. 2008, Coulibaly et al. 2009), antioxidants are often added together with protectants. Although studies showed the protective effects of a popular antioxidant sodium ascorbate (Kurtmann et al. 2009a, Jalali et al. 2012), the incorporation of sodium ascorbate was reported to decrease the viability of microencapsulated *L. rhamnosus* GG (LGG) powder during storage. The negative effects are supposedly because of the difference in abilities of various probiotics to catabolize ascorbate, and some products resulted from the catabolism may be detrimental to cell viability (Ying et al. 2010).

Different protectants variably affect not only the viability of probiotics, but also their physiological functionalities. Cells of *B. animalis* subsp. *lactis* INL1 freeze-dried with different cryoprotectants (PBS, lactose, sucrose or skim milk) were stored at 25°C for 4 weeks. No changes in cell counts were observed for any of the cryoprotectants used. However, only lactose and skim milk were effective for minimizing cell death during simulated gastric digestion (SGD) (Vinderola et al. 2012). Three vaginal lactobacilli (*L. acidophilus* CRL 1259, *L. paracasei* subsp. *paracasei* CRL 1289, and *L. salivarius* CRL 1328) were freeze-dried with lactose, skim milk, ascorbic acid, and the combinations of them, packed into gelatin capsules, and stored at 5°C under darkness up to 12 months. *Lactobacillus* dried with ascorbic acid or combined excipients conserved high viability, whereas the probiotics dried with lactose or skim milk significantly decreased their viability. Abilities to produce lactic acid, $H_2O_2$ and bacteriocin were affected to different extents, while the capacity of lactobacilli to adhere to vaginal epithelial cells was reduced (Zarate and Nader-Macias 2006). In contrast to the negative impact of drying and storage on the probiotic functionalities shown previously, *L. lactis* CRL 1584, *L. lactis* CRL 1827, *L. garvieae* CRL 1828, and *L. plantarum* CRL 1606 were freeze-dried with lactose, sucrose, skim milk, whey protein concentrate, and the combinations of them. Each dried probiotics were filled in glycogelatin capsules, packed in plastic bottles with silica gel to maintain the dry state of the capsules, and stored at 4°C for 18 months. The degree of bacterial cell surface hydrophobicity and autoaggregation of the strains and their inhibitory activity against pathogenic bacteria of the all dried probiotics could be maintained during the storage period and were not different from the cells before freeze-drying (Montel Mendoza et al. 2014).

## 3.2 Other Factors Related to the Storage Stability

### 3.2.1 Intrinsic Factors

Robustness to drying and stability of probiotic powder varies among probiotic strains, and therefore is a critical factor in preservation and storage of probiotic products because it seems there is no single preservation strategy that can apply for all probiotics. One possibility to obtain robust probiotics is the isolation of survivors from numerous cycles of stresses related to the preservation used, e.g., the application of repeated freezing and thawing cycles to obtain cryo-tolerance strains for frozen storage (Monnet et al. 2003). It is recommended that in addition to the aforementioned selection procedures, probiotic should be screened and selected based on specific purposes or expected probiotic functions using insight into intestinal microbiota, nutrition, immunity and mechanisms of action and genomic data (Gueimonde and Salminen 2006). Similarly, the technological function such as stability to preservation and storage should be included at the beginning as a selective pressure for potential probiotics. For example, Simpson et al. (2005) identified a group of closely related *Bifidobacterium* species with heat and oxygen tolerance assay into a distinctive tolerance to heat and oxygen. These species had high initial survival following spray drying, which is the drying process involving high temperature and large volume of air. The cells also retained high viability during storage at refrigerated temperatures. In addition, viability at ambient temperatures was greatly improved when powders were produced using a fluidized-bed spray dryer.

### 3.2.2 Stress Induction

Mild or moderate stress may protect probiotics to subsequent lethal stresses during preservation and storage. The stresses can be induced during fermentation by applying an additional pre-incubation step before the preservation process. However, a drawback of the stress induction is the possible reduced cell growth rate or cell concentration. Moreover, too severe induction may negatively affect cell viability after the preservation (Schoug et al. 2008, Mozzetti et al. 2013). Freezing stability of *L. johnsonii* was improved by 19% when heat shocking was done at 55°C for 45 min, but the number of cells were reduced by 1.5-log unit for the heating (Walker et al. 1999). Examples of the stresses are high osmolarity (Desmond et al. 2002), low (Broadbent and Lin 1999) or high temperatures (Desmond et al. 2004), and low pH (Streit et al. 2008). Effects of stress induction depend on the nature and degree of the stresses applied. For instance, high survival rates during freeze-drying (but not during storage) were obtained when 0.1 mol $l^{-1}$KCl was added at the beginning of fermentation, without any change in membrane properties and betaine accumulation. This condition maintained a high fermentation activity throughout the process. In contrast, the addition of 0.6 mol $l^{-1}$KCl at the beginning of fermentation led to a high survival rate during storage that was related to high intracellular betaine levels, low membrane fluidity and high cycC19:0 concentrations. However, these modifications reduced the fermentation activity during storage. When a moderate stress was applied by combining 0.1 mol $l^{-1}$KCl and 0.6 mol $l^{-1}$KCl at the beginning and at the end of fermentation, the betaine accumulated in the cells without any membrane alteration. This combination could maintain both high fermentation activity and survival rate during storage (Louesdon et al. 2014). In addition to the increased stability during storage of probiotic powder, stress induction also seems to improve stability of probiotics in the products. Fermenter-grown *L. rhamnosus* cells were freeze-dried and added into apple juice (pH 3.4, 3.6 and 3.8). The best survivors during the 6-week storage at 4°C were cells grown at pH 5.0, especially during storage in juice with pH 3.4 (Saarela et al. 2009).

For the screening of the optimal sub-lethal stress conditions, a two-stage continuous culture tool was investigated by Mozzetti et al. (2013). The first reactor was operated under fixed conditions to produce cells with controlled physiology, and the second reactor was used to test the stress pre-treatment combinations. In this setup, the effects of stress pretreatments with up to four different conditions can be examined at simultaneously per day.

### *3.2.3 Growth Phase and Cell Cultivation*

Entry into the stationary phase increases tolerance of cells to lethal stresses because of the production of general stress proteins (van de Guchte et al. 2002). For example, comparative viability after spray drying of probiotic lactobacilli cells in the stationary, early exponential, and lag phase in a study by Corcoran et al. (2004) was over 50%, 14%, and ca. 2%, respectively. It is generally considered that young cells are more sensitive than late logarithmic or stationary phase cells (Saarela et al. 2005), and the effects of stress induction seem to be less pronounced when being applied to stationary phase cells. In a study by Teixeira et al. (1995a), heat shock increased the survival of exponential cells but did not result in normal levels found with un-shocked stationary phase cells. Furthermore, heat shock had no effect on stationary phase cells. Many studies used either cells that had just entered stationary phase (i.e., late exponential phase) (Champagne et al. 1996, Beal et al. 2001) or cells that were already in the stationary phase cells (Corcoran et al. 2004, Lee 2004, Zayed and Roos 2004), but only few studies examined the difference between these two phases, which have only small age gap. A study compared storage stability of the late exponential (15 h) and stationary phase (22 h) of freeze-dried *B. animalis* ssp. *lactis* cells and no difference was observed. Sucrose-formulated *B. animalis* ssp. *lactis* of both 15 and 22 h grown cells had an excellent stability during storage at refrigerated and frozen temperatures for 5–6 months and good stability during storage at 37°C for 2 months. It should be noted also that difference in the cultivation time can influence the probiotic's functionalities such as immunostimulatory activity (Maassen et al. 2003, Sashihara et al. 2007) and resistance to gastric digestion (Vinderola et al. 2012). However, little information is available on the interplay between storage and cultivation time on the stability of probiotics in term of these functionalities.

### *3.2.4 Drying Processes*

Different drying technologies pose different lethal or sub-lethal stresses to probiotic cells. It may also influence physical properties, morphology, microstructure, and the sorption characteristics of probiotic powder. Freeze drying is the mildest drying process in the preparation of probiotic powder, but it is energy intensive and still a batch wise process. Alternative drying processes have been investigated to reduce costs and increase efficiency in the production of probiotic powder. These alternative processes were reviewed and discussed by Santivarangkna et al. (2007) and Meng et al. (2008). Examples of the processes are spray drying, fluidized bed drying, and vacuum drying. Among them, spray drying is highly attractive because it is a continuous process that is well-established in food industries. However, spray drying is not yet commercially applied for the production of probiotic powder due to the low viability after drying (To and Etzel 1997a), low storage stability (Teixeira et al. 1996, Ananta et al. 2005), and poor wetting dispersion properties (Peighambardoust et al. 2011). For example, the survival and fermentation activity of spray-dried *L. lactis* ssp. *cremoris* decreased significantly with storage time in the powders irrespective of type of atomizers or scale of dryers used (Ghandi et al. 2013). The application of spray drying for the dry preservation of lactic acid bacteria was thoroughly described by Peighambardoust et al. (2011). Another attractive and promising alternative drying process is vacuum drying. Vacuum-dried probiotic powder was reported to have better storage stability than the freeze dried one, especially at ambient environment. *L. paracasei* ssp. *paracasei* vacuum-dried with sorbitol was highly stable at storage temperature of 20ºC even at high humidity. The high stability could even be achieved at high temperature of 37ºC, with low $a_w$ of the powder (0.07) (Foerst et al. 2012). The storage stability of the freeze dried powder may be lower than powder dried with other technologies because of the possible higher oxidation rate during storage from the high surface area and porosity of the freeze-dried powder. In the freeze drying process, increased content and uniform ice formation can improve the efficacy of the subsequent drying process. During the annealing step of freeze drying, freezable water entrapped in the amorphous regions (due to rapid freezing) is crystallized into larger and more uniform ice crystals. The rapid freezing occurs during freezing of probiotics as well because frozen probiotics are commonly produced by using liquid nitrogen. With liquid nitrogen, a pellet of 2 mm diameter is cooled from 0 to –50ºC in approximately 10 s or at a freezing rate of approximately 300ºC/min (Oetjen and Haseley 2003). The positive impact of annealing on activity and stability of pharmaceutical products is well characterized. However, information from studies in probiotics is very limited (Santivarangkna et al. 2011b), although the improved storage stability and reduced browning reaction of freeze-dried *L. acidophilus* was reported by Ekdawi-Sever et al. (2003).

### *3.2.5 Encapsulation*

Encapsulation is a process of coating or entrapping of an active component or mixture of active components within the matrix (encapsulant). Encapsulation of probiotics consists of the mixing of a cell suspension with an appropriate encapsulant to form the capsule or gel globule. Common encapsulants for the encapsulation of probiotics are alginate, chitosan, carrageenan, gums (locust bean, gellan gum, xanthan gum, etc.), gelatin, whey protein, and starch (Riaz and Masud 2013). Studies on encapsulation have focused largely on its protective roles on the viability and stability of probiotics in food products and during the transit of probiotics through the gastrointestinal tract. These roles were comprehensively reviewed by Burgain et al. (2011) and Riaz and Masud (2013) and are described in Chapter 16 of this book. Nevertheless, some studies have also reported on the protective effects of the encapsulation during storage of probiotics. Both uncoated and microencapsulated *L. acidophilus* ATCC 43121 cells (single-microencapsulated with Sureteric™, double-microencapsulated using Sureteric™ with: fructooligosaccharide, lactulose or raffinose) were stored at 4, 25 and 37°C. The survival rates of the uncoated cells declined by as much as four orders of magnitude after 16 days at all storage temperatures. For the microencapsulated cells, after 20 days, all encapsulated samples stored at 4°C had the highest survival rates. At 25 and 37°C, *L. acidophilus* double-microencapsulated with lactulose and raffinose exhibited higher survival rates than that which was single-microencapsulated with Sureteric™ and that was double-microencapsulated with fructooligosaccharide (Ann et al. 2007).

### *3.2.6 Genetic Modification*

Stress tolerances and storage stability of probiotics can be specifically enhanced by genetic manipulation. The trehalose biosynthesis genes *ots* BA from *E. coli* was introduced in *L. lactis* NZ9000 (Termont et al. 2006). Following nisin induction, the intracellular accumulation of trehalose reached maxima at approx. 50 mg $g^{-1}$ wet cell weight after 2 to 3 h. The viability of induced NZ9000 was markedly higher than that of the non-induced NZ9000 after freeze drying with skim milk (94.0 vs. 56.7%) and without skim milk (ca. 100 vs. 20%). The induced cells had also higher storage stability and retained almost 100% viability for at least 1 month at 8°C and 10% relative humidity. *S. thermophilus* S4 expressing a small heat-shocked protein encoded by the plasmid pSt04 is able to carry out fermentation at elevated temperature, i.e., at 50°C. In yoghurt culture, together with *L. delbrueckii* subsp. *bulgaricus*, fermentation at elevated temperature results in a mild yoghurt with low post-acidification and improved stability of the starter bacteria during storage at 4°C (El Demerdash et al. 2006).

## 4. Storage Stability of Probiotic in Drying- and Product Matrices

### *4.1 Storage Stability of Probiotics in Drying Matrix*

The addition of protectants significantly reduces the vitality losses during drying and storage (Carvalho et al. 2004b, Santivarangkna et al. 2008a). Therefore, protectants are commonly added and mixed with probiotic cell concentrate before the drying. During drying processes, upon the removal of water, the protectant may experience a transition from the liquid to the glassy state and consequently probiotic cells are embedded in a glassy drying matrix. The glassy state is an amorphous metastable state that resembles a solid, but without any long-range lattice order. The most common parameter describing the glassy state and its transition is the glass transition temperature ($T_g$) below which materials exhibit extremely high viscosity (typically $\geq 10^{12}$ Pa s) and shows temperature-dependent transition (Slade and Levine 1991). The ability of protectants to form glasses is thought to be an important factor for the stabilization of cells during storage. It is believed that a glass matrix can improve storage stability by raising the $T_g$ of biomaterial powder (Crowe et al. 1998, Sun et al. 1998). Diffusion limited deterioration reactions are inhibited at storage temperature lower than $T_g$, and cells therefore survive better.

The $T_g$ may not be constant over long storage time. Depending on package and environment, dry powder of low $a_w$ can absorb water from the surrounding until an equilibrium state is reached. The uptake of water lowers the $T_g$ of the powder and the protective matrix can turn from the glassy into the rubbery state. Therefore, information on the sorption isotherm of dried probiotics is necessary to avoid inactivation due to moisture gain during storage. However, in comparison to foods, very little information on sorption isotherm of probiotics is available (Santivarangkna et al. 2011b). Only a minor difference was found between the water uptake of corresponding matrices with and without probiotic cells at 33% RH but differences were observed at higher RH (Hoobin et al. 2013). From the result, $T_g$ of probiotic powder seems to be governed by the matrix, but the prediction of storage stability or the determination of storage conditions based on only $T_g$ of the drying matrix is not fully relevant. The $T_g$ of the matrix is not an absolute threshold for cell inactivation. Specific interactions between the protectant and the bacterial cell as well between the protectant and its plasticizer water may play a role (Aschenbrenner et al. 2012). Recent studies have shown that chemical reactions are not completely halted in the glassy state during storage of probiotic powder. The inactivation still occurs during storage at a temperature that equals $T_g$ ($T–T_g = 0$) (Higl et al. 2007, Kurtmann et al. 2009b, Hoobin et al. 2013). Results suggest that dry probiotic powder should be stored at a temperature far below $T_g$ to achieve complete stability. In studies with *L. coryniformis* (Schoug et al. 2010) and with *L. rhamnosus* (Miao et al. 2008), viability of cells clearly improved at $T–T_g$ higher than –50ºC and ca. –30ºC, respectively. Glassy state does not essentially protect dry cells from lipid oxidation, which is not a diffusion limited reaction (Kurtmann et al. 2009a). Sucrose, which is a non-reducing sugar, inherited better stability than the reducing sugar lactose. This is because lactose is prone to cause browning reaction, which is related to viability loss in freeze-dried cells (Kurtmann et al. 2009c). Although it was suggested that carbohydrates with high glass transition temperatures probably provide stabilization via fixation of the cells in a glassy powder (Perdana et al. 2014), crystallization of solid carriers during drying also occurred and it can be either beneficial (e.g., with mannitol or sorbitol) or detrimental (e.g., with lactose) to cell survival, which most probably depends on the crystal shape. Moreover, a bacterial cell is not a unique entity but contains an outside compartment, which is embedded in drying matrix and the inside compartment, which contains cell components with different affinities to water. Therefore, not all components or functions of cells are equally sensitive to moisture or temperature (Higl et al. 2007).

As shown from studies above, the inactivation during storage of probiotic powder bacteria shows a correlation but does not solely depend on the $T_g$ of the surrounding matrices. The role of molecular mobility on the stability of the powder is system and temperature dependent. Although the relevance of a glassy matrix strongly depends on the underlying absolute storage temperature, it is not clear whether the relevance is due to factual differences in the respective molecular mobility or to a distinct temperature effect (Aschenbrenner et al. 2014). It was proposed that at low storage temperatures, a potential rate-limiting effect is not apparent due to the lack of detrimental chemical and physical reactions. However, with increasing storage temperature, the restricted mobility is shown to become a rate limiting bottleneck. Thus, the harsher the environmental condition, the more relevant is the protective effect of a surrounding glassy matrix (Aschenbrenner et al. 2014).

### 4.2 Storage Stability of Probiotics in Product Matrix

The matrix and type of products have a large impact on the viability of probiotics (Klu et al. 2012, 2014, Endo et al. 2014). In general, foods have $a_w$ levels ranging from around 0.1 for very dry foods such as potato chips and crunchy snacks; 0.2 for dry foods such as milk powder, cookies, and peanut butter; 0.5 for semi-dry foods such as chocolate bar, chewing gum, and dry fruits; to 0.9 for moist foods such as milk, juices, and yoghurt (Schmidt and Fontana 2008). As mentioned in the previous section, dry probiotics should be stored at low temperatures and water activities (e.g., 0.1–0.3). Since dry probiotic bacteria are normally added and mixed with dry foods at very low doses, typically 0.01–1% of product weight, depending on the concentration of viable cells in the probiotic powder. Hence, at equilibrium, the $a_w$ of products dominate over the $a_w$ of the probiotic powder (Abe et al. 2009, Makinen et al. 2012). As a result, the stability during storage of probiotics decreases proportionally with the increasing $a_w$ and temperature of the probiotic products (Abe et al. 2009, Klu et al. 2012). Although storage at elevated temperatures should be avoided, refrigerated storage is not practically relevant for many foods such as cereal, milk powder, infant formula, and chewing gum. However, the very dry or dry foods have typical water activities in the stable range ($\leq$0.2) for dried probiotic and with the viability loss according to Fig. 2, some probiotics such as *B. longum*, *B. brevis*, *B. breve* have the log loss between ca. 0.01–0.1 per month at $a_w$ 0.2–0.3. This means that dry probiotic products with initial viable cells of, e.g., $10^9$ cfu $g^{-1}$ will have the shelf-life of at least two years at room temperature, provided final viable cells of $10^6$–$10^7$ cfu $g^{-1}$ in the products is satisfactory. Alternatively, overdosing of probiotic powder in the products is unavoidable for the strains that are sensitive to non-refrigerated storage temperatures.

Food components impact stability of probiotics. The long-term survival of *L. rhamnosus* GG in canola oil, extra-virgin olive oil, olive pomace oil, flaxseed oil was monitored at room temperature. Freeze-dried cells in oils kept viable conditions after 4 months, and loss of the viability was only 0.3 to 0.6 log cfu $ml^{-1}$ (Endo et al. 2014). In a study by Nebesny et al. (2007), *L. casei* and *L. paracasei* cells were incorporated in chocolate, which is a product with high fat content, and high viability was retained at elevated temperatures. The survival during the storage for 12 months at 4, 18, and 30ºC was 89–94, 80–87, 60–67% of the initial viable cells. The protective effects of oil can be useful for probiotics that are sold in oil suspension. For example, the oil suspension makes them easy to administer to infants (Saxelin 2008). Moreover, together with drying process, this can be a strategy to produce stable probiotic powder, although more research is still required.

Liquid or moist probiotic products have high $a_w$ ($\geq$0.90) and are mostly chilled products. Probiotic products that can store at ambient environment benefit all stakeholders connected to probiotic products. However, the metabolic activities and detrimental reactions that are not halted or delayed by low temperatures allow only very short shelf-life of the products. In addition to the conventional liquid probiotic products like yoghurt and fermented milk, there are also non-dairy liquid probiotic products such as fermented soymilk, fermented tomato juices, and probiotic fruit and vegetable juices (Rivera-Espinoza and Gallardo-Navarro 2010). These dairy and non-dairy probiotic products have the storage time at refrigerated temperatures between several weeks to months at most. For instance, Modzelewska-KapituŁA et al. (2008) investigated various combinations of commercial yoghurt starter cultures with probiotic strains *L. plantarum* and *L. fermentum* in modified yoghurt production. These probiotic strains did not additionally acidify milk during 24 h and 72 h fermentation at 37°C, but grew well and remained at the level of $10^8$cfu $ml^{-1}$ during 21 days during storage at 4°C. In chestnut mousse, the viability of spray-dried *L. rhamnosus* added as an anhydrous basis remained stable over $10^8$cfu $g^{-1}$ during storage for 3 months at 15°C (Romano et al. 2014). Probiotic oat drink was developed by the fermentation of oat milk with *L. reuteri* and *S. thermophilus*, and the starter survival above $10^7$cfu $ml^{-1}$ could be maintained during the 28 days storage at 4°C (Bernat et al. 2014b). In the fermented tomato juice, the viability of *L. acidophilus* LA39, *L. plantarum* C3, *L. casei* A4, and *L. delbrueckii* D7 ranged from $10^6$ to $10^9$ cfu $ml^{-1}$ after storage for 4 weeks at 4°C (Yoon et al. 2004). In fruit juices, the tested probiotics (*L. salivarius* ssp. *salivarius* UCC118 and UCC500, *B. animalis* ssp. *lactis* Bb-12, *L. casei* DN-114 001, *L. rhamnosus* GG and *L. paracasei* NFBC43338) survived better in orange juice and pineapple juice compared with cranberry juice. The viability of all strains was above $10^7$cfu $ml^{-1}$ in orange juice and above $10^6$ cfu $ml^{-1}$ in pineapple juice for at least 12 weeks. The viability of the probiotics in cranberry decreased drastically within a number of days, depending on pH of the cranberry juice, e.g., viability was below $10^6$cfu $ml^{-1}$ within only four days at pH 3.5 (Sheehan et al. 2007). As shown by the results above, high storage stability is difficult to obtain in some food products. A strategy to increase storage stability is by using a package that protects probiotics from the food matrix until the moment of consumption. For example, probiotics can be combined with oral rehydration salts for the treatment of acute diarrhea. The salts and the probiotic strain are packed in separate sachets, or the salt solution is put in a carton and the probiotic in a straw (i.e., suspended in oil and dried inside the straw) (Saxelin 2008). The straw as such is produced by BioGaia®, in which probiotics cells are immobilized on the inner wall of the straw.

In addition to traditional probiotic products such as yoghurt and fermented milk, there are increasing market opportunities for probiotic niche products like ice cream, cheese, candy, dessert and chewing gum. Consequently, there are also increasing needs for studies and knowledge about storage stability of probiotics in food or product matrices. Examples of the probiotic foods developed recently and their shelf-life are listed in Table 1.

In addition to probiotic foods, a wide variety of single or mixed probiotic strains is used in dietary supplements. These product forms are, for example, capsules, tablets with or without enteric coating, and sachets. The forms have advantages over probiotic foods that by incorporation of appropriate polymers commonly used in drug formulations or by enteric coating, the probiotic can be better protected from harsh conditions of the gastrointestinal tract and can be specifically delivered to

**Table 1.** Stability of probiotics in some probiotic foods recently developed.

| **Probiotic Foods** | **Probiotics** | **Storage temperature (°C)** | **Storage Time** | **Viability cfu $g^{-1}$ or $ml^{-1}$** | **Remarks** | **References** |
|---|---|---|---|---|---|---|
| Raspberry powder | *L. rhamnosus* NRRL B-4495 and *L. acidophilus* NRRL B-442 | 23, 4 | 30 d | 66.8–86.2%, 88.8–91.4% | Raspberry powder was obtained from spray drying lactobacilli and raspberry juice with maltodextrin. | (Anekella and Orsat 2014) |
| Fermented almond milk | *L. reuteri* ATCC 55730, *S. thermophilus* CECT 986 | 4 | 28 d | $10^7$, $3.7x10^6$ | Starter was 1:1 volume ratio of the *L. reuteri* and *S. thermophilus* | (Bernat et al. 2014a) |
| Fermented oat drink | *L. reuteri* ATCC 55730, *S. thermophilus* CECT 986 | 4 | 28 d | $2.6x10^7$, $3.9x10^7$ | Starter was 1:1 volume ratio of the *L. reuteri* and *S. thermophilus* | (Bernat et al. 2014b) |
| Fresh cheese | *L. salivarius* CECT5713 and PS2 | 4 | 28 d | $5x10^6$, $3.9x10^6$ | Pasteurized milk with 0.01% (w/v) $CaCl_2$ was inoculated with *L. lactis* as starter culture. Rennet was added to milk. freeze-dried probiotics were added to the curd, to reach a final concentration of ~8 log10 cfu $g^{-1}$ | (Cardenas et al. 2014) |
| Chest nut mousses | *L. rhamnosus* RBM526 | 15 | 3 mo | $10^8$ | *L. rhamnosus* added as an anhydrous basis | (Romano et al. 2014) |
| Dry-fermented sausages | *L. casei* ATCC 393 | RT, 4–6 | 10–12, 66 d | $\geq 10^6$ | Ripening: RT (40 and 85% RH) for 10–12 days and then decreased to 4–6°C at 2–4°C/day (50 and 75% RH) for up to 66 days | (Sidira et al. 2014) |
| Fermented pineapple juice | *L. casei* NRRL B-442 | 4 | 42 d | $10^6$ | Viabilty was $10^6$ and $5.8x10^4$cfu $ml^{-1}$ in the non-sweetened and sweetened sample, respectively. | (Costa et al. 2013) |
| Mutton fermented sausage | *L. acidophilus* CCDM 476, *B. animalis* CCDM 241a. | 20 | 60 d | $3.4x10^6$, not detected | The sausages were cold smoked at 25°C for 96 h and fermented at 20°C for another 10 days. | (Holko et al. 2013) |

*Table 1. contd....*

*Table 1. contd....*

| Probiotic Foods | Probiotics | Storage temperature (ºC) | Storage Time | Viability cfu $g^{-1}$ or $ml^{-1}$ | Remarks | References |
|---|---|---|---|---|---|---|
| Fruit-based ice cream | *L. acidophilus* (LA 5) | –18 | 10 w | $10^7$ | Ice cream mix was ice cream and yogurt mix, and wood apple puree | (Senanayake et al. 2013) |
| Soy protein bar | *L. acidophilus* LA-2 | 4, RT | 14 w | $6x10^8$, $6x10^5$ | Soy protein bar matrix: soy flour, soy proteinisolate, non-fat dry milk, high fructose corn syrup, dark brown sugar, Gum guar bland, Canola oil, citric acid, and vanilla extract | (Chen and Mustapha 2012) |
| Ice cream | *L. acidophilus* DOWARU™ | –18 | 60 d | $\geq 10^6$ | Higher overrun levels negatively influenced cell viability. | (Ferraz et al. 2012) |
| Chewing gum | *Weissella cibaria* CMU | 20 | 4 w | $2.6x10^8$ | No change in viability at 4ºC for 5 mo (initial viability $8.8x10^9$cfu $g^{-1}$) | (Kang et al. 2012) |
| Goat's milk yoghurt | *L. acidophilus* LA-5, *B. animalis* subsp. *lactis* BB-12, *P. jensenii* 702 | 4 | 4 w | $<10^6$, $10^7$, $10^8$ | Probiotics were used together with commercial dried ABY-1 culture | (Senaka Ranadheera et al. 2012) |
| Fresh-cut apple | *L. rhamonosus* GG | 5, 10 | 28 | $10^7$ | Apples were cut into wedges and dipped in a solution wih *L. rhamnosus* GG ($10^8$ cfu $ml^{-1}$) | (Alegre et al. 2011) |
| Guava mousses | *L. acidophilus* La-5 | 4–18 | 28,112 | $10^6$–$10^7$, $>10^7$ | The enriched fermented milk with the *L. acidophilus* was added to mousses mixture and aerated in a planetary beater. | (Buriti et al. 2010) |
| Yoghurt | *L. fermentum* 4a and *L. plantarum*14 | 4 | 21 d | $10^7$ | Combinations with commercial yoghurt starter YC-X11 Yo-Flex | (Modzelewska-KapituŁA et al. 2008) |

reach the target site. Probiotic capsules contain mainly probiotic powder filled in the moisture-tight capsule. Bruno and Shah (2003) assessed viability of freeze-dried probiotic capsules at various temperatures during prolonged storage. The samples included capsules filled with freeze-dried *B. longum* 1941 and *B. longum* 536, and five commercial probiotic capsule products. The capsules were stored at –18°C, 4°C, and 20°C. The viability at least $10^7$cfu $g^{-1}$ of more than 6 months could be maintained only at –18 and 4°C in all samples. At 20°C, the commercial probiotic capsules could retain this viability level for 5 months, whereas the capsules of *Bifidobacterium* produced *ad hoc* did for only 1–2 months.

For a probiotic tablet, dry probiotic must be mixed and compressed with polymers to get the required shape. The decrease in viability during storage varies among probiotics. In a study by Brachkova et al. (2009), the probiotic stability of tablets containing spores of *Bacillus subtilis* ATCC 6633 was very high, while the tablets with either *L. plantarum* ATCC 8040 or *L. bulgaricus* CIP 101027, and *L. rhamnosus* LGG showed considerable, and low stability for the storage period of observation (6 months) respectively. The compaction force and polymers used in the tablet formulation affect the viability of probiotics. The decrease in viability of 1–2 log cycles usually occur from the compression (Brachkova et al. 2009, Klayraung et al. 2009, e Silva et al. 2013). For instance, the number of viable cells of *L. paracasei* L26 decreased one log cycle from $6.0\times10^9$ to $6.2\times10^8$cfu $g^{-1}$ in the tablets containing the probiotic microparticles, croscarmellose sodium, and cellulose acetate phthalate (CAP) compressed with 9.8 kN (e Silva et al. 2013). However, the inactivation rate during storage is independent from the compression pressure. The initial shock caused by the pressure had no effect on the remaining bacteria and did not make them more sensitive in storage conditions (Muller et al. 2014). The tablet's formulation also affects the stability during storage. For example, the decrease in the viabilities of *L. bulgaricus* CIP101027 and *L. rhamnosus* LGG during compression and storage was more pronounced when microcrystalline cellulose was incorporated in the formulation (Brachkova et al. 2009). Moreover, Muller et al. (2014) reported in a study with *L. rhamnosus* Lcr35(R) that the biological properties of the strain were maintained after compaction (genetic profile, pathogen inhibition) and even improved (gastric resistance).

## 5. Concluding Remarks and Future Prospects

Probiotic powder has advantages over the liquid and frozen forms in term of transportation and storage costs, as well as convenience for probiotic manufacturers, food industries, and consumers. Probiotic powder can be applied in a wide range of products, including probiotic supplements and novelty probiotic products such as cereals, confectionery, and infant formulae. However, storage stability of the probiotic powder and products still require improvements, especially at ambient storage conditions. In addition to viability and fermentation activity, storage stability regarding physiological functionalities of probiotic powder should be taken into consideration as well. Moreover, the functional stability is important not only during storage of the powder, but also during shelf life of the products containing the probiotics added.

Although there are increasing studies on the changes in the functionalities upon the storage, more insight is still required to catch up with the growing demand of new probiotic products.

Stability of probiotic powder depends on many factors. Insight into some factors is well established and rather universal. For example, probiotic powder should have low moisture content and be stored in a hermetic package at low temperature and humidity. Excluding results from the intrinsic nature of probiotics, effects of some factors are varying among studies, and optimization is necessary. Some of the factors are the types and the concentration of protectants and sub-lethal stresses added or applied to probiotic cells before the drying processes. During the process of drying, the protectants may experience a transition from liquid to glassy state. The protective roles and mechanisms of the glassy state are not fully clear, and detrimental reactions of the dry cells embedded in the amorphous protective matrix are not completely halted during storage. However, it is still recommended and practical to store probiotic powder at a temperature below the $T_g$ of the protective matrix. A sub-lethal stress can predominantly improve storage stability of probiotics. However, the responses of cells depend on, for example, stress concentration, time, and growth phase of cells being treated, and therefore neutral or negative effects to viability and stability may occur. Furthermore, a troublesome finding is that the viability and stability of probiotic in drying and storaging is intrinsic to species or strains. It means that storage conditions of probiotic powder and products must be determined and optimized from case to case. As a result, screening for a robust probiotic strain is still essential to get the suitable product-probiotic pair. This means also that the screening procedure should be specific and connected to the intended use of the probiotics, and selective pressures (functional, technological and product properties) should be included early in the screening procedure.

Ultimately, the storage-stable probiotic powder and shelf-stable probiotic products will dramatically reduce transportation and storage costs, promote long-distance trade, and increase convenience for consumers. Moreover, it will open and expand inventing and marketing possibilities of novelty probiotic products.

## References

Abe, F., H. Miyauchi, A. Uchijima, T. Yaeshima and K. Iwatsuki. 2009. Effects of storage temperature and water activity on the survival of bifidobacteria in powder form. International Journal of Dairy Technology 62(2): 234–239.

Alegre, I., I. Vinas, J. Usall, M. Anguera and M. Abadias. 2011. Microbiological and physicochemical quality of fresh-cut apple enriched with the probiotic strain *Lactobacillus rhamnosus* GG. Food microbiology 28(1): 59–66.

Ananta, E., M. Volkert and D. Knorr. 2005. Cellular injuries and storage stability of spray-dried *Lactobacillus rhamnosus* GG. Int. Dairy J. 15(4): 399–409.

Anekella, K. and V. Orsat. 2014. Shelf life stability of lactobacilli encapsulated in raspberry powder: insights into non-dairy probiotics. International journal of food sciences and nutrition 65(4): 411–418.

Ann, E.Y., Y. Kim, S. Oh, J.-Y. Imm, D.-J. Park, K.S. Han and S.H. Kim. 2007. Microencapsulation of *Lactobacillus acidophilus* ATCC 43121 with prebiotic substrates using a hybridisation system. International Journal of Food Science & Technology 42(4): 411–419.

Aschenbrenner, M., E. Grammueller, U. Kulozik and P. Foerst. 2014. The Contribution of the Inherent Restricted Mobility of Glassy Sugar Matrices to the Overall Stability of Freeze-Dried Bacteria

Determined by Low-Resolution Solid-State 1H-NMR. Food and Bioprocess Technology 7(4): 1012–1024.

Aschenbrenner, M., U. Kulozik and P. Foerst. 2012. Evaluation of the relevance of the glassy state as stability criterion for freeze-dried bacteria by application of the Arrhenius and WLF model. Cryobiology 65(3): 308–318.

Ashraf, R. and N.P. Shah. 2011. Selective and differential enumerations of *Lactobacillus delbrueckii* subsp. *bulgaricus*, Streptococcus thermophilus, Lactobacillus acidophilus, *Lactobacillus casei* and *Bifidobacterium* spp. in yoghurt—A review. International journal of food microbiology 149(3): 194–208.

Beal, C., F. Fonseca and G. Corrieu. 2001. Resistance to freezing and frozen storage of *Streptococcus thermophilus* is related to membrane fatty acid composition. Journal of dairy science 84(11): 2347–2356.

Bensch, G., M. Ruger, M. Wassermann, S. Weinholz, U. Reichl and C. Cordes. 2014. Flow cytometric viability assessment of lactic acid bacteria starter cultures produced by fluidized bed drying. Applied microbiology and biotechnology 98(11): 4897–4909.

Bernat, N., M. Chafer, A. Chiralt and C. Gonzalez-Martinez. 2014a. Development of a non-dairy probiotic fermented product based on almond milk and inulin. Food science and technology international = Ciencia y tecnologia de los alimentos internacional.

Bernat, N., M. Chafer, C. Gonzalez-Martinez, J. Rodriguez-Garcia and A. Chiralt. 2014b. Optimisation of oat milk formulation to obtain fermented derivatives by using probiotic Lactobacillus reuteri microorganisms. Food science and technology international = Ciencia y tecnologia de los alimentos internacional.

Borges, S., P. Costa, J. Silva and P. Teixeira. 2013. Effects of processing and storage on Pediococcus pentosaceus SB83 in vaginal formulations: lyophilized powder and tablets. BioMed. research international 2013: 680767.

Brachkova, M.I., A. Duarte and J.F. Pinto. 2009. Evaluation of the viability of *Lactobacillus* spp. after the production of different solid dosage forms. Journal of Pharmaceutical Sciences 98(9): 3329–3339.

Broadbent, J.R. and C. Lin. 1999. Effect of heat shock or cold shock treatment on the resistance of Lactococcus lactis to freezing and lyophilization. Cryobiology 39(1): 88–102.

Bruno, F.A. and N.P. Shah. 2003. Viability of Two Freeze-dried Strains of Bifidobacterium and of Commercial Preparations at Various Temperatures During Prolonged Storage. Journal of Food Science 68(7): 2336–2339.

Bunthof, C.J. and T. Abee. 2002. Development of a flow cytometric method to analyze subpopulations of bacteria in probiotic products and dairy starters. Applied and environmental microbiology 68(6): 2934–2942.

Burgain, J., C. Gaiani, M. Linder and J. Scher. 2011. Encapsulation of probiotic living cells: From laboratory scale to industrial applications. Journal of Food Engineering 104(4): 467–483.

Buriti, F.C., I.A. Castro and S.M. Saad. 2010. Viability of Lactobacillus acidophilus in synbiotic guava mousses and its survival under *in vitro* simulated gastrointestinal conditions. International journal of food microbiology 137(2-3): 121–129.

Bylund, G. 2003. Dairy processing handbook. Tetra Pak Processing Systems AB.

Cardenas, N., J. Calzada, A. Peiroten, E. Jimenez, R. Escudero, J.M. Rodriguez, M. Medina and L. Fernandez. 2014. Development of a Potential Probiotic Fresh Cheese Using Two Lactobacillus salivarius Strains Isolated from Human Milk. BioMed. research international 2014: 801918.

Carvalho, A.S., J. Silva, P. Ho, P. Teixeira, F.X. Malcata and P. Gibbs. 2002. Survival of freeze-dried Lactobacillus plantarum and Lactobacillus rhamnosus during storage in the presence of protectants. Biotechnol. Lett. 24(19): 1587–1591.

Carvalho, A.S., J. Silva, P. Ho, P. Teixeira, F.X. Malcata and P. Gibbs. 2003. Effects of addition of sucrose and salt, and of starvation upon thermotolerance and survival during storage of freeze-dried *Lactobacillus delbrueckii* ssp. *bulgaricus*. J. Food Sci. 68(8): 2538–2541.

Carvalho, A.S., J. Silva, P. Ho, P. Teixeira, F.X. Malcata and P. Gibbs. 2004a. Effects of various sugars added to growth and drying media upon thermotolerance and survival throughout storage of freeze-dried *Lactobacillus delbrueckii* ssp. *bulgaricus*. Biotechnol. Prog. 20(1): 248–254.

Carvalho, A.S., J. Silva, P. Ho, P. Teixeira, F.X. Malcata and P. Gibbs. 2004b. Relevant factors for the preparation of freeze-dried lactic acid bacteria. Int. Dairy J. 14(10): 835–847.

Celik, O.F. and D.J. O'Sullivan. 2013. Factors influencing the stability of freeze-dried stress-resilient and stress-sensitive strains of bifidobacteria. Journal of dairy science 96(6): 3506–3516.

Chaluvadi, S., A.T. Hotchkiss, Jr., J.E. Call, J.B. Luchansky, J.G. Phillips, L. Liu and K.L. Yam. 2012. Protection of probiotic bacteria in a synbiotic matrix following aerobic storage at 4 degrees C. Benef Microbes 3(3): 175–187.

Champagne, C.P., F. Mondou, Y. Raymond and D. Roy. 1996. Effect of polymers and storage temperature on the stability of freeze-dried lactic acid bacteria. Food Res. Int. 29: 555–562.

Champagne, C.P., R.P. Ross, M. Saarela, K.F. Hansen and D. Charalampopoulos. 2011. Recommendations for the viability assessment of probiotics as concentrated cultures and in food matrices. International journal of food microbiology 149(3): 185–193.

Chan, E.S. and Z. Zhang. 2005. Bioencapsulation by compression coating of probiotic bacteria for their protection in an acidic medium. Process Biochemistry 40(10): 3346–3351.

Chávez, B.E. and A.M. Ledeboer. 2007. Drying of Probiotics: Optimization of Formulation and Process to Enhance Storage Survival. Drying Technology 25(7-8): 1193–1201.

Chen, M. and A. Mustapha. 2012. Survival of freeze-dried microcapsules of α-galactosidase producing probiotics in a soy bar matrix. Food microbiology 30(1): 68–73.

Corcoran, B.M., R.P. Ross, G.F. Fitzgerald and C. Stanton. 2004. Comparative survival of probiotic lactobacilli spray-dried in the presence of prebiotic substances. Journal of applied microbiology 96(5): 1024–1039.

Costa, M.G.M., T.V. Fonteles, A.L.T. de Jesus and S. Rodrigues. 2013. Sonicated pineapple juice as substrate for L. casei cultivation for probiotic beverage development: Process optimisation and product stability. Food Chemistry 139(1-4): 261–266.

Coulibaly, I., A.Y. Amenan, G. Lognay, M.L. Fauconnier and P. Thonart. 2009. Survival of freeze-dried leuconostoc mesenteroides and Lactobacillus plantarum related to their cellular fatty acids composition during storage. Appl. Biochem. Biotechnol. 157(1): 70–84.

Crowe, J.H., J.F. Carpenter and L.M. Crowe. 1998. The role of vitrification in anhydrobiosis. Annual Review of Physiology 60: 73–103.

Davis, C. 2014. Enumeration of probiotic strains: Review of culture-dependent and alternative techniques to quantify viable bacteria. Journal of Microbiological Methods 103(0): 9–17.

Desmond, C., G.F. Fitzgerald, C. Stanton and R.P. Ross. 2004. Improved stress tolerance of GroESL-overproducing Lactococcus lactis and probiotic Lactobacillus paracasei NFBC 338. Appl. Environ. Microbiol. 70(10): 5929–5936.

Desmond, C., C. Stanton, G.F. Fitzgerald, K. Collins and R.P. Ross. 2002. Environmental adaptation of probiotic lactobacilli towards improvement of performance during spray drying. International Dairy Journal 12(2-3): 183–190.

Ding, W.K. and N.P. Shah. 2009. An Improved Method of Microencapsulation of Probiotic Bacteria for Their Stability in Acidic and Bile Conditions during Storage. Journal of Food Science 74(2): M53–M61.

e Silva, J.P., S.C. Sousa, P. Costa, E. Cerdeira, M.H. Amaral, J.S. Lobo, A.M. Gomes, M.M. Pintado, D. Rodrigues, T. Rocha-Santos and A.C. Freitas. 2013. Development of probiotic tablets using microparticles: viability studies and stability studies. AAPS Pharm. Sci. Tech. 14(1): 121–127.

Ekdawi-Sever, N., L.A. Geontoro and J.J. de Pablo. 2003. Effects of Annealing on Freeze-Dried Lactobacillus acidophilus. Journal of Food Science 68(8): 2504–2511.

El Demerdash, H.A., J. Oxmann, K.J. Heller and A. Geis. 2006. Yoghurt fermentation at elevated temperatures by strains of Streptococcus thermophilus expressing a small heat-shock protein: application of a two-plasmid system for constructing food-grade strains of Streptococcus thermophilus. Biotechnology journal 1(4): 398–404.

Endo, A., J. Terasjarvi and S. Salminen. 2014. Food matrices and cell conditions influence survival of Lactobacillus rhamnosus GG under heat stresses and during storage. International journal of food microbiology 174: 110–112.

Ferraz, J.L., A.G. Cruz, R.S. Cadena, M.Q. Freitas, U.M. Pinto, C.C. Carvalho, J.A. Faria and H.M. Bolini. 2012. Sensory acceptance and survival of probiotic bacteria in ice cream produced with different overrun levels. J. Food Sci. 77(1): S24–28.

Foerst, P., U. Kulozik, M. Schmitt, S. Bauer and C. Santivarangkna. 2012. Storage stability of vacuum-dried probiotic bacterium Lactobacillus paracasei F19. Food and Bioproducts Processing 90(2): 295–300.

Foligné, B., C. Daniel and B. Pot. 2013. Probiotics from research to market: the possibilities, risks and challenges. Current Opinion in Microbiology 16(3): 284–292.

Fonseca, F., M. Marin and G.J. Morris. 2006. Stabilization of frozen *Lactobacillus delbrueckii* subsp. *bulgaricus* in glycerol suspensions: Freezing kinetics and storage temperature effects. Appl. Environ. Microbiol. 72(10): 6474–6482.

Fonseca, F., J.P. Obert, C. Beal and M. Marin. 2001. State diagrams and sorption isotherms of bacterial suspensions and fermented medium. Thermochimica Acta 366(2): 167–182.

Foschino, R., E. Fiori and A. Galli. 1996. Survival and residual activity of Lactobacillus acidophilus frozen cultures under different conditions. Journal of Dairy Research 63(2): 295–303.

Fu, N. and X.D. Chen. 2011. Towards a maximal cell survival in convective thermal drying processes. Food Research International 44(5): 1127–1149.

Ghandi, A., I.B. Powell, M. Broome and B. Adhikari. 2013. Survival, fermentation activity and storage stability of spray dried Lactococcus lactis produced via different atomization regimes. Journal of Food Engineering 115(1): 83–90.

Golowczyc, M.A., J. Silva, A.G. Abraham, G.L. De Antoni and P. Teixeira. 2010. Preservation of probiotic strains isolated from kefir by spray drying. Letters in applied microbiology 50(1): 7–12.

Gueimonde, M. and S. Salminen. 2006. New methods for selecting and evaluating probiotics. Digestive and liver disease : official journal of the Italian Society of Gastroenterology and the Italian Association for the Study of the Liver 38 Suppl. 2: S242–247.

Higl, B., L. Kurtmann, C.U. Carlsen, J. Ratjen, P. Forst, L.H. Skibsted, U. Kulozik and J. Risbo. 2007. Impact of water activity, temperature, and physical state on the storage stability of Lactobacillus paracasei ssp. paracasei freeze-dried in a lactose matrix. Biotechnol. Prog. 23(4): 794–800.

Holko, I., J. Hrabě, A. Šalaková and V. Rada. 2013. The substitution of a traditional starter culture in mutton fermented sausages by Lactobacillus acidophilus and Bifidobacterium animalis. Meat science 94(3): 275–279.

Hoobin, P., I. Burgar, S. Zhu, D. Ying, L. Sanguansri and M.A. Augustin. 2013. Water sorption properties, molecular mobility and probiotic survival in freeze dried protein-carbohydrate matrices. Food & function 4(9): 1376–1386.

Ishibashi, N., T. Tatematsu, S. Shimamura, M. Tomota and S. Okonogi. 1985. Fundamentals and application of freeze drying to biological materials, dyes and foodstuffs. International Institute of Refrigeration, Paris pp. 227–232.

Jalali, M., D. Abedi, J. Varshosaz, M. Najjarzadeh, M. Mirlohi and N. Tavakoli. 2012. Stability evaluation of freeze-dried *Lactobacillus paracasei* subsp. *tolerance* and *Lactobacillus delbrueckii* subsp. *bulgaricus* in oral capsules. Research in pharmaceutical sciences 7(1): 31–36.

Juárez Tomás, M.a.S., V.S. Ocaña and M.a.E. Nader-Macías. 2004. Viability of vaginal probiotic lactobacilli during refrigerated and frozen storage. Anaerobe 10(1): 1–5.

Juarez Tomas, M.S., E. Bru, G. Martos and M.E. Nader-Macias. 2009. Stability of freeze-dried vaginal Lactobacillus strains in the presence of different lyoprotectors. Canadian journal of microbiology 55(5): 544–552.

Kang, M.S., Y.S. Kim, H.C. Lee, H.S. Lim and J.S. Oh. 2012. Comparison of Temperature and Additives Affecting the Stability of the Probiotic Weissella cibaria. Chonnam medical journal 48(3): 159–163.

Klayraung, S., H. Viernstein and S. Okonogi. 2009. Development of tablets containing probiotics: Effects of formulation and processing parameters on bacterial viability. Int J Pharm 370(1-2): 54–60.

Klu, Y.A.K., R.D. Phillips and J. Chen. 2014. Survival of four commercial probiotic mixtures in full fat and reduced fat peanut butter. Food microbiology 44(0): 34–40.

Klu, Y.A.K., J.H. Williams, R.D. Phillips and J. Chen. 2012. Survival of Lactobacillus rhamnosus GG as Influenced by Storage Conditions and Product Matrixes. Journal of Food Science 77(12): M659–M663.

Kurtmann, L., C.U. Carlsen, J. Risbo and L.H. Skibsted. 2009a. Storage stability of freeze–dried Lactobacillus acidophilus (La-5) in relation to water activity and presence of oxygen and ascorbate. Cryobiology 58(2): 175–180.

Kurtmann, L., C.U. Carlsen, L.H. Skibsted and J. Risbo. 2009b. Water Activity-Temperature State Diagrams of Freeze-Dried Lactobacillus acidophilus (La-5): Influence of Physical State on Bacterial Survival during Storage. Biotechnology progress 25(1): 265–270.

Kurtmann, L., L.H. Skibsted and C.U. Carlsen. 2009c. Browning of Freeze-Dried Probiotic Bacteria Cultures in Relation to Loss of Viability during Storage. Journal of Agricultural and Food Chemistry 57(15): 6736–6741.

Lamboley, L., D. St-Gelais, C.P. Champagne and M. Lamoureux. 2003. Growth and morphology of thermophilic dairy starters in alginate beads. The Journal of general and applied microbiology 49(3): 205–214.

Lee, K. 2004. Cold shock response in *Lactococcus lactis* ssp. *diacetylactis*: a comparison of the protection generated by brief pre-treatment at less severe temperatures. Process Biochemistry 39(12): 2233–2239.

Lee, Y.K. 2008. Selection and Maintenance of Probiotic Microorganisms. Handbook of Probiotics and Prebiotics. John Wiley & Sons, Inc. 177–187.

Louesdon, S., S. Charlot-Rougé, V. Juillard, R. Tourdot-Maréchal and C. Béal. 2014. Osmotic stress affects the stability of freeze-dried Lactobacillus buchneri R1102 as a result of intracellular betaine accumulation and membrane characteristics. Journal of applied microbiology 117(1): 196–207.

Maassen, C.B.M., W.J.A. Boersma, C. van Holten-Neelen, E. Claassen and J.D. Laman. 2003. Growth phase of orally administered Lactobacillus strains differentially affects IgG1/IgG2a ratio for soluble antigens: implications for vaccine development. Vaccine 21(21–22): 2751–2757.

Makinen, K., B. Berger, R. Bel-Rhlid and E. Ananta. 2012. Science and technology for the mastership of probiotic applications in food products. J Biotechnol. 162(4): 356–365.

Manojlović, V., V. Nedović, K. Kailasapathy and N. Zuidam. 2010. Encapsulation of Probiotics for use in Food Products. Encapsulation Technologies for Active Food Ingredients and Food Processing. N.J. Zuidam and V. Nedovic, ed. Springer New York. 269–302.

Meng, X.C., C. Stanton, G.F. Fitzgerald, C. Daly and R.P. Ross. 2008. Anhydrobiotics: The challenges of drying probiotic cultures. Food Chemistry 106(4): 1406–1416.

Miao, S., S. Mills, C. Stanton, G. Fitzgerald, Y. Roos and R.P. Ross. 2008. Effect of disaccharides on survival during storage of freeze dried probiotics. Dairy Sci. Technol. 88(1): 19–30.

Mille, Y., J.P. Obert, L. Beney and P. Gervais. 2004. New drying process for lactic bacteria based on their dehydration behavior in liquid medium. Biotechnol. Bioeng. 88(1): 71–76.

Miyamoto-Shinohara, Y., J. Sukenobe, T. Imaizumi and T. Nakahara. 2008. Survival of freeze-dried bacteria. J. Gen. Appl. Microbiol. 54(1): 9–24.

Modzelewska-KapituŁA, M., L. KŁĘBukowska and K. Kornacki. 2008. Evaluation of the possible use of potentially probiotic Lactobacillus strains in dairy products. International Journal of Dairy Technology 61(2): 165–169.

Monnet, C., C. Beal and G. Corrieu. 2003. Improvement of the resistance of *Lactobacillus delbrueckii* ssp. *bulgaricus* to freezing by natural selection. Journal of dairy science 86(10): 3048–3053.

Montel Mendoza, G., S.E. Pasteris, M.C. Otero and M.E. Fatima Nader-Macias. 2014. Survival and beneficial properties of lactic acid bacteria from raniculture subjected to freeze-drying and storage. Journal of applied microbiology 116(1): 157–166.

Mozzetti, V., F. Grattepanche, B. Berger, E. Rezzonico, F. Arigoni and C. Lacroix. 2013. Fast screening of Bifidobacterium longum sublethal stress conditions in a novel two-stage continuous culture strategy. Beneficial Microbes 4(2): 167–178.

Muller, C., V. Mazel, C. Dausset, V. Busignies, S. Bornes, A. Nivoliez and P. Tchoreloff. 2014. Study of the Lactobacillus rhamnosus Lcr35(R) properties after compression and proposition of a model to predict tablet stability. European journal of pharmaceutics and biopharmaceutics : official journal of Arbeitsgemeinschaft fur Pharmazeutische Verfahrenstechnik e.V.

Muller, J.A., C. Stanton, W. Sybesma, G.F. Fitzgerald and R.P. Ross. 2010. Reconstitution conditions for dried probiotic powders represent a critical step in determining cell viability. Journal of applied Microbiology 108(4): 1369–1379.

Nebesny, E., D. Żyżelewicz, I. Motyl and Z. Libudzisz. 2007. Dark chocolates supplemented with Lactobacillus strains. Eur. Food Res. Technol. 225(1): 33–42.

Oetjen, G. and P. Haseley. 2003. Freeze-Drying. Vol. 2. Wiley-VCH, Weinheim.

Passot, S., S. Cenard, I. Douania, I.C. Tréléa and F. Fonseca. 2012. Critical water activity and amorphous state for optimal preservation of lyophilised lactic acid bacteria. Food Chemistry 132(4): 1699–1705.

Pehkonen, K.S., Y.H. Roos, S. Miao, R.P. Ross and C. Stanton. 2008. State transitions and physicochemical aspects of cryoprotection and stabilization in freeze-drying of Lactobacillus rhamnosus GG (LGG). J. Appl. Microbiol. 104(6): 1732–1743.

Peighambardoust, S.H., A. Golshan Tafti and J. Hesari. 2011. Application of spray drying for preservation of lactic acid starter cultures: a review. Trends in Food Science & Technology 22(5): 215–224.

Perdana, J., M.B. Fox, C. Siwei, R.M. Boom and M.A.I. Schutyser. 2014. Interactions between formulation and spray drying conditions related to survival of Lactobacillus plantarum WCFS1. Food Research International 56(0): 9–17.

Quirós, C., M. Herrero, L.A. García and M. Díaz. 2009. Quantitative Approach to Determining the Contribution of Viable-but-Nonculturable Subpopulations to Malolactic Fermentation Processes. Applied and environmental microbiology 75(9): 2977–2981.

Riaz, Q.U. and T. Masud. 2013. Recent trends and applications of encapsulating materials for probiotic stability. Critical reviews in food science and nutrition 53(3): 231–244.

Rivera-Espinoza, Y. and Y. Gallardo-Navarro. 2010. Non-dairy probiotic products. Food microbiology 27(1): 1–11.

Romano, A., G. Blaiotta, A. Di Cerbo, R. Coppola, P. Masi and M. Aponte. 2014. Spray-dried chestnut extract containing Lactobacillus rhamnosus cells as novel ingredient for a probiotic chestnut mousse. Journal of applied microbiology 116(6): 1632–1641.

Saarela, M., H. Alakomi, A. Puhakka and J. Matto. 2009. Effect of the fermentation pH on the storage stability of Lactobacillus rhamnosus preparations and suitability of *in vitro* analyses of cell physiological functions to predict it. Journal of applied microbiology 106: 1204–1212.

Saarela, M., I. Virkajarvi, L. Nohynek, A. Vaari and J. Matto. 2006. Fibres as carriers for Lactobacillus rhamnosus during freeze-drying and storage in apple juice and chocolate-coated breakfast cereals. Int. J. Food Microbiol. 112(2): 171–178.

Saarela, M., I. Virkajarvi, H.L. Alakomi, T. Mattila-Sandholm, A. Vaari, T. Suomalainen and J. Matto. 2005. Influence of fermentation time, cryoprotectant and neutralization of cell concentrate on freeze-drying survival, storage stability, and acid and bile exposure of *Bifidobacterium animalis* ssp. *lactis* cells produced without milk-based ingredients. J. Appl. Microbiol. 99(6): 1330–1339.

Sanders, M.E., T.R. Klaenhammer, A.C. Ouwehand, B. Pot, E. Johansen, J.T. Heimbach, M.L. Marco, J. Tennilä, R.P. Ross, C. Franz, N. Pagé, R.D. Pridmore, G. Leyer, S. Salminen, D. Charbonneau, E. Call and I. Lenoir-Wijnkoop. 2014. Effects of genetic, processing, or product formulation changes on efficacy and safety of probiotics. Annals of the New York Academy of Sciences 1309(1): 1–18.

Santivarangkna, C., M. Aschenbrenner, U. Kulozik and P. Foerst. 2011a. Role of Glassy State on Stabilities of Freeze-Dried Probiotics. Journal of Food Science 76(8): R152–R156.

Santivarangkna, C., M. Aschenbrenner, U. Kulozik and P. Foerst. 2011b. Role of glassy state on stabilities of freeze-dried probiotics. J. Food Sci. 76(8): R152–156.

Santivarangkna, C., B. Higl and P. Foerst. 2008a. Protection mechanisms of sugars during different stages of preparation process of dried lactic acid starter cultures. Food Microbiol. 25(3): 429–441.

Santivarangkna, C., U. Kulozik and P. Foerst. 2007. Alternative drying processes for the industrial preservation of lactic acid starter cultures. Biotechnol. Prog. 23(2): 302–315.

Santivarangkna, C., U. Kulozik and P. Foerst. 2008b. Inactivation mechanisms of lactic acid starter cultures preserved by drying processes. J. Appl. Microbiol. 105(1): 1–13.

Sashihara, T., N. Sueki, K. Furuichi and S. Ikegami. 2007. Effect of growth conditions of Lactobacillus gasseri OLL2809 on the immunostimulatory activity for production of interleukin-12 (p70) by murine splenocytes. International journal of food microbiology 120(3): 274–281.

Saxelin, M. 2008. Probiotic Formulations and Applications, the Current Probiotics Market, and Changes in the Marketplace: A European Perspective. Clinical Infectious Diseases 46(Supplement 2): S76–S79.

Schmidt, S.J. and A.J. Fontana. 2008. Appendix E: Water Activity Values of Select Food Ingredients and Products. Water Activity in Foods. Blackwell Publishing Ltd. pp. 407–420.

Schoug, A., J. Fischer, H.J. Heipieper, J. Schnurer and S. Hakansson. 2008. Impact of fermentation pH and temperature on freeze-drying survival and membrane lipid composition of Lactobacillus coryniformis Si3. J. Ind. Microbiol. Biotechnol. 35(3): 175–181.

Schoug, Å., D. Mahlin, M. Jonson and S. Håkansson. 2010. Differential effects of polymers PVP90 and Ficoll400 on storage stability and viability of Lactobacillus coryniformis Si3 freeze-dried in sucrose. Journal of applied microbiology 108(3): 1032–1040.

Schwab, C., R. Vogel and M.G. Ganzle. 2007. Influence of oligosaccharides on the viability and membrane properties of Lactobacillus reuteri TMW1.106 during freeze-drying. Cryobiol. 55: 108–114.

Senaka Ranadheera, C., C.A. Evans, M.C. Adams and S.K. Baines. 2012. Probiotic viability and physico-chemical and sensory properties of plain and stirred fruit yogurts made from goat's milk. Food Chem. 135(3): 1411–1418.

Senanayake, S.A., S. Fernando, A. Bamunuarachchi and M. Arsekularatne. 2013. Application of Lactobacillus acidophilus (LA 5) strain in fruit-based ice cream. Food Science & Nutrition 1(6): 428–431.

Sheehan, V.M., P. Ross and G.F. Fitzgerald. 2007. Assessing the acid tolerance and the technological robustness of probiotic cultures for fortification in fruit juices. Innovative Food Science & Emerging Technologies 8(2): 279–284.

Sidira, M., A. Karapetsas, A. Galanis, M. Kanellaki and Y. Kourkoutas. 2014. Effective survival of immobilized Lactobacillus casei during ripening and heat treatment of probiotic dry-fermented sausages and investigation of the microbial dynamics. Meat science 96(2 Pt A): 948–955.

Silva, J., A.S. Carvalho, P. Teixeira and P.A. Gibbs. 2002. Bacteriocin production by spray-dried lactic acid bacteria. Letters in applied microbiology 34(2): 77–81.

Simpson, P.J., C. Stanton, G.F. Fitzgerald and R.P. Ross. 2005. Intrinsic tolerance of Bifidobacterium species to heat and oxygen and survival following spray drying and storage. Journal of applied microbiology 99(3): 493–501.

Slade, L. and H. Levine. 1991. Beyond water activity: recent advances based on an alternative approach to the assessment of food quality and safety. Crit. Rev. Food Sci. Nutr. 30(2-3): 115–360.

Strasser, S., M. Neureiter, M. Geppl, R. Braun and H. Danner. 2009. Influence of lyophilization, fluidized bed drying, addition of protectants, and storage on the viability of lactic acid bacteria. J. Appl. Microbiol.

Streit, F., J. Delettre, G. Corrieu and C. Beal. 2008. Acid adaptation of *Lactobacillus delbrueckii* subsp. *bulgaricus* induces physiological responses at membrane and cytosolic levels that improves cryotolerance. J. Appl. Microbiol. 105(4): 1071–1080.

Sun, W.Q., P. Davidson and H.S.O. Chan. 1998. Protein stability in the amorphous carbohydrate matrix: relevance to anhydrobiosis. Biochimica et Biophysica Acta-General Subjects 1425(1): 245–254.

Talwalkar, A. and K. Kailasapathy. 2004. The role of oxygen in the viability of probiotic bacteria with reference to L. acidophilus and *Bifidobacterium* spp. Current issues in intestinal microbiology 5(1): 1–8.

Teixeira, P., H. Castro and R. Kirby. 1995a. Spray drying as a method for preparing concentrated cultures of Lactobacillus bulgaricus. J. Appl. Bacteriol. 78(4): 456–462.

Teixeira, P.C., M.H. Castro, F.X. Malcata and R.M. Kirby. 1995b. Survival of *Lactobacillus delbrueckii* ssp. *bulgaricus* Following Spray-Drying. Journal of dairy science 78(5): 1025–1031.

Teixeira, P., H. Castro and R. Kirby. 1996. Evidence of membrane lipid oxidation of spray-dried Lactobacillus bulgaricus during storage. Letters in applied microbiology 22(1): 34–38.

Termont, S., K. Vandenbroucke, D. Iserentant, S. Neirynck, L. Steidler, E. Remaut and P. Rottiers. 2006. Intracellular accumulation of trehalose protects Lactococcus lactis from freeze-drying damage and bile toxicity and increases gastric acid resistance. Appl. Environ. Microbiol. 72(12): 7694–7700.

To, B.C.S. and M.R. Etzel. 1997a. Spray drying, freeze drying, or freezing of three different lactic acid bacteria species. Journal of Food Science 62(3): 576–585.

To, B.C.S. and M.R. Etzel. 1997b. Survival of Brevibacterium linens (ATCC 9174) after spray drying, freeze drying, or freezing. J. Food Sci. 62(1): 167–189.

van de Guchte, M., P. Serror, C. Chervaux, T. Smokvina, S.D. Ehrlich and E. Maguin. 2002. Stress responses in lactic acid bacteria. Antonie Van Leeuwenhoek 82(1-4): 187–216.

Vinderola, G., M.F. Zacarias, W. Bockelmann, H. Neve, J. Reinheimer and K.J. Heller. 2012. Preservation of functionality of *Bifidobacterium animalis* subsp. *lactis* INL1 after incorporation of freeze-dried cells into different food matrices. Food microbiology 30(1): 274–280.

Walker, D.C., H.S. Girgis and T.R. Klaenhammer. 1999. The groESL chaperone operon of Lactobacillus johnsonii. Applied and environmental microbiology 65(7): 3033–3041.

Wang, Y.C., R.C. Yu and C.C. Chou. 2004. Viability of lactic acid bacteria and bifidobacteria in fermented soymilk after drying, subsequent rehydration and storage. Int. J. Food Microbiol. 93(2): 209–217.

Yang, S.H., S.H. Seo, S.W. Kim, S.K. Choi and D.H. Kim. 2006. Effect of ginseng polysaccharide on the stability of lactic acid bacteria during freeze-drying process and storage. Arch. Pharm. Res. 29(9): 735–740.

Yao, A.A., I. Coulibaly, G. Lognay, M.L. Fauconnier and P. Thonart. 2008. Impact of polyunsaturated fatty acid degradation on survival and acidification activity of freeze-dried Weissella paramesenteroides LC11 during storage. Applied microbiology and biotechnology 79(6): 1045–1052.

Ying, D.Y., M.C. Phoon, L. Sanguansri, R. Weerakkody, I. Burgar and M.A. Augustin. 2010. Microencapsulated Lactobacillus rhamnosus GG Powders: Relationship of Powder Physical Properties to Probiotic Survival during Storage. Journal of Food Science 75(9): E588–E595.

Yoon, K.Y., E.E. Woodams and Y.D. Hang. 2004. Probiotication of tomato juice by lactic acid bacteria. Journal of microbiology (Seoul, Korea) 42(4): 315–318.

Zarate, G. and M.E. Nader-Macias. 2006. Viability and biological properties of problotic vaginal lactobacilli after lyophilization and refrigerated storage into gelatin capsules. Process Biochemistry 41(8): 1779–1785.

Zayed, G. and Y.H. Roos. 2004. Influence of trehalose and moisture content on survival of Lactobacillus salivarius subjected to freeze-drying and storage. Process Biochemistry 39(9): 1081–1086.

# PART IV

# Applications of Probiotics

# CHAPTER 16

# Encapsulation in Milk Protein Matrices and Controlled Release

*Würth Rebecca,* Foerst Petra* and *Kulozik Ulrich*

## 1. Introduction

The trend of functional food and drinks has increased considerably during the last few years. The enrichment of food and drinks with bioactive ingredients such as probiotics is one of these concepts. By 2015, the functional food and drink segment is expected to reach US $ 130 billion. The share of probiotic food products is estimated to reach US $ 28.8 billion by 2015, which would imply a growth of US $ 12.9 billion since 2008 (Hernández-Rodríguez et al. 2014). All this current data shows the continuing high demand for probiotic products with a positive effect on the consumer's health. To be effective, probiotics have to reach the human intestine in a high number of viable and active cells (Hörmannsperger et al. 2009). The required minimum concentration should be ~$10^6$–$10^7$ CFU/mL during the expected shelf life of the product (Krasaekoopt et al. 2003). Before the probiotic bacteria can reach their destination in the body in available form, a couple of hurdles have to be overcome, as schematically shown in Fig. 1 (Agrawal 2005, Burgain et al. 2011, Champagne et al. 2005, Heidebach et al. 2012, Krasaekoopt et al. 2003).

For industrial applications, after fermentation, the fresh cell concentrate is preserved for distribution. Typical preservation operations include freezing and/or drying. However, both processes generally involve a reduction in viable cell numbers, especially drying which can affect survival due to different stressors. One determining factor known to critically affect viability is dehydration. In the case of thermal processes

Chair for Food Process Engineering and Dairy Technology, TU München, Weihenstephaner Berg 1, 85350 Freising.
E-mail: rebecca.wuerth@tum.de
* Corresponding author

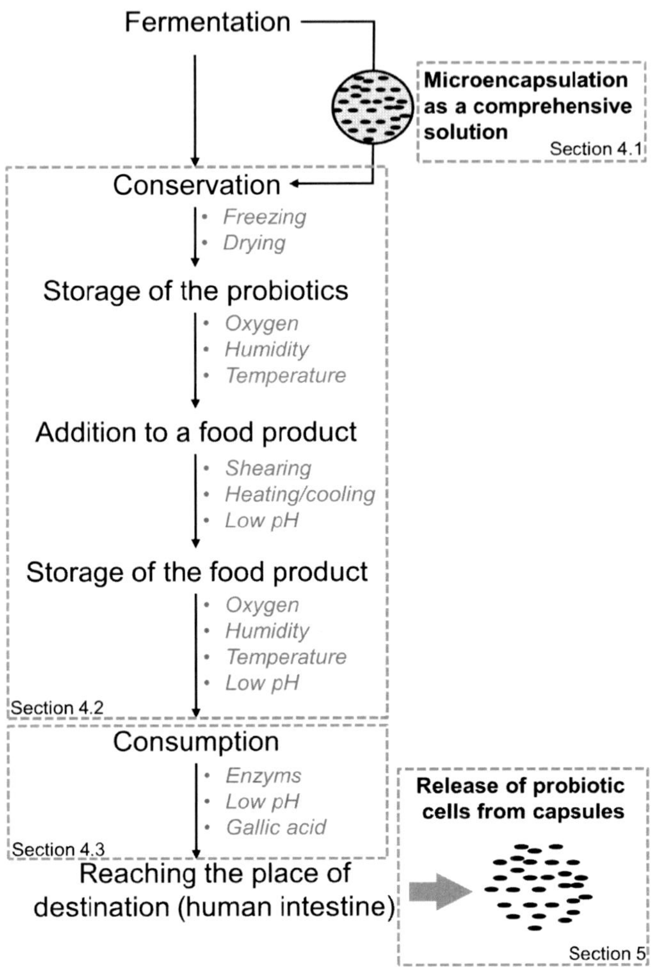


**Fig. 1.** The way of probiotics from fermentation to the intestine.

like spray drying, higher processing temperatures will have an additional negative influence. A further detrimental step is the storage of the preserved bacteria until their application as protective culture in food products. Oxygen, humidity, temperature and also storage time can lead to a reduction in the amount of viable cells. Incorporation into the food product is another critical step. For example, the bacteria may be exposed to shearing and rapidly changing temperatures depending on the individual food product. Moreover, the pH values of the surrounding food matrix are often low, which can negatively affect the active cell number (Burgain et al. 2011, Vos et al. 2010). After consumption, the last barrier of the probiotics is the gastro-intestinal transit. The various digestive enzymes, low pH values and the bile salts are unfavorable factors that can result in a reduction of the viable cell number (Krasaekoopt et al. 2003, Picot and Lacroix 2004). With this in mind, microencapsulation is a method used to prevent a reduction in the viability and active structure of the bacteria. The main idea of microencapsulation is to create a barrier additional to the product matrix. So a reduction of mass transport phenomena is achieved and consequently chemical and biological reactions are slowed down, which are well known to reduce the viable cell counts.

In addition due to stricter regulatory requirements on the probiotic market, there is a need for encapsulation of probiotics. To promote a functional food with benefits for consumer health, a claim is required. As a result, many bacterial strains without proven probiotic effects have disappeared from the market. It is therefore currently essential to protect proven, but less robust strains in their activity and viability. In this context, microencapsulation can become an important aid for the stabilisation of sensitive probiotic strains.

Also some general requirements have to be met for their successful application in food products. The size of the microcapsule has to be large enough for a good protective effect and small enough so that the capsules are not sensorially noticeable. The optimal size of the capsules as well as their structural strength depends on the food product in which they are used (Heidebach et al. 2011). While probiotic microcapsules with diameters around 1 mm may not be sensorially detected in sausages or cheese, these capsule sizes have an adverse sensorial effect in products with lower viscosity like yoghurt (Heidebach et al. 2011). For dairy foods, a size of ~100 μm is generally seen as the threshold to avoid any negative sensorial impact of microcapsules in most products (Hansen et al. 2002, Heidebach et al. 2011). Another indispensable requirement is the water-insolubility of the capsules. Otherwise they would dissolve and lose their protective effect. Furthermore, the functionality and viability of the probiotics must not be destroyed by the encapsulation process. For a final successful use of microencapsulation, it is also necessary that the capsules release the bacteria in the human intestine in a controlled manner.

## 2. Encapsulation Technologies

A variety of available methods for the encapsulation of bioactive substances exists. Different technologies such as spray drying, emulsion and extrusion techniques, liposomes, microemulsions and cyclodextrins have been used in this field (Onwulata 2012). However, for probiotics the methods are on the one hand limited because of the high sensitivity of the bacteria and on the other hand because of the large size of the bacteria compared to other bioactive molecules. Usually extrusion, emulsion techniques or spray drying are used. The following section focuses on current research on these main techniques. All of these methods share the same main principle as shown in Fig. 2. In the first step, the probiotics are mixed homogeneously in the matrix material. Afterwards the matrix material is spread in small drops and finally the drops are hardened. Depending on the method used, the size and shape of the produced capsules differ between 0.1 and 10000 μm.

### *2.1 Extrusion*

Extrusion is the major process for the creation of capsules in the area of immobilized cell technology (ICT), which lays the foundation of the microencapsulation of

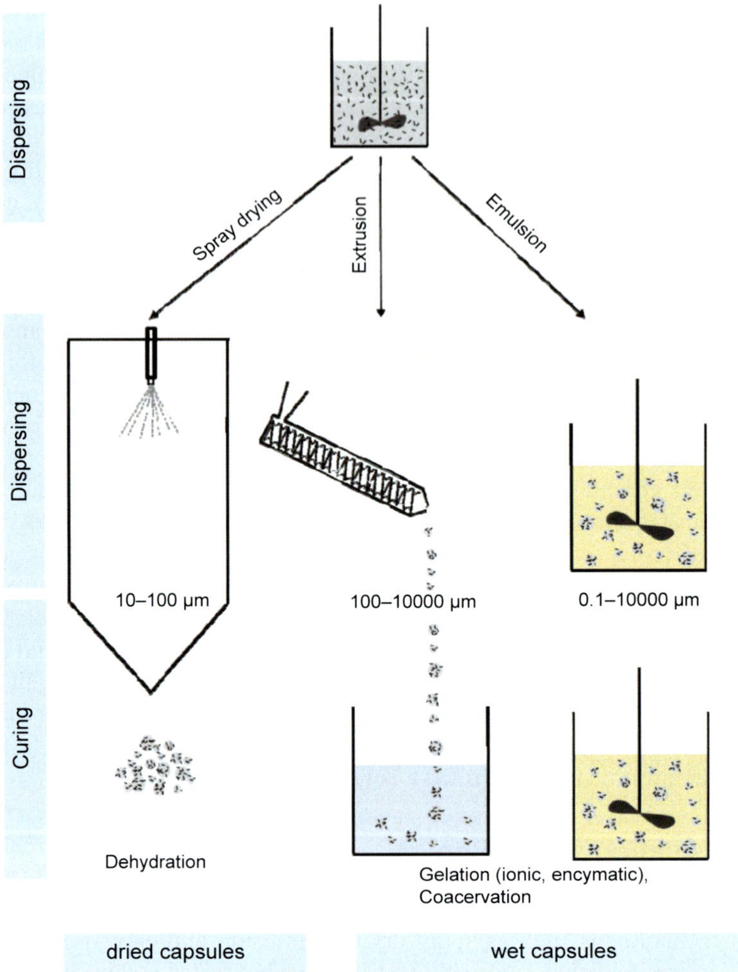


**Fig. 2.** Main encapsulation methods for probiotics.

probiotics for food applications. In the 1980s, intense research on the immobilization of lactic acid bacteria with the goal of an improved biomass production during continuous fermentation processes was performed. Commonly used materials were biopolymers like alginate, gellan gum, xanthan, carrageenan or mixtures thereof. An important requirement for these capsules was a low barrier for the mass transport of substrates and products as well as enough flexibility of the matrix to allow the growth of the immobilized bacteria. By adopting this method for food applications, with the main goal of protecting the bacteria from the surroundings, new demands arose and have become subject of intensive research (Heidebach et al. 2012).

For the microcapsule production by extrusion, the cell-matrix-mixture is mixed homogeneously and then pressed through a nozzle at high pressure to produce droplets.

These droplets are usually directly transferred into a hardening solution for curing. In most cases, divalent cationic solutions are used, which induce ionotropic gelation (Riaz and Masud 2013). From a process engineering point of view, this process can better be characterized as a dropping method. The obtained capsule size is dependent on the viscosity of the encapsulation material, the nozzle diameter and the drop height. Typical encapsulation materials for the extrusion method in food application are: alginate, κ-carrageenan, xanthan or denatured whey protein (Doherty et al. 2011) and combinations of these with other polysaccharides (Burgain et al. 2011, Heidebach et al. 2012, 2011, Krasaekoopt et al. 2003).

In general, extrusion is described in the literature as a gentle method which is frequently used for encapsulation (Vos et al. 2010). The produced capsules are regular in size and shape and are well-suited for lab scale production (Krasaekoopt et al. 2003). However, the resulting capsules are large (around 0.1–10 mm), wet and consequently not stable during long-term storage. The performance of this method is low, due to the slow capsule curing process that limits possible scaling-up of the process (Burgain et al. 2011). To increase the production rate, different developments have been implemented including the use of multiple-nozzle systems, spinning disc atomizers or jet cutter techniques (Burgain et al. 2011, Kailasapathy 2009, Vos et al. 2010).

## 2.2 Emulsion Technology

Emulsification is a common method for the encapsulation of probiotics, which also has its origin in the ICT. Therefore, after creating a homogeneous aqueous mixture of the probiotic cells and the matrix material, emulsification in commercial vegetable oil takes place to form a water-in-oil-emulsion as a first step. Then, the gelation is triggered to transform the droplets into water-insoluble microcapsules. The gelled particles are then separated from the oil by filtration or centrifugation, which then requires an additional washing step.

In this process, the oil phase is only an auxiliary material for the capsule production. To realize smaller droplet sizes, various food-grade surfactants such as polysorbates or phospholipids can be added. Besides the surfactant type and concentration, process parameters such as the shear intensity (Heidebach et al. 2009a), the volume fraction of continuous and discontinuous phase as well as the homogenization methods can be varied to control the particle size (Rathore et al. 2013). Various possible hardening mechanisms can be applied, which can be differentiated based on whether they are an internal or an external trigger. As an external trigger, hardening can be induced by the addition of an agent to the oil phase whereas internal gelation is activated from within the dispersed phase. The possible hardening mechanisms are explained in detail in the section 3.

## 2.3 Spray Drying

In recent years, spray drying has gained importance as an encapsulation process as an alternative to the methods developed for ICT. For encapsulation by spray drying, the mixture of bacteria and wall material is atomized into hot air followed by the rapid evaporation of water. The capsule formation can be influenced by various process parameters like feed rate, air flow rate and air temperature. The composition of the fluid matrix system also determines the final capsule morphology. Typical materials include gum arabic, starch (Burgain et al. 2011), reconstituted skim milk (Gardiner et al. 2000) or whey proteins (Picot and Lacroix 2004). The end product is a dry microcapsule powder. According to the used matrix, the capsules are either water-soluble or water-insoluble. Particularly in the field of dairy-based probiotic products, simple milk protein probiotic powders are sometimes declared as microencapsulated probiotics although there is no other hardening mechanism than the dehydration. Hence, the protecting effects of these powders for food application are limited. Such particles are water soluble and only have a barrier effect during dry storage, but not in liquid food products and during digestion (Heidebach et al. 2010). In contrast to the emulsion and extrusion technologies, the microcapsules are produced on a large scale and by a rapid, continuous process. Furthermore spray drying is a technology abundant in food industry.

The negative effect of relatively high temperature during processing on bacterial viability has to be considered as the main disadvantage of spray drying. A more sensitive alternative is the spray freeze drying. This method works at low temperatures, but current applications are limited. During spray freeze drying, the mixture is atomized in a cold atmosphere (e.g., liquid nitrogen) and afterwards the frozen droplet can be freeze-dried in a separate freeze dryer. This technique is gentler than the classical spray drying, but requires high energy consumption (Semyonov et al. 2010).

## 3. Encapsulation Mechanisms

The focus of most of the works in the area of microencapsulation of probiotics is, along with the different technologies, the wide field of the materials used for encapsulation of probiotics (Burgain et al. 2011, Heidebach et al. 2012). These articles give a good summary on the properties of the materials and so the approach in this section is a detailed classification regarding the underlying encapsulation and curing mechanisms of the different materials independent of the bacterial survival. The influence on the viability of the encapsulated cells is mentioned separately in section 4.1 in this chapter. Furthermore, this section only deals with the gelation and hardening mechanisms of milk proteins and milk protein-polysaccharide mixture. Livney (2010) described the range of structural properties and functionalities which make them highly suitable for their use as delivery systems for bioactive compounds. In addition, they are widely available, natural and GRAS (generally recognized as safe). Very important for the use of milk proteins is the ability of binding ions, the buffer capacity, excellent surface activity, gelation and self-assembly properties as well as the pH- and ionic strength-responsive swelling behaviour (Livney 2010). Furthermore, highly concentrated aqueous solutions of proteins have a relatively low viscosity compared to classical hydrocolloids, which enables the formation of dense matrices (Heidebach et al. 2012). In Table 1 an overview of different encapsulation systems based on milk protein matrices is shown.

**Table 1.** Overview on encapsulation mechanism based on milk protein matrices.

| **Encapsulation mechanism** | **Encapsulation technology** | **Matrix material** | **Encapsulated strain** | **Reference** |
|---|---|---|---|---|
| Cold-set gelation | Extrusion | Pre-denatured WPI | *Lactobacillus rhamnosus* GG | Doherty et al. (2011) |
| Complex coacervation | Emulsion | Pectin, WPI | *Lactobacillus acidophilus* La5 | Gebara et al. (2013) |
| Complex coacervation | Emulsion | Pectin, WPI | *Lactobacillus rhamnosus* CRL 1505 | Gerez et al. (2012) |
| Complex coacervation | Extrusion | Sterilised milk, alginate | *Lactobacillus bulgaricus* | Shi et al. (2013a) |
| Complex coacervation | Extrusion | Sterilised milk, carrageenan, locust bean gum | *Lactobacillus bulgaricus* | Shi et al. (2013b) |
| Dehydration | Spray drying | Skim milk, sweet whey | *Lactobacillus acidophilus* La5 | Maciel et al. (2014) |
| Dehydration | Spray drying | $WPC_{50}$ | *Lactobacillus acidophilus* Ki, *Lactobacillus paracasei* L26, *Bifidobacterium animalis* BB-12 | Rodrigues et al. (2011) |
| Dehydration & Cold-set gelation | Spray drying | Pre-denatured WPI | *Bifidobacterium breve* R070 (BB R070), *Bifidobacterium longum* (BB R023) | Picot and Lacroix (2004) |
| Enzymatic gelation | Emulsion | Micellar casein, WPI (native and denatured) | *Lactobacillus rhamnosus* GG | Burgain et al. (2013a) |
| Enzymatic gelation | Emulsion | Sodium caseinate | *Lactobacillus paracasei* ssp. *paracasei* F19, *Bifidobacterium lactis* Bb12 | Heidebach et al. (2009a) |
| Enzymatic gelation | Emulsion | Skim milk | *Lactobacillus paracasei* ssp. *paracasei* F19, *Bifidobacterium lactis* Bb12 | Heidebach et al. (2009b) |
| Enzymatic gelation | Emulsion | Pre-denatured WPI | *Bifidobacterium bifidum* F-35 | Zou et al. (2012) |

### 3.1 Gelation Mechanisms

In general, gels can be classified as physical or chemical gels. Chemical gels are formed by covalent binding whereas the physical gels are formed by non-covalent bindings via electrostatic, van der Waals, hydrophobic interactions or hydrogen bonding. The formation of physical gels is strongly dependent on time, temperature and milieu conditions. An example of a physical gel is the cold-set gel formed by beta-lactoglobulin (section 3.1.1). Chemical gels form a three dimensional structure induced by a chemical reaction like polycondensation, polyaddition or polymerisation, usually in the presence of crosslinking agents. For forming such structures, biopolymers have to be trifunctional, i.e., that they must have three or more intermolecular covalent groups. These gels are irreversible and usually have a high mechanical stability. The following section describes the gelation of milk proteins and biopolymers and their advantages/disadvantages for use as microencapsulation matrices (Guyot 2013).

#### *3.1.1 Cold-set Gelation*

Cold-set gels constitute an example of physical gels. The unique feature of cold-set gelation of milk proteins is that the activation step of the proteins, i.e., temperature-based exposure of reactive groups, is uncoupled from the subsequent gelation step (Alting et al. 2003) and, therefore, it is an excellent gelation mechanism for the encapsulation of probiotics. The addition of the core material, probiotics in this case, occurs after the thermal activation step and so there is no negative influence on the survival of the bacteria. Specifically, globular protein solutions are heated to denaturation temperatures under conditions at which a strong repulsion between the protein molecules exist. Thus, the molecules unfold but do not extensively aggregate. The solution conditions are set to optimize the strong repulsion during the heat treatment at pH values far from the pI and at low ionic strength. For the induction of gelation, the solution conditions are altered to promote the self-association of the proteins by changing the pH close to pI or by the addition of mono- or divalent cations like $Na^+$ or $Ca^{2+}$ (Matalanis et al. 2011).

To use the cold-set gelation process for the encapsulation of probiotics, two different approaches are possible. The first one was used by Doherty et al. (2011), who produced wet hydrogel capsules out of a whey protein solution. In their work they studied capsule formation in detail concerning the protein concentration and curing media. A pre-denatured (pH 7, 78°C, 45 min) whey protein isolate (WPI) solution with *Lactobacillus rhamnosus* GG was dropped into a sodium acetate buffer (0.5 M, pH 4.6 and 35°C) for curing. The structure of the capsules can be influenced by the protein concentration. Optimal results were obtained with a concentration of 9% (w/v). Capsule shape was improved to form rounder capsules by addition of polysorbate 20 to reduce interfacial tension.

The other possibility for applying this kind of gelation is a two-step process, as developed by Picot and Lacroix (2004). Probiotic bacteria were added to a heat-denatured WPI-solution (10% w/w, 80°C, 30 min), and the mixture was spray dried. The final capsule formation took place during the next step, the rehydration in a $CaCl_2$ solution, to induce the cold-set gelation of the whey proteins.

### *3.1.2 Enzymatic Gelation*

Enzymatic gelation is a rather new hardening method for microencapsulation of probiotics. Transglutaminase (TGase) is an enzyme, which forms inter- and intramolecular iso-peptide bonds between the amino acid residual lysine and glutamine (DeJong and Koppelman 2002). Heidebach et al. (2009b) used Tgase to covalently crosslink sodium caseinate. This kind of gelation takes some time according to the kinetics after TGase addition, so that the protein-cell mixture is still liquid some time to allow the following process steps, in particular the emulsion step. Full gelation depends on the enzyme concentration as well as protein content and requires ~2 h, depending on environmental conditions and composition.

Inspired by the method of Heidebach et al. (2009b), Zou et al. (2012) developed a microencapsulation process by TGase gelation of pre-denatured whey protein isolate (8 and 10% protein content). The pre-denaturation step was necessary because the globular whey proteins are a poor substrate for TGase. Via heat-treatment, the proteins unfold and the active groups become more susceptible to the polymerization, as the authors assume. They showed that the optimal capsule properties concerning gel strength and storage modulus were reached at native pH, 10 units TGase/g WPI and a protein concentration of 10 wt %. The storage modulus (G′) is an indicator of the degree of elasticity of a material (Norton et al. 2011) and in this case, it is one indicator for more stable capsules during processing, storage in food products and digestion. The G′ measured by Heidebach et al. (2009b) after 180 min for sodium caseinate gels was 10x higher than the gels from denatured whey protein.

Alternative to TGase, another enzymatic gelation method for encapsulation was also developed by Heidebach et al. (2009a) using the enzyme chymosin. Based on the classical so called rennet coagulation process from cheese manufacturing, a new method for encapsulation was generated. Skim milk concentrate was incubated with chymosine for 1 h at 5°C. During this time, the hydrolysis of the hairy layer of the casein micelles takes place and the steric repulsion is greatly reduced. In the second step, aggregation is induced by increasing the temperature of the emulsified and enzymatically treated skim milk concentrate resulting in microparticle gelation. For this encapsulation mechanism, only the hydrolysis step was enzymatically driven as gelation took place due to mainly hydrophobic interaction induced and increased by the raised temperature. Based on this work, Burgain et al. (2013a) used the same encapsulation process but refined the matrix system. Instead of skim milk concentrate, they used micellar casein alone or in combination with native or denatured WPI or a mixture of both. These recent examples show that the enzymatic gelation of milk proteins has a high potential as a mild encapsulation process of probiotics.

### *3.2 Complex Coacervation*

Complex coacervation is the formation of electrostatic complexes between proteins and polysaccharides, which commonly occurs between oppositely charged biopolymers (Goh et al. 2009, Kruif et al. 2004). For the formation of the complexes, attractive interactions are necessary which may result from non-specific electrostatics, van der Waals interactions, hydrophobic interactions and/or hydrogen bonding. However, the most important forces for complex coacervation are electrostatic forces. The manipulation of these interactions and hence the production of different microstructures can be achieved by controlling internal (pH, ionic strength, biopolymer ratio, molecular weight and charge of biopolymers) and external (temperature, pressure and shear rate) factors. The formation of such complexes often leads to better techno-functional properties than by using the single proteins or polysaccharides alone. Because of these properties, in the last decades it has become an important trend in encapsulation science to use protein-polysaccharide complexes for the microencapsulation of bioactive components. The coacervation is a good encapsulation mechanism for probiotics in view of encapsulation capacity and controlled liberation of the core material by mechanical stress, temperature or pH change (Rathore et al. 2013). However, detailed information on the molecular structure is still lacking and so it is a challenge to describe the creation of such encapsulation systems and the later release properties (Goh et al. 2009). Milk proteins and especially whey proteins are commonly used in this context in combination with polysaccharides like alginate, chitosan and pectin.

One encapsulation matrix based on the complex coacervation is a alginate-milk system developed by Shi et al. (2013a). In their work, the formulation of the matrix was varied by adding different ratios of alginate to sterilised milk. For complexation, droplets of the matrix with *Lactobacillus bulgaricus* were dropped in a 100 mM $CaCl_2$ solution where they remained for 30 min. The pH value of the solution was not mentioned. A validation as to which of the two materials had a greater influence on the capsule morphology was difficult because capsules out of pure alginate or milk were not considered in this study (Shi et al. 2013a).

Complex coacervation can also be used to create biopolymer coatings on biopolymer matrixcapsules. A pectin-whey protein system was established by Gerez et al. (2012). The pectin capsules were formed by spraying a pectin-butter emulsion containing *L. Rhamnosus* CRL 1505 in a $CaCl_2$ (2%, pH 4) curing bath. The gelation mechanism in this case was the ionotropic gelation of pectin. Afterwards the pectin capsules were transferred to a whey protein solution with a pH value below the pI to form a continuous layer on the pectin capsules. This method was also used by Gebara et al. (2013), who studied the behaviour of *L. acidophilus* La5. Another example of complex coacervation to prepare coated microcapsules was given by Shi et al. (2013b). They dropped sterilized milk containing *L. Bulgaricus* into a $CaCl_2$-solution (16 g/100 g). For the final coating, the formed capsules were transferred into carrageenan and locust bean gum mixture (pH 7.0) for 10 min.

### 3.3 Dehydration

A commonly used encapsulation process, as mentioned in section 2.3, is spray drying. The main hardening mechanism in this case is the dehydration process *per se* and, consequently the resulting capsules/powders are mostly water-soluble. So they are not suitable for applications in moist food products. The solubility properties can be improved by using a matrix material which can be made insoluble during drying or rehydration. A suitable strategy in this context is the pre-denaturation of a protein solution before drying and afterwards gelation is induced during rehydration of the powder in a curing media.

Maciel et al. (2014) produced water-soluble capsules by spray drying at 160/85–95°C (air inlet temperature/outlet temperature) with *L. acidophilus* La-5 in skim milk or sweet whey matrix. The dehydration for the formation of water-soluble capsules was also used to encapsulate three different lactic acid strains (Rodrigues et al. 2011). As matrices, whey protein concentrate 50 ($WPC_{50}$) with and without L-cysteine-HCl was used and dried at 160/75°C. In contrast, a process for water insoluble capsules was established by Picot and Lacroix (2004). During spray drying, dehydration was used as the first capsule forming step. During the final capsule formation, water insoluble capsules were formed by cold-set gelation in a $CaCl_2$ solution.

In general, drying conditions influence the capsule/powder formation and later morphology. From the drying technological point of view, there is ample knowledge regarding the surface composition of powders. Kim et al. (2009) found out that for higher feed solid contents and drying temperatures less fat and protein appear on the surface. The knowledge about these influences could be used to modify the barrier properties by specific application of the individual components under controlled process conditions. However, studies in this field for microencapsulation technology are limited.

## 4. Influence of Production, Storage and Incubation in Simulated Gastric Fluids on the Viability of Encapsulated Probiotic Bacteria

The main goal of microencapsulation is the protection of encapsulated material from harsh environmental stress factors (Burgain et al. 2011, Heidebach et al. 2012). Nevertheless, production processes, storage conditions and consumption are detrimental factors (Fig. 1) that may lead to a reduction in the viability of the encapsulated bacteria. The impact of these processes should be compared with free (i.e., not encapsulated) bacteria. These aspects are addressed in the next section, which focuses on bacteria behavior during microcapsule production, storage and incubation in simulated gastric fluids.

### *4.1 Influence of Microencapsulation Process on the Viability of Probiotic Bacteria*

As described above, the encapsulation methods for probiotics have to be gentle to maintain a high viability of the probiotics without changing the active structure of probiotics. Hence, the most important evaluation criterion for the encapsulation process is the encapsulation yield (EY), which describes the percentage of living encapsulated probiotics after production. Most commonly, EY is defined as a combination of the efficacy of entrapment and survival of viable cells during microencapsulation (Annan et al. 2008, Doherty et al. 2011, Vodnar and Socaciu 2012). However mostly, differences between the loss of cells during production and/or the inactivation during processing cannot be made. Depending on the encapsulation technology and mechanism used, the values may show a large variability. In many cases, the EY values are not even explicitly mentioned (Doherty et al. 2012, Gerez et al. 2012) or the reported reductions are not explained in detail. A further limiting factor is the complete recovery of the encapsulated probiotics. Therefore it has to be ensured that the capsules are completely disintegrated while maintaining the full viability of the probiotic bacteria. Additionally in many cases, the swelling behavior or reduction in water content during production is not considered in the calculation of EY. Although it is well known that hydrocolloid gels will swell or shrink and as a result, the mass and volume of the resulting capsules will change thereby altering the mass balance calculation and potential EY. Different ways to consider these changes have been described in literature. Burgain et al. (2013a) used the total quantity of the capsules after the whole process whereas Heidebach et al. (2009a,b) considered the protein content values of the starting protein-cell mixture and of the capsules after production. For encapsulation by spray drying with parallel water removal, the change in relative dry matter content also has to be taken in account. In most cases, therefore, the survival rate is specified relative to the dry matter content, which makes it easily comparable (Maciel et al. 2014, Picot and Lacroix 2004).

For the comparison of the different processes and process steps regarding the capability of the encapsulation, the definition of Cook et al. (2012) is helpful. They subdivide the ideal microencapsulated probiotic product into two categories. The microcapsules are either wet, water-insoluble capsules for an immediate application in food products or they are a dried powder with long-term storage stability. Depending on the particular demand, both categories have their application in food products. It should be mentioned that for both kind of capsules, different stress conditions occur during production. For instance, extrusion or emulsion techniques are mild encapsulation processes with a gentle hardening mechanism. This type of capsule commonly leads to high EYs and viabilities after production, but the capsules are compromised on storage-stability. On the other hand, dried capsules can further be separated by either a one-step production of dried capsules or via a sequential process involving separately produced capsules with a downstream drying step. For dried capsules, the stress on the probiotics is much higher because of dehydration and/or high temperatures. However, the final product has the added value that the capsules are of long-term storage stability. In Table 2 gives an overview on the different processes with the related matrix material and the resulting Yield/Viability.

### *4.1.1 Extrusion*

The low stress potential for bacteria during an extrusion process was demonstrated by Doherty et al. (2012). They showed that the extrusion through a 150 μm nozzle with and without capsule formation had no significant influence on the viability of the encapsulated bacteria. They analysed these by cell enumeration and visualized the results by confocal microscopy with live-dead staining of the bacteria. The resulting EY was not mentioned in this publication. Shi et al. (2013a) analysed the influence of polymer concentration in alginate-milk capsules on EY. The mixture of sterilized milk and alginate was dropped in 100 mM $CaCl_2$ for hardening. The obtained capsules had an EY from 90–95%, with no significant influence of the polymer concentrations.

In a study by Gerez et al. (2012) double layer pectin-whey protein (native or denatured) capsules were produced by extrusion. The EY for both matrices was around 84.35 ± 0.60%. An explanation as to whether the decrease resulted from

**Table 2.** Influence of the microencapsulation process on the viability of probiotics.

| Encapsulation technology | Matrix material | Encapsulated strain | Viability after processing | Reference |
|---|---|---|---|---|
| Emulsion | Mixtures of micellar casein, WPI (native and denatured) | *L. rhamnosus* GG | 64 ± 4% up to 97 ± 2% | Burgain et al. (2013a) |
| Emulsion | Mixture of pectin and WPI | *L. acidophilus* La5 | 84.35 ± 0.60% | Gebara et al. (2013) |
| Emulsion/freeze drying | Pre-denatured WPI | *B. bifidum* F-35 | 0.27 ± 0.06% | Zou et al. (2012) |
| Emulsion | Sodium caseinate | *L. paracasei* ssp. *paracasei* F19<br>*B. lactis* Bb12 | 70 ± 15%<br>93 ± 22% | Heidebach et al. (2009a) |
| Emulsion | Skim milk | *L. paracasei* ssp. *paracasei* F19<br>*B. lactis* Bb12 | 105 ± 20%<br>115 ± 13% | Heidebach et al. (2009b) |
| Extrusion | Mixtures of sterilised milk and carrageenan-locust bean gum | *L. bulgaricus* | ~60% | Shi et al. (2013b) |
| Extrusion | Mixtures of sterilised milk and alginate | *L. bulgaricus* | 90–95% | Shi et al. (2013a) |
| Spray drying | Pre-denatured WPI | *B. breve* R070<br>*B.longum* R023 | 25.67 ± 0.12%<br>1.44 ± 0.16% | Picot and Lacroix (2004) |
| Spray drying | Skim milk<br>Sweet whey | *L. acidophilus* La5 | 77.73 ± 5.95%<br>75.43 ± 4.03% | Maciel et al. (2014) |
| Spray drying | $WPC_{50}$ with[a] and without[b] L-cysteine-HCl | *L. acidophilus* Ki,<br>*L. paracasei* L26<br>*B. animalis* BB-12 | 41.7%[a], 44.0%[b]<br>45.7%[a], 30.7%[b]<br>39.0%[a], 22.7%[b] | Rodrigues et al. (2011) |

a loss in viability due to the process or from a loss of bacteria in the surrounding curing media was not given (Gebara et al. 2013). Another bilayer capsule type was made from milk with a layer consisting of a carrageenan-locust bean gum mixture (Shi et al. 2013b). Sterilized milk was dropped into a $CaCl_2$ solution for gelling and afterwards the capsules were transferred into the carrageenan-locust bean gum solution. For this kind of capsules, an EY of about 60% was reached. This relatively low EY was explained by the high $CaCl_2$ concentration (16 g/100 mL) in the hardening bath that was harmful for the probiotics, as already shown by Reid et al. (2005) for whey protein microspheres. The film-forming ability of the milk protein is poor compared to alginate (Shi et al. 2013b).

### *4.1.2 Emulsion Technology*

Higher EYs are normally reached by using the emulsion technique. Heidebach et al. (2009a,b) have encapsulated *L. paracasei* ssp. *paracasei* F19 and *B. lactis* Bb12 in a sodium caseinate and skim milk matrix. The EYs for F19 and Bb12 were 70 ± 15% and 93 ± 22%, respectively using sodium caseinate capsules (Heidebach et al. 2009b). With skim milk capsules, even higher EYs of 105 ± 20% (F19) and 115 ± 13% (Bb12) were reached. EY values above 100% are within the range of variability in assessing the survival rate of microorganisms in capsules. These results are explained by the fact that no detrimental steps occur during this process. It is also mentioned that the missing viable cells for the sodium caseinate encapsulation possibly occurred as a result of cells not being released during mechanical disintegration, as also noted by Annan et al. (2008).

The emulsion process of Heidebach et al. (2009a) for a pure milk protein system mixed with *L. rhamnosus* GG was also used by Burgain et al. (2013a). For four different matrices with the same protein content, they obtained EYs ranging from 64 ± 4% up to 97 ± 2% with the highest yield obtained for the matrix consisting of casein and denatured whey proteins. These findings were explained by the interaction between the strain and whey proteins, namely the adhesive forces that helped better retain the bacteria in the gel network. The influence of the mechanical release process on the yield for the different matrices was not explained.

### *4.1.3 Spray Drying*

The encapsulation technology with the highest loss in viability is spray drying. However, in contrast to the emulsion and extrusion methods, the produced microcapsules can have better storage stability and thus a further drying step is unnecessary as already mentioned (Cook et al. 2012). Regarding the influence of drying on bacterial survival, in particular of spray drying, there is a large field of research for the preservation of bacterial cultures available (Peighambardoust et al. 2011, Santivarangkna et al. 2007). This is described in detail in chapter 12 (Alternative Drying Processes for Probiotics and Starter Cultures).

For microencapsulation by spray drying, the suitability of different matrix materials is an important factor. *Lactobacillus acidophilus* La-5 was dried in two different matrices of skim milk and sweet whey at the same total solids content at

180/85–95°C (Maciel et al. 2014). The post-drying EYs of 75.43 ± 4.03% for sweet whey and 77.73 ± 5.95% for skim milk microparticles were obtained. The moisture content and particle sizes were not significantly different. These high survival rates were explained by the presence of lactose and milk proteins, which act as protectants during drying. The differences in the concentration of proteins (casein/whey protein) and lactose seemed to have no effect.

Picot and Lacroix (2004) encapsulated *B. breve* R070 (BB R070) and *B. longum R023* (BL R023) by spray drying (160/80°C) in denatured whey protein suspensions directly or after emulsifying in milk fat. Against the expectations of the authors, the best results were obtained for drying in the pre-denatured whey protein suspension and not for the whey protein/milk fat matrix. BB R070 showed with 25.67 ± 0.12% better heat stability compared to BL R023 with 1.44 ± 0.16%.

The presence of oxygen during spray drying is known as an additional detrimental factor for anaerobic bacteria (Ghandi et al. 2012). Based on this knowledge, Rodrigues et al. (2011) have investigated the effect of the oxygen scavenger L-cysteine-HCl on the survival of *L. acidophilus* Ki, *L. paracasei* L26 and *B. animalis* BB-12 after spray-drying with 160°C/75°C. The survival rate was calculated considering the protein content before and after drying. The influence of the oxygen scavenger was shown to be strain-dependent.

For the sake of completeness freeze dried capsules have to be mentioned in this context, too, because this process may lead to a reduction in viable cell count in addition to the capsule forming step. Although freeze drying is considered a gentle drying process, the survival rates of viable cells after drying were <35% for F19 and <45% for Bb12 (Heidebach et al. 2010). Combining this with the EY of these capsules, F19 only showed ~25% survival whereas that of Bb12 was ~40%. Hence, such secondary processes should be considered when evaluating EY.

### 4.2 Viability of Probiotics after Production or in Food Products as a Function of Storage Time

The general aim of microencapsulation is the protection of the bacteria during storage to obtain viable cell numbers as high as possible in the final product. Two main methods for the investigation of storage behaviour are used: The assessment of viability of encapsulated bacteria stored directly after capsule production and of encapsulated bacteria added to a food product or model systems for food product. For all studies, the viable cell numbers of encapsulated bacteria are compared with non-encapsulated ones under the same conditions.

In general, for encapsulated cells in dried state the same behaviour regarding the influence of temperature, humidity and the other environmental conditions occur like for dried cells. These influences and mechanisms are explained in detail in Chapter 15 (Storage Stability of Probiotic Powder). In contrast to free cells an additional barrier is formed by encapsulation, which may further protect the bacteria during storage. Hence, in the following section the presented results are focused on the effect of encapsulation.

### 4.2.1 Storage Stability of Probiotics after Production

Maciel et al. (2014) studied the storage stability of spray-dried *L. acidophilus* La-5 in a skim milk or sweet whey matrix in sealed PE-bags at 4 and 25°C for 90 d. The powders were classical skim milk and sweet whey powders loaded with bacteria. The results showed a better survival in the skim milk than in the sweet whey matrix and for the same matrix a better survival at lower temperatures. However, a comparison with free cells and an explanation for the different stabilities regarding the matrix was missing (Maciel et al. 2014).

Also with milk protein as matrix Rodrigues et al. (2011) encapsulated *L. acidophilus* Ki, *L. paracasei* L26 and *B. animalis* BB-12 in $WPC_{50}$ with and without L-cysteine-HCl. The samples were stored either at 5°C under oxygen and without controlled humidity to simulate refrigeration conditions over 180 d or at 22°C with defined humidity of 12, 32 and 45%. For the refrigeration conditions under oxygen, all survival rates were better than for the higher temperature, and the encapsulated cells showed in all cases a higher stability than the free cells. The stability behavior and the effect of the scavenger were strain-dependent. The storage of the spray dried encapsulated cells at 22°C appeared feasible for storing probiotic bacteria over a longer time period without being below the recommended number of bacteria required for a food application.

Heidebach et al. (2010) studied the storage stability of dried probiotic microcapsules based on protein hydrogels and showed that the loss of living cells was strongly dependent on the storage conditions and the probiotic strain. Under optimized conditions (4°C/11% RH) after 3 m, there was a 1 log loss of encapsulated *B. lactis* Bb12 (~$3.8 \times 10^9$ CFU/g powder) compared to the free cells. For *L. paracasei* ssp. *paracasei* F19, the reduction was ~2 log higher and at the end of storage only $1.7 \times 10^8$ CFU/g powder living cells were detected.

To promote protective effects, the combination of various materials with different properties is promising. Alginate-milk microcapsules have shown a high potential for protecting probiotics during storage for 30 d at 4°C. At the end of a refrigeration period, full viability of encapsulated *L. bulgaricus* was preserved but an explanation for this excellent result is missing. In comparison, the number of free cells dropped down from 10 to 2.3 log CFU/mL (Shi et al. 2013a). The same strain was also encapsulated in carrageenan-locust bean gum coated milk microspheres and stored under 30 d at 4°C (Shi et al. 2013b) and showed again no loss of viability. Both examples show that polysaccharide-protein capsules may have a high potential for the stabilisation of probiotics during storage (Shi et al. 2013a,b).

### 4.2.2 Storage Stability of Probiotics in Food Products

Another common strategy to evaluate the efficacy of encapsulation is via addition of free and microencapsulated probiotics directly to food products. Typical products which are used for these tests are dairy products and fruit juices because of their low pHvalues. In this regard, spray-dried whey protein microcapsules with *B. breve* R070 (BB R070) and *B. longum* R023 (BL R023) were tested by Picot and Lacroix (2004), who assessed the stability in yoghurt at 4°C for 28 d. The stability for BL R070 in yoghurt after 28 d was higher than for the free bacteria but nevertheless the final cell number was below the required therapeutic effect. Unlike the survival of encapsulated BB R023 which was worse compared with free bacteria.

In contrast, Ying et al. (2013) tested the storage stability of free and spray dried WPI/resistant starch (RS) encapsulated *L. rhamnosus* GG in apple juice at 4 and 25°C for 5 weeks. They investigated the influence of different WPI to RS ratios on the pH value and viability of *L. rhamnosus* GG and found much higher storage stability for encapsulated cells, i.e., free bacteria were no longer detectable after 1 w at 4 and 25°C. Also at both temperatures, the protection by the simple RS capsules was worst. At 4°C over the whole 5 w, a constant decrease in the cell number was observed where as at 25°C only in the first two days the cell number decreased followed by a bacterial growth up to more than 1 log higher than the initial cell number for capsules containing WPI. This growth accompanied the decreasing pH of the samples (Ying et al. 2013). These results further show the high capability of milk protein based matrices for the stabilisation of probiotic bacteria.

Despite the huge amount of data clear conclusions regarding the suitability of a given encapsulation process or matrix regarding probiotic storage stability are difficult to derive. Besides the obvious susceptibility of different bacterial strains, the microencapsulation technique, addition of protective substances or the food matrix in which probiotics are delivered are without exception important parameters (Maciel et al. 2014). All studies mentioned above show that the "perfect" encapsulation system does not exist. It is crucial to define for which application and especially for which product the probiotics should be encapsulated. So the selection of an encapsulation process and material should be contingent on the food product, storage time and conditions and additionally the bacterial strain.

### 4.3 Viability of Probiotics at Simulated Gastric Conditions

Different media and set-ups have been described in the literature to evaluate the gastric tolerance of encapsulated bacteria. An overview about the different used fluids with the detailed formulations is shown in Table 3. Heidebach et al. (2009a,b) used a simple simulated gastric fluid (SGF) according to the Parmacopedia (2008) without digestive enzymes for evaluating the stability of probiotics encapsulated in rennet or sodium caseinate gels at low pH. For the rennet capsules, pH 2.5 was tested for the survival of the two strains *L. paracasei* ssp. *paracasei* F19 and *B. lactis* Bb12 (Heidebach et al. 2009a). Additionally the sodium caseinate encapsulation of these strains at two pH values (2.5 and 3.6) was compared (Heidebach et al. 2009b). In all cases, there was an improvement in the survival rate by encapsulation depending on strain and pH value. In contrast to the sodium caseinate capsules, the rennet capsules showed structural degradation depending on the incubation pH when studied under the microscope. The high stability of the sodium caseinate capsules was explained with the covalent network formed by the TGase whereas the rennet capsules were mainly stabilised by hydrophobic interactions.

Burgain et al. (2013a) studied the digestion of rennet casein-based microcapsules *in situ* with regard to particle size and shape and additionally the survival rate of *L. rhamnosus* GG. Four different matrix compositions of emulsion rennet capsules were

**Table 3.** Viabilty of probiotics at simulated gastric conditions.

| Encapsulation technology and matrix material | Encapsulated strain | Fluid composition | Improvement in survival by encapsulation | Reference |
|---|---|---|---|---|
| Emulsion capsules; micellar casein, WPI (native and denatured) | *L. rhamnosus* GG | 34 mM NaCl<br>3.2 mg/L pepsin<br>pH 2.5 with HCl<br>120 min | 2.0–5.0 CFU/g | Burgain et al. (2013a) |
| Pectin-whey protein and Pectin-whey protein/whey protein capsules | *L. rhamnosus CRL 1505* | KCl (1.1g/L), NaCl (2.0 g/L), $CaCl_2$ (0.1 g/L) and $K_3H_2PO_4$ (0.4 g/L) Mucin (3.5 g/L), pepsin (0.2 g/L)<br>pH 1.2[a] or 2.0[b]<br>120 min | 8.0[a] and 3.0[b] log CFU/mL<br>6.0[a] and 6.0[b] log CFU/mL | Gerez et al. (2012) |
| Emulsion capsules; Sodium caseinate | *L. paracasei* F19<br>*B. lactis* Bb12 | 0.08 M HCl<br>containing 0.2% (w/v) NaCl<br>pH 2.5[c] and 3.6[d]<br>90 min | 2.0[c] and 2.0[d]CFU/g<br>3.0[c] and 0.1[d] CFU/g | Heidebach et al. (2009a) |
| Emulsion capsules; Skim milk | *L. paracasei* F19<br>*B. lactis* Bb12 | 0.08 M HCl<br>containing 0.2% (w/v) NaCl<br>pH 2.5<br>90 min | 0.8 log CFU/g<br>2.8 log CFU/g | Heidebach et al. (2009b) |
| Spray dried capsules; Skim milk, sweet whey | *L. acidophilus* La5 | KCl (1.12 g/L), NaCl (2.0 g/L), $CaCl_2$(0.11 g/L) and $K_3H_2PO_4$ (0.4 g/L) Mucin (3.5 g/L), pepsin (0.26 g/L)<br>pH 2.0 with HCl<br>120 min | 2.5–3.0 log CFU/mL | Maciel et al. (2014) |
| Spray dried capsules; Pre-denatured WPI | *B. breve* R070 ,<br>*B.longum* R023 | 0.1 N HCl<br>0.26 g/L pepsin<br>pH 1.9 with NaOH<br>30 min | 1 log CFU/mL<br>No protection | Picot and Lacroix (2004) |

| | | | | |
|---|---|---|---|---|
| Spray dried capsules; $WPC_{50}$ with and without L-cysteine-HCl | *L. acidophilus* Ki,<br>*L. paracasei* L26,<br>*B. animalis* BB-12 | 0.1 M HCl<br>25 mg/mLpepsin<br>pH 2.0 with HCl<br>60 min | 3.16–3.45 log CFU/g<br>0.23–0.28 log CFU/g<br>0.01–0.16 log CFU/g | Rodrigues et al. (2011) |
| Extrusion capsules; denatured WPI | *L. rhamnosus* GG | *Ex vivo* porcine gastric juice<br>pH 2.0, 2.4, 3.4<br>180 min | 8.5–9.0 log CFU/mL | Doherty et al. (2012) |
| Extrusion capsules; Sterilised milk, alginate | *L. bulgaricus* | 2.0 g/L NaCl<br>pH 2.0 or 2.5<br>120 min | 8.0 log CFU/g<br>10.0 log CFU/g | Shi et al. (2013a) |
| Extrusion capsules; Sterilised milk, carrageenan, locust bean gum | *L. bulgaricus* | 2.0 g/L NaCl<br>pH 2.0 or 2.5<br>120 min | 8.0 log CFU/g<br>10.0 log CFU/g | Shi et al. (2013b) |

compared. After addition to the SGF (with pepsin) they observed an increase in particle size because of capsule aggregation induced by the acidic environment, which was then followed by a reduction in the particle size. The slowest decrease was observed for the casein/denatured whey protein matrix, which also showed the best survival rates, compared to the other matrices and free *L. rhamnosus* GG. The behaviour of the encapsulated bacteria in casein/denatured whey protein matrix was explained by the specific interaction settling between the strain and whey proteins (Burgain et al. 2013b). They showed by AFM force spectroscopy that a specific adhesion between bacteria and whey proteins and non specific interactions with micellar casein exist. Due to these forces, bacteria can be more strongly retained in the gel network and so the survival rates after incubation in SGF are higher. The interactions with the dairy matrix are influenced by the nature of protein also from the nature of strain and the pH in the surrounding (Burgain et al. 2013b). By addition of denatured whey proteins the matrix stability was improved compared to the capsules of Heidebach et al. (2009a). However, it is hard to directly compare these studies, because Heidebach et al. (2009a) used skim milk concentrate as a raw material instead of micellar casein with/without whey proteins as well as different strains and SGJ compositions. Nevertheless both studies showed that by using this kind of matrix, a higher stability in the acid surrounding of the stomach could be achieved.

The stability of whey protein capsules and their protective effect was tested by Doherty et al. (2012). They used *ex vivo* porcine gastric fluid to realize more realistic conditions and analysed the survival of *L. rhamnosus* GG as a function of pH (2.0–3.4) and time for free cells, cell-whey protein mixtures and encapsulated cells. For free bacteria, they detected a reduction of 9 $\log_{10}$ cfu $mL^{-1}$ living cells within the first 30 min for all pH values. By mixing the bacteria with the native WPI suspension, they could improve the survival by 2.5 to 6 $\log_{10}$ cfu $mL^{-1}$ depending on the pH. In the case of the encapsulated bacteria, there was no reduction during the whole 180 min test period at pH 3.4 and the reduction at pH 2.4 and 2.0 was less than 0.5 $\log_{10}$ cfu $mL^{-1}$.

Spray dried $WPC_{50}$ capsules with and without L-cysteine-HCl were also studied for three different strains: *L. acidophilus* Ki, *L. paracasei* L26 and *B. animalis* BB-12 (Rodrigues et al. 2011). Gastrointestinal transit was simulated by a four-step model of the mouth, oesophagus, duodenum and ileum with the typical enzymes, pH levels and mechanical impact for free and encapsulated cells with L-cysteine-HCl. After a 60 min exposure to simulated gastric conditions with pepsin for encapsulated *L. acidophilus* Ki and *L. paracasei* L26, the survival was improved compared to free cells. Whether the improvement in survival for the two strains resulted from the physical stabilizing matrix effect of the capsules or mainly by the buffering capacity was not assessed in the paper. A comparison of the stability of spray-dried *Lactobacillus acidophilus* La-5 in skim milk or sweet whey matrix during exposure to simulated gastric conditions was made in modified SGF (pH 2.0, with pepsin) for 30, 60 and 90 min (Maciel et al. 2014). As reference, free cells without matrix material were treated under the same conditions. The results showed that both encapsulation matrices protected the bacteria compared to the free cells, but a significant difference between the matrices was not found (Maciel et al. 2014).

Gerez et al. (2012) compared pectin/whey protein (P/WP) and pectin-whey protein/whey protein (PWP/WP) capsules and their protective effect on *L. rhamnosus*

CRL 1505 in a SGJ at pH 1.2 and 2.0 after 0, 60 and 120 min. As environment NaCl and pepsin, and also the salts KCl, $CaCl_2$, $K_2HPO_4$ and mucin were used. At pH 2.0, they showed that both kinds of matrices protected the bacteria completely (around 8 $log_{10}$ cfu $mL^{-1}$). At pH 1.2 the survival with P/WP was excellent, whereas with P-WP/WP reductions of 4 $log_{10}$ cfu $mL^{-1}$ after 60 min and of 5 $log_{10}$ cfu $mL^{-1}$ after 120 min were detected. In the P/WP capsules, the carboxyl groups of the inner pectin capsule interacted with the coating layer of whey protein, whereas for the P-WP/WP, parts of the carboxyl groups already interact with pectin from inside the capsule which implies a weaker network.

Two other different polysaccharide/protein capsules were tested by Shi et al. (2013a,b) for the stabilisation of *L. bulgaricus.* In both studies, the same test conditions, i.e., an incubation in SGF at pH 2.0 and 2.5 for 120 min, were applied. After 1 min in SGF, the free cells were not detectable anymore. The carrageenan-locust bean gum coated milk capsules protected the bacteria at pH 2.5 completely and at pH 2.0 only a reduction from 10 to 8 $log_{10}$ cfu $g^{-1}$ was observed. This was explained with the protein buffering capacity in combination with the low porosity of the surface. A similar behaviour was shown for the alginate-milk protein capsules. The cell numbers were the same after 120 min. The only difference was that incubated samples also showed full viability at pH 2.0 within the first 30 min and afterwards a reduction to 8 $log_{10}$ cfu $g^{-1}$. They also showed that with increasing milk protein concentration, the viability of the bacteria improved, which was explained by the higher density of the hydrogel matrix and the resulting lower diffusion rate.

It is difficult to establish a clear conclusion regarding the stability of the different encapsulation systems during exposure to simulated gastric conditions as there are too many different parameters that have an impact on the results. On the one hand there is great variation in the parameters and recipes of the different *in vitro* digestion models and on the other hand there are also a wide range of encapsulation processes and materials as well as encapsulated bacterial strains, which make it hard to compare the results. Based on this complexity, Minekus et al. (2014) developed a harmonised static, *in vitro* digestion method for food to allow the production of comparable data in the future. Moreover, it should be the subject of further research to evaluate independent criteria like diffusion coefficients or the mechanical stability to describe the capsules and to make different encapsulation techniques and materials more easily assimilable. To date, all results have shown that milk proteins—alone or in combination—as encapsulation material have a high potential for the stabilization of bacteria under gastric conditions among other things because of their excellent barrier-forming and buffering properties.

## 5. Evaluation of the Release Properties

To ensure sufficient probiotic efficacy, the bacteria must reach the human intestine and retain the functionality and viability. Therefore, validation of the release properties of encapsulated bacteria is necessary but only little work has directly explored the release. It is generally understood that the capsules have to disintegrate to release their content in a controlled manner, i.e., not too early but in the intestine. The main mechanisms, as described by Matalanis et al. (2011) are diffusion, erosion, fragmentation and swelling/shrinking. The release is also driven by factors such as pH changes, proteolytic enzyme breakdown and osmotic stress. Additionally, the release process is determined by the residence time of the capsules in the human body.

As noted earlier, the stability of spray dried *L. acidophilus* La-5 in skim milk or sweet whey matrix after exposure to simulated intestinal conditions was investigated by Maciel et al. (2014). After incubation in simulated gastric juice (SGJ), the solutions were adjusted to pH 7.0 and the enzyme pancreatine was added to simulate the intestinal surroundings. The viability of free and encapsulated cells was determined after 180 min. In all cases a growth was detected after the SGJ without a significant difference of the matrix material. Like in most other studies, only the viability and not the release characteristics were observed (Maciel et al. 2014).

Gebara et al. (2013) investigated the viability of the same strain, *L. acidophilus* La5, encapsulated in pectin and pectin-whey protein microcapsules during exposure to simulated gastrointestinal conditions. They incubated the free and encapsulated bacteria for 2 h in simulated gastric fluid (SGF) with mucin and pepsin at pH 1.2 and 3.0 followed by 3 h in simulated intestinal fluid (SIF) with pancreatin. In this work after 3 h in the SIF, a complete degradation of the coated capsules took place, whereas the uncoated pectin capsules remained stable. Following incubation in pH 3.0 SGF, the free cell count dropped to 0.14% of the initial load whereas that of the encapsulated cells was only reduced to 12.3–14.8% depending on the type of encapsulation. It is unclear whether this reduction was the result of the negative influence of the SIF or an incomplete release of bacterial cells. For the lower pH-value of the SGF before incubation in the SIF for free and pectin-whey protein encapsulated bacteria a bacterial growth was measured. Similar growth was also observed by Maciel et al. (2014) during incubation in SIF but also after incubation in a SGJ at pH 2.0.

In the studies of Shi et al. (2013a,b) in a simple model solution without enzymes (pH 6.8, 50 mM, $KH_2PO_4$), the release properties for two different sterilised milk matrix systems (alginate and carrageenan) were directly investigated. For alginate-milk capsules, within the first 30 min a very slow release took place, afterwards a sharp increase in the release curve was detected. At 60 min, the release of the whole content was determined. They also mentioned that with increasing alginate concentration, the release rate was significantly slowed down due to a higher matrix density. In their other study, carrageenan/locust bean gum milk protein microcapsules were studied. The bacteria were completely released for this capsule type after 45 min with 50% of them liberated after just 10 min (Shi et al. 2013a,b).

Finally, Doherty et al. (2012) tested the release properties of a *Lactobacillus* strain in whey protein isolate capsules under *ex vivo* gastrointestinal conditions. They found that during an incubation time of 180 min, the release in the gastric fluid was negligibly low, whereas the release in the following *ex vivo* intestinal fluid was very fast. Two-thirds of the bacteria were released during the first 5 min in the intestinal fluid and after 30 min the release was complete. They also tested capsule degradation in phosphate buffer saline (PBS), but saw no change in either capsule stability or bacteria release. This showed that a neutral pH alone was insufficient to promote release and that enzymatic action was necessary.

This chapter presents that there is a huge diversity of investigated encapsulation technologies, mechanisms and used bacterial strains. Therefore, it is necessary to develop new methods for a better comparability regarding storage and gastric stability tests. At the moment, additionally a great variation in the used testing parameters exists. An important step in the right direction was done by Minekus et al. (2014). They developed a harmonised static, *in vitro* digestion method for food with strict regulations not only concerning the recipe of the fluids, but also for the activity tests of the used enzymes. Only by such strict regulations a direct comparison and evaluation of different systems is possible. Besides these analytical aspects, a deeper knowledge about the barrier properties of the capsules is necessary. Therefore, future research should focus on the investigation of capsule structure and morphology like it was done by Burgain et al. (2013b). They investigated, by AFM force spectroscopy, the adhesion between bacteria and matrix and showed that depending on the composition of the matrix the interactions are different. By combination of these results with the survival rates it is possible to gain a deeper understanding of underlying mechanisms. Further possibilities to obtain independent criteria can be the determination of diffusion coefficients or the characterisation of the mechanical stability of the capsules.

Nevertheless the current research shows, that milk protein based microcapsules have a high potential for protecting probiotics from the unfavourable surroundings because of their excellent barrier-forming and buffering properties. Especially the combination with polysaccharides for the creation of new systems is promising and should be a subject of future research.

## References


Parmacopedia. 2008. European pharmacopoeia: Published in accordance with the Convention on the Elaboration of a European Pharmacopoeia (European treaty series no. 50). 6th ed. Strasbourg: Council of Europe.

Agrawal, R. 2005. Probiotics: An emerging food supplement with health benefits. Food Biotechnology 19(3): 227–46.

Alting, A.C., R.J. Hamer, de C.G. Kruif and R.W. Visschers. 2003. Cold-Set Globular Protein Gels: Interactions, Structure and Rheology as a Function of Protein Concentration. Journal of Agricultural and Food Chemistry 51(10): 3150–6.

Annan, N.T., A.D. Borza and L.T. Hansen. 2008. Encapsulation in alginate-coated gelatin microspheres improves survival of the probiotic *Bifidobacterium adolescentis* 15703T during exposure to simulated gastro-intestinal conditions. Food Research International 41(2): 184–93.

Burgain, J., C. Gaiani, C. Cailliez-Grimal, C. Jeandel and J. Scher. 2013a. Encapsulation of *Lactobacillus rhamnosus* GG in microparticles: Influence of casein to whey proteins ratio on bacterial survival during digestion. Innovative Food Science & Emerging Technologies (19): 233–42.

Burgain, J., C. Gaiani, G. Francius, A. Revol-Junelles, C. Cailliez-Grimal, S. Lebeer, H. Tytgat, J. Vanderleyden and J. Scher. 2013b. *In vitro* interactions between probiotic bacteria and milk proteins probed by atomic force microscopy. Colloids and Surfaces B: Biointerfaces 104: 153–62.

Burgain, J., C. Gaiani, M. Linder and J. Scher. 2011. Encapsulation of probiotic living cells: From laboratory scale to industrial applications. Journal of Food Engineering 104(4): 467–83.

Champagne, C.P., N.J. Gardner and D. Roy. 2005. Challenges in the Addition of Probiotic Cultures to Foods. Critical Reviews in Food Science and Nutrition 45(1): 61–84.

Cook, M.T., G. Tzortzis, D. Charalampopoulos and V.V. Khutoryanskiy. 2012. Microencapsulation of probiotics for gastrointestinal delivery. Journal of Controlled Release 162(1): 56–67.

DeJong, G.A.H. and S.J. Koppelman. 2002. Transglutaminase Catalyzed Reactions: Impact on Food Applications. Journal of Food Science 67(8): 2798–806.

Doherty, S.B., M.A. Auty, C. Stanton, R.P. Ross, G.F. Fitzgerald and A. Brodkorb. 2012. Survival of entrapped *Lactobacillus rhamnosus* GG in whey protein micro-beads during simulated *ex vivo* gastrointestinal transit. International Dairy Journal 22(1): 31–43.
Doherty, S.B., V.L. Gee, R.P. Ross, C. Stanton, G.F. Fitzgerald and A. Brodkorb. 2011. Development and characterisation of whey protein micro-beads as potential matrices for probiotic protection. Food Hydrocolloids 25(6): 1604–17.
Gardiner, G.E., E. O'Sullivan, J. Kelly, M.A.E. Auty, G.F. Fitzgerald, J.K. Collins, R.P. Ross and C. Stanton. 2000. Comparative Survival Rates of Human-Derived Probiotic *Lactobacillus paracasei* and *L. salivarius* Strains during Heat Treatment and Spray Drying. Applied and Environmental Microbiology 66(6): 2605–12.
Gebara, C., K.S. Chaves, M.C.E. Ribeiro, F.N. Souza, C.R. Grosso and M.L. Gigante. 2013. Viability of *Lactobacillus acidophilus* La5 in pectin–whey protein microparticles during exposure to simulated gastrointestinal conditions. Food Research International 51(2): 872–8.
Gerez, C., G. Font de Valdez, M. Gigante and C. Grosso. 2012. Whey protein coating bead improves the survival of the probiotic *Lactobacillus rhamnosus* CRL 1505 to low pH. Letters in Applied Microbiology 54(6): 552–6.
Ghandi, A., I.B. Powell, T. Howes, X.D. Chen and B. Adhikari. 2012. Effect of shear rate and oxygen stresses on the survival of *Lactococcus lactis* during the atomization and drying stages of spray drying: A laboratory and pilot scale study. Journal of Food Engineering 113(2): 194–200.
Goh, K.K.T., A. Sarkar and H. Singh. 2009. Milk protein-polysaccharide interactions. pp. 347–76. *In*: M. Boland, H. Singh and A. Thompson (eds.). Milk proteins: From expression to food. London: Academic.
Guyot, C. 2013. Transglutaminase-induzierte und Transglutaminase-unterstützte Gele aus Milchproteinen. München: Dr. Hut.
Hansen, L.T., P.M. Allan-Wojtas, Y. Jin and A.T. Paulson. 2002. Survival of Ca-alginate microencapsulated *bifidobacterium* spp. in milk and simulated gastrointestinal conditions. Food Microbiology 19(1): 35–45.
Heidebach, T., P. Först and U. Kulozik. 2009a. Microencapsulation of probiotic cells by means of rennet-gelation of milk proteins. Food Hydrocolloids 23(7): 1670–7.
Heidebach, T., P. Först and U. Kulozik. 2009b. Transglutaminase-induced caseinate gelation for the microencapsulation of probiotic cells. International Dairy Journal 19(2): 77–84.
Heidebach, T., P. Först and U. Kulozik. 2010. Influence of casein-based microencapsulation on freeze-drying and storage of probiotic cells. Journal of Food Engineering 98(3): 309–16.
Heidebach, T., P. Först and U. Kulozik. 2012. Microencapsulation of Probiotic Cells for Food Applications. Critical Reviews in Food Science and Nutrition 52(4): 291–311.
Heidebach, T., E. Leeb, P. Först and U. Kulozik. 2011. Microencapsulation of Probiotic Cells. pp. 293–312. *In*: M. Fanun (ed.). Colloids in Biotechnology. Boca Raton, FL: CRC Press.
Hernández-Rodríguez, L., C. Lobato-Calleros, D. Pimentel-González and E. Vernon-Carter. 2014. *Lactobacillus plantarum* protection by entrapment in whey protein isolate: κ-carrageenan complex coacervates. Food Hydrocolloids 36: 181–8.
Hörmannsperger, G., T. Clavel, M. Hoffmann, C. Reiff, D. Kelly, G. Loh, M. Blaut, G. Hölzlwimmer, M. Laschinger, D. Haller and B. Zhang. 2009. Post-Translational Inhibition of IP-10 Secretion in IEC by Probiotic Bacteria: Impact on Chronic Inflammation. PLoS ONE 4(2): e4365.
Kailasapathy, K. 2009. Encapsulation technologies for functional foods and nutraceutical product development. CAB Reviews 4(033).
Kim, E.H., X.D. Chen and D. Pearce. 2009. Surface composition of industrial spray-dried milk powders. 2. Effects of spray drying conditions on the surface composition. Food Powder Technology 94(2): 169–81.
Krasaekoopt, W., B. Bhandari and H. Deeth. 2003. Evaluation of encapsulation techniques of probiotics for yoghurt. International Dairy Journal 13(1): 3–13.
Kruif, C.G. de, F. Weinbreck and de R. Vries. 2004. Complex coacervation of proteins and anionic polysaccharides. Current Opinion in Colloid & Interface Science 9(5): 340–9.
Livney, Y.D. 2010. Milk proteins as vehicles for bioactives. Current Opinion in Colloid & Interface Science 15(1-2): 73–83.
Maciel, G., K. Chaves, C. Grosso and M. Gigante. 2014. Microencapsulation of *Lactobacillus acidophilus* La-5 by spray-drying using sweet whey and skim milk as encapsulating materials. Journal of Dairy Science.

Matalanis, A., O.G. Jones and D.J. McClements. 2011. Structured biopolymer-based delivery systems for encapsulation, protection, and release of lipophilic compounds: 25 years of Advances in Food Hydrocolloid Research. Food Hydrocolloids 25(8): 1865–80.

Minekus, M., M. Alminger, P. Alvito, S. Ballance, T. Bohn, C. Bourlieu, F. Carrière, R. Boutrou, M. Corredig, D. Dupont, C. Dufour, L. Egger, M. Golding, S. Karakaya, B. Kirkhus, S. Le Feunteun, U. Lesmes, A. Macierzanka, A. Mackie, S. Marze, D.J. McClements, O. Ménard, I. Recio, C.N. Santos, R.P. Singh, G.E. Vegarud, M.S.J. Wickham, W. Weitschies and A. Brodkorb. 2014. A standardised static *in vitro* digestion method suitable for food—an international consensus. Food Funct. 5(6): 1113.

Norton, I.T., F. Spyropoulos and P. Cox. 2011. Practical food rheology: An interpretive approach. Ames, Iowa: Blackwell.

Onwulata, C. 2012. Encapsulation of New Active Ingredients. Annu. Rev. Food Sci. Technol. 3(1): 183–202.

Peighambardoust, S., A. Golshan Tafti and J. Hesari. 2011. Application of spray drying for preservation of lactic acid starter cultures: a review. Trends in Food Science & Technology 22(5): 215–24.

Picot, A. and C. Lacroix. 2004. Encapsulation of bifidobacteria in whey protein-based microcapsules and survival in simulated gastrointestinal conditions and in yoghurt. International Dairy Journal 14(6): 505–15.

Rathore, S., P.M. Desai, C.V. Liew, L.W. Chan and P.W.S. Heng. 2013. Microencapsulation of microbial cells. Journal of Food Engineering 116(2): 369–81.

Reid, A.A., J.C. Vuillemard, M. Britten, Y. Arcand, E. Farnworth and C.P. Champagne. 2005. Microentrapment of probiotic bacteria in a $Ca^{2+}$-induced whey protein gel and effects on their viability in a dynamic gastro-intestinal model. Journal of Microencapsulation 22(6): 603–19.

Riaz, Q.U.A and T. Masud . 2013. Recent Trends and Applications of Encapsulating Materials for Probiotic Stability. Critical Reviews in Food Science and Nutrition 53(3): 231–44.

Rodrigues, D., S. Sousa, T. Rocha-Santos, J.P. Silva, J.M. Sousa Lobo, P. Costa, M.H. Amaral, M.M. Pintado, A.M. Gomes, F.X. Malcata and A.C. Freitas. 2011. Influence of l-cysteine, oxygen and relative humidity upon survival throughout storage of probiotic bacteria in whey protein-based microcapsules. International Dairy Journal 21(11): 869–76.

Santivarangkna, C., U. Kulozik and P. Foerst. 2007. Alternative Drying Processes for the Industrial Preservation of Lactic Acid Starter Cultures. Biotechnol. Progress 23(2): 302–15.

Semyonov, D., O. Ramon, Z. Kaplun, L. Levin-Brener, N. Gurevich and E. Shimoni. 2010. Microencapsulation of *Lactobacillus paracasei* by spray freeze drying. Food Research International 43(1): 193–202.

Shi, L., Z. Li, D. Li, M. Xu, H. Chen, Z. Zhang and Z. Tang. 2013a. Encapsulation of probiotic *Lactobacillus bulgaricus* in alginate–milk microspheres and evaluation of the survival in simulated gastrointestinal conditions. Journal of Food Engineering 117(1): 99–104.

Shi, L., Z. Li, Z. Zhang, T. Zhang, W. Yu, M. Zhou and Z. Tang. 2013b. Encapsulation of *Lactobacillus bulgaricus* in carrageenan-locust bean gum coated milk microspheres with double layer structure. LWT—Food Science and Technology 54(1): 147–51.

Vodnar, D.C. and C. Socaciu. 2012. Green tea increases the survival yield of Bifidobacteria in simulated gastrointestinal environment and during refrigerated conditions. Chemistry Central Journal 6(1): 61.

Vos P de, M.M. Faas, M. Spasojevic and J. Sikkema. 2010. Encapsulation for preservation of functionality and targeted delivery of bioactive food components. International Dairy Journal 20(4): 292–302.

Ying, D., S. Schwander, R. Weerakkody, L. Sanguansri, C. Gantenbein-Demarchi and M.A. Augustin. 2013. Microencapsulated *Lactobacillus rhamnosus* GG in whey protein and resistant starch matrices: Probiotic survival in fruit juice. Journal of Functional Foods 5(1): 98–105.

Zou, Q., X. Liu, J. Zhao, F. Tian, H. Zhang, H. Zhang and W. Chen. 2012. Microencapsulation of *Bifidobacterium bifidum* F-35 in Whey Protein-Based Microcapsules by Transglutaminase-Induced Gelation. Journal of Food Science 77(5): M270.

# CHAPTER 17

# Novel Dairy Probiotic Products

*Shah Nagendra Prasad*

## 1. Introduction

A number of dairy products have evolved over the past decades including cultured buttermilk, sour cream, yoghurt, ymer, taetmojolk, folkjolk, dahi, zabadi, kefir, kumys, and cheeses such as Cheddar cheese and cottage cheese. These products use a number of organisms including *Lactococcuslactis* ssp. *lactis*, *L. lactis* ssp. *cremoris*, *Lc. lactis* ssp. *lactis* biovar. *diacetylactis, Leuc. mesenteroides* ssp. *cremoris, Lactobacillus delbrueckii* ssp. *bulgaricus,* and *Streptococcus thermophilus* are used for making various dairy products. However, the microorganisms used for making these dairy foods do not survive in the gastrointestinal tract and do not provide probiotic effects. Hence, the recent trend is to add probiotic bacteria including *Lactobacillus acidophilus*, *Bifidobacterium* spp., *L. casei, Lb. casei* (strain Shirota), *L. reuteri*, and *L. rhamnosus* GG to dairy foods. Probiotic organisms are incorporated in yogurt, sour cream, milk powder, buttermilk and frozen fermented desserts. Cheddar cheese, cottage cheese, Mozzarella cheese and cheese based spreads are relatively new products in which probiotic organisms are incorporated. Yet, a more recent trend has been to use starter bacteria or probiotic bacteria to release bioactive peptides from milk proteins in probiotic products for additional health benefits. Examples of such products include Calpis and Evolus. Both products contain bioactive peptides such as angiotensin I-converting enzyme (ACE) inhibitory peptides including isoleucine-proline-proline (IPP) or valine-proline-proline (VPP) for lowering blood pressure. A number of dairy based functional foods containing bioactive peptides have already been commercialized.

## 2. Fermented Dairy Products

Fermentation of milk is one of the ancient techniques for food preservation. Although no records confirm the origin of fermented dairy products, it is commonly believed that

School of Biological Sciences, The University of Hong Kong.
E-mail: npshah@hku.hk

the roots of fermented dairy products date back to the Persian times in the Middle East (Ross, Morgan and Hill 2002). 'Yoghurt' was the first name given to the fermented milk product by the ancient Turkish nomads residing in Asia, in the 8th century (Prajapati et al. 2003). With the invasion of the Tartars, Hans and Mongols, the skills of making fermented milk products was also carried far and wide, extending to Russia and Europe. With the growth of civilization, the production of various fermented milk products such as Laban Rayeb or Laban Khad, became a common practice in ancient Egypt. Danone, the Europe based food giant, undertook the first commercial production of yogurt in 1922 in Spain and the art of yogurt production has been advancing ever since. Dairy products are now an important part of the diet in most parts of the world, and are manufactured in many regions around the world. West European countries lead the world market in the production and per capita consumption of yogurt and other fermented products. The consumption of fermented dairy products, especially yogurt, is increasing in Canada, Australia and USA.

The Southeast Asian belt including Iran continues to be the significant region for the production of fermented dairy products. Some of the fermented milk products and their origin are listed in Table 1.

Metchnikoff pioneered the research on the health benefits of the fermented milk at the start of the 20th century (Metchnikoff 2004). He observed that the average life span of the Bulgarian peasants consuming large quantity of yogurt and fermented milks, was exceptionally high and this finding stimulated the commencement of an incessant research on the health benefits of fermented dairy products. Growing concern about healthy diet in the global population has given a great impetus to the production of functional food. In particular, dairy products have contributed significantly to the functional food market, where they account for around 60% of the sales of functional food in Europe. Dairy products are also among the popular functional foods in the USA, where they are ranked second, with consumers spending $5.0 billion on functional dairy foods in 2004 (Vierhile 2006).

Owing to the increasing awareness on the health benefits of the fermented dairy products today, there are a variety of them in the market. The dairy industry has come up with innovative variations in fermented products, on the basis of their nutritional and/or textural characteristics (Shah 2003). Not only does milk fermentation conserve the vital nutrients of milk, but also enhances the nutritive quality thereby providing additional health benefits to the consumers. In different parts of the world, these fermented foods are used as an ingredient in cooked product or are consumed as condiment, snack, dessert, drink or spread. Factors such as source of milk (from different animals), diverse microflora, addition of supplements and preservative and the processing techniques like freezing, drying and concentrating contribute to the variations in fermentation of milk.

A number of organisms are used in making fermented dairy foods including *Lactococcuslactis* ssp. *lactis*, *L. lactis* ssp. *cremoris*, *Lc. lactis* ssp. *lactis* biovar. *diacetylactis, Leuc. mesenteroides* ssp. *cremoris, Lactobacillus delbrueckii* ssp. *bulgaricus,* and *Streptococcus thermophilus* (Table 2). Most of these starter organisms produce lactic acid as the sole product of fermentation leading to milk coagulation. Lactic acid also prevents spoilage and increases the shelf life of dairy products

**Table 1.** Some important fermented milk products.

| Product | Country of origin | | Characteristics |
|---|---|---|---|
| Airan | Central Asia, Bulgaria | | Milk soured by *Lb. bulgaricus* and used as refreshing beverages |
| Bulgarian milk | Bulgaria | | Very sour milk fermented by *Lb. bulgaricus* |
| Dahi | India | | Milk soured by using previous day soured milk as starter |
| Kefir | Caucasusian China | | Milk fermented with kefir grains containing lactobacilli and yeast. Lactic acid, alcohol, and $CO_2$ give sparkling characteristics |
| Kishk | Egypt, Arab world | | Fermented milk mixed with par boiled wheat and dried |
| Kumys or Kumiss | Central Asia, Mongol, Russia | | Mare milk is fermented by lactobacilli and yeast. Lactic acid, alcohol, and $CO_2$ give sparkling characteristics |
| Laban | Egypt | | Soured milk coagulated in earthenware utensils |
| Langfil or Tattemjolk | Sweden | | Milk fermented with slime producing lactococci |
| Leben | Iraq | | Milk soured with yogurt bacteria and whey is partially drained by hanging the curd in clothes |
| Mast | Iran | | Natural yogurt with firm consistency and cooked flavor |
| Skyr | Iceland | | Fermented milk made from ewe milk with the help of rennet and starter |
| Taette | Norway | | Viscous fermented milk |
| Trahana | Greece | | Fermented milk made by mixing wheat flour followed by drying |
| Villi | Finland | | Viscous milk fermented with lactic acid bacteria and mold |
| Yakult | Japan | 1935 AD | Highly heat-treated milk fermented by *L. casei* Shirota strain |
| Ymer | Denmark | | Protein fortified milk fermented with leuconostocs and lactococci |
| Yogurt or yoghurt | Turkey | 800 AD | Custard like sour fermented milk |

substantially. Health benefits of fermented dairy foods are also increased due to fermentation. Some health benefits are listed and discussed in Shah (2006a,b, 2007, 2010, 2012a,b) and Prajapati and Shah (2011).

Lactic acid bacteria used as starter culture and other organisms used for making fermented products as well as other dairy foods do not survive in gastrointestinal tract; hence the trend has been to add probiotic bacteria for providing additional health benefits beyond what is provided by fermented dairy foods.

## 3. Probiotic Organisms

A number of bacteria are used as probiotic, including *Lactobacillus acidophilus*, *L. casei*, *Leuconostoc*, *Pediococcus*, *Bifidobacterium*, and *Enterococcus*, but the main organism believed to have probiotic characteristics are *L. acidophilus*, *Bifidobacterium*

**Table 2.** Some fermented milks and their starter cultures.

| **Product** | **Starter organisms** |
|---|---|
| Butter milk (Bulgarian) | *Lactobacillus delbrueckii* ssp. *bulgaricus* |
| Butter milk (cultured) | *Lactococcus lactis* ssp. *lactis, Lactococcus lactis* ssp. *cremoris, Lactococcus lactis* ssp. *lactis* biovar. *diacetylactis, Leuconostoc mesenteroides* ssp. *cremoris* |
| Cultured cream | *Lc. lactis* ssp. *lactis, Lc. lactis* ssp. *cremoris, Lc. lactis* ssp. *lactis* biovar. *diacetylactis*, and *Leu. mesenteroides* ssp. *cremoris* |
| Dahi | *Lb. delbrueckii* ssp. *bulgaricus, Streptococcus thermophilus* or *Lc. lactis* ssp. *lactis, Lc. lactis* ssp. *cremoris, Lc. lactis* ssp. *lactis* biovar. *diacetylactis,* and *Leu. mesenteroides* ssp. *cremoris* |
| Filjolk | *Lc. lactis* ssp. *lactis, Lc. lactis* ssp. *cremoris, Lc. lactis* ssp. *lactis* biovar. *diacetylactis* |
| Kefir | *Lc. lactis* ssp. *lactis, Lc. lactis* ssp. *cremoris, Lc. lactis* ssp. *lactis* biovar. *diacetylactis,* and *Leu. mesenteroides* ssp. *dextranicum, Str. thermophilus, Lb. delbrueckii* ssp. *bulgaricus, Lb. acidophilus, Lb. helveticus, Lb. kefir, Lb. kefiranofaciens, Kluyveromycesmarxianus, Sacchamyces* spp. |
| Kumys | *Lb. acidophilus, Lb. delbrueckii* ssp. *bulgaricus, Sacchamyces lactis, Torula koumiss* |
| Tatmjolk | *Lc. lactis* ssp. *lactis, Lc. lactis* ssp. *cremoris, Lc. lactis* ssp. *lactis* biovar. *diacetylactis* |
| Viilli | *Lc. lactis* ssp. *lactis, Lc. lactis* ssp. *cremoris, Lc. lactis* ssp. *lactis* biovar. *diacetylactis, Geotrichumcandidum* |
| Yakult | *Lb. paracasei* ssp. *casei* |
| Yogurt | *Lb. delbrueckii* ssp. *bulgaricus, Str. thermophilus* |
| Cheddar cheese | *Lc. lactis* ssp. *lactis*, *L. lactis* ssp. *cremoris* |
| Cottage cheese | *Lc. lactis* ssp. *lactis*, *L. lactis* ssp. *cremoris* |

spp., and *L. casei*. *Lactobacillus* and *Bifidobacterium* have a long and safe history in the manufacture of dairy products and have been initially isolated from the gastrointestinal tract.

At present, there are 56 species of the genus *Lactobacillus*. Details of these organisms can be found in Shah (2006b, 2012b). *L. acidophilus* tends to grow slowly in milk, taking 20–24 h to complete the fermentation process. Ironically, most strains of *L. acidophilus* do not survive well in low pH of fermented milk and it is difficult to maintain large numbers in the product. The poor growth is partly related to low concentration of small peptides and free amino acids in milk, which would be insufficient to support the bacterial growth.

*Bifidobacterium* are normal inhabitants of the human gastrointestinal tract. Selected strains survive stomach and intestinal transit and reach the colon in abundant numbers. Presently, 32 species in the genus *Bifidobacterium* are recognized, many of which have been isolated from dental caries, faeces and vagina, animal intestinal tracts or rumen, and honeybees. The species found in humans are: *B. adolescentis*, *B. angulatum*, *B. bifidum*, *B. breve*, *B. catenulatum*, *B. dentium*, *B. infantis*, *B. longum*, and *B. pseudocatenulatum*. *B. breve*, and *B. infantis* (Shah and Lankaputhra 2002). Further details about these organisms can be found in Shah (2002, 2011a,b).

A number of health benefits are claimed in favour of products containing probiotic organisms including prevention of gastrointestinal infections, improvement in lactose metabolism, antimutagenic and anticarcinogenic properties, reduction in serum cholesterol, anti-diarrhoeal properties, immune system stimulation, improvement in inflammatory bowel disease and suppression of *Helicobacter pylori* infection (Shah 2004). Some of the health benefits are well established, while other benefits have shown promising results in animal models. Well documented health benefits include relief of lactose maldigestion symptoms, shortening of rotavirus diarrhoea and immune modulation. Promising results available for selected strains are management of inflammatory bowel disease, irritable bowel syndrome, urinary tract infection and allergies. Health benefits not substantiated include lowering of serum cholesterol level, lowering of blood pressure and management of constipation.

Health benefits provided by probiotic bacteria are strain specific, and not species- or genus-specific. It is important to note that no strain will provide all claimed benefits, not even strains of the same species, and not all strains of the same species will be effective against defined health conditions. The strains *L. rhamnosus* GG (Valio), *S. cerevisiae* Boulardii (Biocodex), *L. casei* Shirota (Yakult), and *B. animalis* Bb -12 (Chr. Hansen) have the strongest human health efficacy data with respect to management of lactose malabsorption, rotaviral diarrhoea, antibiotic-associated diarrhoea, and *Clostridium difficile* diarrhoea (Shah 2006a,b, 2010, 2012a,b). Further details about the health benefits can be found in Shah (2006a,b, 2010, 2012a,b, 2013).

There are currently over 70 probiotic products that are produced worldwide including sour cream, buttermilk, yoghurt, powdered milk, and frozen desserts. More than 53 different types of milk products containing probiotic organisms are marketed in Japan alone. Probiotics are also incorporated in drinks, and marketed as supplements which include tablets, capsules and freeze-dried preparations. Though the European market is rich in dairy based probiotic products, it is dominated mainly by yogurt.

## 4. Probiotic Dairy Foods

Although yogurts and fermented milks still remain the main media for incorporation of probiotic cultures, current trend has been to incorporate probiotic organisms in other products such as milk-based desserts, cheeses based-dips and spreads, powdered milk for newborn infants, ice cream, butter, mayonnaise, and various types of cheese (Champagne et al. 2005, Saad et al. 2008).

Fermented beverages are the most traditional and widely consumed probiotic products. Fermentation is usually an inexpensive process, requiring a low-cost technology. Moreover, many bases can be used for these types of products: milk, cereals, fruits, roots, or even mixture of these. From country to country, different mixtures of ingredients and technologies are employed to develop many types of probiotic dairy beverages, as outlined by Prado et al. (2008).

In fermented probiotic products, it is common to use probiotic bacteria mixed together with other starter bacteria suited for the fermentation of the specific product. In many cases consumers find products fermented with *L. delbrueckii* spp. *Bulgaricus* to be too acidic and with too strong acetaldehyde flavor (typically found in yogurt). Therefore, in several probiotic formulations, *L. delbrueckii* spp. *Bulgaricus* is omitted. Examples of such cultures are the so-called ABT cultures (ABT standing for *L. acidophilus*, *Bifidobacterium*, and *S. thermophilus*) (Shah 2003).

The carbonation as a method for improving bacterial viability in fermented milk of probiotic pasteurized milk incorporated with *Lactobacillus acidophilus* and *Bifidobacterium bifidum* was studied by Vinderola et al. (2000a). The higher acidity levels of carbonated and lactic acidified samples enhanced growth and metabolic activity of the starter during fermentation, thus reducing incubation time.

Hernandez-Mendoza et al. (2007) and Zoellner et al. (2009) developed a whey-based probiotic product with *Lactobacillus reuteri* and *Bifidobacterium bifidum*.

Antunes et al. (2009) developed and evaluated the acceptability of a probiotic buttermilk-like fermented milk product flavored with different fruit flavors (strawberry, vanilla, mint, graviola, and cupuaçu) with added sucrose or sucralose.

Table 3 lists probiotic based fermented products and probiotic organisms contained therein. Most products contain one or more strains of probiotic organisms. These organisms grow very slowly in milk taking up to 24 h to complete the fermentation process and the trend has been to use *S. thermophilus* or *L. delbrueckii* ssp. *bulgaricus* or both, to speed up the fermentation process. If only *S. thermophilus* alone is added in addition to probiotic organism(s), the fermentation time could be 7–8 h, however, when both organisms (*S. thermophilus* or *L. delbrueckii* ssp. *bulgaricus*) are added the fermentation time could be as little as 4 h.

**Table 3.** Fermented probiotic based products and probiotic organisms therein.

| | | |
|---|---|---|
| ACO-yogurt | Switzerland | *S. thermophilus, L. bulgaricus, L. acidophilus* |
| Cultura-AB | Denmark | *L. acidophilus, B. bifidum* |
| AB-yogurt | Denmark | *L. acidophilus, B. bifidum, S. thermophilus* |
| Biogarde | Germany | *L. acidophilus, B. bifidum, S. thermophilus* |
| Bifighurt | Germany | *B. longum, S. thermophilus* |
| Gefilac | Finland | *L. casei* GG (rhamnosus) |
| Yakult | Japan | *L. casei* |
| MiruMiru | Japan | *L. acidophilus, L. casei* |
| Biokys | Slovakia | *B. bifidum, L. acidophilus, Pediococcus acidilactici* |
| Ofilus | France | *B. bifidum, B. longum, L. acidophilus, S. lactis, S. cremoris* |
| Gaio | Denmark | *E. faecium, S. thermophilus* |
| LC1 | Europe | *L. acidophilus* La1 |
| Symbalance* | Switzerland | *L. reuteri, L. casei, L. acidophilus* |
| Probiotic plus oligofructose | Germany | *L. acidophilus, L. bifidus, LA7* |
| ProCult3 | Germany | *B. longum BB536* |
| Actimel Orange | Germany | *L. acidophilus* |
| Fysiq* | Netherlands | *L. acidophilus Gilliland* |
| DanActive Immunity | USA | *L. casei Immunitas TM* |
| Activia | USA | *B. animalis* |
| Morinaga BB536 NomuDrinkingyogurt | Japan | *B. longum BB536* |
| Megumi series | Japan | *L. gasseri* sp., *Bifidobacterium* sp. |
| Actimel (Danone) | France | *L. casei Immunitass DN-114001* |
| Yakult Light | Netherlands | *L. casei Shirota* |
| Vitality | UK | *Bifidobacterium* sp., *L. acidophilus* |

Some milk-based probiotic products that have been developed by a number of researchers worldwide are listed in Table 4. Although a large number of products are developed, it is important to understand that most of them are not commercially available and they have been tested for research purposes on a small or laboratory scale.

**Table 4.** Some dairy probiotic products developed worldwide.

| Food Product | Reference |
|---|---|
| Acidophilus milk drink | Itsaranuwat et al. 2003 |
| Synbiotic acidophilus milk | Amiri et al. 2008 |
| Acidophilus "sweet" drink | Speck 1978 |
| Biogarde, mil-mil, and acidophilus milk with yeasts | Gomes and Malcata 1999 |
| Fermented lactic beverages supplemented with oligofructose and cheese whey | Castro et al. 2009 |
| Fermented goat's milk | Martín-Diana et al. 2003 |
| Regular full-fat yogurts | Aryana and Mcgrew 2007 |
| Açai yogurt | Almeida et al. 2008, 2009 |
| Mango soy fortified probiotic yogurt | Kaur et al. 2009 |
| Peanut milk yogurt | Isanga and Zhang 2009 |
| Traditional Greek yogurt | Maragkoudakis et al. 2006 |
| Corn milk yogurt | Supavititpatana et al. 2008 |
| Banana-based yogurt | Sousa et al. 2007 |
| Cheddar cheese | Ong and Shah 2008a,b, 2009 |
| Minas fresco cheese | Souza and Saad 2009 |
| Feta cheese | Kailasapathy and Masondole 2005 |
| Cheese from caprine milk | Kalavrouzioti et al. 2005 |
| Kazar cheese | Özer et al. 2008 |
| Semi-hard reduced-fat cheese | Thage et al. 2005 |
| White-brined cheese | Yilmaztekin et al. 2004 |
| Cottage cheese | Blanchette et al. 1996 |
| Canestrato pugliese hard cheese | Corbo et al. 2001 |
| Argentine fresco cheese | Vinderola et al. 2000b |
| Goat semi-solid cheese | Gomes and Malcata 1999 |
| Petit-Suisse cheese supplemented with inulin and/or oligofructose | Cardarelli et al. 2008 |
| Crescenza cheese | Gobbetti et al. 1997 |
| Manufacture of Turkish Beyaz cheese | Kiliç et al. (2009) |
| Synbiotic ice cream | Homayouni et al. (2008), Caglar et al. (2008) |
| Probiotic ice cream | Akin et al. (2007), Akalin and Erisir (2008), Hekmat and McMahon (1992), Cruz et al. (2009) |
| Low-fat ice cream | Akalin and Erisir (2008) |
| Acidophilus milk-based ice cream | Andrighetto and Gomes (2003) |
| Acidophilus butter and progurt | Gomes and Malcata (1999) |
| Guava-based mousse | Saad et al. (2007) |
| Coconut flan | Saad et al. (2008) |

Adapted from Granato et al. (2010)

Yogurt is one of the most popular dairy products world wide in which probiotic organisms are incorporated; with more innovations in the product formulations, its popularity has increased in the recent times. Often, probiotic bacteria such as, *L. acidophilus*, *L. casei*, *L. rhamnosus* GG, *L. johnsonii* LA1 and *Bifidobacterium* spp. are added in yogurt.

The growth of probiotic bacteria in milk is slower than the typical yogurt starter bacteria; the fermentation time for making yogurt with the yogurt starter culture is around 4 h, whereas, with only the probiotic bacteria, it could take as long as 24 h. Therefore, the trend is to use both, the typical yogurt starter bacteria as the main starter culture and probiotic bacteria as an adjunct starter (Shah 2000b).

Kefir is another product that is becoming popular in the Western world. Kefir is fermented milk made from kefir grains and is a self-carbonated refreshing drink prepared from lactic acid and yeast fermentation of milk. Common lactic acid bacteria involved in kefir fermentation include *L. delbrueckii* ssp. *bulgaricus*, *L. acidophilus, L. kefir*, *L. kefirogranum*, *Enterococcus faecalis*, *E. faecium*, *S. thermophilus*, and *Lactococcus lactis* ssp. *cremoris*. The commonly identified yeasts from kefir are *Candida kefir*, *Kluyveromycesmarxianus*, *K. lactis*, *Saccharomyces kefir*, and *S. cerevisiae*. Probiotic kefir has been marketed recently with added probiotic organisms (Fanworth 2005).

Kumys is yet another product that is becoming popular in the Western world (Koroleva and Robinson 1991). Kumys is produced by a mixed yeast and lactic acid fermentation of mare's milk. However, due to limited availability of mare's milk, cow's milk is used for industrial production of kumys. To level the differences between these two milks, cow's milk is often fortified by the addition of sucrose or whey, enabling fermentation of cow's milk that is comparable to that of mare's milk. *L. delbrueckii* ssp. *bulgaricus*, *L. acidophilus* and *Saccharomyces lactis* commonly constitute the starter culture for kumys production. Kumys possesses a characteristic sour flavor, attributed to the production of ethyl alcohol and lactic acid, along with fizziness that is imparted by the carbon dioxide produced. The alcohol content in kumys varies from 1.0 to 2.5%, whereas the lactic acid content ranges from 0.7 to 1.8% (Kosikowski and Mistry 1966).

Skyr is a concentrated cultured dairy product produced from skim milk and is very popular in Iceland. The product is traditionally concentrated by draining whey using cheese-cloth. *Lb. delbrueckii* ssp. *bulgaricus*, and the *Lb. helveticus* cultures as well as a lactose fermenting yeast are used for fermentation. The fermentation is conducted in two stages: 40ºC for 4–5 h to favour the growth of thermophilic lactic starter culture, and 18ºC for 18 h to promote the growth of the yeast (Tamime and Marshall 1997).

Viiliis another cultured dairy product popular in Finland. The fermentation is primarily achieved by the culture consisting of *Lc. lactis* ssp. *lactis* biovar. *diacetylactis* and *Leuconostoc mesenteroides* ssp. *cremoris* and lactose fermenting mold *Geotrichumcandidum*. The cream layer is usually covered with the mold and the product is eaten with a spoon. Pasteurized milk is fermented at approximately 20ºC until a final acidity of 0.9% is reached (Tamime and Marshall 1997).

## 5. Probiotic Ice Cream and Desserts

Ice creams have shown great potential for use as media for probiotic cultures. One of the advantages is that ice cream is consumed by all age groups. Although several steps in ice cream processing are optimized in order to maintain the viability of microorganisms in order to provide therapeutic benefits to consumers. Probiotic cultures are added to ice creams by either adding them directly to the pasteurized mix prior to freezing. Alternatively, milk may be used as a substrate for fermentation, followed by mixing with based mix in order to produce frozen fermented dairy desserts.

Probiotic ice cream with inulin and sugar added with *Lactobacillus acidophilus* and *Bifidobacterium lactis* was studied by Akin et al. (2007). An increase in sugar concentration improved the quality of probiotic ice cream, but the addition of inulin had no effect.

Aragon-Alegro et al. (2007) developed a synbiotic chocolate mousse with *Lactobacillus paracasei* subsp. *paracasei* and inulin. Results suggested that probiotic organisms can be incorporated in chocolate mousse.

Probiotic coconut flan with *Bifidobacterium lactis* and *Lactobacillus paracasei* was studied by Saad et al. (2008). Probiotic coconut flans were preferred as compared with the control product, showing its great potential as a probiotic functional food.

In studies by Fávaro-Trindade et al. (2006), ice creams containing acerola pulp were formulated with the use of different starter cultures (*Bifidobacterium longum, B. lactis, Streptococcus thermophilus,* and *Lactobacillus delbrueckii* spp. *Bulgaricus*). The acceptance of the product was lower than that containing probiotic bacteria.

## 6. Probiotic Cheeses

Cheeses are relatively new products in which probiotic organisms are being incorporated. Its versatility offers opportunities for many marketing strategies as a probiotic food carrier. In order to develop probiotic cheese, it is important to understand the all their processing steps, as well as their level of impact (positive or negative) on the survival of probiotic organisms. It should be kept in mind that for the manufacture of probiotic cheese, the changes in processing line be kept to a minimum when compared to traditional non-probiotic products in order to exploit the product commercially.

Vinderola et al. (2009) examined the potential of using *Lactobacillus paracasei* A13, *Bifidobacterium bifidum* A1, and *L. acidophilus* A3 in a probiotic fresco cheese, a popular cheese in Argentina, during its manufacture and refrigerated storage at 5°C and 12°C for 60 days. *L. paracasei* A13 was reported to grow a half log order at 43°C during the manufacture of cheese and another half log order during the first 15 days of storage at 5°C.

Kasımoglu et al. (2004) examined the effects of adding *L. acidophilus* 593 N to Turkish white cheese. The cheese was ripened in vacuum packs or in brine at 4°C for 90 days. On ripening in the vacuum pack, *L. acidophilus* survived to above $10^7$ CFU/g.

Cardarelli et al. (2008) examined the influence of inulin, oligofructose and oligosaccharides on *Bifidobacterium animalis* subsp. *lactis* and *Lactobacillus acidophilus* in synbiotic petit-Suisse cheeses. The viable number of organisms increased from 6.08 up to 6.99 log CFU/g for *L. acidophilus* 7.20 up to 7.69 log CFU/g for *B. animalis* subsp. *lactis*. Cheeses containing oligofructose were most preferred.

Blanchette et al. (1996) studied the potential of adding *Bifidobacterium infantis* to cottage cheese to develop probiotic cottage cheese by adding cream dressing fermented by *Bifidobacterium infantis* to the dry curd. Probiotic cottage cheeses were made at two pH levels (pH 4.5 and pH 5.5) and one without any probiotic organism. Control and the product at pH 5.5 were acceptable.

Kiliç et al. (2009) studied effects of supplementing *Lactobacillus fermentum* and *L. plantarum* to Turkish Beyaz probiotic cheese. A control cheese was made with commercial starter culture containing *Lactococcus lactis* spp. *Cremoris* and *L. lactis* spp. *lactis* and probiotic culture. Probiotic cheeses were similar in preference with the control.

## 7. Dairy based Probiotic Products Containing Bioactives

In addition to exceptional nutritional attributes, milk and milk derived products such as fermented milk contain components that possess a range of different bioactivities (Shah 2000a, Ramchandran and Shah 2008). Bioactive molecules are compounds that have active biological role in behavioural, gastrointestinal, hormonal, immunological, and nutritional response.

Milk proteins in their native form exert a range of physiological activities. Hydrolysis of these proteins can also liberate bioactive peptides from milk proteins and enhance bioactivity (Shah 2000a). This can be achieved with the help of proteolytic dairy starter cultures which possess appreciable level of proteolytic activity, which is required for their rapid growth in milk. Some bioactive peptides are also released during the process of fermentation through proteolysis of milk caseins. There has been an increasing interest in some important groups of bioactive inhibitory peptides, particularly, angiotensin I-converting enzyme (ACE, EC 3.4.15.1) inhibitory (ACE-I) peptides, which have been extensively studied due to their hypotensive role. A number of probiotic strains have also been found capable of producing different bioactive peptides with ACE-inhibitory activity in yogurt and soy based yogurt (Donkor et al. 2005, 2006a,b).

A number of probiotic products containing ACE-I peptides have been commercialized successfully. Some examples include Calpis® and Evolus®, which contain the ACE-inhibitory tri-peptides, valine-proline-proline (VPP) and isoleucine-proline-proline (IPP). Calpis® is prepared by fermentation with *Lb. helveticus* and *Saccharomyces cerevisiae* (Takano 2002); whereas Evolus®, is manufactured using a *Lb. helveticus* LBK-16 H strain (Seppo et al. 2002). Studies have revealed that these products help in lowering hypertension via renin-angiotensin system, thereby improving the vascular function in hypertensive subjects (Jauhiainen et al. 2007).

Table 5 lists dairy based functional products containing bioactive peptides and their claimed bioactivities. Most of these products have claimed blood pressure lowering effects due to release of ACE-I peptides. However, some products have claimed improvement in athletic performance and muscle recovery and aid in relaxation and sleep. Clinical trials have confirmed blood pressure lowering effects of some products.

**Table 5.** Functional products containing bioactive peptides and their claimed bioactivities.

| Brand name | Type of product | Claimed functional bioactive peptides | Health claims | Manufacturer |
|---|---|---|---|---|
| Calpis | Sour milk | β-casein, κ-casein VPP, LPP | Reduction of blood pressure | Calpis Co., Japan |
| Evolus | Fermented milk drink | β-casein VPP, LPP | Reduction of blood pressure | Valio, Finland |
| BioZate | Hydrolysed WPI | β-lactoglobulin, f(142–148) | Reduction of blood pressure | Davisco Foods Int. USA |
| Prodiet F200/ Lactium | Flavored milk drink | $\alpha_{s1}$-casein f(91–100) TLGTLGGLLA | Reduction of stress effects | Ingredia, France |
| Festivo | Fermented low-fat hard cheese | $\alpha_{s1}$-casein f(1-9), $\alpha_{s1}$-casein f(1-7), $\alpha_{s1}$-casein f(1-6) | No health claims | MTT Agrifood Research Finland |
| Cysteine Peptide | Ingredient/hydrolysate | Milk protein derived peptide | Reduction of blood pressure | DMV Int., the Netherlands |
| C12 | Ingredient/hydrolysate | Casein derived peptide | Reduction of blood pressure | DMV Int., the Netherlands |
| Capolac | Ingredient/hydrolysate | Casein derived peptide | Improves athletic performance and muscle recovery | DSM Food Specialities, the Netherlands |
| Vivinal Alpha | Ingredient/hydrolysate | Whey derived peptide | Aids relaxation and sleep | Borculo Domo Ingredients, the Netherlands |

## 8. Technological Challenges in the Development of Probiotic Dairy Products

Probiotic organisms are expected to be delivered to the desired sites in an active and viable form in a large number in order to exert a health benefit. The viability and activity of probiotic organisms in the products is important for achieving health benefits. The recommended levels of probiotic organisms ranges from $10^6$ cfu/mL to $10^7$– $10^8$ cfu/mL (Shah 2000b). These high numbers are required to compensate for the possible losses in the number of probiotic organisms as their numbers decline during processing and storage of a probiotic product as well as during passage through the gastrointestinal tract. In Japan, the Fermented Milks and Lactic Acid Bacteria Beverages Association has developed a standard which requires that the level of probiotic be present at $\geq 10^7$ viable cells per gram of a product (Shah 2000b). However, numerous studies have demonstrated that probiotic organisms grow poorly in milk, resulting in low final concentrations in yogurt and even the loss of the viability during prolonged cold storage.

A number of commercial yogurt type products have been analyzed in Australia and Europe for the presence of *L. acidophilus* and *Bifidobacterium* over the years (Shah 2000b, Tharmaraj and Shah 2003). Most of the products contained very low concentrations of probiotic organisms, especially *Bifidobacterium*.

Similarly, it is important to use proper enumeration techniques in order to assess the viability and survival of probiotic bacteria. Several media for selective enumeration of *L. acidophilus*, *Bifidobacterium* spp. and *L. casei* have been suggested; however, most of these methods are reported to be based on pure cultures. The picture will be different when these micro-organisms are enumerated from a product that contains 5–6 different probiotic organisms, in addition to the starter culture used for the manufacture of a particular product (Shah 2000b). Tharmaraj and Shah (2003) recommended media for selective enumeration of *S. thermophilus*, *L. delbrueckii* ssp. *bulgaricus*, *L. acidophilus*, *Bifidobacterium* spp., *L. casei*, *L. rhamnosus* and *Propionibacterium* in a mixture of probiotic bacteria. More details about selective and differential enumeration of probiotic organisms can be found in Dave and Shah (1996), Tharmaraj and Shah (2003), and Ashraf and Shah (2011).

The viability and activity of probiotic organisms could be affected during processing, storage, and passage through the gastro-intestinal tract. In general, probiotic organisms are susceptible to environmental conditions such as water activity, redox potential, temperature, and acidity (Shah 2000b) and during industrial scale production such as freeze drying, spray drying or freeze concentration. Increased solute concentration and intracellular ice formation in case of freezing and freeze drying, and exposure to elevated temperatures during spray drying may have detrimental effects on viability and survival of probiotic organisms. Cryoprotectants are added either to the growth medium or before the freezing or dehydration step in order to improve survival of probiotic bacteria.

The viability of probiotic organisms in a delivery system (i.e., food matrix) depends on a strain, interactions between microbial species present, production of hydrogen peroxide, and final acidity of the product. The viability would be also affected due to availability of nutrients, growth promoters and inhibitors, concentration of sugars, dissolved oxygen and oxygen permeation through package (especially for *Bifidobacterium* spp.), inoculation level, and fermentation time (Shah 2000b). *Bifidobacterium* is not as acid tolerant as other probiotic organisms. *B. animalis* subsp. *lactis* is better acid tolerant then other *Bifidobacterium* species; hence, this species is widely used. The use of *L. delbrueckii* ssp. *bulgaricus* in fermented products, particularly in yogurt, may affect the survival of *L. acidophilus* and *Bifidobacterium* due to acid and hydrogen peroxide produced during fermentation. However, due to its proteolytic nature, *L. delbrueckii* ssp. *bulgaricus* may liberate essential amino acids such as valine, glycine, and histidine, which are required to support the growth of *Bifidobacterium* (Shihata and Shah 2002). Additionally, *S. thermophilus* may also stimulate the growth of certain probiotic organisms due to uptake of oxygen during growth (Dave and Shah 1997c).

The presence of oxygen (positive redox potential) in probiotic products may have a detrimental effect on the viability and survival of probiotic organisms during storage of a product as *L. acidophilus* is microaerophilic and *Bifidobacterium* spp. is anaerobic. They lack an electron-transport chain which results in the incomplete reduction of oxygen to hydrogen peroxide. Furthermore, they do not produce catalase, thus are incapable of converting hydrogen peroxide into oxygen and water. *Bifidobacterium* spp. is generally more susceptible to the deleterious presence of oxygen than *L. acidophilus*. For better survival of probiotic organisms, exclusion of oxygen is very important (Ahn et al. 2001, Dave and Shah 1997c). Oxygen permeation through packaging materials during storage can also have detrimental effect on the survival of probiotic organisms (Dave and Shah 1997a,b,c, Tamime et al. 2005). Similarly, *L. delbrueckii* ssp. *bulgaricus* is known to produce hydrogen peroxide in the presence of oxygen, which may affect probiotic organisms indirectly (Dave and Shah 1997a, Joseph et al. 1998). A synergistic inhibition of probiotic cultures due to acid and hydrogen peroxide was also observed by Lankaputhra et al. (1996).

The final pH of a product may also influence the survival of probiotic organisms. Below pH 4.4, probiotic organisms do not survive well and a substantial decrease in number of probiotic bacteria is usually observed. Having pH below 4.5 is a legal requirement for yogurt in some countries. Acid is also produced during storage of a product. This process, frequently referred to as post-acidification, usually occurs during production of yogurt due to the high acid producing nature of *Lactobacillus delbrueckii* ssp. *bulgaricus* (Dave and Shah 1997a).

The numbers of probiotic bacteria in frozen fermented dairy desserts or frozen yogurt are reduced significantly by acid, freeze-injury, sugar concentration of the product and oxygen toxicity (Davidson et al. 2000). For this reason, protecting cells with the help of microencapsulation is promising. Microencapsulation is a process where the cells are retained within the encapsulating membrane in order to reduce the cell injury or cell loss. The use of gelatine or vegetable gum as encapsulating materials has been reported to provide protection to acid sensitive probiotic organisms. Encapsulated probiotic organisms when incorporated in fermented frozen dairy desserts, yogurt or freeze-dried yogurt showed improved viability in comparison to non-encapsulated control organisms (Ravula and Shah 2000, Capela et al. 2006).

The survival of probiotic organisms in the product and subsequently in the gastrointestinal tract can be improved by addition of an appropriate prebiotic, which is a non-digestible food ingredient that beneficially affects the host by selectively stimulating the growth and/or activity of one or a limited number of bacteria in the colon in order to improve host health (Gibson and Roberfroid 1995). While their role is the selective stimulation of a limited number of colonic and preferable beneficial bacteria, a range of prebiotics has been used as a tool for improvement of probiotic activity and survival in fermented foods during growth and storage (Bruno et al. 2002, Liong and Shah 2005a,b, 2006, Dave and Shah 1998).

## 9. Conclusions

A number of lactic cultures are used to make fermented milks and other milk based dairy foods. However, these lactic cultures do not survive in the gastrointestinal tract and the recent trend is to add probiotic organisms to dairy foods. This has resulted in a number of probiotic based dairy foods. The popularity of probiotic products is increasing tremendously with the increasing concern of a healthy diet. Several probiotic products based on bioactive peptides released from milk proteins have already been commercialized. These products provide health benefits from probiotic organisms as well as from bioactive peptides.

## References


Akalin, A.S. and D. Erisir. 2008. Effects of inulin and oligofructose on the rheological characteristics and probiotic culture survival in low-fat probiotic ice cream. J. Food Sci. 73: 184–188.

Ahn, J.B., H.J. Hwang and J.H. Park. 2001. Physiological responses of oxygen-tolerant anaerobic *Bifidobacterium longum* under oxygen. J. Ind. Microbiol. Biotechnol. 11: 443–451.

Akin, M.B., M.S. Akin and Z. Kirmaci. 2007. Effects of inulin and sugar levels on the viability of yogurt and probiotic bacteria and the physical and sensory characteristics in probiotic ice cream. Food Chem. 104: 93–99.

Almeida, M.H.B., A.G. Cruz, J.A.F. Faria, M.R.L. Moura, L.M.J. Carvalho and M.C.J. Freitas. 2009. Effect of the açai pulp on the sensory attributes of probiotic yogurts. Int. J. Prob. Preb. 4: 41–44.

Almeida, M.H.B., S.S. Zoellner, A.G. Cruz, M.R.L. Moura, L.M.J. Carvalho and A.S. Sant'ana. 2008. Potentially probiotic açaí yogurt. Int. J. Dairy Tech. 61: 178–182.

Amiri, Z.R., P. Khandelwal, B.R. Aruna and N. Sahebjamnia. 2008. Optimization of process parameters for preparation of synbiotic acidophilus milk via selected probiotics and prebiotics using artificial neural network. J. Biotech. 136: 460.

Andrighetto, C. and M.I.F.V. Gomes. 2003. Produção de picolés utilizando leite acidófilo. Brazilian Journal of Food Technology 6: 267–271.

Antunes, A.E.C., T.F. Cazetto and H.M.A. Bolini. 2005. Viability of probiotic micro-organisms during storage, postacidification and sensory analysis of fat-free yogurts with added whey protein concentrate. Int. J. Dairy Technol. 58(3): 169–173.

Antunes, A.E.C., E.R.A. Silva, A.G.F. Van Dender, E.T.G. Mascara, I. Moreno, E.V. Faria, M. Padula and A.L.S. Lerayer. 2009. Probiotic buttermilk-like fermented milk product development scale in a semi-industrial scale: Physicochemical, microbiological and sensory acceptability. Int. J. Dairy Technol. 62: 556–561.

Aragon-Alegro, L.C., J.H.A. Alegro, H.R.C. Cardarelli, M.C. Chiu and S.M.I. Saad. 2007. Potentially probiotic and synbiotic chocolate mousse. LWT—Food Sci. Technol. 40(4): 669–675.

Aryana, K.J. and P. Mcgrew. 2007. Quality attributes of yogurt with *Lactobacillus casei* and various prebiotics LWT—Food Sci. Tech. 40: 1808–1814.

Ashraf, R. and N.P. Shah. 2011. Selective and differential enumeration of *Lactobacillus delbrueckii* ssp. *bulgaricus, Streptococcus thermophilus, Lactobacillus acidophilus*, *L. casei*, and *Bifidobacterium* spp. in yoghurt: A review. International J. Food Microbiology 149(3): 194–208.

Blanchette, L., D. Roy, G. Belanger and S.F. Gauthier. 1996. Production of cottage cheese using dressing fermented by bifidobacteria. J. Dairy Sci. 79: 8–15.

Bolduc, M.P., Y. Raymond, P. Fustier, C.P. Champagne and J.C. Vuillemard. 2006. Sensitivity of bifidobacteria to oxygen and redox potential in nonfermented pasteurized milk. Int. Dairy J. 16: 1038–1048.

Bruno, F., W.E.V. Lankaputhra and N.P. Shah. 2002. Growth, viability and activity of *Bifidobacterium* spp. in skim milk containing prebiotics. Journal of Food Science 67(7): 2740–2744.

Capela, P., T.K.C. Hay and N.P. Shah. 2006. Effect of cryoprotectants, prebiotics and microencapsulation on survival of probiotic organisms in yoghurt and freeze-dried yoghurt. Food Research International 39: 203–211.

Cardarelli, H.R., F.C.A. Buriti, I.A. Castro and S.M.I. Saad. 2008. Inulin and oligofructose improve sensory quality and increase the probiotic viable count in potentially synbiotic petit-suisse cheese. LWT—Food Sci. Tech. 41: 1037–1046.

Çaglar, E., O.O. Kuscu, S.S. Kuwetli, S.K. Cildir, N. Sandalli and S. Twetman. 2008. Short-term effect of ice-cream containing *Bifidobacterium lactis* Bb-12 on the number of salivary mutans streptococci and lactobacilli. Acta Odon Scand 66: 154–158.

Castro, F.P., T.M. Cunha, P.J. Ogliari, R.F. Teófilo, M.M.C. Ferreira and E.S. Prudêncio. 2009. Influence of different content of cheese whey and oligofructose on the properties of fermented lactic beverages: Study using response surface methodology LWT—Food Sci. Tech. 42: 993–997.

Champagne, C.P., D. Roy and N.J. Gardner. 2005. Challenges in the addition of probiotic cultures to foods. Crit. Rev. Food Sci. Nut. 45: 61–84.

Corbo, M.R., M. Albenzio, M. De Angelis, A. Sevi and M. Gobbetti. 2001. Microbiological and biochemical properties of Canestrato Pugliese hard cheese supplemented with bifidobacteria. J. Dairy Sci. 84: 551–561.

Cruz, A.G., A.E.C. Antunes, A.L.O.P. Sousa, J.A.F. Faria and S.M.I. Saad. 2009. Ice cream as a probiotic food carrier. Food Res. Int. 42: 1233–1239.

Dave, R. and N.P. Shah. 1996. Evaluation of media for selective enumeration of *S. thermophilus*, *L. delbrueckii* ssp. *bulgaricus*, *Lactobacillus acidophilus* and *Bifidobacterium* species. Journal of Dairy Science 79: 1529–1536.

Dave, R. and N.P. Shah. 1997a. Effectiveness of cysteine as redox potential reducing agent in improving viability of probiotic bacteria in yogurts made with commercial starter cultures. International Dairy Journal 7: 537–545.

Dave, R. and N.P. Shah. 1997b. Effectiveness of ascorbic acid as oxygen scavenger in improving viability of probiotic bacteria in yogurts made with commercial starter cultures. International Dairy Journal 7: 435–443.

Dave, R. and N.P. Shah. 1997c. Effect of level of starter culture on viability of yogurt and probiotic bacteria in yogurts. Food Australia 49(4): 32–37.

Dave, R.I. and N.P. Shah. 1998. Ingredient supplementation effects on viability of probiotic bacteria in yogurt. J. Dairy Sci.: 2804–2816.

Davidson, R.H., S.E. Duncan, C.R. Hackney, W.N. Eigel and J.W. Boling. 2000. Probiotic culture surival and implications in fermented frozen yogurt characteristics. J. Dairy Sci. 83: 666–673.

Donkor, O.N., A. Henriksson, T. Vasiljevic and N.P. Shah. 2005. Probiotic strains as starter cultures improve angiotensin-converting enzyme inhibitory activity in soy yogurt. J. Food Sci. 70(4): M375–M381.

Donkor, O.N., A. Henriksson, T. Vasiljevic and N.P. Shah. 2006a. Activity of probiotic strains in soy yoghurt during prolonged cold storage. Int. Dairy J. 16: 1181–1189.

Donkor, O.N., A. Henriksson, T. Vasiljevic and N.P. Shah. 2006b. Effect of acidification on the activity of probiotics in yoghurt during storage. Int. Dairy J. 16(10): 1181–1189.

Donkor, O.N., A. Henriksson, T. Vasiljevic and N.P. Shah. 2007a. Proteolytic activity of dairy lactic acid bacteria and probiotics as determinant of viability and angiotensin-converting enzyme inhibitory activity in fermented milk. Le Lait 87: 21–38.

Donkor, O.N., S.L.I. Nilmini, P. Stolic, T. Vasiljevic and N.P. Shah. 2007b. Survival and activity of selected probiotic organisms in set-type yoghurt during cold storage. International Dairy Journal 17(6): 657–665.

Farnworth, E.R. 2005. Kefir—a complex probiotic. Food Sci. Technol. Bull.: Functional Foods 2: 1–17.

Favaro-Trindade, C.S., S. Bernardi, R.B. Bodini, J.C.C. Balieiro and E. Almeida. 2006. Sensory acceptability and stability of probiotic microorganisms and vitamin C in fermented acerola (*Malpighia emarginata* DC) ice cream. J. Food Sci. 71: 492–495.

Gibson, G.R. and M.B. Roberfroid. 1995. Dietary modulation of the human colonic microbiota: introducing the concept of prebiotics. J. Nutr. 125(6): 1401–12.

Gobbetti, M., A. Corsetti, E. Smacchi, A. Zocchetti and M. De Angelis. 1997. Production of Crescenza cheese by incorporation of bifidobacteria. J. Dairy Sci. 81: 37–47.

Gomes, A.M.P. and F.X. Malcata. 1999. *Bifidobacterium* spp. and *Lactobacillus acidophilus*: biological, biochemical, technological and therapeutical properties relevant for use as probiotics. Trends Food Sci. Technol. 10: 139–157.

Granato, D., G.F. Branco, A.G. Cruz, J.A.F. Faria and N.P. Shah. 2010. An overview of functional foods and strategies for developing dairy probiotic foods. Comprehensive Reviews in Food Sci. and Food Safety 9: 455–470.

Hekmat, S. and D.J. McMahon. 1992. Survival of *Lactobacillus acidophilus* and *Bifidobacterium bifidum* in ice cream for use as a probiotic food. J. Dairy Sci. 75(6): 1415–1422.

Hernandez-Mendoza, A., V.J. Robles, J.O. Angulo, J. Cruz and H. Garcia. 2007. Preparation of a whey-based probiotic product with *Lactobacillus reuteri* and *Bifidobacterium bifidum*. Food Tech. Biotechnol. 45(1): 27–31.

Homayouni, A., A. Azizi, M.R. Ehsani, M.S. Yarmand and S.H. Razavi. 2008. Effect of microencapsulation and resistant starch on the probiotic survival and sensory properties of synbiotic ice cream. Food Chem. 111: 50–55.

Isanga, J. and G. Zhang. 2009. Production and evaluation of some physicochemical parameters of peanut milk yogurt. LWT—Food Sci. Technol. 42: 1132–1138.

Itsaranuwat, P., K.S.H. Al-Haddad and R.K. Robinson. 2003. The potential therapeutic benefits of consuming 'health-promoting' fermented dairy products: a brief update. Int. J. Dairy Technol. 56: 203–210.

Jauhiainen, T., M. Rönnback, H. Vapaatalo, K. Wuolle, H. Kautiainen and R. Korpela. 2007. < i> *Lactobacillus helveticus*</i> fermented milk reduces arterial stiffness in hypertensive subjects. International dairy journal 17(10): 1209–1211.

Joseph, P.J., R.I. Dave and N.P. Shah. 1998. Antagonism between yogurt bacteria and probiotic bacteria isolated from commercial starter cultures, commercial yogurts, and a probiotic capsule. Food Aust. 50: 20–23.

Kailasapathy, K. and L. Masondole. 2005. Survival of free and microencapsulated *Lactobacillus acidophilus* and *Bifidobacterium lactis* and their effect on texture of feta cheese. Aust. J. Dairy Technol. 60: 252–258.

Kalavrouzioti, I., M. Hatzikamari, E. Litopoulou-Tzanetaki and N. Tzanetakis. 2005. Production of hard cheese from caprine milk by use of probiotic cultures as adjuncts. Int. J. Dairy Technol. 58: 30–38.

Kasımoglu, A., M. Göncüoglu and S. Akgün. 2004. Probiotic white cheese with *Lactobacillus acidophilus*. Int. Dairy J. 14: 1067–1073.

Kaur, H., H.N. Mishra and P. Umar. 2009. Textural properties of mango soy fortified probiotic yogurt: optimisation of inoculum level of yogurt and probiotic culture. Int. J. Food Sci. Technol. 44: 415–424.

Kiliç, G.B., H. Kuleansan, I. Eralp and A.G. Karahan. 2009. Manufacture of Turkish Beyaz cheese added with probiotic strains. LWT—Food Sci. Technol. 42(5): 1003–1008.

Koroleva, N.S. and R.K. Robinson. 1991. Products prepared with lactic acid bacteria and yeasts. Therapeutic properties of fermented milks 159–179.

Kosikowski, F. and V.V. Mistry. 1966. Cheese and fermented milk foods: Edwards Bros., Ann Arbor, Mich.

Lankaputhra, W.E.V., N.P. Shah and M. Britz. 1996. Survival of bifidobacteria during refrigerated storage in presence of acid and hydrogen peroxide. Milchwissenschaft 51(2): 65–70.

Liong, M.T. and N.P. Shah. 2005a. Optimization of cholesterol removal by probiotics in presence of prebiotics using response surface methodology. Applied and Environmental Microbiology 71: 1745–1753.

Liong, M.T. and N.P Shah. 2005b. Roles of probiotics and prebiotics on cholesterol: The hypothesized mechanisms. Nutrafoods 4(4): 45–57.

Liong, M.T. and N.P. Shah. 2006. The application of response surface methodology to optimize removal of cholesterol and to evaluate growth characteristics and production of organic acids by *Bifidobacterium infantis* ATCC 17930 in the presence of prebiotics. International J. of Probiotics and Prebiotics 1(1): 41–56.

Lourens-Hattingh, A. and B.C. Viljoen. 2001. Yogurt as probiotic carrier food. Int. Dairy J. 11: 1–17.

Maragkoudakisa, P.A., C. Miarisa, P. Rojeza, N. Manalisb, F. Magkanarib, G. Kalantzopoulosa and E. Tsakalidou. 2006. Production of traditional Greek yogurt using *Lactobacillus* strains with probiotic potential as starter adjuncts. Int. Dairy J. 16(1): 52–60.

Martín-Diana, A.B., C. Janer, C. Peláez and T. Requena. 2003. Development of a fermented goat's milk containing probiotic bacteria. Int. Dairy J. 13: 827–833.

Metchnikoff, II. 2004. The prolongation of life: Optimistic studies (reprinted edition 1907): New York, NY, USA: Springer.

Ong, L. and N.P. Shah. 2008a. Release and identification of angiotensin converting enzyme-inhibitory peptides as influenced by ripening temperatures and probiotic adjuncts in Cheddar cheeses. LWT—Food Science and Technology 41(9): 1555–1566.

Ong, L. and N.P. Shah. 2008b. Influence of Probiotic *Lactobacillus acidophilus* and *Lb. helveticus* on proteolysis, organic acid profiles and ACE-inhibitory activity of Cheddar cheeses ripened at 4, 8 and 12°C. Journal of Food Science 73(3): M111–M120.

Ong, L. and N.P. Shah. 2009. Probiotic Cheddar cheese: Influence of ripening temperatures on survival of probiotic microorganisms, cheese composition and organic acid profiles. LWT—Food Sci. Technol. 42: 1260–1268.

Özer, B., Y.S. Uzun and H.A. Kirmaci. 2008. Effect of microencapsulation on viability of *Lactobacillus acidophilus* La-5 and *Bifidobacterium bifidum* Bb-12 during kasar cheese ripening. Int. J. Dairy Technol. 61: 237–244.

Prajapati, Jashbhai. B., Nair and M. Baboo. 2003. The history of fermented foods. Handbook of Fermented Funct. Foods 2: 1–24.

Prajapati, J.B. and N.P. Shah. 2011. Probiotics and health claims—An Indian perspective. Probioitics and Health Claims. Edited by W. Kneifel, and S. Salminen. Blackwell Publishing Ltd., Oxford pp. 134–145.

Prado, F.C., J.L. Parada, A. Pandey and C.R. Soccol. 2008. Trends in non-dairy probiotic beverages. Food Res. Int. 41: 111–123.

Ramchandran, L. and N.P. Shah. 2008. Proteolytic profiles, and angiotensin-I converting enzyme and alpha-glucosidase inhibitory activities of selected lactic acid bacteria. J. of Food Science 73(2): M75–M81.

Ravula, R. and N.P. Shah. 2000. Influence of water activity on fermentation, organic acids production and viability of yogurt and probiotic bacteria. Australian Journal of Dairy Technology 55(3): 127–131.

Ross, R.P., S. Morgan and C. Hill C. 2002. Preservation and fermentation: past, present and future. Int. J. Food Microbiol. 2002. 79(1-2): 3–16.

Saad, S.M.I, F.C.A. Buriti and T.R. Komatsu. 2007. Activity of passion fruit (*Passiflora edulis*) and guava (*Psidium guajava*) pulps on *Lactobacillus acidophilus* in refrigerated mousses. Braz J. Microbiol. 38: 315–317.

Saad, S.M.I., S.B.M. Corrêa and I.A. Castro. 2008. Probiotic potential and sensory properties of coconut flan supplemented with *Lactobacillus paracasei* and *Bifidobacterium lactis*. Int. J. Food Sci. Technol. 43: 1560–1568.

Seppo, L., O. Kerojoki, T. Suomalainen and R. Korpela. 2002. The effect of a *Lactobacillus helveticus* LBK-16 H fermented milk on hypertension: a pilot study on humans. Milchwissenschaft 57(3): 124–127.

Shah, N.P. 2000a. Natural bioactive substances in milk and colostrum. British J. Nutrition 84 (Supp. 1) S3–S10.

Shah, N.P. 2000b. Probiotic bacteria: selective enumeration and survival in dairy foods. J. Dairy Sci. 83(4): 894–907.

Shah, N.P. 2001. Functional foods from probiotics and prebiotics. Food Technol. 55(11): 46–53.

Shah, N.P. 2002. *Bifidobacterium* spp.: Applications in fermented milks. Encyclopedia of Dairy Science, Academic Press, London pp. 147–151.

Shah, N.P. 2003. Yogurt: The product and its manufacture. Encyclopedia of food science and nutrition 10: 6252–6260.

Shah, N.P. 2004. Probiotics and prebiotics. AgroFood Industry HiTech 15(1): 13–16.

Shah, N.P. 2006a. Health benefits of yogurt and fermented milks. pp. 327–340. *In*: R.C. Chandan (ed.). Manufacturing Yogurt and Fermented Milks. Blackwell Publishing Professional, Iowa, USA.

Shah, N.P. 2006b. Microorganisms and health attributes (probiotics). pp. 341–354. *In*: R.C. Chandan (ed.). Manufacturing Yogurt and Fermented Milks. Blackwell Publishing Professional, Iowa, USA.

Shah, N.P. 2007. Functional cultures and health benefits. Int. Dairy J. 17: 1262–1277.

Shah, N.P. 2010. Probiotics: Health benefits, efficacy and safety. Edited by D. Bagchi, F.C. Lau and D.K. Ghosh, Taylor and Francis Publisher, USA pp. 485–496.

Shah, N.P. 2011a. *Bifidobacterium* spp.: Morphology and Physiology. Encyclopedia of Dairy Science, Academic Press, London pp. 381–387.

Shah, N.P. 2011b. *Bifidobacterium* spp.: Applications in fermented milks. Encyclopedia of Dairy Science, Academic Press, London 388–394.

Shah, N.P. 2012a. Health benefits of yogurt and fermented milks. pp. 433–450. *In*: R.C. Chandan (ed.). Manufacturing Yogurt and Fermented Milks. Blackwell Publishing Professional, Iowa, USA.

Shah, N.P. 2012b. Microorganisms and health attributes (probiotics). pp. 451–468. *In*: R.C. Chandan (ed.). Manufacturing Yogurt and Fermented Milks. Blackwell Publishing Professional, Iowa, USA.

Shah, N.P. 2013. Microorganisms and health attributes (probiotics). pp. 451–466. *In*: R.C. Chandan and A. Kilara (eds.). Manufacturing Yogurt and Fermented Milks. Blackwell Publishing Professional, Iowa, USA.

Shah, N.P. and W.E.V. Lankaputhra. 2002. *Bifidobacterium* spp.: Morphology and Physiology. Encyclopedia of Dairy Science, Academic Press, London pp. 141–146.

Shihata, A. and N.P. Shah. 2002. Influence of addition of proteolytic strains of *Lactobacillus delbrueckii* spp. *bulgaricus* to Commercial ABT starter cultures on texture of yogurt, exopolysaccharide production and survival of bacteria. International Dairy Journal 12(9): 765–772.

Sousa, R.C.S., R.A. Lira, F.C. Oliveira, D.O. Santos and O.A.P. Sierra. 2007. Desenvolvimento e aceitação sensorial de iogurte probiótico *light* de banana. *In*: Proceedings of III Encontro Regional Sul de Ciência e Tecnologia de Alimentos. pp. 549–553. Curitiba: Anais do IX ERSCTA.

Souza, C.H.B. and S.M.I. Saad. 2009. Viability of *Lactobacillus acidophilus* La-5 added solely or in co-culture with a yogurt starter culture and implications on physico-chemical and related properties of Minas fresh cheese during storage. LWT—Food Sci. Technol. 42: 633–640.

Speck, M.L. 1978. The development of sweet acidophilus milk. Dairy Ice Cream Field J. 70: A-D.

Supavititpatana, P., T.I. Wirjantoro, A. Apichartsrangkoon and P. Raviyan. 2008. Addition of gelatin enhanced gelation of corn–milk yogurt. Food Chem. 106: 211–216.

Takano, T. 2002. Anti-hypertensive activity of fermented dairy products containing biogenic peptides. Antonie Van Leeuwenhoek 82(1-4): 333–340.

Tamime, A.Y. and V.M.E. Marshall. 1997. Microbiology and technology of fermented milks. pp. 57e152. *In*: B.A. Law (ed.). Microbiology and biochemistry of cheese and fermented milk (2nd ed.). London: Chapman and Hall.

Tamime, A.Y., M. Saarela, K. Sondergaard, V.V. Mistry and N.P. Shah. 2005. Production and maintaining viability of probiotics (chapter 3). pp. 39–72. *In*: A.Y. Tamime (ed.). Probiotic Dairy Products. Blackwell Publishing, Oxford, United Kingdom.

Thage, B.V., M.L. Broe, M.H. Petersen, M.A. Petersen, M.A. Bennedsen and Y. Ardö. 2005. Aroma development in semi-hard reduced-fat cheese inoculated with *Lactobacillus paracasei* strains with different aminotransferase profiles. Int. Dairy J. 15: 795–805.

Tharmaraj, N. and N.P. Shah. 2003. Selective enumeration of *Lactobacillus delbrueckii* ssp. *bulgaricus*, *Streptococcus thermophilus*, *Lactobacillus acidophilus*, bifidobacteria, *Lactobacillus casei*, *Lactobacillus rhamnosus* and Propionibacteria. Journal of Dairy Science 86: 2288–2296.

Vasiljevic, T. and N.P. Shah. 2007. Fermented Milk: Health Benefits Beyond Probiotic Effect. Handbook of Food Products Manufacturing: Health, Meat, Milk, Poultry, Seafood, and Vegetables 2: 99.

Vasiljevic, T. and N.P. Shah. 2008. Cultured milk and yogurt. *In*: R.C. Chandan, A. Kilara and N.P. Shah (eds.). Dairy Processing and Quality Control. Blackwell Publishing, Ames, Iowa, USA.

Vierhile, T. 2006. Functional 'add-ins' boost yogurt consumption. Food Technol. 60: 44–48.

Vinderola, C.G., M. Gueimonde, T. Delgado, J.A. Reinheimer and C.G. Reyes-Gavilán. 2000a. Characteristics of carbonated fermented milk and survival of probiotic bacteria. Int. Dairy J. 10(3): 213–220.

Vinderola, C.G., W. Prosello, T.D. Ghiberto and J.A. Reinheimer. 2000b. Viability of probiotic (*Bifidobacterium*, *Lactobacillus acidophilus* and *Lactobacillus casei*) and non probiotic microflora in Argentinean fresco cheese. J. Dairy Sci. 83: 1905–1911.

Vinderola, G., W. Prosello, F. Molinari, D. Ghilberto and J. Reinheimer. 2009. Growth of *Lactobacillus paracasei* A13 in Argentinian probiotic cheese and its impact on the characteristics of the product. Intl. J. Food Microbiol. 135: 171–174.

Yilmaztekin, M., B.H. Özer and F. Atasoy. 2004. Survival of *Lactobacillus acidophilus* La-5 and *Bifidobacterium bifidum* BB-02 in white-brined cheese. Int. J. Food Sci. Nutr. 55: 53–60.

Zoellner, S.S., A.G. Cruz, J.A.F. Faria, H.M.A. Bolini, M.R.L. Moura, L.MJ. Carvalho and A.S. Sant´ana. 2009. Whey beverage with açai pulp as a food carrier of probiotic bacteria. Aust. J. Dairy Technol. 64: 165–169.

# CHAPTER 18

# Non-Dairy Probiotic Products

*Nissreen Abu-Ghannam*[1,*] and *Gaurav Rajauria*[2]

## 1. Introduction

The term probiotic was technically defined as "live microorganisms which upon ingestion in certain numbers exert health benefits beyond inherent nutrition" (FAO/WHO 2001). This definition requires that the microorganisms must be alive and present in high numbers, generally more than $10^9$ cells per daily ingested dose. Probiotic food products are considered as functional foods which are defined to contain health-promoting components beyond traditional nutrients and the addition of probiotic cultures is one approach in which foods could be modified to become functional.

The market for functional food is one of the fastest growing segments of the global food industry and estimated to be worth a current value around US$ 25 billion with a compound growth rate of 7.4%. There are a number of key drivers behind this unprecedented growth rate, including the increase in world population and changes in the demographics of that population (increase in aging population), advances in the understanding of the relationship between diet and health, and the demand for health and wellness food products from childhood to old age (Espin et al. 2007). Probiotic products have been widely promoted in the last decade for multitude of health benefits, and are well recognized by consumers as being "good for you" type products. The global market of probiotic ingredients, supplements and foods (dairy and non-dairy) reached nearly US$ 23.1 billion in 2012 and is expected to reach US$ 36.7 billion, with a compound growth rate of 6.2%, over the five-year period from 2013–2018. Up to 78% of current probiotic sales world-wide are mainly delivered through dairy products and the remaining 22% are mostly accounted for by non-dairy probiotic foods and beverages (bccresearch.com 2014, Hui and Özgül Evranuz 2012).

[1] School of Food Science and Environmental Health, Dublin Institute of Technology, Cathal Brugha Street, Dublin 1, Ireland.
E-mail: nissreen.abughannam@dit.ie

[2] School of Agriculture and Food Science, University College Dublin, Lyons Research Farm, Newcastle, Co. Dublin, Ireland.
E-mail: gaurav.rajauria@ucd.ie

* Corresponding author

There is a growing interest in the development of non-dairy probiotic products due to issues such as lactose intolerance in many populations around the world and the unfavorable cholesterol content typically associated with fermented dairy products. The ongoing trend of vegetarianism and the requirements of cold storage environments for fermented dairy products are additional drivers for the consideration and development of non-dairy products as carriers of probiotic agents. It is worth noting that the annual dairy consumption in Asia is considerably low ranging from 8–100 kg/capita in comparison to 251 kg in the USA, 310 kg in Australia and up to 330 kg for some European countries, thus suggesting the potential development of non-dairy probiotic products as another avenue for the growth of the functional food sector (Sharma and Mishra 2013). Some commercial examples of non-dairy probiotic products are listed in Table 1.

Fruits, vegetables, legumes, cereal and meat have been utilized world-wide in the development of many traditional fermented food products as means of food preservation or nutritional enhancement. Additionally, plant products are characterized as being rich in minerals, vitamins, antioxidants and fibres, thus further enhancing their characteristics as candidates for the development of non-dairy probiotic products.

This chapter will provide an overview on the utilization of botanical and meat sources as substrates for probiotic growth highlighting, throughout, challenges and opportunities in utilizing such sources.

## 2. Cereal Grains and Soy Based Probiotic Fermentation

Cereal grains and legumes, often called staple crops are the world's largest food source, and constitute a major source of dietary nutrients for human and animal consumption. The global annual production of cereal grains is more than 2.3 billion metric tons and is grown on 73% of the world's total harvested area and remains the world's largest food yielding source (Serna-Saldivar 2010, Charalampopoulos et al. 2002). Although, in contrast to milk and meat, the nutritional quality of cereal grains is inferior due to essential amino acid deficiency such as lysine in cereals and methionine in legumes and the presence of antinutrients such as phytic acid and tannins, they are however considered as good substrates for the proliferation of probiotic bacteria due to their high content of non-digestible carbohydrates thereby acting as prebiotics. Lactic acid fermentation of cereals has been traditionally considered as the most simple and economical way for improving their nutritional value, shelf life, safety, sensory properties, and functional qualities (Charalampopoulos et al. 2002). Lactic acid bacteria (LAB) fermentation of cereals has been reported to enhance the degradation of phytase and release minerals such as manganese, iron, zinc and calcium that supports the growth of probiotic bacteria (Blandino et al. 2003). Additionally, fermentation of cereal grains by LAB cultures not only enhances the bio-availability of essential vitamins and minerals, but also reduces the level of non-digestible carbohydrates typically associated with cereals (Soccol et al. 2010).

Cereal grains are rich in native prebiotics like non-digestible carbohydrates which enhance acid and bile tolerance levels, resulting in better survival and protection of probiotic bacteria in the extreme environment of the gastrointestinal tract

**Table 1:** Commercial examples of non-dairy probiotic products (Sharma and Mishra 2013, Gupta and Abu-Ghannam 2012, Prado et al. 2008).

| **Brand/Trade Name** | **Description** | **Producer** |
|---|---|---|
| Vita Biosa© | Mixture of aromatic herbs and other plants, which are fermented by a combination of lactic acid bacteria and yeast cultures | Biosa, Denmark |
| Jovita Probiotisch© | Blend of cereals, fruit and probiotic yogurt | H&J Bruggen, Germany |
| Proviva© | Fermented oatmeal gruel, malted barley containing *Lactobacillus plantarum*. The first probiotic food that does not contain milk, or milk constituents (1994) | Skane Mejerier, Sweden |
| Provie© | Fruit drink containing *Lb. plantarum* | Skane Mejerier, Sweden |
| Rela© | Fruit juice containing *Lactobacillus reuteri MM53* | Biogaia, Sweden |
| Grainfields Wholegrain Liquid© | dairy-free, no genetically modified ingredients and no added sugar containing *Lactobacillus acidophilus, Lactobacillus delbruekii, Saccharomyces boulardii and Sc. cerevisiae* | AGM Foods Pty, Australia |
| Soytreat© | Kefir type product with six probiotics | Lifeway, USA |
| SOYosa© | Range of products based on soy and oats and includes a refreshing drink | Bioferme, Finland |
| Yosa© | Oat product flavoured with natural fruits and berries containing probiotic bacteria (*Lb. acidophilus, Bifidobacterium lactis*) | Bioferme, Finland |
| Gefilus© | Fruit drinks containing *Lactobacillus rhamnosus, Propionibacterium freudenreichii* | Valio Ltd., Finland |
| Proflora© | Freeze-dried product containing *Lb. acidophilus, Lb. delbrueckii, Steptococcus thermophilus* and Bifidobacterium | Chefaro, Belgium |
| Bactisubtil© | Freeze-dried product containing *Bacillus* sp. *strain IP5832* | Synthelabo, Belgium |
| Snack Fibra© | Snacks and bars with natural fibers and extra minerals and vitamins | Celig″ueta, Spain |
| Bififlor© | Freeze-dried product containing *Lb. acidophilus, Lb. rhamnosus,* Bifidobacterium | Eko-Bio, The Netherlands |
| Hardaliye© | from the natural fermentation of the red grape, or grape juice with the addition of the crushed mustard seeds and benzoic acid | CDS Agro Ltd., Turkey |
| Biola© | Probiotic juice drinks containing *Lb. rhamnosus GG*, available in orange–mango and apple–pear flavors | Tine BA, Norway |
| GoodBelly© | Organic fruit juice-based probiotic beverage contains a patented *Lb. plantarum* 299v culture | Nextfoods, UK |

(Lamsal and Faubion 2009, Patel et al. 2004). Additionally, due to the presence of non-digestible carbohydrates, cereal grains can also act as synbiotics (exhibiting probiotic and prebiotic properties) which selectively stimulate the growth of probiotic bacteria in the colon (Charalampopoulos et al. 2002).

A multitude of non-dairy traditional fermented cereals products have been created throughout history for human nutrition worldwide, but only recently the probiotic activity of such products has been investigated (Rivera-Espinoza and Gallardo-Navarro 2010). Some of the interesting characteristics of the probiotic strains isolated from fermented cereal products are their ability to survive bile toxicity during their passage into the gastrointestinal system. Due to milk shortages in many parts of the world, cereals constitute the principal raw material for the development of multitude of non-dairy probiotic beverages, gruels and porridge. *Idli* (steam cooked fermented paste) and *Dosa* (fermented paste fried as pan-cake) are the most popular fermented foods in India made from black gram and rice respectively. Some of the LAB found to be responsible for the fermentation process are *Leuconostoc mesenteroides*, *Lactobacillus delbrueckii*, *Lactobacillus lactis* and *Pediococcus cerevisiae* (Blandinao et al. 2003).

Prado et al. (2008) provided a comprehensive review on fermented cereal beverages from around the world. For example *Boza* is a very popular colloid beverage consumed in Bulgaria, Albania, Turkey and Romania and made from wheat, rye, millet, maize and other cereals mixed with sugar. It mainly consists of yeast and lactic acid bacteria in an average LAB/yeast ratio of 2:4. *Mahewu* is a sour beverage made from corn meal and is common in Africa and some Arabian Gulf countries where a maize porridge, sorghum, millet, malt are added for spontaneous fermentation by the natural flora, predominantly by *Lb. lactis*, at ambient temperature. *Pozol* is a beverage made from fermented corn dough consumed mainly in Mexico.

Charalapompoulos et al. (2002) identified a number of parameters that could influence the growth of probiotic strains on cereals such as: the composition and processing of cereal grains, the substrate formulation, the growth capability and productivity of the starter culture, the stability of the probiotic strain during storage and the nutritional value of the final product. Variations in the capacity of different cereals, or their mixtures, to support the growth of probiotics have been reported, for example malt medium support a wide range of LAB bacteria including *Lactobacillus plantarum*, *Lactobacillus fermentum*, *Lactobacillus acidophilus* and *Lactobacillus reuteri* better than barley and wheat media. Also wheat and barley extracts exhibited a significant protective effect on the viability of *Lb. plantarum*, *Lb. acidophilus* and *Lb. reuteri* under a pH of 2.5 (Charalapompoulos 2002).

In a recent study by Rathore et al. (2012), malt, barley and a mixture of both was utilized as a substrate for the fermentation with *Lb. plantarum* (NCIMB 8826) and *Lb. acidophilus* (NCIMB 8821). The study reported increased growth in the malt substrate resulting in significant amounts of lactic acid production (0.5–3.5 g/L) along with 7.9 to 8.5 log CFU/mL cell populations being reached within 6 h of fermentation. The study concluded that the mixing of microorganism strains could give rise to the production of more flavour attributes thus offering potential opportunities for the developments of novel probiotic beverages from cereals.

Survival, functionality and competition with other microorganisms are a number of challenges for the probiotic strain in any non-dairy substrate. In cereal-based probiotics, yeasts which could naturally be present on dried cereals or could be added deliberately to the fermentation process could interact with the growth potential of the probiotic strains. The co-cultured organism may compete for growth nutrients or could produce metabolic products with inhibitory effects. Kedia et al. (2007) reported that the introduction of yeast in malt-based substrate fermentation had increased the growth of *Lb. reuteri* as compared to the pure LAB culture as yeast may produce vitamins that enhance the growth of LAB.

Probiotic cultures may require some growth promoters to initiate growth in cereals as reported by Helland et al. (2004). In these two studies, a malted brown rice media (cereal porridge) and cereal pudding of rice and maize were fermented with *Bifidobacterium longum* BB536 and *Bifidobacterium animalis* Bb12, respectively and the results obtained showed that the growth of Bifidobacterium is not possible unless the substrate was enriched with a growth promoter like milk. Helland et al. (2004) fermented a maize porridge made of a mixture of maize flour and barley malt with various probiotic strains and suggested that maize porridge supplemented with barely malt is a better medium for probiotic growth.

Oat-based substrates have been utilized successfully in probiotic fermentations due to their compositional structure such as the presence of special mixed links (1→3) and (1→4) β-glucan, arabinoxylans and cellulose dietary fibers along with comparatively high level of unsaturated lipids, proteins, vitamins and phenolic antioxidants (Awaisheh 2012). Whole-grain oat was fermented with *Lb. plantarum* to obtain a fermented drink that combined the health benefits of a probiotic culture with the oat prebiotic β-glucan and maintained a viable cell count of 10 log CFU/mL after 24 d of refrigerated storage at 4–6°C, higher than that required for considerations as a probiotic drink (Angelov et al. 2006). Gupta et al. (2010) reported negligible changes in the β-glucan level after 21 d of refrigerated storage of a *Lb. plantarum* oat-based fermented drink. Oat-bran breakfast cereals were found to support the growth of *Lb. plantarum* within a moisture content of 50–58%, indicating potential development of probiotic breakfast cereals (Patel et al. 2004). LAB fermentation of oat-based substrates combines the effects of β-glucan for cholesterol reduction and the effect of LAB benefits to maintain and improve the intestinal balance.

Soybean is one of the most important legumes in the traditional Asian diet and is considered to be rich in high quality protein and oligosaccharides. However, wider utilization of soybeans has been limited by its typical beany flavour and the presence of raffinose and stachyose oligosaccharides which are not easily digested by the human digestive system (Soccol et al. 2010). Traditionally, fermentation has been the most suitable option to increase the digestibility of soy products and to reduce their unacceptable off-flavour (Han et al. 2001). For example, a variety of fermented soybean based foods such as Turangbai, Hawaijar, Aakhuni, Bekangum, Pruyaan and Kinema have been produced and consumed by the ethnic people of North-Eastern India, Nepal and Bhutan (Tamang 2005). The utilization of probiotic bacteria was reported to enhance the level of free isoflavones in soy products, reduce the sugars that cause flatulence and also the level of off-flavour n-hexanal and pentanal organic compounds which are typically associated with the "beany" flavour of soy products (Champagne et al. 2005). Soymilk is considered as a good substrate for supporting the survival of a range of probiotic strains including *Lb. rhamnosus, Lb. reuteri*, *Lb. acidophilus* and *Lb. Fermentum.* This is possibly attributed to the oligosaccharides of soybeans acting as carbon sources (Wang et al. 2006). Several studies indicated that soy based products such as frozen soy desert, soy tempeh, soy yogurt, soy beverage and soy cheese worked as good substrates for the growth of probiotic bacteria showing total viable cell count as high as 8–9 log CFU/mL (Rivera-Espinoza and Gallardo-Navarro 2010, Wang et al. 2006). In particular *Lb. rhamnosus* strain seems to withstand a range of processing conditions of soybeans, suggesting its potential in the development of soy-based probiotic products.

Beans contain phenolic compounds that exhibit antioxidant activity with possible benefits to human health. LAB of soyfoods shows that this activity gets remarkably stronger in comparison to un-fermented beans (Wu and Chou 2009). Combining the health benefits of probiotic fermentation and the increased antioxidant capacity places soy-based products as a strong candidate for the development of non-dairy based probiotic products.

## 3. Fruits and Vegetables Based Probiotic Fermentation

Fruits and vegetables could also be considered possible substrates for probiotic fermentation, being rich in several beneficial bioactive ingredients such as vitamins, dietary fibres and minerals that could support the growth of probiotic strains (Rößle et al. 2010). In the process of developing such products, fruits and vegetables are typically minimally processed (such as peeling and shredding) which results in the release of nutrients from the cellular component thus making a favourable environment for the growth of probiotic microorganisms without compromising much of the bioactive content of fruits and vegetables (Oliveira et al. 2011, Soccol et al. 2010).

The structural composition of fruits and vegetable is typically characterized as having wide intercellular spaces thus allowing an easier entry and colonization by probiotic bacteria. In addition, fruits and vegetables are considered rich in indigestible cellulosic fiber thus providing a protective shield for probiotic bacteria to sustain their probiotic integrity while passing through the gastrointestinal tract resulting in augmented benefits for the host (Kourkoutas et al. 2006). In some cases, the rate of survival of probiotic bacteria in fruit and vegetable-based substrates was better than that observed in dairy-based probiotic products, with the additional advantage of lacking certain undesirable components associated with dairy products such as lactose and cholesterol (Prado et al. 2008).

However, it should be noted that the lack of certain essential amino acids and vitamins in some types of fruits and vegetables and their acidic nature could limit the growth of probiotic bacteria (Saad et al. 2011). The probiotic cell viability has been reported to depend upon the strains used, the characteristics of the substrate, oxygen content and final acidity of the product (Shah 2001). Based on the acid resistance capacity of probiotic strains, Sheehan et al. (2007) reported that Lactobacillus and Bifidobacterium strains survived for longer in orange and pineapple juice compared to cranberry juice. The same study reported that *Lb. casei*, *Lb. rhamnosus* and *Lactobacillus paracasei* displayed great robustness surviving at levels above 7 log CFU/mL in orange juice and above 6 log CFU/mL in pineapple juice for at least 12 wk. However, as juices are generally submitted to a thermal pasteurization step to enhance their safety and quality, this will subsequently reduce the viability of the probiotic bacteria when incorporated in such juices. This was demonstrated in the

Sheehan et al. (2007) study were a combination of thermal pasteurisation at 76ºC for 30 s, combined with high-pressure treatment of 400 MPa for 5 min showed that *Lb. casei*, *Lb. rhamnosus* and *Lb. paracasei* were not capable of withstanding the treatments required to achieve a stable juice at levels > 6 log CFU/mL. Shah (2007) reported that the optimum probiotic growth temperature is between 35–40ºC, and the best pH is between 6.4 and 4.5, ceasing when a pH of 4.0–3.6 is reached. A number of suggested solutions to provide barriers against unfavourable conditions during probiotic fermentation included the immobilization of probiotic strains on agar, polyacrylamide, calcium pectate gel, chemically modified chitosan beads and alginates (Kourkoutas et al. 2005). Fruit juices have the added benefit of their sucrose content as demonstrated by Saarela et al. (2006) where sucrose-protected cells were reported to survive better than reconstituted skim-milk protected cells. Additionally, some of the techniques practiced to enhance the viability of probiotic bacteria include microencapsulation and vacuum impregnation. The various approaches described earlier were designed with the aim of not only protecting the probiotic bacteria from the extreme external processing conditions or the internal environment of the digestive system but also to increase their viability in the finished products (Prado et al. 2008, Yoon et al. 2006, Nedovic et al. 2011).

Nutrient supplementation was reported to be an important factor for initiating probiotic growth in some fruit and vegetable based substrates. In particular enriching with brewer's yeast autolysate prior to fermentation resulted in better survival of probiotic cultures in the typically low pH environment of fruits and vegetables. Rakin et al. (2007), reported that *Lb. plantarum* and *Lb. delbrueckii* were capable of surviving the low pH of beetroot juice that was supplemented with brewer's yeast autolysate, remaining at 5–7 log CFU/mL after 4 wk of storage at 4ºC. *Lb. casei*, recognised for its low tolerance for low pH environments, lost cell viability completely after 2 wk.

The application of a heat treatment to the vegetable or fruit substrate prior to fermentation could enhance the growth rate and the viability of probiotic bacteria. For example, Yoon et al. (2005) reported that *Lb. plantarum*, *Lb. casei* and *Lb. delbrueckii* grew rapidly on sterilized cabbage juice without nutrient supplementation and reached 8 log CFU/mL after 48 h of fermentation at 30ºC. *Lb. plantarum* and *Lb. delbrueckii* produced significantly more titratable acidity than *Lb. casei*, suggesting that *Lb. casei* requires some essential growth nutrients that are deficient in cabbage juice. Jaiswal et al. (2012) studied the effects of blanching York cabbage at 95ºC for 12 min as a means to inactivate surface microflora, reduce microbial competition, prior to fermentation with *Lb. plantarum.* The study showed that a growth of 9 log CFU/mL was attained after 36 h of fermentation and was sustained for a storage period of 15 d at 4ºC. This study highlighted that the application of a relatively mild heat treatment to fruits and vegetables prior to fermentation could significantly improve the chances of probiotic growth and survival without compromising the nutritional content of such products. The Jaiswal et al. (2012) study also highlighted that probiotic fermentation retained most of the polyphenolic content and the antioxidant activity of York cabbage unlike other processing methods known to reduce and negatively influence of such health related properties. Similar approaches should be relevant to other types of fruits and vegetables where the initial heat treatment could be optimized to maximize growth and viability of probiotic strains.

Apart from pH reduction and lactic acid production, probiotic fermentation of plant-based products have the potential for enhancing the prebiotic characteristics of the developed product, thus further enhancing the overall health benefits. Vergara et al. (2010) fermented cashew apple juice to produce prebiotic oligosaccharides. The prebiotic effect of cashew apple juice fermented with *Lactobacillus mesenteroides* was demonstrated by the better growth of *Lactobacillus johnsonii* in fermented cashew apple juice when compared to the observed growth in culture media containing only glucose and fructose as the carbon source.

A number of innovative vegetable and fruit-based fermented products have been recently investigated. di Cagno et al. (2011) developed a protocol for the manufacture of fermented smoothies, in which white grape juice and aloevera extract were mixed with red fruits (cherries, blackberries prunes and tomatoes) or green vegetables (fennels, spinach, papaya and kiwi) and were fermented using a mixed culture of *Lb. plantarum*, *Weissella cibaria* and *Lactobacillus pentosus* strains. The previous study reported an enhancement in the antioxidant capacity and the sensory properties of the developed product.

The sensory properties of probiotic based products are an essential component of the success of such products and it is worth noting that not all plant-based products would show an enhancement in their sensory properties upon probiotic fermentation. For example the sensory impact study by Luckow and Delahunty (2004) showed that consumers prefer the sensory characteristics of conventional orange juices to those containing probiotics, but if the associated health benefits were provided, then consumer preference increases over the conventional orange juices. The same study suggested the addition of tropical fruits in percentages not exceeding 10% (v/v) to mask perceptible off-flavours of probiotics when incorporated in juices.

Overall, reports have been indicating that tomatoes, carrot, cabbage, artichokes and red beet juices were proven to be particularly suitable for probiotic fermentation, allowing rapid growth of the strains in addition to the production of a viable cell population above 8 log CFU/mL (Rivera-Espinoza and Gallardo-Navarro 2010). Probiotic growth can be further enhanced by the incorporation of essential nutrients.

Daily intake of fruits and vegetables is estimated to be lower than the doses (400 g) recommended by the World Health Organization (WHO), and the Food and Agriculture Organization (FAO). Fruits and vegetables can either be consumed as fresh or industrially processed. Minimal processing, while attractive to consumers due to freshness attributes and high sensory properties, have short shelf life due to rapid microbial growth. Other methods of preservation such as canning or pasteurization do enhance the shelf-life of fruits and vegetables but can result in alterations in the physical and chemical properties of such products. The consumer trend towards fresh-like, health-promoting and rich flavour ready-to-eat or drink foods and beverages is increasing. Lactic acid fermentation presents one of the most suitable approaches for increasing daily consumption of fresh-like vegetables and fruits (di Cagno et al. 2011). Probiotic juices have become the most studied and suitable substrate for the growth and viability of probiotic strains. Due to their taste, flavour and nutrient content, probiotic juices have the potential to be widely accepted and consumed by a wide range of consumers.

## 4. Meats and Seafood Based Probiotic Fermentation

Meat and seafood based products are considered as basic sources of high quality proteins and amino acids, numerous minerals and a good source of vitamins such as A, D, E and B complex (Kołożyn-Krajewska and Dolatowski 2012). Due to its composition and structure, meat serves as an excellent medium for probiotic growth and has been reported to protect the probiotic bacteria against bile during the passage through the GI tract (Khan et al. 2011).

The preservation of meat by LAB fermentation is an ancient practice that provides considerably stable meat products with acceptable quality and sensory characteristics. Meat fermentation typically relies on native lactic acid bacteria that are usually present at low numbers on raw meat surfaces prior to fermentation. LAB fermentation is known for favourable modifications in flavour and texture in addition to substantial improvements in the product shelf-life and consumer convenience (Xu et al. 2008, Tu et al. 2010). In particular, dry fermented type sausages, without thermal processing, lend themselves to be appropriate candidates for the development of probiotic meat-based products. Such products tend to acquire their characteristics fermentation flavours earlier on during the preparation stage, so that the later incorporation or addition of probiotic cultures provides better chances for survival without altering the sensorial properties of the final product. As part of its manufacturing process, dry fermented sausages contain high numbers of LAB, but they are not regarded as probiotics (Leroy et al. 2006).

Typical LAB strains commonly utilized in meat starter cultures to develop the taste, flavour and texture, associated with fermented meats, include *Lb. casei*, *Lb. plantarum*, *Lb. sakei*, *Lactobacillus pentosus* and *Pediococcus acidilactici*. The inclusion of functional starter cultures in fermented meats such as *Lb. reuteri* and *B. longum* could offer further benefits in addition to those offered by classical processing starter cultures. This approach could open up a new range of meat products characterized as being healthier and without compromising their traditional sensory characteristics (Amor and Mayo 2007). Research has highlighted so far that probiotic organisms tend to survive poorly in fermented foods in general, and encapsulation has been suggested as a mean for protection especially in high acidic environments that typically characterize acid foods. In the case of fermented meats, the meat and fat matrix could provide a natural "encapsulate" for probiotic organisms, thus rendering meat products as better candidates to support probiotic organisms than fruits, vegetables and cereal products. Muthukumarasamy and Holley (2006) reported that encapsulation could reduce the inhibitory action of probiotic organisms against pathogenic bacteria. Accordingly, products intended for probiotic inclusion should be of a very good microbiological quality in order to realize the full health potential of probiotic organisms. The suitability of probiotic cultures will have to be verified for each individual sausage type with the emphasis on using dominant strains of probiotics due to their adaptation to the meat environment. For example, *Lb. plantarum* and *Lb. pentosus*, which are naturally present in Scandinavian-type fermented sausage, were considered as appropriate candidates for probiotic meat starter cultures (Klingberg et al. 2005).

Despite the popularity of fermented meat products, the commercial production of probiotic meat products is still not common and only very few manufacturers offer fermented meat products enriched with probiotics (Kołożyn-Krajewska and Dolatowski 2012). Germans and Japanese producers were the pioneers in incorporating probiotic bacteria into meat products launching for example a probiotic salami (fermented with intestinal *Lb. acidophilus*, *Lb. casei*, and *Bifidobacterium* spp.) and probiotic meat spread (fermented with *Lb. rhamnosus* FERM P-15120) (Arihara 2006). Currently, probiotic meat products are still a relatively new product concept for the meat industry despite a high demand by health conscious consumers. Meat products such as dry fermented sausages, ham or loin have the advantage of being considered as an appropriate substrate for probiotic growth as their production usually requires slight or no heat processing, thus providing the appropriate conditions essential for the survival of probiotic cultures (Ammor and Mayo 2007). However, because of the mild processing conditions these products are subjected to, other indigenous meat micro-flora could grow on meat surfaces and subsequently compete with the potential survival and growth of probiotic bacteria on such products and the overall progress of the fermentation process. Accordingly, there is a need for the selection of appropriate probiotic strains with a high viability when incorporated in fermented meat matrixes (Tu et al. 2010). Starter cultures are modified to withstand the anaerobic atmosphere, high salt concentrations, low temperatures and low pH that prevail in fermented meat products. Immobilization of probiotic cultures in order to enhance their chances of survival and also to withstand certain heat treatments that meat is typically exposed to has been reported. Sidira et al. (2013) studied the effect of immobilization on the cell viability of *Lb. casei* ATCC 393 during probiotic fermentation of dry-fermented sausages. The sausages were fermented with free or immobilized *Lb. casei* ATCC 393 bacteria on wheat, and after 66 d of ripening it was observed that the viable cell counts in sausages produced with immobilized culture were higher than that required for the characterization as a probiotic product. The results of the previous study also indicated that the same product had an improved profile of aroma-related compounds with a total of 124 volatile compounds including esters, organic acids and carbonyl compounds being identified.

The incorporation of probiotic cultures in other types of meat products is receiving an increasing interest. In this case, raw meat products are now believed to be an appropriate matrix for the growth of probiotic microorganisms. For instance, Neffe-Skocińska et al. (2011) studied the impact of probiotic culture incorporation in pork sirloin meat upon storage at different temperatures (16, 20, and 24°C) and a sensory evaluation was conducted after the completion of an aging period of 21 d. During this aging period, the viable counts of probiotic bacteria were at their highest (7–8 log CFU/g) at 20°C while the lowest viable counts (4–6 log CFU/g) were recorded at 16°C. The results concluded that the probiotic pork sirloins meat product, stored at 20°C with 21 d aging period were considered to be the higher quality product from a microbiological and sensorial perspective.

In the case of seafood, the application of probiotic bacteria is only limited to increase the shelf life with little or no impact on the final sensory characteristics of the product. In contrast to meat, LAB are usually not considered as natives of fish or seafood, but certain strains of Lactobacillus and Lactococcus, have been found to be associated with fishery products (Ghanbari et al. 2013). The antimicrobial compounds released by LAB such as lactic acid, hydrogen peroxide ($H_2O_2$), carbon dioxide ($CO_2$), diacetyl (2,3-butanedione) and bacteriocins are basically responsible for killing the harmful bacteria and extending the shelf-life of seafood (Nes 2011). For example, Fall et al. (2010) inhibited the growth of *Brochothrix thermosphacta* bacteria in cooked shrimp and improved its sensory properties by using the probiotic strain *Lactococcus piscium* CNCM I-4031. The incorporation of probiotic strains of lactic acid bacteria in meat products will open up new opportunities for manufacturers and consumers and could alleviate some of the health concerns typically associated with meat consumption.

## 5. New Possible Sources for Probiotic Fermentation

Several plant food processors produce high levels of agro-industrial by-products globally. Disposal of this waste incurs considerable costs to processors and represents a significant environmental hazard. However there is a growing awareness that this material could represent a valuable resource in the development of value-added food products. Brewers' spent grain (BSG) is a classical example of an agro-industrial waste resulting from brewing, where millions of tons of this waste material are produced annually and typically ending up either in animal feed or in landfills with very little application for human consumption. Due to the presence of polysaccharides and proteins, BSG has been used as a substitute to expensive carbon sources for industrial production of lactic acid. Production of 4 g/L lactic acid was reported by *Lb. delbrueckii* using BSG (Mussatto et al. 2007). Interest in the addition of BSG as a means to enhance the quality of food products for human consumption has increased due to its high content of oligosaccharides and phenolic compounds. In a study reported by Gupta et al. (2013) BSG was utilized in the development of a fermented nutraceutical liquid product using *Lb. plantarum* ATCC 8014 in a 7L bioreactor. BSG in water was autoclaved at 121°C for 1 min in order to release nutrients and breakdown its lignocellulosic material. A growth of 10 log CFU/mL was reported after 19 h of fermentation and was maintained for a storage period of 30 d at 4°C. *Lb. plantarum* of BSG additionally resulted in the release of sugars, antioxidants and phenolic compounds in the broth medium. These attributes opens up new opportunities and possibilities for the development of new probiotic beverages that utilizes agro-industrial by-products which would be both economically and environmentally attractive and viable.

Marine underutilized resources, mainly algal materials were also investigated as possible substrate candidates in the development of probiotic products, as virtually most of the research has been limited to terrestrial plants. Gupta et al. (2011) utilized *Lb. plantarum* in the fermentation of three species of edible Irish brown seaweeds; *Himanthalia elongata*, *Laminaria digitata* and *Laminaria saccharina*. Heat treatment at 95°C for 15 min considerably enhanced the growth of *Lb. plantarum* due to sugar release as compared to non-heat-treated seaweeds. In particular, the *Laminaria* species is rich in laminaran polysaccharides which can be fermented by LAB. A maximum 10 log CFU/mL was achieved at the end of 16–24 h of fermentation for *L. digitata* and *L. saccharina*. The Gupta et al. (2011) study indicated the fermentative capability of seaweeds as a sole source of nutrition for the growth of *Lb. plantarum*. Seaweeds are a vast untapped resource of nutraceuticals and its potential incorporation in the development of probiotic plant-based products merits the importance of further needed research.

Table 2 illustrates a list of non-dairy probiotic products recently explored from cereals, legumes, soy, fruits, vegetables, meat and other possible sources.

## 6. Enhancing the Viability of Probiotics in Plant Based-Products

International standards stipulate that fermented products claiming health benefits should contain a minimum of $10^7$ viable probiotic bacteria per gram of the product when sold (Shah 2001). Probiotic viability could be defined as surviving environmental and processing conditions and reaching the site of action and producing a beneficial health effect to the host (De Vos et al. 2010). Generally, probiotic bacteria suffer from poor survival in probiotic products mainly attributed to nutrients composition, low pH conditions and the type of the probiotic strain utilized. These factors have to a greater extent been controlled and optimized so survival rates of probiotics could be enhanced during the storage of such products. In fact, survival of probiotics in the extreme acidic pH of the human gastro-intestinal system is the actual challenge for developing probiotic products generally, and particularly plant-based probiotic products which tend to be more on the acidic side thus limiting the types of probiotic strains that could be utilized to acid-resistant types, and potentially losing out on strains that could contribute better to flavour and texture or could confer more health benefits to the host. Probiotic encapsulation, or providing probiotic living cells with a physical barrier, has been proposed as an efficient technology to improve viability during long-term storage and to preserve the metabolic activity in the gastrointestinal tract (Zuidam and Nedovic 2010). Encapsulation is defined as a process that entraps a substance into another substance, producing particles in the nanometer (nanoencapsulation), micrometer (microencapsulation) or millimetre scale (Burgain et al. 2011). Carrier material should isolate and protect bacterial cells from the effects of hostile environments, be safe for consumption with Generally Recognised As Safe (GRAS) status and cost effective in order to minimize the influence on the final product. Low cost carrier materials include starches, inulin, pectin and most carbohydrates (De Vos et al. 2010). Food-grade polymers such as alginate, chitosan, carboxymethyl cellulose (CMC), carrageenan, gelatine and pectin have been widely used for various microencapsulation techniques.

Currently there is a range of well-established microencapsulation technologies for the protection of probiotic bacteria (Burgain et al. 2011). The selection of an encapsulation method depends on a number of factors such as the required particle size, physical and chemical properties of the carrier material, the applications of the encapsulated material, and the required release mechanism and cost. Recently, research is directed towards the development of carrier materials that offer multiple delivery and other benefits, in addition to protecting probiotic bacteria, including functional, nutraceutical and prebiotic properties. Nanoencapsulation has the potential to provide delivery of probiotic bacteria to certain parts of the gastro-intestinal tract where they could interact with specific receptors to maximize the delivery of their health capacity. *In vivo* studies are required using human subjects to confirm the efficacy of encapsulation in delivering probiotic bacteria and their controlled release in the gastro-intestinal system.

**Table 2.** Potential non-dairy probiotic fermented products recently developed.

| Products | Probiotic culture | Substrate/Matrix | Viability of the probiotics in the product | | References |
|---|---|---|---|---|---|
| | | | **Abundance** | **Storage time** | |
| ***Cerealgrains and soy based*** | | | | | |
| Beverage | *Lb. plantarum* | Rice, barley, oats, wheat, soy flour and red grape juice | 8.4 log CFU/g | 30 d | Coda et al. (2012) |
| Beverages | *Lb. plantarum*<br>*Lb. acidophilus* | Malt, barley and a mixture of both | Above 8 log CFU/mL | 6 h | Rathore et al. (2012) |
| Soybean bar | *Lb. acidophilus* | Soybean | Above 8 log CFU/g | 8 wk | Chen and Mustapha 2012 |
| Oat based drink | *Lb. plantarum* | Oat | Above 10 log CFU/mL | 21 d | Gupta et al. (2010) |
| ***Fruit and vegetable based*** | | | | | |
| Pineapple beverage | *Lb. casei* | Pineapple juice | 6.03 log CFU/mL | 42 d | Costa 2013 |
| Vegetable drinks | *Lb. acidophilus*<br>*Lb. plantarum* | Bitter gourd, bottle gourd and carrot | 8 log CFU/mL | 72 h | Sharma and Mishra 2013 |
| Carrot juice | LAB | Pasteurized carrot juice | 9–10 log CFU/mL | 4 wk | Tamminen et al. (2013) |
| Snack product | *Lb. acidophilus* | Apple, mandarin and pineapple grape juice | Above 7 log CFU/g | 3–15 d | Betoret et al. (2012) |
| Apple beverage | *Lb. casei*<br>*Lb. acidophilus* | Fuji and Gala apples | Above 7.6 log CFU/mL | 28 d | de Souza Luciana Neves Ellendersen et al. (2012) |
| Cantaloupe beverage | *Lb. casei* | Cantaloupe juice | 8.3 log CFU/mL | 42 d | Fonteles et al. (2012) |
| Table olives | *Lb. pentosus* | Olives | 6 log CFU/ mL | 7–14 d | Aponte et al. (2012) |
| Pear Juice | *Lb. acidophilus* | Pear | 6 to 7 log CFU/mL | 72 h | Ankolekar et al. (2012) |
| Beverage | *Lb. plantarum*<br>*Lb. paracasei*<br>*Lb. fermentum* | Shalgam<br>(black carrot, turnip) | 6 to 8 log CFU/mL | 10 d | Tanguler and Erten 2012 |
| Probiotic products | *Lb. plantarum* | York cabbage | 10.3 log CFU/mL | 15 d | Jaiswal et al. (2012) |
| ***Meat and seafood based*** | | | | | |
| Dry-fermented sausages | *Lb. casei* | Dry-sausages | 8 log CFU/g | 6 mon | Wójciak et al. (2012) |

| | | | | | |
|---|---|---|---|---|---|
| Dry-fermented sausages | *Lb. fermentum*<br>*P. acidilactici* | Iberian dry-sausages | Above 7 log CFU/g | 50 d | Ruiz-Moyano et al. (2011) |
| Meat product | *Lb. casei* | Pork sirloins | 7 to 8 log CFU/g | 21 d | Neffe-Skocińska et al. (2011) |
| Som-fug | LAB | Fish | Above 8 log CFU/g | 15 d | Riebroy et al. (2007) |
| ***Others sources*** | | | | | |
| Nutraceutical drink | *Lb. plantarum* | Brewers' spent grain | 10.4 log CFU/mL | 30 d | Gupta et al. (2013) |
| Seaweed beverage | *Lb. plantarum* | Brown seaweed | 10 log CFU/mL | 72 h | Gupta et al. (2011) |

## 7. Health Claims of Probiotic Products

The European Food Safety Authority (EFSA) has not released a favourable opinion in relation to live organisms other than for live cultures in yoghurt, which were shown to improve the digestion of lactose in yoghurt in individuals with lactose maldigestion. The unfavourable opinion of EFSA was mainly because microorganisms were not properly characterized in the health claims submitted or due to the poor evidence of beneficial effect. The current situation as stated by EFSA requires further research to support a beneficial physiological effect of probiotics specifically in humans (Pravst 2012). It is expected that the same argument would be also applied to non-dairy probiotic products.

## 8. Concluding Remarks

- The application of non-thermal methodologies should be considered for the elimination/reduction of surface microflora of non-dairy substrates in order to reduce competition upon inoculation with the probiotic culture. In the production of fermented dairy products, the primary raw material milk is typically pasteurized to eliminate pathogens and reduce spoilage microorganisms. Non-thermal treatments will enhance the viability of the probiotic cultures with minimal losses in the nutritional attributes of the non-dairy substrates.
- Lactic acid bacteria, including some probiotic strains, could inhibit the growth of pathogenic bacteria resulting in an extension of the microbiological shelf-life, thus adding an element of safety in addition to nutrition for non-dairy fermented products.
- Mixed strains of probiotics could offer a range of flavours, thus creating opportunities for novel probiotic products particularly masking the "beany flavour" that is typically associated with cereal and legume-based products.
- There is a significant potential for the incorporation of probiotics in bakery products given their wide range and high consumption rates. In this case, the probiotic cultures should be encapsulated for protection against the high temperature applied to baked products.
- While encapsulation could significantly help towards increasing the viability and availability of probiotic cultures, on the negative side, encapsulation would restrain the inhibitory action of probiotic cultures against some pathogenic microorganisms. However, this could be counteracted by utilizing good microbiological quality raw material.
- Combining genetic engineering of microorganisms with novel processing technologies will be expected to create new fermented products in which desired properties would be emphasized. New products should be developed and traditional ones should be maintained and improved.
- Probiotic fermentation of plant-based products has the potential to improve the nutraceutical properties of the final products by enhancing the levels of phenolic content and antioxidant activity.

- The exploitation of lactic acid fermentation through the selection of controlled fermentation processes and starter cultures could be considered as an approach for enhancing the consumption of fresh-like vegetable and fruits among the world population.

## References


Ammor, M.S. and B. Mayo. 2007. Selection criteria for lactic acid bacteria to be used as functional starter cultures in dry sausage production: an update. Meat Science 76: 138–146.

Angelov, A., V. Gotcheva, R. Kuncheva and T. Hristozova. 2006. Development of a new oat-based probiotic drink. International Journal of Food Microbiology 112: 75–80.

Ankolekar, C., M. Pinto, D. Greene and K. Shetty. 2012. *In vitro* bioassay based screening of antihyperglycemia and antihypertensive activities of *Lactobacillus acidophilus* fermented pear juice. Innovative Food Science & Emerging Technologies 13: 221–230.

Aponte, M., G. Blaiotta, F. La Croce, A. Mazzaglia, V. Farina, L. Settanni and G. Moschetti. 2012. Use of selected autochthonous lactic acid bacteria for Spanish-style table olive fermentation. Food Microbiology 30: 8–16.

Arihara, K. 2006. Strategies for designing novel functional meat products. Meat Science 74: 219–229.

Awaisheh, S.S. 2012. Probiotic Food Products Classes, Types, and Processing. *In*: E.C. Rigobelo (ed.). Probiotics, InTech, Rijeka, Croatia.

BCC Research. 2014. The Probiotics Market: Ingredients, Supplements, Foods. Report code FOD035D. BCC Research, Wellesley, MA, USA (Website: www.bccresearch.com accessed on 30 May 2014).

Betoret, E., E. Sentandreu, N. Betoret, P. Codoñer-Franch, V. Valls-Bellés and P. Fito. 2012. Technological development and functional properties of an apple snack rich in flavonoid from mandarin juice. Innovative Food Science & Emerging Technologies 16: 298–304.

Blandino, A., M.E. Al-Aseeri, S.S. Pandiella, D. Cantero and C. Webb. 2003. Cereal-Based Fermented Foods and Beverages: Review. Food Research International 36: 527–543.

Burgain, J., C. Gaiani, M. Linder and J. Scher. 2011. Encapsulation of probiotic living cells: From laboratory scale to industrial applications. Journal of Food Engineering 104: 467–483.

Champagne, C.P., N.J. Gardner and D. Roy. 2005. Challenges in the addition of probiotic cultures to foods. Critical Reviews in Food Science and Nutrition 45: 61–84.

Charalampopoulos, D., R. Wang, S.S. Pandiella and C. Webb. 2002. Application of cereals and cereal components in functional foods: a review International Journal of Food Microbiology 79: 131–141.

Chen, M. and A. Mustapha. 2012. Survival of freeze-dried microcapsules of α-galactosidase producing probiotics in a soy bar matrix. Food Microbiology 30: 68–73.

Coda, R., A. Lanera, A. Trani, M. Gobbetti and R. Di Cagno. 2012. Yogurt-like beverages made of a mixture of cereals, soy and grape must: Microbiology, texture, nutritional and sensory properties. International Journal of Food Microbiology 155: 120–127.

Costa, M.G.M., T.V. Fonteles, A.L.T. De Jesus and S. Rodrigues. 2013. Sonicated pineapple juice as substrate for *L. casei* cultivation for probiotic beverage development: process optimization and product stability. Food Chemistry 139: 261–266.

de Souza Neves Ellendersen, L., D. Granato, K.B. Guergoletto and G. Wosiacki. 2012. Development and sensory profile of a probiotic beverage from apple fermented with *Lactobacillus casei*. Engineering in Life Sciences 12: 475–485.

di Cagno, R., G. Minervini, C.G. Rizzello, M. De Angelis and M. Gobbetti. 2011. Effect of lactic acid fermentation on antioxidant, texture, color and sensory properties of red and green smoothies. Food Microbiology 28: 1062–1071.

Espin, J.C., M.T., Garcia-Consesa and F.A. Thomas-Barberan. 2007. Nutraceuticals: facts and fiction, Phytochemistry 68: 2986–3008.

Fall, P.A., F. Leroi, M. Cardinal, F. Chevalier and M.F. Pilet. 2010. Inhibition of *Brochothrix thermosphacta* and sensory improvement of tropical peeled cooked shrimp by *Lactococcus piscium* CNCM I-4031. Letters in Applied Microbiology 50: 357–361.

FAO/WHO (Food and Agriculture Organization of the United Nations, World Health Organization). 2001. Evaluation of the health and nutritional properties of probiotics in food including powder milk with live lactic acid bacteria. Report of the Joint FAO/WHO Expert Consultation, Cordoba, Argentina.

Fonteles, T.V., M.G.M. Costa, A.L.T. de Jesus and S. Rodrigues. 2012. Optimization of the Fermentation of Cantaloupe Juice by *Lactobacillus casei* NRRL B-442. Food and Bioprocess Technology 5: 2819–2826.

Ghanbari, M., M. Jami, K.J. Domig and W. Kneifel. 2013. Seafood biopreservation by lactic acid bacteria—A review. LWT—Food Science and Technology 54: 315–324.

Gupta, S., A.K. Jaiswal and N. Abu-Ghannam. 2013. Optimization of fermentation conditions for the utilization of brewing waste to develop a nutraceutical rich liquid product. Industrial Crops and Products 44: 272–282.

Gupta, S., N. Abu-Ghannam and A.G.M. Scannell. 2011. Growth and kinetics of *Lactobacillus plantarum* in the fermentation of edible Irish brown seaweeds. Food and Bioproducts Processing 89: 346–355.

Gupta, S., S. Cox and N. Abu-Ghannam. 2010. Process optimization for the development of a functional beverage based on lactic acid fermentation of oats. Biochemical Engineering Journal 52: 199–204.

Gupta, S. and N. Abu-Ghannam. 2012. Probiotic Fermentation of Plant Based Products: Possibilities and Opportunities, Critical Reviews in Food Science and Nutrition 52: 183–199.

Han, B.-Z., F.M. Rombouts and M.J. Robert Nout. 2001. A Chinese fermented soybean food. International Journal of Food Microbiology 65: 1–10.

Helland, M.H., T. Wicklund and J.A. Narvhus. 2004. Growth and metabolism of selected strains of probiotic bacteria, in maize porridge with added malted barley. International Journal of Food Microbiology 91: 305–313.

Hui, Y.H. and E. Özgül Evranuz. 2012. Handbook of Plant-Based Fermented Food and Beverage Technology, Second Edition. CRC Press, Boca Raton, FL, USA.

Jaiswal, A.K., S. Gupta and N. Abu-Ghannam. 2012. Optimization of lactic acid fermentation of York cabbage for the development of potential probiotic products. International Journal of Food Science and Technology 47: 1605–1612.

Kedia, G., R. Wang, H. Patel and S.S. Pandiella. 2007. Use of mixed cultures for the fermentation of cereal-based substrates with potential probiotic properties. Process Biochemistry 42: 65–70.

Khan, M.I., M.S. Arshad, F.M. Anjum, A. Sameen, A. Rehman and W.T. Gill. 2011. Meat as a functional food with special reference to probiotic sausages. Food Research International 44: 3125–3133.

Klingberg, T.D., L. Axelsson, K. Naterstad, D. Elsser and B.B. Budde. 2005. Identification of potential probiotic starter cultures for Scandinavian-type fermented sausages. International Journal of Food Microbiology 105: 419–431.

Kołożyn-Krajewskaa, D. and Z.J. Dolatowskib. 2012. Probiotic meat products and human nutrition. Process Biochemistry 47: 1761–1772.

Kourkoutas, Y., L. Bosnea, S. Taboukos, C. Baras, D. Lambrou and M. Kanellaki. 2006. Probiotic cheese production using *Lactobacillus casei* cells immobilized on fruit pieces. Journal of Dairy Science 89: 1439–1451.

Kourkoutas, Y., V. Xolias, M. Kallis, E. Bezirtzoglou and M. Kanellaki. 2005. *Lactobacillus casei* cell immobilization on fruit pieces for probiotic additive, fermented milk and lactic acid production. Process Biochemistry 40: 411–416.

Lamsal, B. and J. Faubion. 2009. The Beneficial Use of Cereal and Cereal Components in Probiotic Foods. Food Reviews International 25: 103–114.

Leroy, F., J. Verluyten and L. De Vuyst. 2006. Functional meat starter cultures for improved sausage fermentation. International Journal of Food Microbiology 106: 270–285.

Luckow, T. and C. Delahunty. 2004. Which juice is healthier? A consumer study of probiotic non-dairy juice drinks. Food Quality and Preference 15: 751–759.

Mussatto, S.I., M. Fernandes, G. Dragone, I.M. Mancilha and I.C. Roberto. 2007. Brewer's spent grain as raw material for lactic acid production by *Lactobacillus delbrueckii*. Biotechnology Letters 29: 1973–1976.

Muthukumarasamy, P. and R. Holley. 2006. Microbiological and sensory quality of dry fermented sausages containing alginate-microencapsulated *Lactobacillus reuteri*. International Journal of Food Microbiology 111: 164–169.

Nedovic, V., A. Kalusevic, V. Manojlovic, S. Levic and B. Bugarski. 2011. An Overview of Encapsulation Technologies for Food Applications. Procedia Food Science 1: 1806–1815.

Neffe-Skocińska, K., M. Gierejkiewicz and D. KoŁozyn-Krajewska. 2011. Optimization of fermentation conditions for dry-aged sirloins with probiotic bacteria added. Żywność. Nauka. Technologia. Jakość 18: 36–46.

Nes, I.F. 2011. History, current knowledge and future directions on bacteriocin research in lactic acid bacteria. pp. 3–12. *In*: D. Drider and S. Rebuffat (eds.). Prokaryotic Antimicrobial Peptides: From Genes to Applications. Springer Publishing Company, New York, USA.

Oliveira, M.A., V.M. Souza, A.M. Morato Bergamini and E.C. Pereira De Martinis. 2011. Microbiological quality of ready-to-eat minimally processed vegetables consumed in Brazil. Food Control 22: 1400–1403.

Patel, H.M., R. Wang, O. Chandrashekar, S.S. Pandiella and C. Webb. 2004. Proliferation of *Lactobacillus plantarum* in solid-state fermentation of oats. Biotechnology Progress 20: 110–116.

Prado, F.C., J.L. Parada, A. Pandey and C.R. Soccol. 2008. Trends in Non-Dairy Probiotic Beverages. Food Research International 41: 111–123.

Pravst, I. 2012. Functional Foods in Europe: A Focus on Health Claims. *In*: B. Valdez (ed.). Scientific, Health and Social Aspects of the Food Industry. InTech, Rijeka, Croatia.

Rakin, M., M. Vukasinovic, S. Siler-Marinkovic and M. Maksimovic. 2007. Contribution of lactic acid fermentation to improved nutritive quality vegetable juices enriched with brewer's yeast autolysate. Food Chemistry 100: 599–602.

Rathore, S., I. Salmerón and S.S. Pandiella. 2012. Production of potentially probiotic beverages using single and mixed cereal substrates fermented with lactic acid bacteria cultures. Food Microbiology 30: 239–244.

Riebroy, S., S. Banjakul, W. Visessanguan and M. Tanaka. 2007. Effect of iced storage of big-eye snapper (*Priacanthus tayenus*) on the chemical composition, properties and acceptability of Som-fug, a fermented Thai fish mince. Food Chemistry 102: 270–280.

Rivera-Espinoza, Y. and Y. Gallardo-Navarro. 2010. Non-dairy probiotics products. Food Microbiology 27: 1–11.

Rößle, C., M.A.E. Auty, N. Brunton, R.T. Gormley and F. Butler. 2010. Evaluation of fresh-cut apple slices enriched with probiotic bacteria. Innovative Food Science and Emerging Technologies 11: 203–209.

Ruiz-Moyano, S., A. Martín, M.J. Benito, A. Hernández, R. Casquete and M. deG. Córdoba. 2011. Application of *Lactobacillus fermentum* HL57 and *Pediococcus acidilactici* SP979 as potential probiotics in the manufacture of traditional Iberian dry-fermented sausages. Food Microbiology 28: 839–847.

Saad, S.M.I., T.R. Komatsu, D. Granato, G.F. Branco and F.C.A. Buriti. 2011. Probióticos e Prebióticos em Alimentos: Aspectos Tecnológicos, Legislação e Segurança no Uso. pp. 23–49. *In*: S.M.I. Saad, A.G. Cruz and J.A.F. Faria (eds.). Probióticos e Prebióticos em Alimentos: fundamentos e aplicações tecnológicas. Editora Varela, São Paulo, Brazil.

Saarela, M., I. Virkajärvi, H.L. Alakomi, P. Sigvart-Mattila and J. Mättö. 2006. Stability and functionality of freeze-dried probiotic Bifidobacterium cells during storage in juice and milk. International Dairy Journal 16: 121477–121482.

Serna-Saldivar, S.O. 2010. Cereal grains: Properties, processing, and nutritional attributes, *In*: Food Preservation Technology. CRC Press, FL, USA.

Shah, N.P. 2001. Functional foods from probiotics and prebiotics. Food Technology 55: 46–53.

Shah, N.P. 2007. Functional cultures and health benefits. International Dairy Journal 17: 1262–1277.

Sharma, V. and H.N. Mishra. 2013. Fermentation of vegetable juice mixture by probiotic lactic acid bacteria. Nutrafoods 12: 17–22.

Sheehan, V.M., P. Ross and G.F. Fitzgerald. 2007. Assessing the Acid Tolerance and the Technological Robustness of Probiotic Cultures for Fortification in Fruit Juices. Innovative Food Science and Emerging Technology 8: 279–284.

Sidira, M., M. Kanellaki and Y. Kourkoutas. 2013. Profile of Aroma-related Volatile Compounds Isolated from Probiotic Dry-fermented Sausages Produced with Free or Immobilized *L. casei* using SPME GC/MS Analysis. pp. 135–147. *In*: C.-T. Ho, C. Mussinan, F. Shahidi and E.T. Contis (eds.). Nutrition, Functional and Sensory Properties of Foods. RSC Publishing Company, Cambridge, UK.

Soccol, C.R., L.P. de Souza-Vandenberghe, M.R. Spier, A.B. Pedroni-Mederios, C.T. Yamaguishi, J.D.D. Lindner, A. Pandey and V. Thomaz-Soccol. 2010. The potential of probiotics: a review. Food Technology and Biotechnology 48: 413–434.

Tamang, J.P. 2005. Ethnic fermented foods of the Eastern Himalayas, 2nd International conference on fermented foods, health status and social wellbeing, Anand, India. SASNET.

Tamminen, M., S. Salminen and A.C. Ouwehand. 2013. Fermentation of Carrot Juice by Probiotics: Viability and Preservation of Adhesion. International Journal of Biotechnology for Wellness Industries 2: 10–15.

Tanguler, H. and H. Erten. 2012. Occurrence and growth of lactic acid bacteria species during the fermentation of shalgam (salgam), a traditional Turkish fermented beverage. LWT—Food Science and Technology 46: 36–41.

Tu, R.-J., H.-Y. Wu, Y.-S. Lock and M.-J. Chen. 2010. Evaluation of microbial dynamics during the ripening of a traditional Taiwanese naturally fermented ham. Food Microbiology 27: 460–467.
Vergara, C.M.A.C., T.L. Honorato, G.A. Maia and S. Rodrigues. 2010. Prebiotic effect of fermented cashew apple (*Anacardium occidentale* L.) juice, LWT—Food Science and Technology 43: 141–145.
Wang, Y.C., R.C. Yu, H.Y. Yang and C.C. Chou. 2006. Antioxidatives activities of soymilk fermented with lactic acid bacteria and bifidobacteria. Food Microbiology 23: 128–135.
Wójciak, K.M., Z.J. Dolatowski, D. Kołożyn-Krajewska and M. Trząskowska. 2012. The Effect of the *Lactobacillus Casei* Lock 0900 Probiotic Strain on the Quality of Dry-Fermented Sausage During Chilling Storage. Journal of Food Quality 35: 353–365.
Wu, C.H. and C.C. Chou. 2009. Enhancement of Aglycone, Vitamin K2 and Superoxide Dismutase Activity of Black Soybean through Fermentation with *Bacillus subtilis* BCRC 14715 at Different Temperatures. Journal of Agricultural and Food Chemistry 57: 10695–10700.
Xu, H.-Q., W.-W. Wang, Y.-S. Jiang and Z.-J. Wang. 2008. Study on fermentation properties of lactic acid bacteria isolated from traditional fermented meat products. Science and Technology of Food Industry (Chinese) 1: 88–92.
Yoon, K.Y., E.E. Woodams and Y.D. Hang. 2005. Fermentation of beet juice by beneficial lactic acid bacteria. LWT—Food Science and Technology 38: 73–75.
Yoon, K.Y., E.E. Woodams and Y.D. Hang. 2006. Production of Probiotic Cabbage Juice by Lactic Acid Bacteria. Bioresource Technologies 97: 1427–1430.
Zuidam, N.J. and N. Viktor. 2010. Encapsulation Technologies for Active Food Ingredients and Food Processing, Springer, USA.

# Index

## I



## L



## M



## N



## O



## P



## R



## S



## T



## V



## Y